Micro-Small Fixed-Wing UAV Control

Quan Quan · Wenhan Gao · Runxiao Liu ·
Xinquan Chen · Xunhua Dai · Shuli Lv · Lin Xu ·
Yue Li

Micro-Small Fixed-Wing UAV Control

Design and Practice

Springer

Quan Quan
Beihang University
Beijing, China

Wenhan Gao
Beihang University
Beijing, China

Runxiao Liu
Beihang University
Beijing, China

Xinquan Chen
Beihang University
Beijing, China

Xunhua Dai
Central South University
Changsha, Hunan, China

Shuli Lv
Beihang University
Beijing, China

Lin Xu
Tiangong University
Tianjin, China

Yue Li
Beihang University
Beijing, China

ISBN 978-981-95-1953-8 ISBN 978-981-95-1954-5 (eBook)
https://doi.org/10.1007/978-981-95-1954-5

This Springer imprint is published by the registered company Springer Nature Singapore Pte Ltd.
The registered company address is: 152 Beach Road, #21-01/04 Gateway East, Singapore 189721, Singapore

Preface

Micro-small fixed-wing unmanned aerial vehicles (UAVs) are increasingly being used in both military and civilian fields. In the military field, they can perform tasks that traditional aircraft find challenging, including reconnaissance, surveillance, rescue, and special operations. For example, using micro-small fixed-wing UAVs for reconnaissance and surveillance on the battlefield can provide real-time intelligence support, thereby enhancing operational command decision-making possibilities. In the civilian field, due to their flexibility, efficiency, and safety, they hold irreplaceable advantages in certain special scenarios, including aerial photography, agricultural plant protection, environmental monitoring, logistics distribution, and rescue. For instance, using micro-small fixed-wing UAVs for aerial photography in urban areas can provide high-definition aerial views, which facilitate urban planning and tourism promotion.

The widespread application of UAVs drives the need to cultivate more full-stack UAV flight control engineers. Full-stack UAV engineers are professionals skilled in UAV software and hardware design, control systems development, data processing, and applications. They can effectively integrate these different technologies, improve efficiency, reduce costs, and drive technological innovation, thereby providing their institutions with more competitive advantages in the market.

To better cultivate full-stack UAV flight control engineers, we have written this book. The aim is to enable readers with zero foundation to quickly get started, master advanced development methods, and keep up with the cutting edge of technology. This book strives for foundational, comprehensive, systematic, and technically advanced aspects, focusing on breakthroughs in the following areas:

1. Rich Cases: Currently, this book includes eight theme experiments, involving level flight performance evaluation experiment, controlled dynamic system modeling experiment, system identification experiment, low-level flight control experiment, path following and planning experiment, trajectory planning and differential flatness control experiment, image-based visual servo docking control experiment, and swarm flight control experiment. Each experiment can serve as an independent case.

2. Depth and Breadth: The eight theme experiments cover UAV design, modeling, single-UAV control and planning, swarm control and planning, and also involve the development of visual perception algorithms. Each theme experiment generally includes basic experiments, analysis experiments, design experiments, and hardware-in-the-loop simulation experiments. The experiments gradually deepen and are closely connected to current engineering practice.
3. Interactivity and Practicality: All experiments include complete code. Basic experiments and analysis experiments help readers get started, while design experiments require readers to complete independently. Hardware-in-the-loop simulation experiments use autopilot hardware (CubePilot, PixHawk) to implement the entire flight control task. Readers will encounter software such as MATLAB/Simulink, Python, QGroundControl, and ground stations, as well as hardware like CubePilot or PixHawk autopilots and remote controllers, using the RflySim toolchain to achieve rapid development, deployment, and experimentation from algorithms to real machines.
4. Development Platform: This book conveys the idea and methods of model-based development, including platforms, development methods, and a series of key technologies. Based on this book, readers can further develop their own algorithms and share them, achieving up-to-date effects.

The writing of this book has lasted for more than five years, and many students from the Reliable Flight Control Research Group at Beihang University have participated in its writing. Wenhan Gao was responsible for the early work of Chaps. 2, 4, 6, and 7; Runxiao Liu has been responsible for Chaps. 5 and 9; Xinquan Chen has been responsible for Chaps. 1 and 3, and the later revisions of Chap. 7; Lin Xu has been responsible for the later revisions of Chaps. 4 and 6; Shuli Lv has been responsible for Chap. 8; Yue Li has been responsible for Chap. 10; Assoc. Prof. Xunhua Dai of Central South University wrote Chap. 2 and guided the hardware-in-the-loop simulation experiments for the entire book; Hao Sha participated in the revision of Chap. 4; Dongyuan Zhang participated in the revision of Chap. 6; Hanli Chen participated in the revision of the hardware-in-the-loop part of Chap. 9; and Qingpeng Chen participated in the revision of Chap. 10. I am responsible for the architecture and experimental content design of the entire book, organizing and participating in the writing and revision of the entire book.

The writing of this book is inseparable from the support of predecessors, colleagues, friends, and students. Special thanks to Profs. Zongji Chen, Gongzhang Shen, Ping Zhang, Yuhua Fei, Jingbao Lu, Yunfei Zhang, Honglun Wang, Qingzhen Zhang, Zhenghua Liu, Weiqi Li, Lingyu Yang, Dawei Li, Jiong Yang, Jing Zhang from Beihang University and Prof. Wenhai Wu from the Naval Aviation University for their valuable suggestions. Thanks to Yunjie Ma-Li, Jing Li from Aerospace Knowledge magazine for their help with illustrations. The main content of this book has been tested many times in the "Course Design and Comprehensive Experiment" courses I led at Beihang University, and I appreciate the feedback and supplements from the students. Finally, this work was also supported in part by the National Natural Science Foundation of China under Grant 62450127.

Due to limited expertise, there may be errors in the book. Readers are kindly requested to send emails to qq_buaa@buaa.edu.cn for criticism and correction. In addition, if you want to use it as a school textbook, you can also contact us and we will send you more complete materials for reference.

Beijing, China
July 2025

Quan Quan

Contents

1 Introduction ... 1
Quan Quan, Xinquan Chen, and Wenhan Gao
1.1 Basic Concept ... 1
1.1.1 Composition and Classification ... 1
1.1.2 Configuration ... 4
1.1.3 Airfoil and Wing Shape ... 7
1.1.4 Propulsion System ... 11
1.2 Flight Principles ... 14
1.2.1 Aerodynamic Principle ... 14
1.2.2 Aerodynamics Overview ... 17
1.2.3 Force Analysis ... 21
1.2.4 Stability Analysis ... 23
1.3 Development of Design Concept ... 26
1.3.1 Manual Control Technology (Began in Early 20th Century) ... 26
1.3.2 Automatic Control Technology (Began in the 1930s) ... 27
1.3.3 Active Control Technology (Began in 1960–1970s) ... 28
1.3.4 ADA Control Technology (Began in 2010s) ... 29
1.4 Structure of This Book ... 30
1.5 Exercise ... 32
References ... 33

2 Experimental Platform Configuration and Development Process ... 35
Xunhua Dai and Quan Quan
2.1 Overall Introduction ... 35
2.1.1 Software Platform ... 35
2.1.2 Hardware Platform ... 38
2.1.3 Software/Hardware Overall Relationship ... 40
2.1.4 Overview of Development Process ... 42

2.2 Software Platform Installation and Configuration 44
2.2.1 Software Platform Installation Steps 44
2.2.2 Advanced Settings 47
2.2.3 Installation Completion Performance 48
2.2.4 Basic Software Checking 49
2.2.5 Other Features of RflySim 51
2.3 Quick Start Experiment 52
2.3.1 Introduction and Usage of Core Software 52
2.3.2 Introduction to Automatic Code Generation 53
2.3.3 Control Experiment Based on Visual Feedback 58
2.3.4 UAV Swarm Control Experiment 62
References 64

3 Level Flight Performance Evaluation Experiment 65
Xinquan Chen, Wenhan Gao, and Quan Quan
3.1 Experimental Principles of Level Flight Performance Evaluation 65
3.1.1 Level Flight Performance Analysis 66
3.1.2 Propulsion System Modeling 69
3.1.3 Flight Envelope 74
3.2 Basic Experiment 76
3.2.1 Experimental Objective 76
3.2.2 Experimental Procedure 77
3.3 Analysis Experiment 82
3.3.1 Experimental Objective 82
3.3.2 Experimental Procedure 82
3.4 Design Experiment 87
3.4.1 Experimental Objective 87
3.4.2 Experimental Design 87
3.5 Exercise 89
References 91

4 Controlled Dynamic System Modeling Experiment 93
Lin Xu, Wenhan Gao, Hao Sha, Xunhua Dai, and Quan Quan
4.1 Preliminary 94
4.1.1 Coordinate System 94
4.1.2 Kinematics and Dynamics 100
4.1.3 Force and Moment 102
4.1.4 Trimming and Linearization 106
4.1.5 Analysis of Fixed-Wing UAV Lateral Motion 115
4.1.6 Fixed-Wing UAV Modal Analysis 117
4.2 Basic Experiment 119
4.2.1 Experimental Objective 119
4.2.2 Experimental Procedure 120

4.3 Analysis Experiment ... 126
4.3.1 Experimental Objective ... 126
4.3.2 Experimental Procedure ... 127
4.4 Design Experiment ... 134
4.4.1 Experimental Objective ... 134
4.4.2 Experimental Design ... 134
4.4.3 Experimental Procedure ... 135
4.5 Exercise ... 143
References ... 146

5 System Identification Experiment ... 147
Runxiao Liu, Xunhua Dai, and Quan Quan
5.1 Preliminary ... 147
5.1.1 Steps of System Identification ... 148
5.1.2 Guidance Model Identification for Micro-Small Fixed-Wing UAVs ... 149
5.1.3 Aerodynamic Parameter Identification for Micro-Small Fixed-Wing UAVs ... 150
5.2 Basic Experiment ... 156
5.2.1 Experimental Objective ... 156
5.2.2 Experimental Procedure ... 157
5.3 Analysis Experiment ... 162
5.3.1 Experimental Objective ... 162
5.3.2 Experimental Procedure ... 163
5.4 Design Experiment ... 169
5.4.1 Experimental Objective ... 169
5.4.2 Experimental Design ... 169
5.4.3 Experimental Procedure ... 171
5.5 HIL Experiment ... 175
5.5.1 Guidance Model Identification HIL Simulation Experiment ... 175
5.5.2 HIL Simulation for Aerodynamic Parameter Identification ... 180
5.6 Exercise ... 185
References ... 186

6 Low-Level Flight Control Experiment ... 187
Lin Xu, Wenhan Gao, Dongyuan Zhang, Xunhua Dai, and Quan Quan
6.1 Preliminary ... 187
6.1.1 Attitude Control ... 188
6.1.2 Altitude and Airspeed Control ... 201
6.1.3 Principles and Tuning Steps of PID Control ... 208
6.1.4 Gain Scheduling Control ... 210

6.2 Basic Experiment ... 211
6.2.1 Experimental Objective ... 211
6.2.2 Experimental Procedure ... 211
6.3 Analysis Experiment ... 216
6.3.1 Experimental Objective ... 216
6.3.2 Experimental Procedure ... 216
6.4 Design Experiment ... 227
6.4.1 Experimental Objective ... 227
6.4.2 Experimental Design ... 227
6.4.3 Experimental Procedure ... 228
6.5 HIL Experiment ... 235
6.5.1 Experimental Objective ... 235
6.5.2 Experimental Procedure ... 235
6.6 Exercise ... 240
References ... 241

7 Path Following and Planning Experiment ... 243
Wenhan Gao, Xinquan Chen, Xunhua Dai, and Quan Quan
7.1 Preliminary ... 243
7.1.1 L1 Straight-Line Following ... 244
7.1.2 Circular Path Following ... 245
7.1.3 Desired Acceleration Performed by the Low-Level Flight Controller ... 246
7.1.4 Dubins Curves ... 247
7.1.5 Path Length Computation ... 248
7.1.6 Related Development ... 253
7.2 Basic Experiment ... 254
7.2.1 Experimental Objective ... 254
7.2.2 Experimental Procedure ... 254
7.3 Analysis Experiment ... 263
7.3.1 Experimental Objective ... 263
7.3.2 Experimental Procedure ... 265
7.4 Design Experiment ... 269
7.4.1 Experimental Objective ... 269
7.4.2 Experimental Design ... 270
7.4.3 Experimental Procedure ... 274
7.5 Exercise ... 278
References ... 280

8 Trajectory Planning and Differential Flatness Control Experiment ... 281
Shuli Lv, Xunhua Dai, and Quan Quan
8.1 Preliminary ... 281
8.1.1 Coordinated Flight Model ... 282
8.1.2 Differential Flatness Control and Trajectory Tracking ... 283

8.1.3 Trajectory Planning ... 285
8.1.4 B-Spline Trajectory Generation ... 287
8.1.5 Related Development ... 296
8.2 Basic Experiment ... 298
8.2.1 Experimental Objective ... 298
8.2.2 Experimental Procedure ... 298
8.3 Analysis Experiment ... 303
8.3.1 Experimental Objective ... 303
8.3.2 Experimental Procedure ... 303
8.4 Design Experiment ... 310
8.4.1 Experimental Objective ... 310
8.4.2 Experimental Design ... 311
8.4.3 Experimental Procedure ... 316
8.5 Exercise ... 323
References ... 325

9 Image-Based Visual Servo Docking Control Experiment ... 327
Runxiao Liu, Hanli Chen, Xunhua Dai, and Quan Quan
9.1 Preliminary ... 327
9.1.1 Fundamental Principles of Image-Based Visual Servo Control ... 328
9.1.2 Target Recognition and Localization ... 331
9.1.3 Docking Control ... 332
9.1.4 Related Development ... 341
9.2 Basic Experiment ... 343
9.2.1 Experimental Objective ... 343
9.2.2 Experimental Procedure ... 343
9.3 Analysis Experiment ... 349
9.3.1 Experimental Objective ... 349
9.3.2 Experimental Procedure ... 349
9.4 Design Experiment ... 355
9.4.1 Experimental Objective ... 355
9.4.2 Experimental Design ... 355
9.4.3 Experimental Procedure ... 358
9.5 HIL Experiment ... 362
9.5.1 Experimental Objective ... 362
9.5.2 Experimental Design ... 362
9.5.3 Experimental Procedure ... 366
9.6 Exercise ... 375
References ... 375

10 Swarm Flight Control Experiment ... 377
Yue Li, QingPeng Chen, Xunhua Dai, and Quan Quan
10.1 Preliminary ... 377
10.1.1 Basic Concepts ... 377
10.1.2 Controlled Model ... 379
10.1.3 Swarm Control Method ... 380

10.2 Basic Experiment 384
10.2.1 Experimental Objective 384
10.2.2 Experimental Procedure 385
10.3 Analysis Experiment 390
10.3.1 Experimental Objective 390
10.3.2 Experimental Procedure 390
10.4 Design Experiment 395
10.4.1 Experimental Objective 395
10.4.2 Experimental Design 396
10.4.3 Experimental Procedure 400
10.4.4 HIL Experiment 405
10.4.5 Related Development 415
10.5 Exercise 417
10.6 Appendix 418
References 420

Symbols

$\mathbb{R}$, $\mathbb{R}^n$	Real number field, n-dimensional Euclidean space
$\mathbb{R}_+$	Set of positive real numbers
∇	Gradient
$\frac{\partial \mathbf{x}}{\partial \mathbf{z}}$	Partial derivative with respect to the state variable $\mathbf{z}$
$\dot{\mathbf{x}}$, $\frac{\mathrm{d}\mathbf{x}}{\mathrm{d}t}$	First-order derivative with respect to time t
$\lvert * \rvert$	Absolute value
$\lvert\lvert * \rvert\rvert$	Euclidean norm, $\lvert\lvert \mathbf{x} \rvert\rvert \triangleq \sqrt{\mathbf{x}^{\mathrm{T}}\mathbf{x}}$, $\mathbf{x} \in \mathbb{R}^n$
$\mathbf{A}$	Matrix
$\mathbf{A}^{\mathrm{T}}$	Transpose of $\mathbf{A}$
b	Wingspan of the fixed-wing (m)
c	Chord length of the fixed-wing (m)
C_D	Drag coefficient
C_l	Rolling moment coefficient
C_L	Lift coefficient
C_m	Pitching moment coefficient
C_n	Yawing moment coefficient
C_Y	Side force coefficient
$\mathbf{f}$	Total external force on the fixed-wing, ${}^{\mathrm{e}}\mathbf{f}$ and ${}^{\mathrm{b}}\mathbf{f}$ represent the vector $\mathbf{f}$ in the ground coordinate system and body coordinate system, respectively, $f_{x_i}, f_{y_i}, f_{z_i}$ represent the external force components along the coordinate axes $o_i x_i$, $o_i y_i$, $o_i z_i$, $i = \mathrm{e}, \mathrm{b}$
$\mathbf{f}_{\mathrm{a}}$	Aerodynamic force, ${}^{\mathrm{b}}\mathbf{f}_{\mathrm{a}}$ and ${}^{\mathrm{w}}\mathbf{f}_{\mathrm{a}}$ represent the vector $\mathbf{f}_{\mathrm{a}}$ in the body coordinate system and wind coordinate system, respectively, and ${}^{\mathrm{w}}\mathbf{f}_{\mathrm{a}} = \left[f_{\mathrm{ax}}\ f_{\mathrm{ay}}\ f_{\mathrm{az}}\right]^{\mathrm{T}}, f_{\mathrm{ax_b}}, f_{\mathrm{ay_b}}, f_{\mathrm{az_b}}$ represent the aerodynamic force components along the coordinate axes $o_{\mathrm{b}x_{\mathrm{b}}}$, $o_{\mathrm{b}y_{\mathrm{b}}}$, $o_{\mathrm{b}z_{\mathrm{b}}}$
g	Gravitational acceleration $(\mathrm{m/s^2})$
H	Flight altitude of the fixed-wing, namely $p_{z_{\mathrm{e}}}$
$\mathbf{J}$	Moment of inertia matrix of the fixed-wing, J_{xx}, J_{yy}, J_{zz} are the principal moments of inertia
m	Mass of the fixed-wing UAV (kg)

$\mathbf{m}$	Total external torque on the fixed-wing, ${}^{\mathrm{e}}\mathbf{m}$ and ${}^{\mathrm{b}}\mathbf{m}$ represent the vector $\mathbf{m}$ in the ground coordinate system and body coordinate system, respectively, $m_{x_i}, m_{y_i}, m_{z_i}$ represent the external torque components along the coordinate axes o_ix_i, o_iy_i, o_iz_i, $i=\mathrm{e, b}$ (N·m)
$\mathbf{m}_\mathrm{a}$	Aerodynamic torque in the body coordinate system, pitch torque max, roll torque may, yaw torque maz represent the torque components around $o\mathrm{b}x_\mathrm{b}$, $o_\mathrm{b}y_\mathrm{b}$, $o_\mathrm{b}z_\mathrm{b}$ respectively (N·m)
$o_\mathrm{b}x_\mathrm{b}y_\mathrm{b}z_\mathrm{b}$	Body coordinate system
$o_\mathrm{e}x_\mathrm{e}y_\mathrm{e}z_\mathrm{e}$	Ground coordinate system
$o_\mathrm{s}x_\mathrm{s}y_\mathrm{s}z_\mathrm{s}$	Stability coordinate system
$o_\mathrm{w}x_\mathrm{w}y_\mathrm{w}z_\mathrm{w}$	Wind coordinate system
$\mathbf{p}$	Position of the center of the fixed-wing, ${}^{\mathrm{e}}\mathbf{p}$ and ${}^{\mathrm{b}}\mathbf{p}$ are the representations of the vector $\mathbf{p}$ in the ground coordinate system and body coordinate system, respectively. $p_{x_i}, p_{y_i}, p_{z_i}$ represent the position components along the coordinate axes o_ix_i, o_iy_i, o_iz_i, $i=\mathrm{e, b}$ (m)
$\mathbf{R}(*)$	Rotation matrix
$\mathbf{R}_j^i$	Rotation matrix that represents the rotation of a vector from coordinate system j to coordinate system i. $i, j=\mathrm{b, e, s, w}$
S	Wing area of the fixed-wing (m^2)
${}^{\mathrm{b}}\mathbf{t}$	Thrust on the fixed wing T (N)
$\mathbf{u}$	Input, $\mathbf{u}_\mathrm{lon}$: input of the longitudinal channel, $\mathbf{u}_\mathrm{lat}$: input of the lateral-directional channel
$\mathbf{v}$	Velocity of the center of the fixed-wing, ${}^{\mathrm{e}}\mathbf{v}$ and ${}^{\mathrm{b}}\mathbf{v}$ are the representations of the vector $\mathbf{v}$ in the ground coordinate system and body coordinate system, respectively, $v_{x_i}, v_{y_i}, v_{z_i}$ represent the velocity components along the coordinate axes o_ix_i, o_iy_i, o_iz_i, $i=\mathrm{e, b}$ (m/s)
$\mathbf{v}_\mathrm{a}$	Airspeed vector, its vector magnitude is V_a which is the airspeed, ${}^{\mathrm{b}}\mathbf{v}_\mathrm{a}$ is the representation of the airspeed vector in the body coordinate system, and ${}^{\mathrm{b}}\mathbf{v}_\mathrm{a}=\left[v_{\mathrm{a}x_\mathrm{b}}\,v_{\mathrm{a}y_\mathrm{b}}\,v_{\mathrm{a}z_\mathrm{b}}\right]^\mathrm{T}$, ${}^{\mathrm{w}}\mathbf{v}_\mathrm{a}$ is the representation of the airspeed vector in the wind coordinate system, and ${}^{\mathrm{w}}\mathbf{v}_\mathrm{a}=[V_\mathrm{a}00]^\mathrm{T}$ (m/s)
$\mathbf{v}_\mathrm{g}$	Velocity relative to the ground coordinate system (also known as ground speed), ${}^{\mathrm{b}}\mathbf{v}_\mathrm{g}$ is the representation of the ground speed vector in the body coordinate system, and ${}^{\mathrm{b}}\mathbf{v}_\mathrm{g}=\left[v_{\mathrm{g}x_\mathrm{b}}\,v_{\mathrm{g}y_\mathrm{b}}\,v_{\mathrm{g}z_\mathrm{b}}\right]^\mathrm{T}$ (m/s)
$\mathbf{v}_\mathrm{w}$	Wind speed relative to the ground coordinate system (also known as wind speed), ${}^{\mathrm{b}}\mathbf{v}_\mathrm{w}$ is the representation of the wind speed vector in the body coordinate system, and ${}^{\mathrm{b}}\mathbf{v}_\mathrm{w}=\left[v_{\mathrm{w}x_\mathrm{b}}\,v_{\mathrm{w}y_\mathrm{b}}\,v_{\mathrm{w}z_\mathrm{b}}\right]^\mathrm{T}$ (m/s)
$\mathbf{W}$	Matrix representing the relationship between attitude angular rate and body angular velocity
$\mathbf{x}^*$	State variable under the reference motion
$\Delta\mathbf{x}$	Increment of the state variable
$\mathbf{x}_\mathrm{lon}$	State variable of the longitudinal channel
$\mathbf{x}_\mathrm{lat}$	State variable of the lateral-directional channel
$\mathbf{x}_\mathrm{d}$	Desired value of the state variable

$\mathbf{x}_\times$	$\mathbf{x}_\times = \begin{bmatrix} 0 & -x_3 & x_2 \\ x_3 & 0 & -x_1 \\ -x_2 & x_1 & 0 \end{bmatrix}$
α	Angle of attack (rad)
β	Sideslip angle (rad)
δ_e, δ_a, δ_r	Elevator deflection, aileron deflection, rudder deflection (rad)
γ	Flight path angle
Θ	Euler angles, $\Theta = [\phi\,\theta\,\psi]^\mathrm{T}$, where ϕ, θ, ψ are roll angle, pitch angle, and yaw angle, respectively
ρ	Air density (kg/m^3)
ω	Angular velocity of the fixed-wing, ${}^\mathrm{e}\omega$ and ${}^\mathrm{b}\omega$ represent the vector ω in the ground coordinate system and body coordinate system, respectively. ω_{x_i}, ω_{y_i}, ω_{z_i} represent the angular velocity components along the coordinate axes o_ix_i, o_iy_i, o_iz_i, $i = \mathrm{e}, \mathrm{b}$ (rad/s)
ω_c	Cut-off frequency (rad/s)

Abbreviations

ADA	Autonomous, Dependable, and Affordable
ADC	Analog to Digital Conversion
CFD	Computational Fluid Dynamics
CIFER	Comprehensive Identification from Frequency Response
DARPA	Defense Advanced Research Projects Agency
DLL	Dynamic-Link Library
ENU	East-North-Up
ESC	Electronic Speed Controller
FBW	Fly-By-Wire
FC	Flight Controller
FSD	Full Self-Drive
GAASI	General Atomics Aeronautical Systems, Inc.
GPS	Global Positioning Systems
HIL	Hardware-In-the-Loop
IBVS	Image-Based Visual Servo
LQR	Linear Quadratic Regulator
LSC	Loop Shaping Control
MBD	Model-Based Design
MIL	Model-In-the-Loop
MPC	Model Predictive Control
NACA	National Advisory Committee for Aeronautics
NED	North-East-Down
PBVS	Position-Based Visual Servo
PSP	Pixhawk Support Package
PWM	Pulse Width Modulation
QGC	Q Ground Control
RFLY	Reliable Flight Control Group
RMSE	Root Mean Square Error
SDPSO	Hybrid Particle Swarm Optimization Algorithm
SIL	Software-In-the-Loop
TEC	Total Energy Control

TECS	Total Energy Control System
UAV	Unmanned Aerial Vehicle
UDP	User Datagram Protocol
VTOL	Vertical Take-Off and Landing

Chapter 1
Introduction

Quan Quan, Xinquan Chen, and Wenhan Gao

Unmanned Aerial Vehicles (UAVs), also known as *drones*, are aircraft operated without a pilot, using radio remote control devices and autonomous flight control systems. They are widely used in various fields, including military and civilian applications. Fixed-wing UAVs, known for their long endurance and high cruising speeds, have become one of the most popular UAVs being developed worldwide. This chapter introduces the basic concepts of fixed-wing UAVs and discusses their principles of flight. This chapter's exploration of fixed-wing UAVs provides readers with foundational knowledge of micro-small fixed-wing UAVs, laying the groundwork for subsequent chapters and introducing them to the full development and application process of such UAVs.

1.1 Basic Concept

1.1.1 Composition and Classification

Fixed-wing aircraft[1] generate lift via fixed wings attached to the fuselage, are engine-powered, and fly within the atmosphere, weighing more than the same air volume [1]. For example, the commercial, military, and general aviation aircraft, well-known in Fig. 1.1, are all fixed-wing aircraft. Typical fixed-wing aircraft consist mainly or partly of structural components, landing gear, electrical systems, flight control sys-

[1] Fixed-wing aircraft is a broad category that includes all aircraft that generate lift using wings, including airplanes and gliders. An airplane is a subclass of fixed-wing aircraft, typically referring to larger, fixed-wing, powered, pilot-controlled aircraft used primarily for transport, commercial, or military purposes. In short, all airplanes are fixed-wing aircraft, but not all fixed-wing aircraft are airplanes. Airplanes refer specifically to traditional, large, powered aircraft while fixed-wing aircraft cover a wider range of categories.

Q. Quan et al., *Micro-Small Fixed-Wing UAV Control*,
https://doi.org/10.1007/978-981-95-1954-5_1

(a) C919 (b) J-20 (c) RX4E

Fig. 1.1 Fixed-wing aircraft

tems, propulsion systems, control devices, communication systems, human-machine environment, and payloads. *Structural components* typically include wings, fuselage, and tail. Wings and the tail usually have ailerons, elevators, and rudders for control. The wing is the main component for generating lift, and the tail is essential for balancing and controlling the aircraft's attitude. The *flight control system* mainly performs functions such as attitude control, speed and altitude control, automatic stabilization and holding, and flight performance optimization. The *propulsion system* primarily generates thrust or propulsion to move the aircraft forward by overcoming the resistance created by relative motion with the air. Common propulsion units include propeller, turbojet, and turbofan engines. The *control devices* are principally divided into two categories: the primary control device and the secondary control device. In recent years, with the advancement of electronic technology, a proportion of the flight control devices from conventional mechanical control systems have been progressively superseded by more sophisticated teletype control systems. Consequently, pilots are no longer required to manually manipulate the control surfaces of fixed-wing aircraft; instead, they can set movement instructions to control the plane. Fixed-wing aircraft are powered mainly by the chemical energy of aviation fuel and electricity, although some also use solar energy. For example, the RX4E shown in Fig. 1.1c is a four-seat electric fixed-wing aircraft. It is powered by electricity to drive the engine and propeller to generate thrust, while Fig. 1.1a, b rely on aviation fuel as their energy source.

Fixed-wing UAVs are a type of aircraft that can be operated remotely without a pilot on board. They are controlled remotely by a ground station and equipped with advanced computing technology for signal reception and autonomous operation. The "Interim Regulations on the Flight Management of Unmanned Aerial Vehicles" implemented by the State Council and the Central Military Commission in China on January 1, 2024, categorizes UAVs into five types: micro UAVs, light UAVs, small UAVs, medium UAVs, and large UAVs.

Micro UAVs are UAVs under 0.25 kg, with a maximum altitude of 50 m, a level flight speed of 40 km/h, and compliant wireless communication equipment. They can be manually controlled throughout the flight. Figure 1.2a shows a micro UAV. *Light UAVs* refer to UAVs with an empty weight not exceeding 4 kg and a maximum takeoff weight not exceeding 7 kg. They have a maximum speed of 100 km/h, airspace holding, and reliable surveillance for airspace management, and can be manually controlled throughout the flight, excluding micro-UAVs Fig. 1.2b. *Small UAVs*

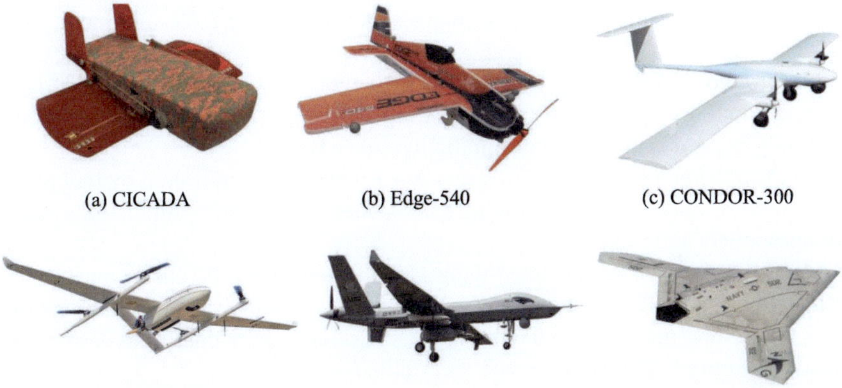

(a) CICADA (b) Edge-540 (c) CONDOR-300

(d) CW-40 (e) Wing Loong (f) X-47B

Fig. 1.2 Fixed-wing UAVs

refer to UAVs with an empty weight not exceeding 15 kg and a maximum takeoff weight not exceeding 25 kg. They possess airspace holding and reliable surveillance capabilities that meet airspace management requirements and can be manually controlled throughout the flight, except for micro and light UAVs, as shown in Fig. 1.2c. *Medium UAVs* refer to UAVs with a maximum takeoff weight not exceeding 150 kg, but excluding micro, light, and small UAVs, shown in Fig. 1.2d. Large UAVs refer to UAVs with a maximum takeoff weight exceeding 150 kg, such as the Wing Loong reconnaissance UAV from China and the X-47B stealth UAV from the United States, as shown in Fig. 1.2e, f.

Micro-small fixed-wing UAVs generally refer to fixed-wing UAVs in the micro to small size range. They are typically smaller in size, with shorter wingspans and fuselage lengths, and usually fly at lower speeds. They differ from larger fixed-wing UAVs in four primary aspects.

(1) *Aerodynamics*: Due to their small size and lower speeds, micro-small fixed-wing UAVs experience smaller inertial forces compared to aerodynamic viscous forces[2]. Consequently, the airflow around micro-small fixed-wing UAVs may tend to turbulence,[3] resulting in flow behaviors distinct from those observed in larger fixed-wing aircraft. Generally, large fixed-wing aircraft are more prone to localized flow

[2] Inertia force is a fictitious force experienced in accelerating or rotating environments, similar to the sensation of being pushed back into the seat of a car during sudden acceleration. Viscous force is the frictional force within a fluid that impedes flow, as seen in the slower flow of honey compared to water due to honey's greater viscosity.

[3] Turbulence is a chaotic, irregular fluid motion, akin to the tumultuous waves in a fast-flowing river or the turbulent sensation experienced by an aircraft encountering unstable air currents. Unlike the orderly flow of calm water, this motion is characterized by random fluctuations in direction and velocity across various scales.

separation during stalls. In contrast, micro-small fixed-wing UAVs are more prone to overall stall[4].

(2) *Structure*: Micro-small fixed-wing UAVs typically experience lower loads, so their airframe materials often consist entirely of composites, with metal materials used only in critical connections, aiming to reduce weight.

(3) *Propulsion System*[5]: Unlike larger fixed-wing aircraft powered by fuel turboprop engines, micro-small fixed-wing UAVs often use electric propulsion systems with propellers and motors.

(4) *Control*: Micro-small fixed-wing UAVs often fly in the speed range of 9–18 m/s, at altitudes of around a hundred meters, where wind speeds usually exceed 4m/s. Therefore, micro-small fixed-wing UAVs often require control to ensure high maneuverability in the wind for safe and stable flight [2].

1.1.2 Configuration

The configuration of fixed-wing aircraft generally refers to the longitudinal arrangement of the horizontal tail relative to the wings. As shown in Fig. 1.3, it typically includes conventional, biplane, monoplane, canard, flying wing, triplane, tiltwing, or tiltrotor configurations. Additionally, there are hybrid aircraft with both fixed-wing and multirotor configurations, including the fixed-wing multirotor hybrid (e.g., quadplane) and lifting-wing multirotor configurations. This section provides a brief introduction to several typical configurations of fixed-wing aircraft.

1. Conventional configuration

Micro-small fixed-wing UAVs with conventional configurations typically consist of a fuselage, wings, tail wings, propulsion systems, control systems, and communication systems. They adopt a single-wing configuration, with the wing located on the upper part of the fuselage, providing the primary source of lift, and equipped with ailerons on the trailing edge. The tail wing is located at the rear of the fuselage and is divided into vertical stabilizers and horizontal stabilizers. The aircraft's stabilizers consist of a vertical stabilizing surface, a rudder, a horizontal stabilizing surface, and an elevator. The conventional configuration is further subdivided based on the arrangement of control surfaces.

Conventional aircraft have control surfaces, including ailerons, elevators, and rudders, as shown in Fig. 1.4. The ailerons are positioned at the outer edges of the wings, with their deflection angle represented by δ_a. The elevators are situated at the rear

[4] Stall occurs when the aircraft's airspeed is too slow, or the angle of attack is too high, causing the wings to generate insufficient lift, resulting in the aircraft's inability to maintain routine flight. This may lead to a loss of altitude or difficulty in controlling the aircraft.

[5] In aircraft, the "power system" generates and distributes electrical power for onboard systems, while the "propulsion system" provides thrust to propel the aircraft. The power system comprises components such as generators and batteries, whereas the propulsion system includes engines, propellers, or jet turbines. Both are essential but serve different functions.

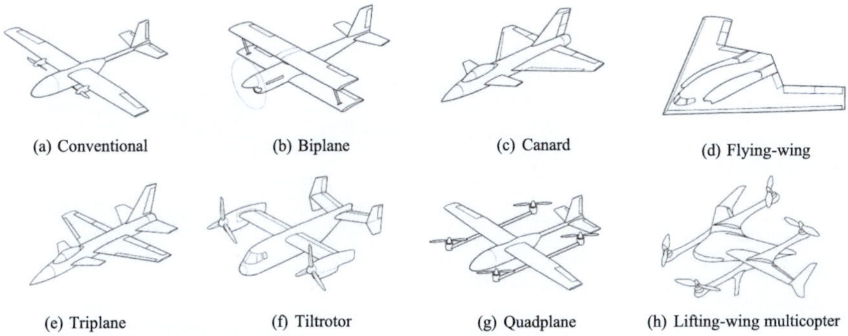

Fig. 1.3 Fixed-wing aircraft of different configurations

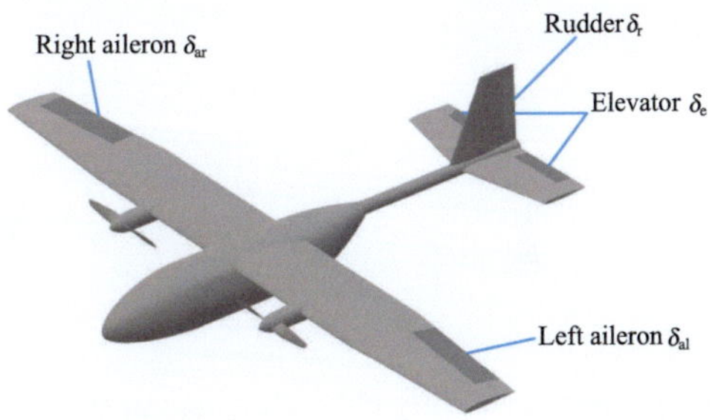

Fig. 1.4 Standard conventional configuration

edge of the horizontal stabilizer, and their deflection angle is indicated by δ_e. Similarly, the rudder is found at the rear edge of the vertical stabilizer, and its deflection angle is denoted by δ_r. The positive direction of rudder variation corresponds to the generation of a negative moment. The ailerons are symmetrically positioned on the left and right wings. Let the deflection of the left aileron be denoted as δ_{al} and that of the right aileron as δ_{ar}. The differential deflection of the ailerons is then expressed as

$$\delta_a = \delta_{al} - \delta_{ar}. \tag{1.1}$$

There are also two special cases in the conventional configuration. The first is the "V-tail" configuration, illustrated in Fig. 1.5, where control surfaces are presented on both sides of the tail. Denoting the deflection of the right tail control surface as δ_{rr} and the left as δ_{rl}, the differential deflection and co-directional deflection of these surfaces function as the elevator and rudder, respectively, as described by

$$\begin{bmatrix} \delta_e \\ \delta_r \end{bmatrix} = \begin{bmatrix} 1 & 1 \\ -1 & 1 \end{bmatrix} \begin{bmatrix} \delta_{rr} \\ \delta_{rl} \end{bmatrix}. \tag{1.2}$$

The second case is the no-tail configuration, characterized by a large dihedral angle at the outer edges of the wing. In this design, the rolling moment is generated

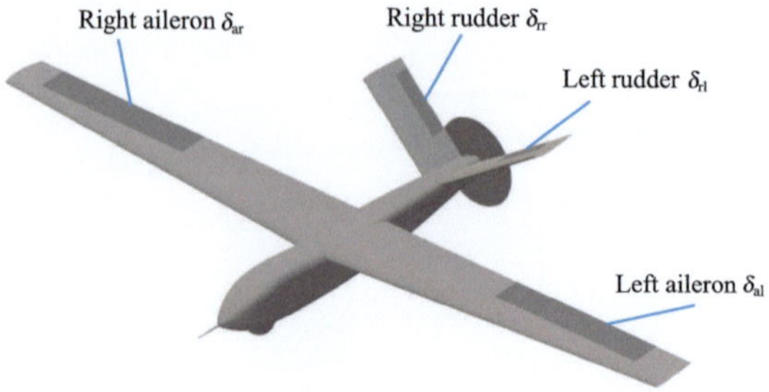

Fig. 1.5 "V-tail" conventional configuration

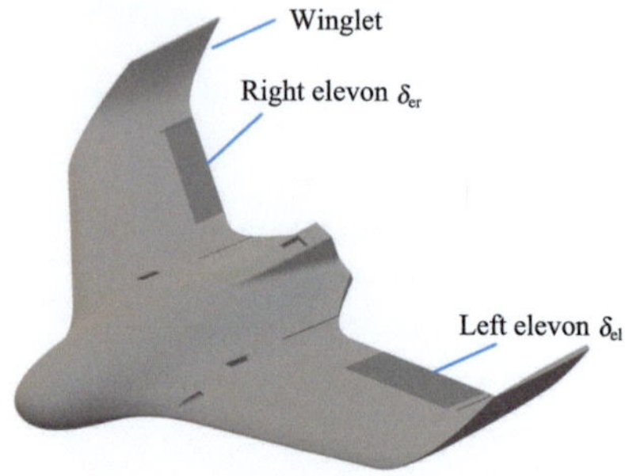

Fig. 1.6 Flying wing configuration

by the lateral stability characteristics of the fixed wing when the rudder is deflected. This configuration is relatively rare and typically exhibits inferior maneuverability compared to fixed-wing UAVs with ailerons. Still, it is favored in military applications due to its ease of maintenance and disassembly.

2. Flying Wing configuration

The flying wing configuration differs from the conventional configuration by integrating the wings and fuselage into a single, airfoil-shaped structure or a form closely resembling it. The entire wing and fuselage together generate lift. To ensure directional stability, winglets are placed on either side of the wing, acting as a tail in conventional configurations, as shown in Fig. 1.6. The control surfaces of the flying wing configuration are relatively simple, consisting of elevons located on either side of the rear edge of the fuselage. The deflection of the right elevon is denoted by δ_{er}, and that of the left by δ_{el}. Differential deflection simulates the aileron function, while co-directional deflection plays the same role as the elevator, described mathematically by

$$\begin{bmatrix} \delta_e \\ \delta_a \end{bmatrix} = \begin{bmatrix} 1 & 1 \\ -1 & 1 \end{bmatrix} \begin{bmatrix} \delta_{er} \\ \delta_{el} \end{bmatrix}. \tag{1.3}$$

3. Composite configuration

In recent years, with the development of multicopters, many fixed-wing UAVs have combined the vertical takeoff and landing advantages of multicopters. The most common is the quadplane configuration, which directly combines a conventionally configured fixed-wing UAV with a quadcopter, as shown in Fig. 1.7a. During takeoff

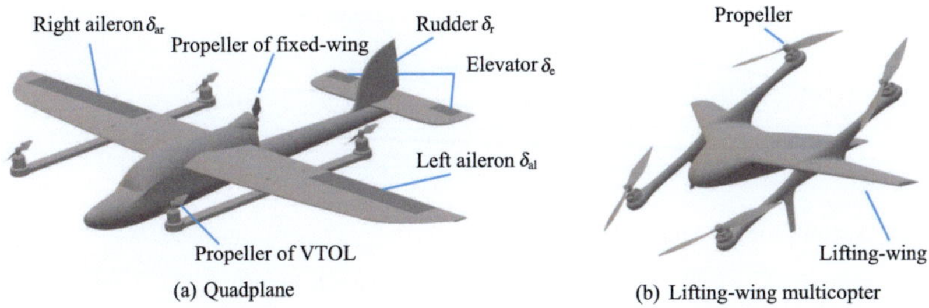

Fig. 1.7 Composite configuration

and landing, this configuration operates similarly to a quadcopter, with flight control managed by the four propellers of the Vertical Takeoff and Landing (VTOL) system. During cruising flight, the propellers of VTOL stop rotating, and thrust is provided by the fixed-wing propeller. The lifting-wing multicopter is a hybrid configuration that combines a blended wing body (also known as a *lifting-wing*) with a multirotor. The rotor disc plane is fixed at an angle to the lifting-wing (e.g., 30–45°). This design significantly improves forward flight efficiency while retaining the original simplicity, reliability, and wind resistance stability of the multirotor. Its unique overall configuration is markedly different from conventional VTOL aircraft, such as tail-sitter and fixed-wing-multirotor hybrid (e.g., quadplane) designs. Compared to traditional fixed-wing and fixed-wing-multirotor hybrid aircraft, the lifting-wing multicopter, with its short wings and powerful rotors, offers superior wind resistance performance [3–5].

1.1.3 Airfoil and Wing Shape

The wing is the primary source of lift for fixed-wing UAVs, and its ability to generate lift is mainly influenced by the parameters[6] of the airfoil profile and the geometric shape of the wing planform. This section provides a brief introduction to the relevant knowledge of airfoils and wing shapes.

1. Airfoil

In general, wings have a symmetrical surface. The airfoil profile, which is obtained by cutting the wing with a plane parallel to the aircraft's symmetrical axis, is commonly referred to as the *airfoil*. The leading edge of the airfoil is the upstream end of the

[6] In the book, there will be discussions on parameters and coefficients, which have some distinctions. Parameters describe the properties of a system, including scalars, vectors, matrices, and other quantities, such as angle of attack, velocity, mass, and the moments of inertia matrix. Coefficients represent the degree of relationship between variables, typically represented by real numbers, such as the relationship between the lift coefficient and the angle of attack. Parameters can vary, while coefficients are usually constants or functions.

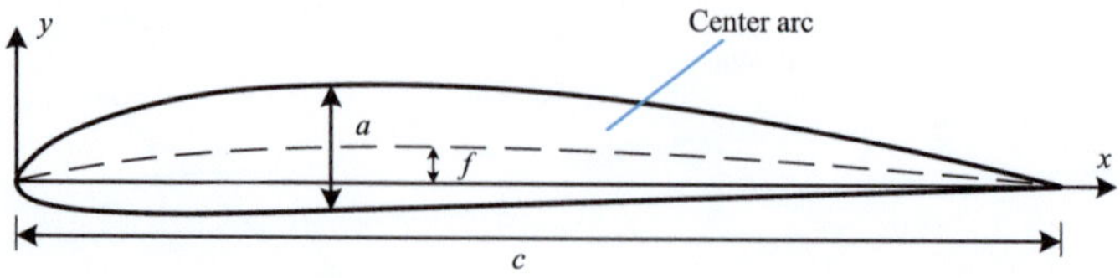

Fig. 1.8 Airfoil geometric parameters

wing, and in subsonic conditions, it often adopts a smooth curve with a large radius of curvature to reduce profile drag. Micro-small fixed-wing UAVs typically employ low-speed airfoil profiles characterized by a rounded leading edge and a pointed trailing edge.

(1) Geometric Parameters of Airfoils

As shown in Fig. 1.8, a circle inscribed within the airfoil profile near the leading edge and tangent to the curve at this point defines the leading edge radius. The line connecting the leading edge point of the airfoil to the trailing edge tip is called the *chord line* of the airfoil. The length of the chord line is referred to as the *geometric chord length*, commonly denoted as c. Setting up a coordinate system with the leading edge point of the airfoil as the origin and the chord line as the x axis, the line joining the midpoint of the top and bottom surfaces of the airfoil in the y direction is called the *mean camber line*. In the y direction, the maximum distance between the mean camber line and the chord line is represented by f. The ratio f/c represents the *camber*. The maximum y directional height of the upper and lower surfaces of the airfoil is denoted by a, and the relative thickness of the airfoil is represented by a/c [1].

(2) Reference Airfoil

With the development of potential flow theory, Kutta-Joukowski theorem, Prandtl's boundary layer theory, and the continuous improvement of wind tunnel experimental methods, the design theory of airfoils has been gradually improved. A large number of lift curves of different airfoils have been obtained through continuous experiments. In subsequent designs, a standard airfoil is selected from the existing empirical curves based on the design requirements, and then modified and optimized to meet the actual design requirements.

Airfoils are generally classified into several types, including flat-convex (Fig. 1.9a), symmetrical (Fig. 1.9b), concave-convex (Fig. 1.9c), double-convex (Fig. 1.9d), S-type (Fig. 1.9e) and supercritical (Fig. 1.9f) airfoils. Standard reference airfoils include the NACA (National Advisory Committee for Aeronautics, now replaced by NASA) series, the RAF-6E airfoil, the Clark-Y airfoil, the ARA-D airfoil, the C4 airfoil, and the supercritical airfoil [1].

2. Wing Shape

The wing shape parameters primarily encompass planar and solid geometric parameters. The key planar geometric parameters include wing area, wingspan, chord length, and sweep angle. When the planar geometric parameters of the wing are described, a

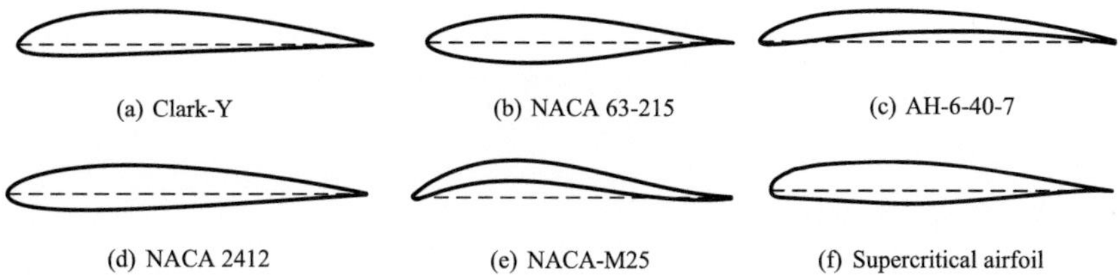

Fig. 1.9 Common reference airfoils

Fig. 1.10 Planar geometric parameters

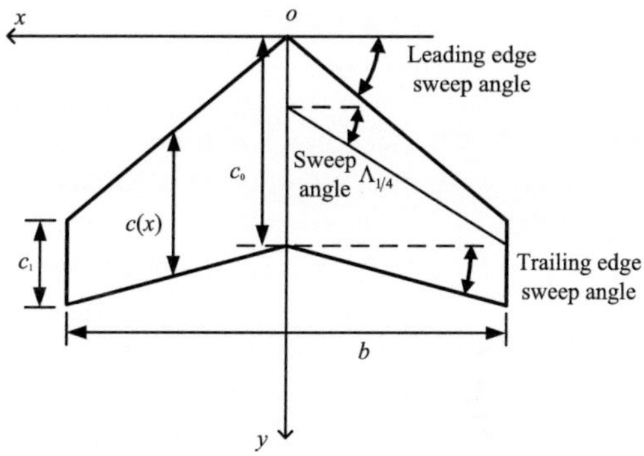

planar coordinate system is established as shown in Fig. 1.10. The *wingspan*, denoted as b, represents the maximum length of the wing along the x-axis. The chord length, $c(x)$, varies along the spanwise direction and is a function of the position x along the span. The chord length at the root ($x = 0$) is denoted by c_0, while at the tip ($x = b/2$), it is denoted by c_1. The wing area, representing the planform area, is calculated by integrating the chord length $c(x)$ along the x-axis, as given by

$$S = 2\int_0^{b/2} c(x)\mathrm{d}x. \tag{1.4}$$

Based on the wing area, the most commonly mentioned chord length is the *mean aerodynamic chord* c_A, defined as follows:

$$c_\mathrm{A} = \frac{2}{S}\int_0^{b/2} c^2(x)\mathrm{d}x. \tag{1.5}$$

This is considered to be the average length from the leading edge to the trailing edge of the wing. Non-dimensional parameters commonly used to characterize the planar shape of the wing include the *aspect ratio* λ (for a rectangular wing, this is the ratio of the span to the chord length) and the *taper ratio* η (the ratio of the root chord to the tip chord of the wing). In general, the aspect ratio is related to the lift coefficient of the wing. The greater the aspect ratio, the greater the lift coefficient, but the drag will increase accordingly. A higher aspect ratio usually means that the wing is longer and narrower. They are represented as follows:

$$\begin{aligned}\lambda &= \frac{b^2}{S} \approx \frac{b^2}{bc} = \frac{b}{c} \\ \eta &= \frac{c_0}{c_1}.\end{aligned} \tag{1.6}$$

The *sweep angle* of the wing refers to the angle at which the wing is tilted backward relative to the direction of flight, and this angle is the angle between the 1/4 chord line and the x axis.

The primary geometric parameters of the wing are the *geometric twist angle* (the angle formed by the chord sections of the airfoil at different spanwise positions which are not in the same plane) and *dihedral* (or *anhedral*). The geometric twist angle refers to the angle between the chord line of the airfoil section at any spanwise position and the chord line of the root section of the wing, as shown in Fig. 1.11a, b. In general, the geometric twist angle of the wing is designed to adjust the spanwise distribution of the aerodynamic loads. Designing negative twist angles of the wingtip relative to the root can transfer aerodynamic loads from the wingtip to the wing root, thereby reducing the bending moment loads on the wing. A negative twist angle design can also prevent wingtip stalls. On tailless aircraft, it can provide a nose-up moment to compensate for the nose-down moment on the wing. Aerodynamic twists can also be used to enhance aerodynamic performance by modifying the local aerodynamic characteristics through changes in wing shape, rather than physically rotating the wing itself.

The *dihedral* (or *anhedral*) of the wing refers to the angle between the chord plane of the wing and the horizontal plane of the aircraft, as shown in Fig. 1.12. A certain amount of dihedral angle for low-speed wings can improve roll stability [6, p. 202].

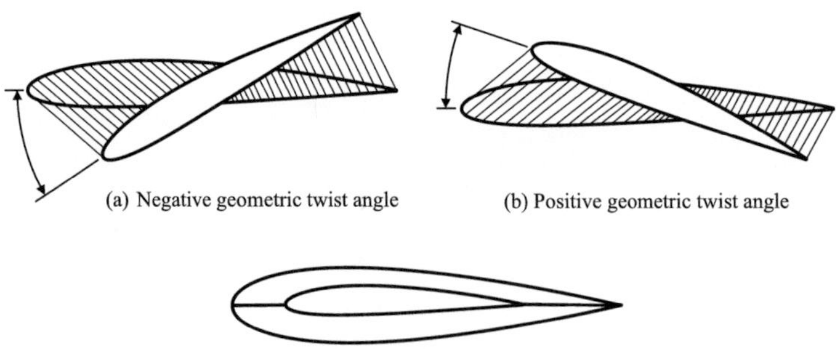

(c) Aerodynamic twist

Fig. 1.11 Geometric twist and aerodynamic twist

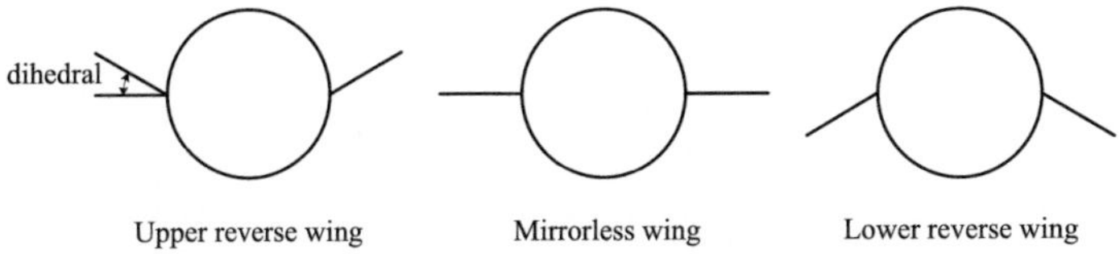

Fig. 1.12 Wing mounting style

1.1.4 Propulsion System

Fixed-wing UAVs typically rely on two primary power sources: fuel and batteries. In this book, micro-small fixed-wing UAVs primarily use motors to drive propellers and generate thrust. The propulsion system of these micro-small fixed-wing UAVs comprises motors, propellers, and servos. Each component of the propulsion system must work in concert with the others; therefore, the optimal combination must be selected to ensure compatibility and optimal functionality. This chapter provides a concise overview of the fundamental concepts underlying each propulsion system component, with detailed mathematical modeling elaborated in Chap. 3.

1. Motor

A motor is a device that converts electrical energy into mechanical energy and vice versa, generally divided into DC motors and AC motors. In the propulsion system of micro-small fixed-wing UAVs, the primary function of the motor is to convert electrical energy from the battery into mechanical energy, thereby rotating the propeller and generating thrust to propel the UAV. Among them, sensorless brushless DC outrunner motors are widely used in micro-small fixed-wing UAVs and other types of aircraft due to their high efficiency, ease of miniaturization, low manufacturing costs, and stable speed. Figure 1.13 shows a standard sensorless brushless DC motor in fixed-wing UAVs.

The circuitry of a sensorless brushless DC motor typically includes star-connected three-phase half-bridge, star-connected three-phase full-bridge, and delta-connected three-phase full-bridge configurations. Among them, the star-connected three-phase full-bridge is more commonly used, as shown in Fig. 1.14. The sensorless brushless DC motor changes the states of the power transistors in the inverter according to

Fig. 1.13 Brushless DC motor

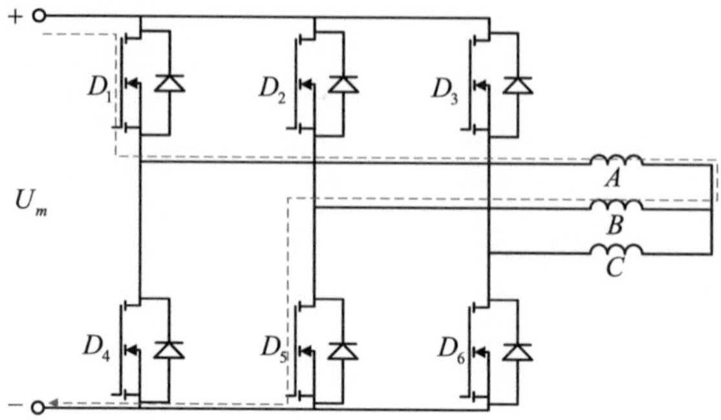

Fig. 1.14 Star-connected three-phase full-bridge main circuit

the observed phase of the back electromotive force, causing the current to flow sequentially through the motor coils. This generates a rotating magnetic field that interacts with the magnetic field generated by the magnets on the rotor, thus driving the motor rotation. For example, as shown in Fig. 1.14, turning on switches D_1 and D_5 allows current to flow from inductor A to inductor B, generating the corresponding magnetic field.

2. Propeller

The propeller is the device that directly generates the required thrust for micro-small fixed-wing UAVs. It is generally divided into fixed-pitch propellers and variable-pitch propellers. In micro-small fixed-wing UAVs, simple fixed-pitch propellers are mainly used, as shown in Fig. 1.15a. Since the efficiency of the motor varies with the output torque (which depends on the propeller type, size, speed, etc.), the matching between the motor and the propeller should be considered when designing the whole propulsion system. Proper matching can enhance the motor's efficiency, producing the same thrust with reduced energy consumption and thereby improving the UAV's endurance.

The working principle of the propeller is very similar to that of the wing. When considering the mechanism by which the propeller generates thrust, the propeller is essentially supposed to be a twisted wing. The main difference between the two is that the wing is translational in the air, whereas the propeller rotates around the shaft as it moves with the aircraft. At present, the design of any cross-section of the propeller is equivalent to the design of the corresponding airfoil, as shown in Fig. 1.15b. In pro-

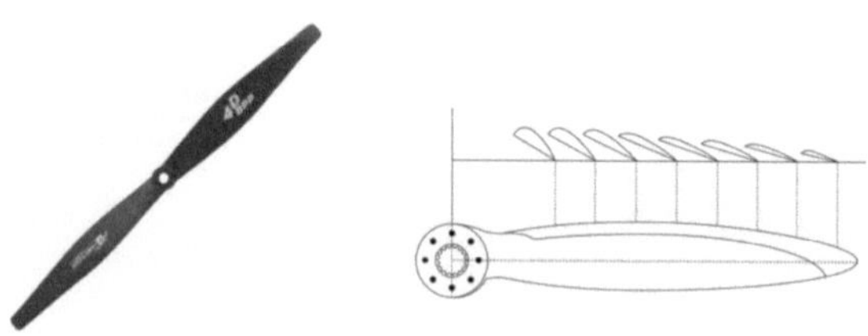

(a) T-motor T8542 F3P 4D (b) Propeller blade cross section

Fig. 1.15 Propeller and blade cross section

peller design and manufacturing, key parameters such as input power, the maximum required voltage, the number of blades, and propeller diameter are primarily considered. The airfoil distribution is then determined based on a suitable airfoil family from existing data and optimized to meet design requirements [1]. When selecting propellers for micro-small fixed-wing UAVs, the main parameters to consider are the diameter, pitch, and number of blades of the propeller.

3. Servo

A servo is an angular position servo driver suitable for systems that require changing angles to achieve control functions. It is divided into digital servos and analog servos according to different input signals. In micro-small fixed-wing UAVs, electric servos are commonly used, as shown in Fig. 1.16a, to control the deflection angles of control surfaces such as ailerons, elevators, and rudders to achieve flight control of micro-small fixed-wing UAVs. An electric servo mainly consists of a motor, a gear reducer, a control board, and a position detection element. Its electrical interface has three wires: positive power supply, negative power supply, and signal interface. The specific internal structure of the servo is shown in Fig. 1.16b.

The motor speed within the servo is high, but the torque is low; therefore, a gear reducer is used to decrease the speed and increase the torque, meeting the design requirements. The interfaces of the servo generally include Pulse Width Modulation (PWM), serial servo, CAN bus servo, Dshot, etc. For traditional analog servos, their deflection angles are controlled by PWM signals input on the signal line, as shown in Fig. 1.17. This is generally a square wave signal with a frequency of 50–400 Hz, and the deflection angle of the servo corresponds to the duty cycle of the square wave. When a PWM signal servos for micro-small fixed-wing UAV propulsion systems, the primary consideration is whether the servo can maintain the deflection angle with the torque it can transmit.

Fig. 1.16 Servo

(a) SG90 (b) Internal structure

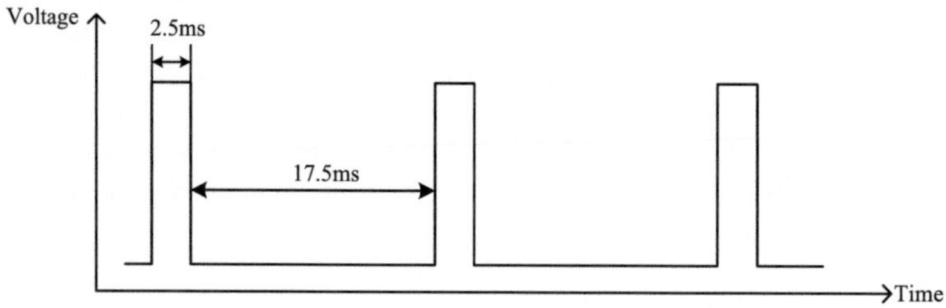

Fig. 1.17 PWM signal

1.2 Flight Principles

1.2.1 *Aerodynamic Principle*

1. Principle of Relative Motion

During flight, an aircraft generates lift and drag through relative motion with the air. The aircraft relies on its power as well as lift and drag to maintain stable flight. As air is a type of fluid with properties similar to other fluids, when an object moves through the air, aerodynamic forces are generated due to relative motion. The principle of relative motion states that when an object moves through stationary air compared to when air flows over a stationary object, if their velocities are the same, the aerodynamic forces experienced by the object are identical.

According to the principle of relative motion, aerodynamic forces depend on the magnitude of the relative velocity between the object and the air. Therefore, an experimental method is designed to maintain the aircraft stationary while allowing air to flow over it at the same velocity in the opposite direction, thereby investigating the aerodynamic characteristics of the aircraft at that velocity. This is known as the *wind tunnel test*. Currently, wind tunnel tests are widely used in industries such as aerospace and have become a crucial method for exploring the aerodynamic characteristics of aircraft. Figure 1.18 shows a fixed-wing aircraft undergoing the wind tunnel test.

2. Continuity Theorem

When a fluid flows through a pipe with a variable cross-section, due to the continuous and uninterrupted nature of fluid flow, according to the principle of mass conservation, the mass of fluid passing through any cross-section of the pipe per unit time remains constant.

As shown in Fig. 1.19, for a steady flow of gas, the mass of gas flowing into Section 1 per unit time and the mass of gas flowing out of Section 2 per unit time are equal. This leads to the continuity equation for steady gas flow:

$$\rho_1 S_1 V_1 = \rho_2 S_2 V_2. \tag{1.7}$$

Fig. 1.18 Wind tunnel test

Fig. 1.19 Variable cross-section pipe

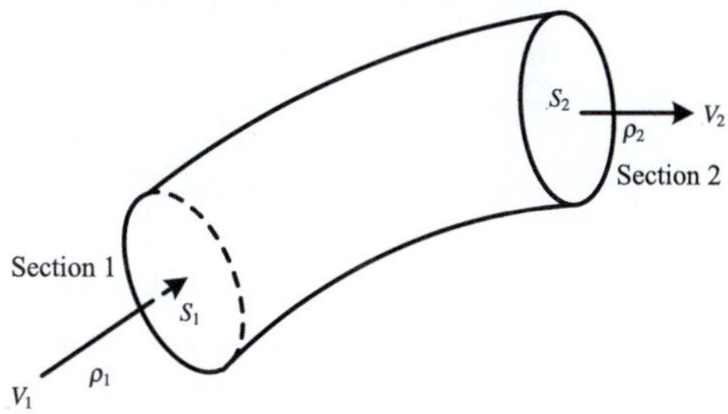

Here, ρ_i represents gas density (unit: kg/m^3); S_i represents cross-sectional area (unit: m^2); V_i represents gas flow velocity (unit: m/s), with $i = 1, 2, ...$ This equation is extended to any cross-section:

$$\rho_1 S_1 V_1 = \rho_2 S_2 V_2 = \rho_3 S_3 V_3. \tag{1.8}$$

Generally, when the velocity of air is less than 0.3 times the Mach number (the ratio of the speed of a fixed-wing aircraft to the local speed of sound), air is considered an incompressible fluid. At this point, the density of air is the same at any cross-section. Thus, the continuity equation is rewritten as:

$$S_1 V_1 = S_2 V_2 = S_3 V_3. \tag{1.9}$$

As is observed from the above equation, within a pipe, the velocity of the low-speed flowing gas is inversely proportional to the cross-sectional area. That is to say, the larger the cross-sectional area of the pipe, the lower the gas flow rate. In the flight process of micro-small fixed-wing UAVs, air can generally be regarded as an incompressible fluid.

3. Bernoulli's principle

Bernoulli's principle explains how aircraft generate lift by describing the relationship among pressure, kinetic energy, and potential energy in a fluid, based on the conservation of energy. The theorem was verbally stated by Daniel Bernoulli in 1738 and mathematically derived by Leonhard Euler in 1755. It applies to ideal fluid cases where the effects of fluid viscosity are neglected. For the case of steady incompressible fluid flow, Bernoulli's equation is:

$$\frac{P_1}{\rho_1} + \frac{{V_1}^2}{2} + gz_1 = \frac{P_2}{\rho_2} + \frac{{V_2}^2}{2} + gz_2. \tag{1.10}$$

From an energy perspective, P_1/ρ_1 represents the pressure energy of the fluid, $V_1^2/2$ represents the kinetic energy of the fluid, and gz_1 represents the potential energy of the fluid. When the fluid moves on the same horizontal plane or the change in height z is small, the change in potential energy of the fluid is neglected, leading to:

$$P + \frac{\rho V^2}{2} = P^*. \tag{1.11}$$

Here P is called the *static pressure*; $\rho V^2/2$ is called the *dynamic pressure*; and P^* is called the *total pressure*. For micro-small fixed-wing UAVs in level flight, static pressure is the pressure exerted by the air on the aircraft at rest, and dynamic pressure is the force exerted on the micro-small fixed-wing UAV due to its relative motion with the air. In simple terms, Bernoulli's law can be summarized as *the sum of kinetic energy, potential energy, and internal energy remains constant*. As shown in Fig. 1.20, when a high-speed train travels, it creates a zone of low pressure that attracts nearby objects, such as people. This is because the high speed of the train reduces the pressure of the air between the train and the person, while the slower-moving air outside creates a higher pressure, resulting in a pressure difference that pushes the person towards the train. This is an example of Bernoulli's law in everyday life. Based on the continuity theorem and Bernoulli's principle, when a fluid flows horizontally through a pipe of variable cross-section: the smaller the cross-sectional area of the pipe, the faster the fluid velocity and the lower the pressure; the larger the cross-sectional area of the pipe, the slower the fluid velocity and the higher the pressure. These are the fundamental principles of aerodynamic generation and variation in micro-small fixed-wing UAVs.

Fig. 1.20 A high-speed train attracts people nearby

1.2.2 Aerodynamics Overview

1. Lift

For a micro-small fixed-wing UAV to fly in the air, it first needs to generate lift to balance the weight of the UAV and utilize propulsion to generate thrust, thereby balancing the resistance encountered during flight. This ensures continuous flight of the micro-small fixed-wing UAV in the air. The lift of a micro-small fixed-wing UAV mainly depends on the relative motion between the wing and the air.

When the air meets the leading edge of the wing, the air velocity relative to the wing is zero, which is called the *stagnation point* of the wing. According to Bernoulli's principle, it is inferred that the pressure at the stagnation point is maximum. The air separates from the stagnation point and flows towards the upper and lower surfaces of the wing. The upper and lower surfaces mentioned here refer to the areas above and below the chord line of the wing profile. Generally, the upper surface of the wing profile is more convex, and the lower surface is flatter. When the upper and lower surfaces of the wing profile are precisely the same, it is called the *symmetric profile*. If the chord line of the wing profile forms an angle with the incoming flow velocity, it is called the *angle of attack* α. At this point, the airflow between the upper and lower surfaces of the wing profile will exhibit different flow line states, and the stagnation point will shift downward towards the lower surface of the wing, as shown in Fig. 1.21. The passage through which the airflow passes over the upper and lower surfaces of the wing profile is called the *pipe*. For a conventional asymmetrical airfoil profile with a convex upper surface and a flat lower surface, or a wing profile with an angle of attack, the pipe becomes narrower (decreased cross-sectional area) as the airflow passes over the upper surface and wider (increased cross-sectional area) as it passes over the lower surface. According to the continuity theorem and Bernoulli's principle, air velocity increases and pressure decreases over the upper surface, while air velocity decreases and pressure increases over the lower surface. At this point, a pressure difference is generated between the upper and lower surfaces, resulting in an

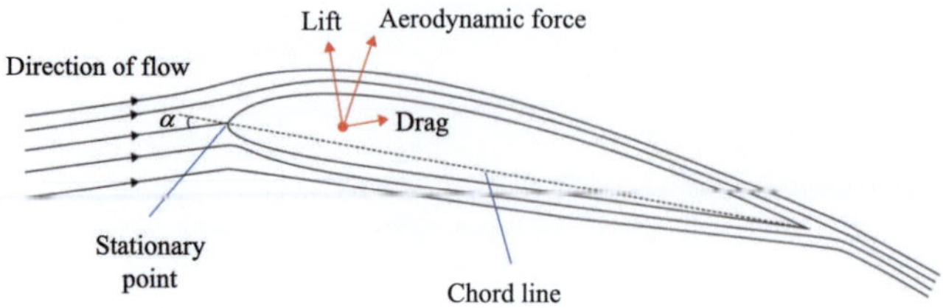

Fig. 1.21 Force on an airfoil

upward lift force on the wing. According to Bernoulli's principle, the lift generated by the wing profile is described as:

$$f = \frac{1}{2}\rho V^2 C_f. \tag{1.12}$$

Here, f represents the lift generated by the wing profile, ρ represents the air density of the low-speed straight and uniform airflow, and C_f is called the *lift coefficient* of the wing profile, mainly related to the angle of attack and wing profile parameters.

Due to the influence of the wing's geometric parameters, including planar and solid geometry, the airfoil profile varies slightly across different sections of the wing. Consequently, the lift coefficient C_L of the wing differs from the lift coefficient C_f of the airfoil profile. Thus, the lift generated by the wing is described as:

$$L = \frac{1}{2}\rho V^2 SC_L. \tag{1.13}$$

Generally, asymmetrical airfoil profiles can generate lift at a zero angle of attack or even a negative angle of attack. In contrast, symmetrical airfoil profiles require a certain angle of attack to generate lift. For the same airfoil, different angles of attack yield varying lift forces. With an increasing angle of attack, the lift force of the wing also increases. However, the angle of attack cannot continue to grow. After reaching a certain angle, under the adverse pressure gradient of the airflow and the effect of air viscosity, the airflow will separate from the leading edge of the wing. A separated flow region will form on the upper surface of the wing at the trailing edge, leading to a sudden decrease in lift force and a rapid increase in drag, resulting in a reduction of aircraft speed, known as *stall*. The angle of attack at this time is called the *critical angle of attack*. During the flight of a micro-small fixed-wing UAV, it is necessary to ensure that the angle of attack of the wing is within an appropriate range to avoid stalling.

As the angle of attack changes, the lift force of the micro-small fixed-wing UAV will change in size. However, there is a fixed point where the combined moment of aerodynamic forces acting on the airfoil remains constant concerning this point. This point is usually referred to as the *aerodynamic center*. According to Fig. 1.22, the aerodynamic center of the airfoil is calculated by converting the lift and moment

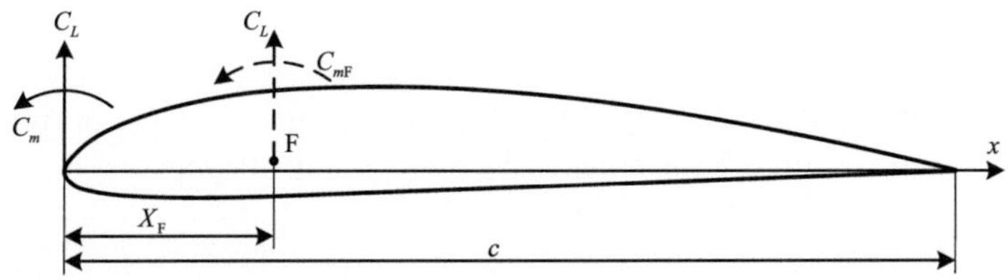

Fig. 1.22 Aerodynamic center

acting on the leading edge to the F point on the chord line. The moment at the F point is then expressed by:

$$\frac{1}{2}\rho V^2 SC_{m\mathrm{F}} = \frac{1}{2}\rho V^2 SC_L\frac{X_\mathrm{F}}{c} + \frac{1}{2}\rho V^2 SC_m. \quad (1.14)$$

Here, $C_{m\mathrm{F}}$ is the moment coefficient of the converted point; C_m is the moment coefficient of the leading edge of the wing. Since the lift coefficient and moment coefficient of the airfoil are both related to the angle of attack, Eq. (1.14) is written as:

$$C_{m\mathrm{F}} = \bar{X}_\mathrm{F}\frac{\partial C_L}{\partial \alpha}\bigg|_{\alpha=\alpha_0}(\alpha-\alpha_0) + C_m(\alpha_0) + \frac{\partial C_m}{\partial \alpha}\bigg|_{\alpha=\alpha_0}(\alpha-\alpha_0). \quad (1.15)$$

Here, $\bar{X}_\mathrm{F} = X_\mathrm{F}/c$, α_0 is the *zero lift angle of attack*. Since the moment coefficient at the aerodynamic center does not change with the angle of attack, that is, it satisfies:

$$\frac{\partial C_{m\mathrm{F}}}{\partial \alpha} = 0. \quad (1.16)$$

Therefore, the distance from the leading edge point to the aerodynamic center is obtained as:

$$\bar{X}_\mathrm{F} = -\left(\frac{\partial C_m}{\partial \alpha}\Big/\frac{\partial C_L}{\partial \alpha}\right)\bigg|_{\alpha=\alpha_0}. \quad (1.17)$$

When the angle of attack is small, $\partial C_L/\partial\alpha$ and $\partial C_m/\partial\alpha$ are essentially constant.

The aerodynamic center of the airfoil is generally located near the 0.25-chord length of the airfoil. In general, when a micro-small fixed-wing UAV flies at an appropriate angle of attack, the pitch moment of the UAV changes linearly with the angle of attack. However, after exceeding the critical angle of attack, due to the separation of airflow, the pitch moment of the wing will change rapidly. In severe cases, it may lead to the loss of control of the entire UAV.

2. Drag

Unlike lift, the composition of drag for fixed-wing UAVs is relatively complex. External components of fixed-wing UAVs, including wings, fuselage, and propulsion systems, all generate drag. Drag mainly consists of four parts: pressure drag, friction drag, induced drag, and interference drag.

(1) *Pressure Drag*, also known as the *form drag*, is caused by the same mechanism as lift. As airflow passes over protruding parts of the aircraft, unfavorable pressure gradients and air viscosity cause the airflow to separate on the leeward side, preventing pressure recovery and resulting in a back pressure differential. Pressure resistance is closely related to the frontal area of the object. The larger the frontal area, the greater the drag. In addition, if the object has a large degree of protrusion or if the trailing edge is not smooth and the cross-sectional area of the aircraft shrinks sharply at the trailing edge, it is also prone to pressure drag. Typically, various parts of a micro-small fixed-wing UAV are streamlined to reduce the maximum frontal area of the UAV and thus reduce drag.

(2) *Friction Drag*, also known as the *viscous drag*, is generated when air comes into contact with the surface of the UAV. When airflow passes over the surface of the UAV, due to the viscosity of the air, air molecules experience friction with the surface, hindering the flow of air and generating friction drag. Both the wing and fuselage of the UAV will generate pressure drag and friction drag.

(3) *Induced Drag*, also known as the *lift-induced drag*, is mainly generated by the wing and tail of the aircraft. The occurrence of induced drag is related to lift. Lift comes from the pressure difference between the upper and lower surfaces of the wing, which creates wingtip vortices. The wingtip vortices generate downwash velocity, causing the airflow direction near the wingtip to tilt downward. This results in the aerodynamic force on the upper surface of the wing near the wingtip tilting backward, thereby producing induced drag. When the lift coefficient of the wing increases, the effect of the vortex becomes more pronounced. Therefore, induced drag is considered to be caused by the appearance of lift. If there is no lift, then the induced drag is zero. Induced drag is mainly related to the planform shape, airfoil profile, and aspect ratio of the wing. It is reduced by increasing the aspect ratio and selecting an appropriate wing planform shape.

(4) *Interference Drag* refers to the additional drag generated at the junction of adjacent components of the UAV. When each component of the UAV flies independently in the airflow, it creates drag. However, when they are combined, the total drag is not the vector sum of the drag experienced by each component. In comparison, the total drag will be larger. Interference drag is mainly related to the relative positions of different components of the UAV. Therefore, when designing the overall installation structure, the drag impact must be taken into consideration. If necessary, vortex generators should be added at component junctions to reduce drag.

When researching a micro-small fixed-wing UAV, the primary source of drag is the wing. Its formula is as follows:

$$D = \frac{1}{2}\rho V^2 S C_D. \tag{1.18}$$

The lift coefficient C_L in Eq. (1.13) is only related to the angle of attack and the lift characteristics of the wing, while the drag coefficient C_D in Eq. (1.18) is related not only to the angle of attack and the drag characteristics of the wing but also to the lift coefficient C_L. In the actual flight process of UAVs, drag is mainly divided into *zero lift drag* D_0 and *lift-induced drag* D_{i}, so the drag generated by the wing is often written in the following form:

$$D = D_0 + D_{\mathrm{i}} = \frac{1}{2}\rho V^2 SC_{D0} + \frac{1}{2}\rho V^2 SAC_L^2. \tag{1.19}$$

Here, A is the lift-induced drag factor or induced drag factor, which is expressed as $A = 1/(\pi\lambda e)$; $\lambda = b/c$ is the aspect ratio; the parameter e is the Oswald efficiency factor, typically taken as 0.7 for a rectangular wing.

1.2.3 Force Analysis

During the flight of a fixed-wing UAV, it is considered to be acted upon by four forces: gravity, thrust from the propulsion system, lift generated by the air, and drag. Under the influence of these four forces, the motion of a fixed-wing UAV is roughly decomposed into four basic flight maneuvers: steady level flight, steady gliding, steady climbing, and steady coordinated turn.

When a UAV is in a state of steady-level flight, the direction of flight velocity is parallel to the horizontal plane, as shown in Fig. 1.23. The lift generated by the air is perpendicular to the horizontal plane and upward, while gravity acts vertically downward. Drag acts opposite to the direction of flight velocity, also parallel to the horizontal plane, and thrust from the propulsion system acts in the same direction as the flight velocity. Gravity and lift balance vertically, while thrust and drag balance horizontally.

When the UAV is in a state of steady gliding flight, the thrust from the UAV's propulsion system is at low power consumption or turned off, and the angle between

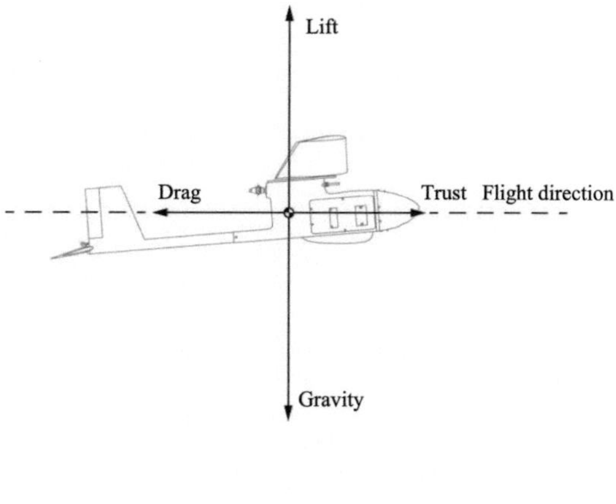

Fig. 1.23 Force analysis in level flight

the flight velocity direction and the horizontal plane is the *glide angle* (*flight path angle*). When the propulsion system is turned off and the UAV glides without power, lift and drag balance the various components of gravity, as shown in Fig. 1.24.

When the UAV is in a state of steady climbing flight, the angle between the flight velocity direction and the horizontal plane is the climbing angle (flight path angle), as shown in Fig. 1.25. The total lift is the combined force of the lift generated by the wings and the thrust from the propulsion system. In contrast, gravity is decomposed into two components: one acting opposite to lift and the other acting opposite to thrust, but in the same direction as drag.

When the UAV is in a state of steady level turning flight, the flight velocity direction is parallel to the horizontal plane. Still, the wing plane of the aircraft is at a certain angle of bank to the horizontal plane. The aircraft performs circular motion in the horizontal plane, as shown in Fig. 1.26. At this time, the lift balances the weight in the vertical direction, and the horizontal component of the lift provides the centripetal force for the circular motion.

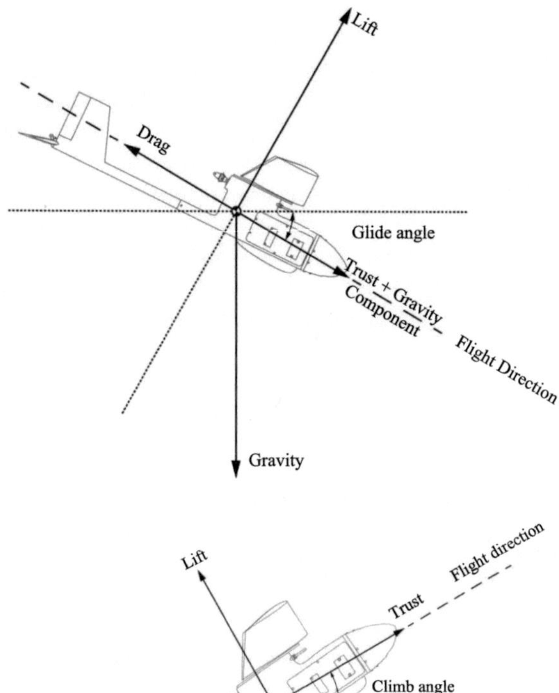

Fig. 1.24 Force analysis in gliding flight

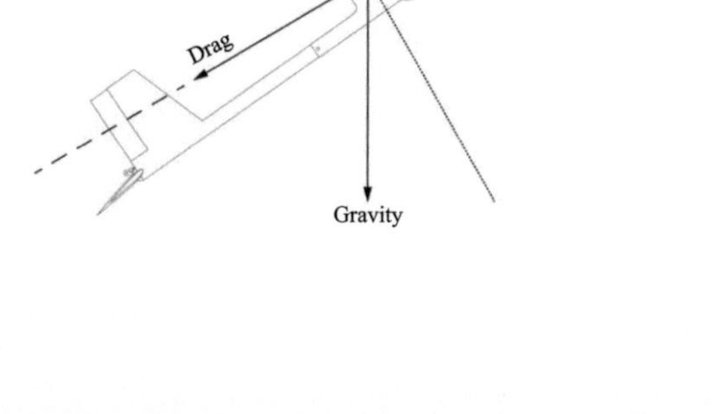

Fig. 1.25 Force analysis in climbing flight

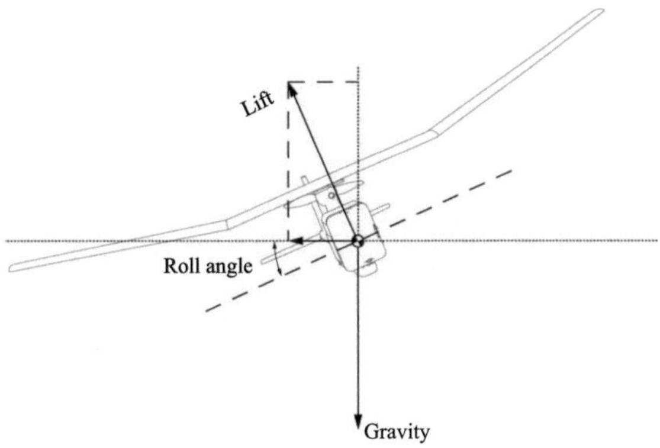

Fig. 1.26 Force analysis in turning flight

1.2.4 Stability Analysis

During flight, micro-small fixed-wing UAVs may be subject to various disturbances. If the UAV can return to a stable flight state after being disturbed and then the disturbance disappears, then the UAV is *dynamically stable*. Dynamic stability refers to the process by which the UAV returns to its initial motion state after being disturbed, such as convergence, oscillation, or divergence. However, analyzing dynamic stability is complex and often infeasible during the initial design of an aircraft. To address this issue, the concept of *static stability* is introduced. It refers to the trend of motion at the initial moment when external disturbances cease to act, such as recovery, deviation, or neither recovery nor deviation from the equilibrium position. The motion of micro-small fixed-wing UAVs in the air mainly includes pitch, yaw, and roll. Based on the motion of micro-small fixed-wing UAVs, the stability of the UAV is divided into longitudinal stability, directional stability, and lateral stability [1].

1. Longitudinal Stability of Fixed-Wing UAVs

The ability of a micro-small fixed-wing UAV to automatically restore its original longitudinal state after a disturbance is known as *longitudinal* or *pitch stability*. As the pitch of the UAV changes, the angle of attack of the wings and tail changes accordingly. For example, as the UAV pitches up, the angle of attack of the wings and tail increases. At this point, both the wings and the tail generate additional lift. It is called the *aerodynamic center* of the aircraft, as shown in Fig. 1.27. Suppose the aerodynamic center of the aircraft is behind the center of gravity, as the angle of attack increases. In that case, the increase in lift acting on the aerodynamic center against the moment of the center of gravity will resist the tendency of the UAV to increase pitch. As the angle of attack decreases, the negative increment of lift acting on the aerodynamic center (lift increment downwards) against the moment of the center of gravity will resist the tendency of the UAV to decrease pitch. In this case, the UAV is longitudinally stable. It is observed that the longitudinal stability of the UAV mainly depends on the relative position of the center of gravity and the aerodynamic center,

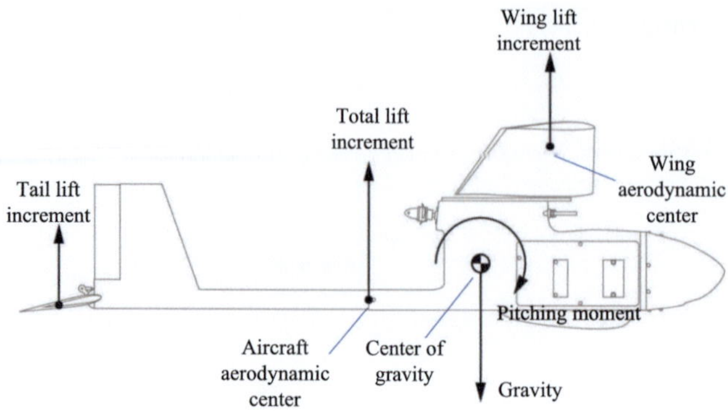

Fig. 1.27 Longitudinal stability analysis

and only when the center of gravity is in front of the aerodynamic center can the longitudinal stability of the UAV be ensured.

2. Directional Stability of Fixed-Wing UAVs

Directional static stability is more accurately termed *weathervane*, as a fixed-wing aircraft with this characteristic will align its nose with a new direction when disturbed by wind, rather than returning to its original heading. Micro-small fixed-wing UAVs mainly rely on the vertical tail to ensure their directional stability. When the UAV is disturbed, and the aircraft's body axis is not parallel to the incoming airflow, the UAV will experience sideslip, as shown in Fig. 1.28. The angle between the direction of UAV motion and the incoming airflow (acute angle) is called the *sideslip angle*. When the sideslip angle occurs, there is an angle between the incoming airflow and

Fig. 1.28 Directional stability analysis

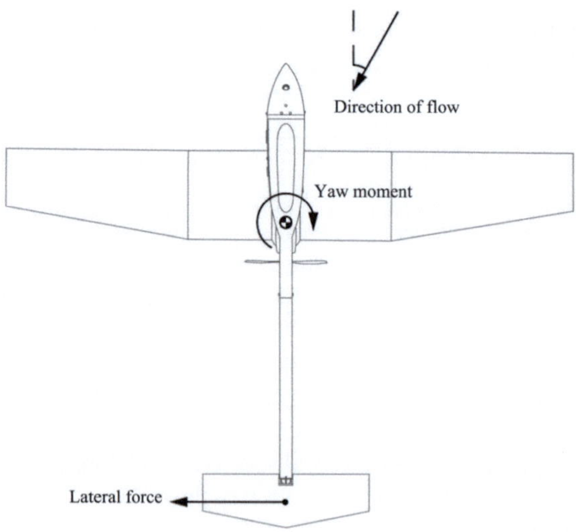

the chord line of the vertical tail, resulting in a side force on the vertical tail. Since the vertical tail is positioned behind the center of gravity, this side force, relative to the center of gravity, will counteract the tendency of the UAV to sideslip, reducing the sideslip angle until it disappears.

3. Roll Stability of Fixed-Wing UAVs

The ability of a micro-small fixed-wing UAV to automatically restore its wing to a horizontal position after deviating from the original horizontal state and producing a roll angle is called the *roll stability*. The lateral stability of the UAV is the inherent ability to resist changes in roll angle during the design of the overall aircraft structure, mainly relying on the roll moment generated by the wing's dihedral angle and wing sweep angle to restore the horizontal state, as shown in Fig. 1.29. Assuming the micro-small fixed-wing UAV is flying out of the page, when the aircraft rolls to the right (from the perspective of the onboard pilot), causing a right sideslip, the relative airflow acts on the wing with a dihedral angle, causing the angle of attack to increase after superposition with the incoming flow ahead. However, due to the dihedral angle of the wing, the angle of attack on the two wings is increased differently. The increase in angle of attack on the right wing is greater than that on the left wing, resulting in a greater lift increment on the right wing than on the left wing. This generates a rolling moment to the left, causing the UAV to roll back to a level state and achieve roll stability. In actual UAV structure design, improving lateral static stability can also be considered by adjusting the sweep angle.

In general, when analyzing micro-small fixed-wing UAVs, roll stability and directional stability are considered together and collectively referred to as *lateral-directional stability*. Their aerodynamic forces and moments are affected by the sideslip angle, as well as by the roll rate, yaw rate, aileron, and rudder control. The longitudinal stability, directional stability, and lateral stability described above represent the static stability of micro-small fixed-wing UAVs. They are the inherent performance of the entire UAV configuration when designing micro-small fixed-wing UAVs. Stability ensures a particular ability of the UAV to resist disturbances, providing the basic conditions for stable flight of fixed-wing UAVs in the air. How-

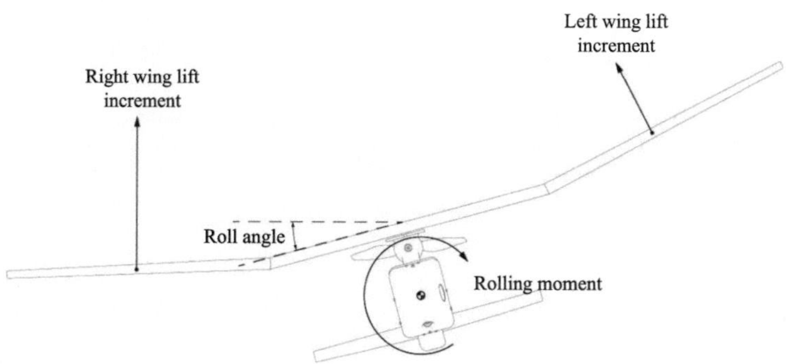

Fig. 1.29 Lateral stability analysis

ever, relying solely on the stability of the aircraft itself is not enough. When external disturbances are significant or persistent disturbances exist (such as continuous crosswinds or vertical gusts), pilot or control programs are still needed to control the UAV in real-time, adjust the UAV's flight attitude, and ensure the UAV's regular flight.

1.3 Development of Design Concept

Looking back at the history of fixed-wing aircraft, as shown in Table 1.1, it is classified according to the development of design concepts.

1.3.1 Manual Control Technology (Began in Early 20th Century)

On December 17, 1903, as shown in Fig. 1.30, the Wright brothers successfully conducted the world's first fully controlled, powered, heavier-than-air flight capable of sustained flight without landing, marking a significant milestone in aviation history.

Table 1.1 Development of fixed-wing UAV design concepts

Starting year	Design principle	Objectives	Command transmission	Social factors	Hardware factors	Control factors
Early 20th century	Manual control technology	Smooth aircraft control	Mechanical	Personal Hobby, World War I	Mechanical gyroscope	–
1930s	Automatic control technology	Reduce flightcontrol difficulty	Mechanical, Hydraulic	World War II	Autopilot	Classical control
1960s-70s	Active control technology	Improve flight control performance and safety	Fly-by-Wire	Cold War	Digital computers, Flight management systems	Modern control
2010s	ADA control technology	More economical, Lower threshold	Fly-by-wire, Optical fiber control, wireless	Rise of personal aviation, New economic growth needs, Continuous local wars	MEMS sensors, High-performance chips, High-power and high-density, Power, Smart bus	Artificial intelligence

Fig. 1.30 Early aircraft

Fixed-wing aircraft were widely used during World War I. Due to technological limitations at the time, aircraft were primarily controlled by pilots manually manipulating control sticks and pedals to regulate the aircraft's heading and attitude, relying on electronic gyroscopes to measure the aircraft's flight attitude. Pilots needed to adjust and balance controls to maintain stable flight precisely. This manual control technology, besides initial pilot training, mainly relied on mechanical control devices to fly the aircraft, with aircraft propulsion, structure, and control systems dominating aircraft design in the early period until the emergence of automatic control technology.

1.3.2 Automatic Control Technology (Began in the 1930s)

The year 1932 saw the introduction of the Boeing Company's Model 247, the first of the all-metal, unbraced-wing airliners. A heroic demonstration was given by yet another flight by Wiley Post, this time alone. Between the 15th and 22nd of July 1933, he flew the Winnie Mae around the world in a total flying time of 115 h, 3.5 min [7]. World War II profoundly impacted aircraft development, leading to more powerful engines, improved wing designs, advanced flight control systems, and enhanced electronic reconnaissance and weaponry, shaping the course of modern military and civilian aviation. During this time, fixed-wing aircraft primarily relied on fully electronic autopilots, which were made up of analog circuits, thereby reducing the workload of pilots. Automatic control technology utilizes sensors, computers, and actuators to monitor and adjust the aircraft's status and heading, enabling the aircraft to stabilize, track routes, and maintain altitude automatically. Besides mechanical transmission, hydraulic systems were also used to assist in controlling high-speed and large aircraft, facilitating maneuverability.

Various classical control technologies emerged in the context of automatic control, including Bode plots, Nyquist curves, root loci, and Nichols charts. Subsequently, in 1948, Wiener published "Cybernetics", a landmark in the birth of control theory [8]. Later, Chinese scientist Hsue-Shen Tsien published "Engineering Cybernetics" in 1954, which extended control theory to engineering technology [9].

1.3.3 Active Control Technology (Began in 1960–1970s)

Influenced by the Cold War, the United States actively developed combat aircraft with air superiority, requiring significant improvements in maneuverability, among other factors. Conventional designs often consider aerodynamics, structural strength, and engine factors, compromising one for the other to meet mission requirements. The design approach using active control technology broke this pattern by elevating flight control systems to equal importance with the factors above, making them one of the four key factors to consider in aircraft selection and playing a proactive role. Active control technology uses Fly-By-Wire (FBW), replacing traditional mechanical connections with electronic signals to transmit aircraft control commands. The primary purpose of this technology was to enhance aircraft maneuverability, stability, and safety while reducing the pilot's workload. These technologies utilized computers and advanced control systems to change the aircraft's configuration and control actively. Active control introduced the concept of relaxed static stability in aircraft design, offering benefits such as reduced flight resistance, increased total lift, reduced weight, and enhanced maneuverability. Additionally, aircraft could actively control maneuvers by adjusting wing shapes, controlling surface positions, changing engine thrust, and implementing algorithms for flight control laws, attitude control, and propulsion distribution. As shown in Fig. 1.31, third-generation fighter jets abroad have widely adopted active control technology, such as the F-16, F/A-18, Su-27, and MiG-29. Civilian aircraft have also adopted active control technology, as seen in the Boeing 777 and Airbus A320. From the mid-1970s onwards, integrated control systems have been developed in the military field, with applications in flight control

(a) F-16 (b) F-18

(c) Su-27 (d) Mig-29

Fig. 1.31 Third-generation fighter jets

that aim to coordinate these key subsystems to ensure aircraft stability, precision, and efficiency under various flight conditions. It integrates firepower, propulsion, and navigation control systems, enabling aircraft to perform multiple missions and cope with different operating environments.

This period was the golden age of modern control theory development, especially in the 1960s-80s, with the work of Kalman, Bellman, and Pontryagin playing a crucial role in the development of state-space models and optimal control theory. These methods found wide applications in aerospace, navigation, and flight control. Subsequently, multivariable control and robustness theory became part of modern control theory. These theories and technologies allow engineers to deal more effectively with multivariable systems, improving the performance of control systems. Meanwhile, adaptive control and intelligent control also developed, with intelligent control at that time typically referring to shallow neural network control, fuzzy logic control, and other approaches.

1.3.4 ADA Control Technology (Began in 2010s)

ADA stands for Autonomous, Dependable, and Affordable [10]. In recent years, the increasing complexity of aircraft sensors and functions has led to more intricate designs, higher production costs, and greater maintenance difficulties. These factors have led to the discontinuation of many high-performance aircraft, such as the F-22 Raptor fighter jet (2015) and the Airbus A380 (2021). In addition, with the rise of personal aviation and the low-altitude economy, eVTOL aircraft represent a rapidly developing field in aviation. These aircraft utilize electric and vertical takeoff and landing technologies, with the potential for revolutionary impact. It directly benefits from the development of high-power and high-density electric systems, lightweight materials, MEMS sensors, high-performance chips, intelligent buses, and other technologies. Whether military, traditional civilian, or personal aviation, autonomy, dependability, and affordability are all critical design factors to consider in aircraft design.

(1) Aircraft's *autonomy* refers to the extent to which an aircraft can perform various tasks and operations without human intervention. This includes flight missions, navigation, control, and decision-making aspects. Specifically, aircraft autonomy is divided into manual control (corresponding to the period of manual control technology), assistance and cooperation (corresponding to the period of automatic control technology), semi-autonomous (roughly corresponding to the period of active control technology), and fully autonomous. Here, full autonomy is the highest level of autonomy, where the aircraft can independently perform tasks in almost all situations, including flight planning, navigation, flight control, obstacle avoidance, and decision-making [11, 12]. Additionally, aircraft autonomy is typically associated with UAVs and autonomous aircraft. Autonomy involves integrating automation, sensing technology, artificial intelligence, and autonomous decision-making systems to ensure that aircraft can perform tasks safely and efficiently. With ongoing technological

advancements, aircraft autonomy will continue to be a key area of research and development in the aviation field.

(2) In the 1980s, *dependability* was first proposed as a term encompassing the availability of products and the factors influencing it, including reliability, safety, and maintenance support. Aircraft dependability refers to the ability of aircraft systems, components, and the overall aircraft to operate reliably and safely under specific operating conditions and within a specified time range according to design requirements and performance parameters. Dependability is a crucial aspect of aircraft safety and performance, ensuring that the aircraft maintains a high level of performance and safety throughout its entire lifecycle, which is essential for flight safety and aircraft availability.

(3) Aircraft's *affordability* refers to the balance between the cost and performance of an aircraft over its entire lifecycle, including purchasing, maintenance, fuel consumption, operation, and termination of operation, including purchase costs, fuel efficiency, maintenance costs, service life, etc. Affordability is a crucial concept in the aviation industry and aviation transportation, as it directly affects the profitability of airlines and the operational efficiency of aircraft.

This period is the golden age of big data and artificial intelligence development, with massive data, deep learning (algorithms), and high-performance chips (computing power) playing a crucial role. In many ways, intelligent algorithms surpass human capabilities and have a continuous impact on various industries. Currently, the fields of electric vehicles, aviation, and aerospace are moving towards more ADA designs. For example, Tesla reduces costs through integrated molding technology; General Atomics Aeronautical Systems, Inc. (GAASI) announces its latest Gambit series unmanned combat aircraft, drawing inspiration from the automotive industry and manufacturing Gambit drones that are converted into four models using methods similar to commercial vehicle assembly lines, thereby reducing manufacturing costs and speeding up development; XSpace reduces costs through engine modularity and standardization; a large number of artificial intelligence technologies represented by deep learning have been introduced, and Tesla's "Full Self-Drive (FSD)" car, based on end-to-end learning, reportedly not only avoids defects introduced by human programming but also accelerates iterative improvements in autonomy and safety performance.

1.4 Structure of This Book

The purpose of this book is to enable readers to design control systems using the knowledge gained from introductory technical courses and professional courses, such as "Control Theory," after completing them. It aims to provide practical exercises in applying professional theories to solve engineering problems. Additionally, due to the strong emphasis on engineering practice in flight control-related courses, it is challenging to fully understand the essence of flight control solely through theoretical study. This book presents a comprehensive mathematical model of fixed-wing UAVs,

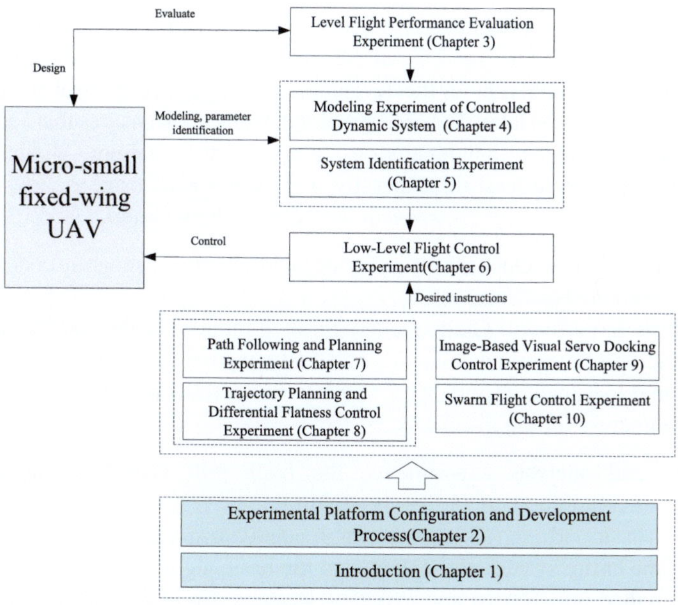

Fig. 1.32 Book structure

along with the design of software and hardware based on this model, in a simulated environment, as illustrated in Fig. 1.32.

The book contains eight theme experiments:

(1) Experiment 1: Level Flight Performance Evaluation Experiment
(2) Experiment 2: Controlled Dynamic System Modeling Experiment
(3) Experiment 3: System Identification Experiment
(4) Experiment 4: Low-Level Flight Control Experiment
(5) Experiment 5: Path Following and Planning Experiment
(6) Experiment 6: Trajectory Planning and Differential Flatness Control Experiment
(7) Experiment 7: Image-Based Visual Servo Docking Control Experiment
(8) Experiment 8: Swarm Flight Control Experiment.

The Level Flight Performance Evaluation Experiment (Experiment 1) is used to evaluate the performance of micro-small fixed-wing UAVs, including the maximum flight speed, cruising speed, endurance, and corresponding flight distance. In turn, the methods in this chapter help to design key parameters of fixed-wing UAVs. Some parameters in *Experiment 1* will be used in *Experiment 2*. The Controlled Dynamic System Modeling Experiment (Experiment 2) will establish full-state non-linear equations of fixed-wing UAVs and analyze the entire model. The acquisition of some parameters in *Experiment 2* will be through *Experiment 3*. The System Identification Experiment (Experiment 3) aims to identify key parameters using observed data, which is particularly important for obtaining the true parameters of fixed-wing

UAVs. Based on the models from the previous experiments, the Low-Level Flight Control Experiment (Experiment 4) can achieve some command tracking and following. This lays the foundation for the development of high-level control and planning (Experiments 5, 6, 7, 8) later. The code provided in this book ensures that each theme experiment is completed individually or in a progressive structure. All eight theme experiments include at least three step-by-step experiments from shallow to deep: *Basic Experiment*, *Analysis Experiment*, and *Design Experiment*.

(1) Basic Experiment: Open the offered code, read and run the program code, observe and record analysis data.
(2) Analysis Experiment: Guide readers through modifying the code, running the modified programs, and collecting and analyzing the data.
(3) Design Experiment: Based on the above two experiments, design an experiment independently for the given task.

For basic and analysis experiments, the book will provide complete code https://rflysim.com/doc/en/C/5.FW_Book.html.

Through basic and analysis experiments, readers can quickly understand the architecture of the entire system, the model, and the basic algorithm. The design experiment is divided into *Design* and *Practice*. Readers can independently design and develop algorithms based on the code architecture in the basic and analysis experiments, replace relevant modules one by one, and conduct Hardware-In-the-Loop (HIL) simulation tests. This book also sets up the Software-In-the-Loop (SIL) and HIL simulations based on the RflySim toolchain, which will be introduced explicitly in Chap. 2.

1.5 Exercise

1. In addition to Bernoulli's principle, other principles contribute to the generation of lift by the wings of a fixed-wing UAV. It is recommended that the reader conduct online research on this topic and present their findings.
2. As a nascent athletic pursuit, Frisbee can traverse a considerable distance when users impart a rotational angular velocity upon it and propel it at a specific angle. Please provide a comprehensive analysis of the flight principle of a Frisbee.
3. Please analyze in detail the force and stability of a kite when it is towed by the user with a thin rope under certain horizontal wind speed conditions.
4. As shown in Fig. 1.33, some fixed-wing model aircraft can hover vertically. Please analyze the forces acting on it.

Fig. 1.33 Vertical hover

(a) Landing

(b) Flight trajectory

Fig. 1.34 Aircraft landing

5. The Red Bull Air Force flies a small aircraft as shown in Fig. 1.34a to perform ultra-short takeoff and landing on the rooftop helipad of the Burj al-Arab in Dubai.[7] Why can this aircraft take off and land in such a short distance? Please describe the entire process of takeoff and landing.

References

1. J.D. Anderson, *Introduction to Flight* (McGraw-Hill Education, 2007)
2. R.W. Beard, T.W. McLain, *Small Unmanned Aircraft: Theory and Practice* (Princeton University Press, 2012)
3. H. Zhang, S. Tan, Z. Song, Q. Quan, Performance evaluation and design method of lifting-wing multicopters. IEEE/ASME Trans. Mechatron. **27**(3), 1606–1616 (2022)
4. K. Xiao, Y. Meng, X. Dai, H. Zhang, Q. Quan, A lifting wing fixed on multirotor UAVs for long flight ranges, in *2021 International Conference on Unmanned Aircraft Systems (ICUAS)*. (IEEE, 2021), pp.1605–1610

[7] https://www.redbull.com/int-en/projects/bullseye-landing.

5. Q. Quan, S. Wang, W. Gao, Lifting-wing quadcopter modeling and unified control. J. Guidance Control Dyn. **48**(3), 689–699 (2025)
6. M. Simons, *Model Aircraft Aerodynamics* (Motorbooks Intl, 2000)
7. D. McRuer, D. Graham, Flight control century: Triumphs of the systems approach. J. Guidance Control Dyn. **27**(2), 161–173 (2004)
8. N. Wiener, *Cybernetics or Control and Communication in the Animal and the Machine* (MIT Press, 1961)
9. H.-S. Tsien, *Engineering Cybernetics* (McGraw-Hill, 1954)
10. K.-Y. Cai, K.S. Trivedi, B. Yin, S-ADA: software as an autonomous, dependable and affordable system, in *2021 51st Annual IEEE/IFIP International Conference on Dependable Systems and Networks—Supplemental Volume (DSN-S)* (2021), pp. 19–20
11. United States, *Department of Defense, Unmanned Aircraft Systems Roadmap: 2005–2030* (Department of Defense, Office of the Secretary of Defense, 2005)
12. Z. Chen, J. Wei, Y. Wang, R. Zhou, UAV autonomous control levels and system structure. Acta Aeronautica et Astronautica Sinica **32**(6), 1075–1083 (2011). in Chinese

Chapter 2
Experimental Platform Configuration and Development Process

Xunhua Dai and Quan Quan

This book's experimental platform comprises two components: the RflySim software platform and the hardware platform, both of which adhere to the Model-Based Design (MBD) concept. The development process includes numerical simulation, Software-In-the-Loop (SIL, also referred to as SITL) simulations, and Hardware-in-the-Loop (HIL, also known as HITL) simulations. The following sections introduce the basic components, installation setup, and development process for each part of the platform.

2.1 Overall Introduction

2.1.1 Software Platform

The RflySim toolchain is the core experimental platform used in this book. It is a new development and testing platform developed by Beihang Reliable Flight Control Group (Rfly[1]) since 2016, led by Prof. Quan and mainly developed by Associate Prof. Dai Xunhua,[2] in collaboration with engineers from Beijing ZhuoYi Intelligent Technology Co. Ltd. Since 2020, FeiSi Laboratory, under Beijing ZhuoYi Intelligent Technology Co. Ltd, is responsible for promoting the development of advanced functions and ensuring long-term growth and maintenance. The platform is based on the MBD concept. It utilizes Pixhawk/PX4, MATLAB/Simulink, ROS, and other technologies to enable unmanned aerial system algorithm simulation, HIL simulation, and seamless transition to real-world implementation. MATLAB/Simulink is used throughout the design phase to achieve functionalities such as controller design, code generation, and autopilot code compilation. With its powerful visual

[1] https://rfly.buaa.edu.cn/.

[2] Xunhua Dai studied in the Reliable Flight Control Group from 2012 to 2020.

Q. Quan et al., *Micro-Small Fixed-Wing UAV Control*,
https://doi.org/10.1007/978-981-95-1954-5_2

processing libraries, Python is the upper-level tool for visual perception and UAV swarm algorithm development. The software package provided with this book includes a one-click installation script, which simplifies the entire software installation and configuration process. Along with the RflySim toolchain provided in this book, the software package constitutes the experimental software platform. The RflySim toolchain comprises numerous software tools essential for modeling, simulation, and algorithm validation processes in unmanned systems [1–3]. Its core dependent components mainly consist of the following parts.

1. **Microsoft Windows**: Microsoft Windows is an operating system with a graphical user interface developed by Microsoft Corporation. It is one of the most widely used operating systems globally, running on computers, smartphones, and other devices. The RflySim toolchain supports Windows 10 and later versions. The software is developed using a batch scripting language, which enables one-click activation of all necessary software and completion of deployment configuration, significantly improving experimental efficiency.
2. **MATLAB/Simulink**: MATLAB/Simulink is a visual simulation tool developed by MathWorks.[3] It allows for the easy creation of simulation systems using modular programming languages. The tool is widely used in the aircraft and automotive industries for modeling dynamic systems, controller design, SIL and HIL simulation, and performance analysis. The simulation software package and example code provided in this book are recommended for use with MATLAB 2022b and later versions. It is essential to note that this book does not include the MATLAB installation package or provide an installation process. Therefore, readers are advised to purchase and install the required toolboxes independently. If possible, it is recommended that MATLAB R2022b be installed with all toolboxes selected for full installation.

   - MATLAB/Simulink: Used for a basic simulation platform of fixed-wing UAVs.
   - Control System Toolbox: Provides tools for systematic analysis, design, and tuning of linear control systems.
   - Curve Fitting Toolbox: Provides tools for data curve and surface fitting.
   - Aerospace Blockset: Simulink reference examples and modules for modeling and simulating high-fidelity aircraft, multicopters, and spacecraft platforms.
   - Aerospace Toolbox: Tools for analyzing aviation and aerospace vehicles' motion, tasks, and environments.
   - MATLAB Coder: Converts MATLAB files into C/C++ code and generates executable files.
   - Simulink Coder: Converts Simulink models into C/C++ code and generates executable files.
   - Stateflow: Toolbox for designing state machines.

[3] https://www.mathworks.com/.

3. **Pixhawk support package (PSPtoolbox)**[4]: MathWorks has officially launched a toolbox for Pixhawk autopilots.The team has made appropriate updates and optimizations to the official toolkit to ensure compatibility with the latest Pixhawk and MATLAB versions.
4. **FlightGear flight simulator**[5]: FlightGear is a widely used open-source flight simulator software that can receive flight status updates sent by Simulink via UDP (User Datagram Protocol). This enables real-time display of the aircraft's flight status during Simulink simulations. It is primarily used to display the attitudes of fixed-wing UAVs during numerical simulations.
5. **PX4 autopilot**[6]: The PX4[7] autopilot is an open-source project that originated from the Pixhawk project at the Computer Vision and Geometry Lab at ETH Zurich in Switzerland. It provides a high-performance autopilot for flight control enthusiasts and research teams worldwide at a low cost.
6. **PX4 firmware compilation environment**: Compiling PX4 firmware on Windows is not possible directly. The recommended approach is to install a Linux emulation environment such as WSL, WSL2 or Cygwin and deploy the compilation toolchain. This enables the compilation of the PX4 source code and controller code generated by Simulink into firmware format '.px4', which can then be downloaded and flashed onto the Pixhawk autopilot hardware, similar to the process of reinstalling the system. After restarting, the control algorithms generated by the PSP toolbox will run automatically.
7. **QGroundControl (QGC) ground station**[8]: QGroundControl is an open-source ground station designed for the PX4 autopilot. It is written in C++ using the QT editor and is mainly used for sensor calibration and parameter adjustment of PX4 autopilots before takeoff. Additionally, it receives the flight status of fixed-wing UAVs and sends control commands in real-time during flight through wireless data transmission modules.
8. **CopterSim real-time motion simulation software**: CopterSim is a core software component of the RflySim toolchain, consisting of two main parts: model and communication. The model performs simulations based on parameters and supports Dynamic-Link Library (DLL) operation. It works with other software to create a software/hardware-in-the-loop simulation. CopterSim serves as a hub for all data communication. The flight control system and CopterSim are connected via a serial port or a network TCP/UDP connection, using MAVLink for data transmission to achieve control loop simulation and simulate outdoor flight scenarios.
9. **Python38Env environment**: Python is a widely used cross-platform programming language in scientific computing and research due to its simplicity, readability, and extensibility. Python38Env is a packaged Python environment based

[4] https://ww2.mathworks.cn/help/uav/px4-spkg.html.

[5] https://www.flightgear.org/.

[6] https://github.com/PX4/PX4-Autopilot.

[7] https://px4.io/.

[8] http://qgroundcontrol.com/.

on Python 3.12 that includes libraries such as NumPy, PyMavlink, OpenCV, and PyUlog, making it convenient for Python-related programs to call specific compilers.
10. **WinWSL compiler**: The WinWSL subsystem is a subsystem on the Windows operating system that allows users to run Linux applications in Windows. In this case, WinWSL is a set of PX4-firmware compilation environments deployed and packaged by the team based on the WSL/Ubuntu22.04 subsystem.
11. **RflySim3D**: RflySim3D is a simulation software for unmanned systems developed using the Unreal Engine 4 (UE4) engine. It communicates with other software in the platform through UDP, enabling a high-fidelity display of fixed-wing UAV attitudes and positions. The HIL simulation software platform is composed of CopterSim simulation software and RflySim3D display software. The platform has a distributed, independent running mechanism that ensures compatibility for future swarm simulations.
12. **Simulink swarm control interface**: The RflySim toolchain provides a rich set of UAV swarm control modules. The Simulink UAV swarm control interface file is implemented through C++ mixed compilation using Simulink S functions, in conjunction with the UDP module provided by Simulink.
13. **RflySim supplementary material files**: The RflySim toolchain provides comprehensive learning materials and example files to help users build and develop their own unmanned systems in a one-stop way, from the low-level control algorithm to the middle-level planning algorithm to the upper-level learning algorithm development and validation. The folder **[*installation directory*]/ PX4PSP/RflySimAPIs/*** contains all source codes and learning materials for the platform. Users can explore the API and PPT files in each chapter to run and simulate the examples provided by the RflySim toolchain, aiding in their familiarization with unmanned systems development. Additionally, online documentation is available at rflysim.com.

2.1.2 Hardware Platform

The recommended hardware platform for this book consists of four main components.

1. **Ground computer**: The ground computer is a personal computer with a high-performance operating system. Its primary function in this experimental platform is to provide a software runtime environment for the simulation software packages provided in this book. It is responsible for controller design, SIL simulation, code generation, HIL simulation, and other functions. To ensure the smooth operation of all software, the computer must meet the basic configuration requirements for both low-level and upper-level flight control systems, as described in this book.

   - Operating System: Windows 10 (version 1809 or later)/Windows 11, 64-bit;
   - Processor: Intel 10th generation i5 processor or higher, or equivalent AMD processor;

- Memory: 16GB RAM or more, with a frequency of DDR3 1600MHz or higher;
- Graphics Card: Intel integrated graphics UHD 620 or higher, or equivalent AMD graphics card;
- Hard Drive: Remaining capacity of 60GB or more, SSD recommended;
- Ports: At least one USB Type-A port, with extension capability;
- Display: Resolution of 1080P or higher; dual monitors recommended.

It is important to note that the above configuration meets the requirements for underlying automatic code generation and HIL simulation validation. For computers with lower-performance graphics cards, running the relevant one-click startup scripts provided with the platform may require executing the "***.bat" scripts with the "LowGPU" suffix located in the RflySimAPIs folder of the installation directory. For more realistic 3D scenes or to run the vision-based control examples in this book, you will need a computer with a higher-performance graphics card, recommending a discrete NVIDIA GTX2060 or higher (or equivalent AMD graphics card).[9]

2. **Autopilot**: The autopilot serves as the operational platform for control algorithms. It possesses powerful computing capabilities and rich sensors to acquire flight status and calculate control commands for the propulsion system, thereby achieving the flight control of fixed-wing UAVs. This book chooses the widely used PX4 series open-source software system, with hardware products such as Pixhawk[10] and CubePilot.[11] This open-source hardware is designed to provide an easy-to-use, high-quality, and affordable flight control system for education, hobbies, and developers. It offers a range of flight control hardware products to meet various performance and cost requirements, promoting the development of the UAV industry. These hardware platforms support open-source flight control software such as PX4 or Ardupilot. This book primarily focuses on "CubePilot/Pixhawk+PX4" software and hardware architecture.
3. **Remote control system**: It is used to transmit control commands from ground operators to the autopilot of fixed-wing UAVs. It includes remote controllers, receivers, chargers, and other necessary components.
4. **Airframe**: It comprises a fuselage, landing gear, and other components for transporting payloads, propulsion systems, and autopilots. These elements must exhibit optimal aerodynamic characteristics and structural integrity to guarantee the dependability of flight operations.
According to the functionalities and requirements of the RflySim toolchain and subsequent experimental needs in this book, the recommended educational hardware mainly includes:

   (1) The recommended desktop computer model is: Refer to Sect. 2.1.2 for recommended computer configurations, detailed computer model recommen-

---

[9] For increased development efficiency, higher computer performance is preferable.

[10] http://pixhawk.org/.

[11] http://cubepilot.org/.

dations can be found at: "[*installation directory*]\RflySimAPIs \1.RflySimIntro\2.AdvExps\e4_ComputerIntro\Readme.pdf".

(2) The hardware required for HIL simulation includes autopilot hardware, remote controllers, remote controller receivers, and several DuPont wires. Flight controllers recommended by the RflySim toolchain include:
   (a) Pixhawk 6x: Detailed configurations are found in "[*installation directory*]\RflySimAPIs\1.RflySimIntro\2. AdvExps\e2_FCUIntro\2.Pixhawk6x\Readme.pdf".
   (b) CubePilot Cube Orange: Detailed configurations are found in "[*installation directory*]\RflySimAPIs \1.RflySimIntro\2.AdvExps\e2_FCUIntro\3.CubePilot\Readme.pdf".

   Recommended remote controllers include:
   (a) WELY ET10: The matching remote controller receiver is WELY RF209S. Detailed configurations are found in "[*installation directory*]\RflySimAPIs\1.RflySimIntro\2.AdvExps\e1_RCIntro \1.WELY_ET10\Readme.pdf".
   (b) FLYSKY FS-i6S, with the matching remote controller receiver being FLYSKY FS-iA6B. Detailed configurations are found in "[*installation directory*]\RflySimAPIs\1.RflySimIntro\2.AdvExps\e1_RCIntro\2FS_i6S\Readme.pdf".

(3) The recommended aircraft is the UltraStick 25E [4], one of the series of electric remote-controlled aircraft launched by HorizonHobby. It is used for research, entertainment, practice, and even small-scale aerial performances. It is particularly suitable for flying in areas without fuel-powered aircraft. The basic configuration of the UltraStick 25E is shown in Fig. 2.1. For more detailed introductions and instructions on using this aircraft, refer to: "[*installation directory*]\RflySimAPIs\2.RflySimUsage\0.ApiExps\e2 _ FWConfig\Readme.pdf".

2.1.3 Software/Hardware Overall Relationship

The preceding sections presented the software architecture of the RflySim toolchain, along with its hardware components. The software cores primarily consist of CopterSim and RflySim3D, while the hardware devices include ground computers, autopilots, and remote controllers. Proficient use of these tools can greatly enhance development efficiency, reduce complexity, and save time. Figure 2.2 shows the interrelationship between the software and hardware components of the experimental platform. Subsequent chapters will provide detailed information on the specific composition of each component and the development process using each part.

The RflySim toolchain's overall framework is illustrated in Fig. 2.2. It comprises modules such as the "Low-Level Controller" motion simulation model, the 3D simulation model, and the "Upper-Level Controller" for swarm control. The "Low-Level

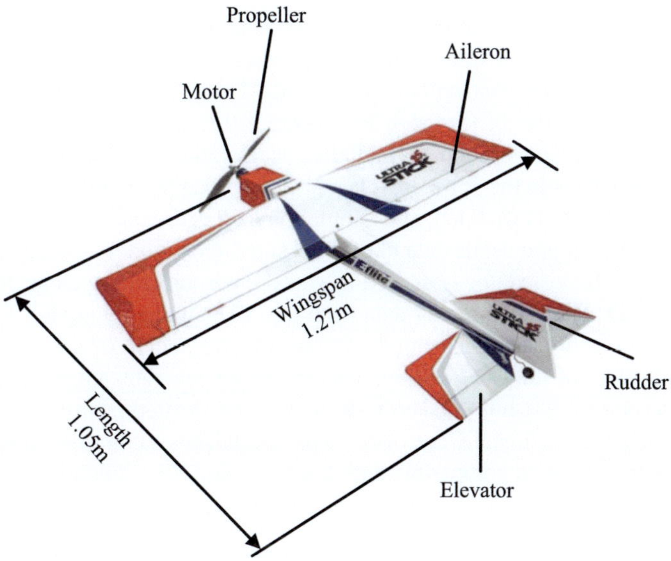

Fig. 2.1 UltraStick 25E

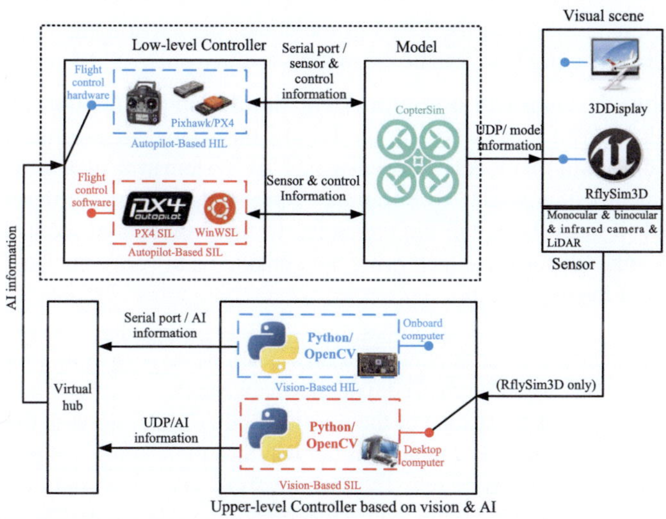

Fig. 2.2 The RflySim toolchain's overall framework

Controller" can switch between SIL and HIL simulations based on PX4. Sensor data and control information are transmitted to CopterSim through UDP or serial communication. CopterSim receives motor control signals from the "Low-Level Controller" and inputs them into the DLL model. The model is then driven forward in real-time at 1 kHz, and the output sensor data (including acceleration, GPS, and magnetometer readings) is sent back to the "Low-Level Controller" to enable regular operation. The sensor hardware data is shielded, and simulation data is used as a substitute. Meanwhile, it sends simulation state data to the "3D simulation model" so that the motion of unmanned vehicles can be observed in visual environments, such as 3DDisplay and RflySim3D. The visual sensor data from the "3D simulation model" (including camera images and laser point clouds) is the primary source of data for "Upper-Level Controller" of the UAV swarm. The controller generates upper-level control commands such as disarming, takeoff, desired speed, desired position, and desired trajectory to the "Low-Level Controller". This establishes a closed-loop simulation and development system compatible with both HIL and SIL simulations.

2.1.4 Overview of Development Process

When developing based on the RflySim toolchain, the experimental development process of a fixed-wing UAV can be cascaded into three phases: MIL, SIL, and HIL.

1. **Algorithm development and numerical simulation verification phase**. This phase, also known as the Model-in-the-Loop Simulation (MIL) verification phase, focuses on building a model of a fixed-wing UAV system in an ideal and simplified environment using MATLAB/Simulink. Various algorithms are developed and validated to meet the initial requirements. In these models of fixed-wing UAV systems, other disturbance factors are not considered to verify the feasibility and accuracy of the algorithms.
2. **SIL simulation verification phase**[12]: SIL simulation requires compiling and executing source code generated on a host machine as a separate process. This allows for testing the numerical equivalence between the model and the generated code by comparing the results of the normal mode simulation with the SIL simulation. The SIL simulation phase on the RflySim toolchain is conducted in the MATLAB/Python environment. The task involves designing control algorithms in Simulink/Python using provided UAV simulation models and code examples. It is important to ensure consistency between the input-output signal and the actual UAV system by correctly connecting the model and controller. Before conducting the SIL simulation, the fixed-wing UAV model built in MATLAB/Simulink must be compiled into a DLL file and then loaded into CopterSim. The fixed-wing UAV

[12] Subsequent chapters of this book do not specifically differentiate between "model-in-the-loop simulation" and "software-in-the-loop simulation" but collectively refer to "software-in-the-loop simulation" to distinguish between "software-in-the-loop simulation" and "hardware-in-the-loop simulation."

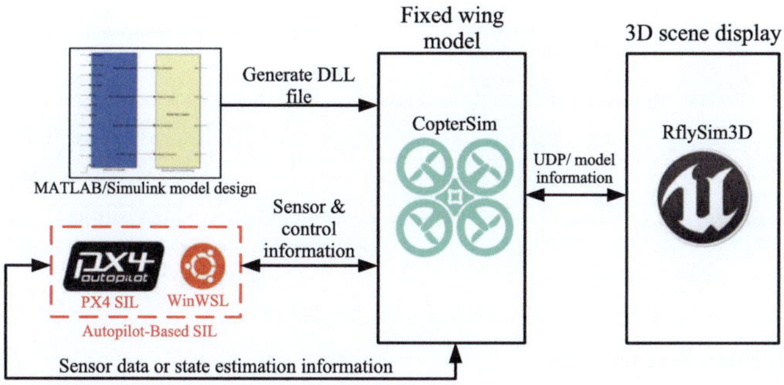

Fig. 2.3 SIL simulation structure

model in CopterSim sends sensor data and state estimation information, such as attitude angles, angular rates, position, and velocity, to the controller. The controller then sends PWM control commands to each motor, creating a closed-loop SIL simulation system. Figure 2.3 illustrates the communication structure during the simulation process. Users can observe control performance and modify or design controllers to meet desired performance requirements.

3. **HIL simulation verification phase**: HIL simulation is a development and testing technique used for real-time embedded systems. It provides a dynamic system model that can simulate a real system environment, incorporates mathematical representations of relevant dynamic systems, and connects them to the simulation system platform through the input and output of the embedded system. The RflySim toolchain enables the importation of Simulink fixed-wing UAV model parameters into CopterSim. It also allows the downloading of the Simulink controller algorithm-generated code to the CubePilot/Pixhawk autopilot. Finally, USB signal cables replace the virtual signal lines in Simulink. CopterSim transmits sensor data, such as accelerometers, barometers, and magnetometers, to the CubePilot/Pixhawk autopilot via USB. The PX4 flight controller software in the CubePilot/Pixhawk autopilot filters and estimates the received sensor data, and sends the estimated state information to the controller through the internal uORB message bus. Figure 2.5 illustrates the controller, which then sends PWM control commands for each motor back to CopterSim via USB, establishing a closed-loop HIL simulation. Figure 2.4 illustrates the communication structure during the simulation process. In the HIL simulation, the fixed-wing UAV model runs at the same speed as the actual clock, ensuring real-time simulation. The control algorithm is also deployed and run on a real embedded system, making it more similar to an actual fixed-wing UAV system. It is important to note that transmission delays may occur in hardware communication, and the simulation models and environments in HIL systems may differ from those in SIL systems. As a result, additional adjustments to controller parameters may be required to meet design requirements and satisfy real-world scenarios.

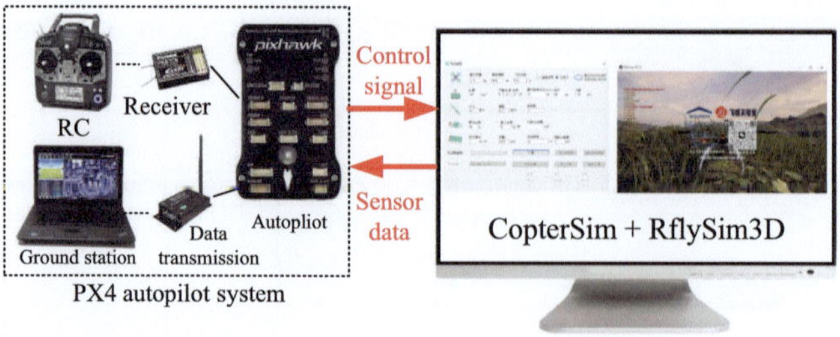

Fig. 2.4 HIL simulation structure

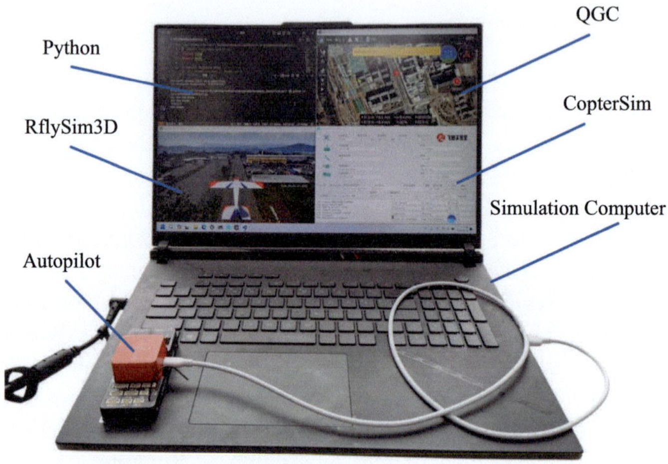

Fig. 2.5 Hardware layout of HIL simulation

2.2 Software Platform Installation and Configuration

2.2.1 Software Platform Installation Steps

The installation process of the entire RflySim toolchain software package has been highly automated, allowing for quick and fully automatic deployment by running a one-click installation script. The overall installation steps are summarized in the following:

(1) Visit the official software package download website (https://rflysim.com/download.html), enter your email and personal information to obtain the latest version's download link. After downloading, the installation package is an image file named "RflySimToolchain-***.iso". Extract the installation folder using Windows Explorer or another virtual drive software, and then read the "HowToInstall.pdf" file for detailed installation instructions. Alternatively, you can visit rflysim.com for the latest online user manual.

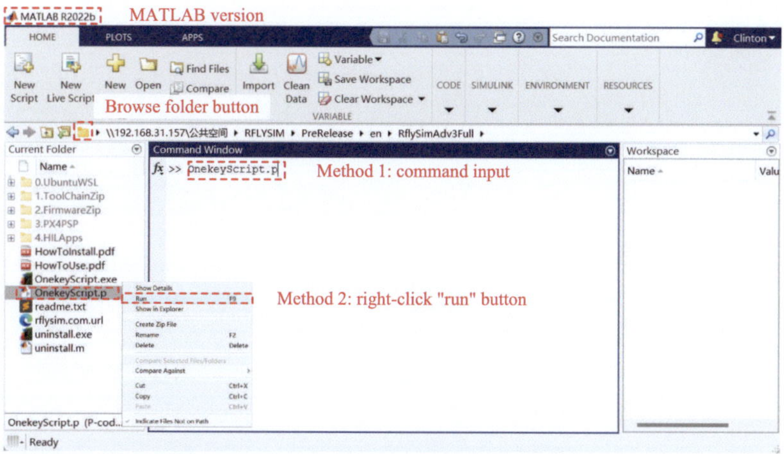

Fig. 2.6 RflySim simulation package and one-click installation script

The experimental code examples of this book are obtained through the following two methods:

(a) After installing the RflySim toolchain, you can find all accompanying materials for this book in "[*installation directory*]\RflySimAPIs\5.RflySimFlyCtrl \1.BasicExps \e10-FixedWingCtrl" folder.
(b) Download via browser link: https://rflysim.com/doc/en/C/5.FW_Book.html.

(2) Enable the WSL (Windows Subsystem for Linux) feature. Double-click "0.UbuntuWSL\EnableWSL.bat" script after extraction (close antivirus software first to prevent interception), then click "Yes" in "User Account Control" window to automatically enable the WSL subsystem.
(3) The .exe method is the default recommended installation approach. Locate the OnekeyScript.exe file within the installation package and double-click it to begin.
(4) Alternatively, you can install using MATLAB (version 2022b or later). As shown in Fig. 2.6, type the command "OneKeyScript" in "Command Window" of MATLAB and press the Enter key on the keyboard to run the one-click installation script.[13]
(5) In the configuration window that pops up, as shown in Fig. 2.7, select the appropriate configuration based on the actual software and hardware situation (here, beginners are recommended to use the default configuration, and the installation directory should have at least about 25GB of free space). Click the "OK" button in the bottom right corner of Fig. 2.7.
(6) Please wait for the software package installation and deployment to complete. This process will take approximately 30 min.

[13] Another method is to select "OneKeyScript.p" file with the mouse, as shown in Fig. 2.6, and right-click the mouse. In the pop-up menu, click the "Run" button to run the one-click installation script.

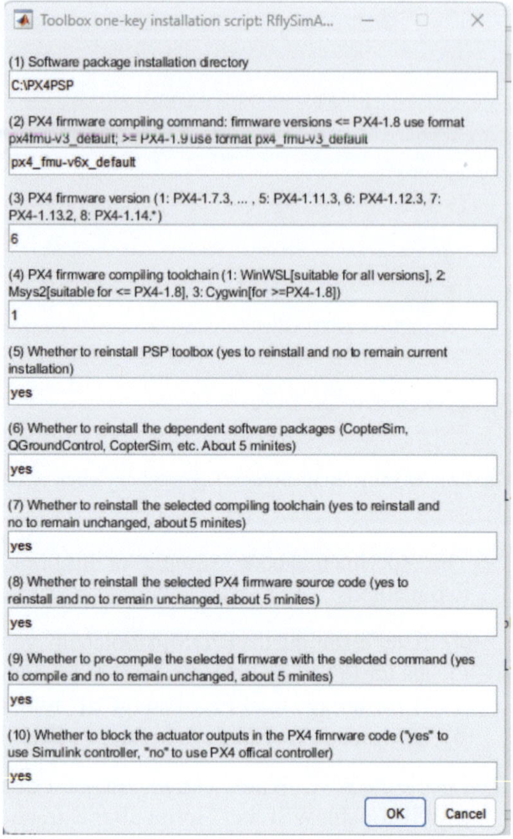

Fig. 2.7 PX4 install script installation option

It is important to note the following.

(a) If the script fails to create shortcuts, it may be due to antivirus software. To generate all software shortcuts, manually click the "GenerateShortcutCMD.bat" script in the installation directory after closing your antivirus software.
(b) If you wish to change the firmware compilation version or restore the compilation environment, there is no need to uninstall the platform. Simply run the "OneKeyScript" command again and modify the corresponding options.
(c) Readers can refer to the "RflySim*******.txt" file in the installation directory for more detailed information and platform update logs.

2.2.2 Advanced Settings

For advanced independent developers, the installation path, CubePilot/Pixhawk hardware version, PX4 firmware version, and compilation environment can be set in the configuration window displayed in Fig. 2.7. The options in Fig. 2.7 are explained in detail below.

(1) **Software package installation path**. The platform's dependent files, which require approximately 50 GB of space, will be installed in the specified path. The default installation path is "C:\PX4PSP". If there is insufficient space on the C drive, you can choose a path on another drive. It is important to note that the path name must be correct and can only consist of pure English characters; otherwise, compilation may fail.

(2) **PX4 firmware compilation command**. The default compilation command is "px4_fmu-v6x_default" corresponding to "Pixhawk 6x" autopilot hardware. After generating the controller code in PSP, the compiler automatically compiles it into a firmware file in the "px4_fmu-v6x_default" format. It then downloads and flashes this ".px4" file onto the supported hardware, thereby deploying the algorithm. For different CubePilot/Pixhawk hardware products, different PX4 firmware compilation commands need to be selected. Figure 2.8 shows some of the CubePilot/Pixhawk series autopilot hardware products. If you need to change to a different compilation command (e.g., changing to "CubePilot_cubeorange_default"), you only need to enter the command in MATLAB: PX4CMD("CubePilot_cubeorange_default") or use the command: PX4CMD CubePilot_cubeorange_default.

- Pixhawk 4/CUAV V5 nano: px4_fmu-v5_default;
- Pixhawk 6x: px4_fmu-v6x_default;

Fig. 2.8 CubePilot/Pixhawk hardware

- Pixhawk 6c: px4_fmu-v6c_defaul;
- Cube Orange: CubePilot_cubeorange_default.

The RflySim toolchain is compatible with autopilots supported by PX4 officials. For a detailed list of autopilots supported by PX4 officials, please refer to: https://docs.px4.io/main/en/flight_controller/.

(3) **PX4 firmware version**. The PX4 source code version is continually updated, and the latest firmware version is currently 1.14. As firmware versions are upgraded, functionality increases, and support for new products is extended. However, compatibility with some older autopilots may be reduced. Over time, some hardware may be discontinued. It is recommended that readers select the available autopilots and then select the appropriate stable PX4 firmware version. The method and steps for querying compilation commands can be found via QGC. Other options in the one-click script are left as defaults. Detailed configuration methods can be found in the "HowToInstall.pdf " file included in the RflySim installation package.

(4) **Other configurations**. For more detailed explanations on options such as the PX4 firmware compiler, please refer to the file " [*installation directory*]\RflySimAPIs\1.RflySimIntro\Readme.pdf " to find whether to reinstall the PSP toolbox, whether to reinstall other dependency packages, whether to re-configure the compilation environment of the firmware compiler, and whether to re-deploy the PX4 firmware code.

2.2.3 *Installation Completion Performance*

After successfully running the one-click installation script above, you should see the following content.

(1) On the desktop, you should see shortcuts for core tools generated as shown in Fig. 2.9.

(2) As shown in Fig. 2.10, all software storage folders appear in the installation path (default is "C:\PX4PSP"). It is important to note that all software on this platform is completely self-contained (also known as portable software, which is

Fig. 2.9 Emulation package desktop shortcut

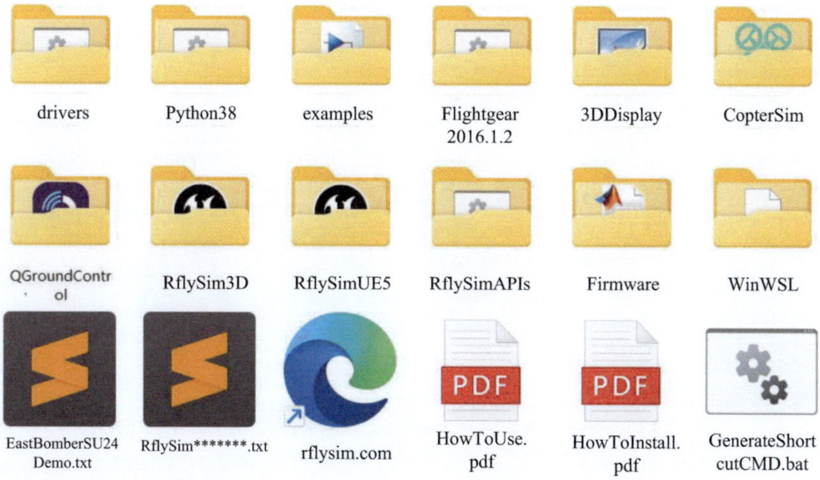

Fig. 2.10 All files under the installation directory of simulation software package

characterized by complete independence from other system software) operates independently of any previously installed software (such as the official version of QGroundControl or FlightGear). In Fig. 2.10, the "Firmware" folder stores the PX4 source code; the "Examples" folder contains Simulink code examples included in the PSP Toolbox; the "Drives" folder stores the drivers for the CubePilot/Pixhawk series autopilots; the "Python38" folder contains a Python environment used to implement the automatic code download functionality in the PSP Toolbox.

(3) As depicted in Fig. 2.11, by opening any Simulink file and clicking on the Library Browser button, you can locate the "Pixhawk Target Blocks" module library of the PSP toolbox.

If you wish to uninstall the software platform, you can double-click the uninstall.exe file or use MATLAB to open the installation package directory and run the "uninstall.m" script to complete the uninstallation process.

2.2.4 Basic Software Checking

(1) **Confirming software startup**. Double-click on the desktop shortcuts shown in Fig. 2.9, namely "CopterSim", "QGroundControl", and "RflySim3D", to open each software interface as depicted in Fig. 2.12. Confirm that each software can run properly.
(2) **Run software simulation startup script to check platform installation**. As shown in Fig. 2.13, follow the steps below. Double-click on the shortcut "SIL-

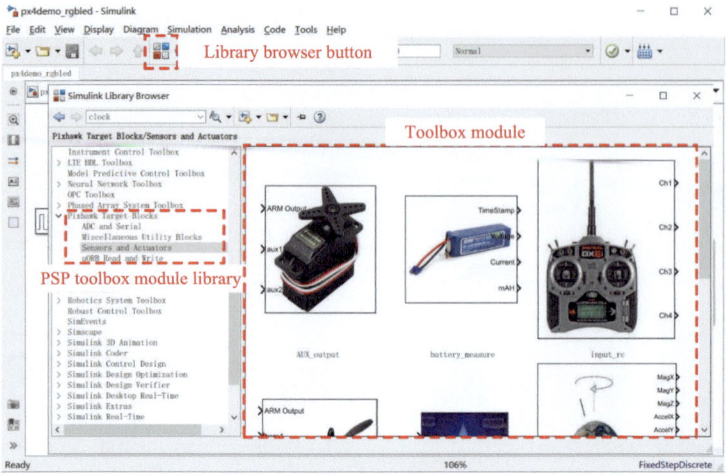

Fig. 2.11 PSP tools module in simulink library browser

Run.bat" (or "SILRunLowGPU.bat") in the RflyTools directory on the desktop. After running the one-click startup script, enter the number of simulation instances as '1' and wait until the text "Copter-Sim/PX4 EKF 3DFixed: 1/1" appears in the top-left corner of the RflySim3D software. This indicates the ini-

(a) FlightGear-F450

(b) CopterSim

(c) QGroundControl

(d) RflySim3D

Fig. 2.12 Software interface

Fig. 2.13 One-button start SIL simulation process

tiation of the multirotor SIL simulation. You can then unlock and take off the multirotor in QGC, thus initiating the one-click startup of the multirotor SIL simulation.

(3) **Confirmation of fixed-wing SIL and HIL simulation functionality**. After the installation of the RflySim toolchain for this book, double-click on the script " [*installation directory*]\RflySimAPIs\SITLRunFw.bat" to open the SIL simulation. Enter "1" to initiate the SIL simulation. Alternatively, double-click on the script " [*installation directory*]\RflySimAPIs\HILRunFw.bat", enter the flight controller port number, to initiate the HIL simulation. You can use the ground station to draw a trajectory for takeoff and flight for fixed-wing UAVs. For specific instructions, please refer to the document: " [*installation directory*]\RflySimAPIs \2.RflySimUsage\0.ApiExps\e2_FWConfig\Readme.pdf ".

2.2.5 Other Features of RflySim

As shown in Fig. 2.14, the RflySim toolchain supports the modeling of other vehicles. The unified modeling framework for other unmanned vehicle systems decomposes the entire unmanned vehicle system into two parts: the controlled system and the control system. Sensor data and control signals are exchanged between the controlled system and the control system. The controlled system is further divided into four subsystems: the airframe subsystem, the actuator subsystem, the 3D environment subsystem, and the sensor subsystem.

RflySim3D, a real-time visual software tailored for unmanned vehicles, is developed based on the Unreal Engine real-time 3D rendering engine. It supports the direct

Fig. 2.14 RflySim highly realistic 3D scene example

use of a large number of 3D scenes and aircraft models from the UE4 scene library. It also supports the creation of 3D scenes and aircraft models using software such as 3D Studio Max and AutoCAD, which can then be imported into Unreal Engine 4 (UE4).

RflySim provides an SDK with a depth camera sensor module, including basic sensor parameters and recommended installation positions. Users can design onboard visual sensors by setting relevant parameters such as orientation, focal length, and field of view, as well as customizing the onboard visual modules. It provides decision system input/output interfaces based on decision system input/output interface protocols.

RflySim supports one-click startup of UAV swarm simulations, MATLAB/Simulink-based UAV swarm simulation development, and multi-SITL, multi-HIL, and virtual-real combined UAV swarm simulations. It also supports distributed UAV swarm simulation across multiple computers within a LAN. For more information on additional features, please visit rflysim.com.

2.3 Quick Start Experiment

2.3.1 Introduction and Usage of Core Software

The RflySim toolchain comprises several software tools essential for developing unmanned systems, including modeling, simulation, and algorithm validation. These tools comprise CopterSim, QGroundControl, RflySim3D/RflySimUE5, Python38Env, WinWSL subsystem, one-click run scripts for SIL/HIL, MATLAB automatic code generation toolbox, Simulink UAV swarm control interface, PX4 Firmware source code, as well as accompanying data files and hardware systems.

Table 2.1 provides an overview of the core software included in the RflySim toolchain, along with their usage methods. For more detailed instructions, refer to: " [*installation directory*]\RflySimAPIs\2.RflySimUsage \0.ApiExps\e1_RflySimSoftwareReadme\Index.pdf".

The RflySim toolchain involves various aspects of unmanned system development, including constructing 3D scenes, establishing dynamic models for unmanned systems, low-level control, intelligent perception, health assessment, network simulation, UAV swarm control, and so on. As shown in Table 2.2, detailed explanations of each subfolder within "[*installation directory*]**RflySimAPIs**" folder are provided.

The internal structure of each tutorial example folder is shown in Table 2.3. The folders contain experiments of varying difficulty levels to help users learn the relevant content of the lecture in a progressive manner.

2.3.2 Introduction to Automatic Code Generation

Figure 2.15 shows the relationship among the PSP toolbox, PX4 software system, and the Cubepilot/Pixhawk hardware system. Key features of the toolkit include:

- The ability to simulate and test different aircraft models and autopilot algorithms in Simulink and automatically deploy the algorithms to CubePilot/Pixhawk autopilots;
- The toolbox provides practical examples, including flight control, remote control data processing, and attitude controllers;
- Numerous interface modules are provided in the toolbox for accessing the software and hardware components of CubePilot/Pixhawk;
- The capability to automatically record flight data of sensors, actuators, and deployed controllers;
- The ability to subscribe and publish uORB topic messages. All data in the PX4 autopilot software is temporarily stored in a uORB message pool. Subscribing

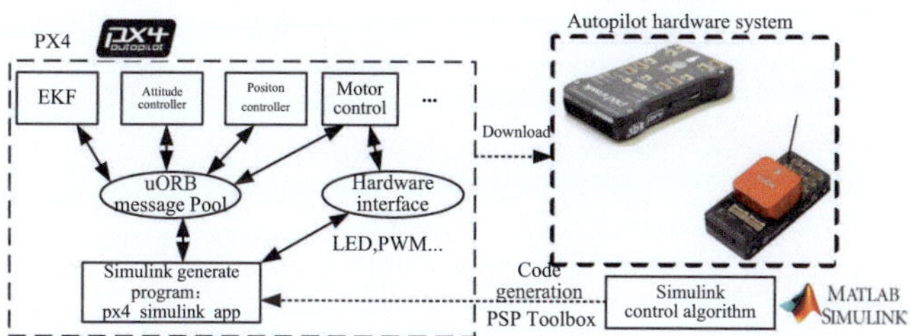

Fig. 2.15 Simulink and CubePilot/Pixhawk autopilot code generation diagram

Table 2.1 Introduction to the main software of RflySim toolchain

Software name	Introduction
CopterSim	CopterSim is a software for HIL simulation developed for the Pixhawk/PX4 autopilot platform. The software configures the multicopter model and connects to the Pixhawk autopilot via a USB serial port to simulate indoor flight conditions for outdoor tests. It consists of two main parts: the model and communication. The simulation is performed immediately after the calculation based on the set model parameters. Additionally, the model can run the dynamic model DLL and form part of a software/hardware in the loop simulation with other software. For more detailed usage methods, see: "[*installation directory*]\RflySimAPIs\2.RflySimUsage\0.ApiExps\e1_RflySimSoftwareReadme\CopterSim\Readme.pdf"
FlightGear 2016.1.2	FlightGear is based on LaRCsim and has been integrated into the Navion aircraft. However, this integration has resulted in several limitations, such as many features being hand-coded instead of being configurable through files, and multiple attempts to develop or add additional flight models. Despite these limitations, FlightGear is now capable of supporting a variety of different flight models that can be selected at runtime. For more detailed usage methods, see: "[*installation directory*]\RflySimAPIs\2.RflySimUsage\0.ApiExps\e1_RflySimSoftware-Readme\FlightGear 2016.1.2 \Readme.pdf"
QGroundControl	QGroundControl is an open-source ground station designed for the latest architecture of PX4 software. It uses the QT editor and the C++ language to write its core code. It supports source code modification and secondary development of functions, which is suitable for UAV ground station research and experimentation, as well as customization and modification of UAV ground station functions. For more detailed usage methods, see: "[*installation directory*]\RflySimAPIs\2.RflySimUsage\0.ApiExps\e1_RflySimSoftware-Readme\QGroundControl\Readme.pdf"
RflySim3D	RflySim3D utilizes UDP communication and can receive external commands, such as switching scenes, creating UAVs, and enabling the built-in physical collision of the UE. The commands will be described in Introduction to the RflySim3D interface and usage,see: "[*installation directory*]\RflySimAPIs\3.RflySim3DUE\Readme.pdf". In summary, RflySim3D can receive UDP commands from CopterSim, Python, and Simulink and return collision/terrain data as well as visual data. For more detailed usage, see: "[*installation directory*]\RflySimAPIs\2.RflySimUsage\0.ApiExps\e1_RflySimSoftware-Readme\RflySim3D\Readme.pdf"

Table 2.2 Search for each lecture under the RflySimAPIs directory

No.	Chapter name	Introduction	Folder
1	Lecture 1: Introduction	This lecture provides an introduction to the RflySim toolchain, covering version differences, installation, and functional features	1. RflySimIntro
2	Lecture 2: Lab Platform Configuration	This lecture focuses on the RflySim toolchain configuration process. The use of core components and experimental procedures	2. RflySimUsage
3	Lecture 3: 3D scene modeling and simulation	This lecture focuses on the architecture of the RflySim3D software, the use of functional 3D modeling and scene development software, and the interface of the RflySim toolchain	3. RflySim3DUE
4	Lecture 4: Modeling and Simulation of Vehicle Motion	This lecture mainly describes the control model construction of unmanned vehicles and the interface and steps of the RflySim toolchain model import to help readers understand the basic theory of unmanned system modeling	4. RflySimModel
5	Lecture 5: Low-Level Control and Filtering Estimation	This lecture contains a large number of low-level development code examples for unmanned systems. It provides code generation and download capabilities. The Simulink control algorithm can be designed to generate C++ code with a single click, compile all the necessary C++ code, and produce the final PX4 firmware. Implement the basic experimental flow of Sim2Real	5. RflySimFlyCtrl
6	Lecture 6: External Control and Trajectory Planning	This lecture mainly describes how the external control interface developed by the unmanned system sends commands to the agent to realize the control functions, such as trajectory planning of the upper-level layer	6. RflySimExtCtrl

(continued)

Table 2.2 (continued)

No.	Chapter name	Introduction	Folder
7	Lecture 7: Safety Testing and Health Assessment	This lecture mainly focuses on the process from software unit and integration verification, embedded software and hardware verification, and software and hardware integration verification to complete machine integration and test verification in unmanned system development. Implement fault injection and safety testing for all phases of the above development	7. RflySimPHM
8	Lecture 8: Visual Perception and Obstacle Avoidance	This lecture primarily discusses visual sensors and related theories, such as coordinate systems, common sensors for vision-based control, etc. It introduces the configuration methods for environments related to visual development, such as Linux, ROS, MAVROS, and the visual interface of the RflySim toolchain	8. RflySimVision
9	Lecture 9: Communication Protocol and Swarm Networking	This lecture primarily discusses the methods and current status of networking for unmanned systems, the system architecture of UAV swarm communication in the RflySim toolchain, and simulation code examples for unmanned system networking	9. RflySimComm
10	Lecture 10: Swarm Control and Game	This lecture primarily focuses on developing UAV swarm control for multi-agent unmanned systems. It introduces technologies such as UAV swarm, task planning, and game theory. The distributed control framework for the UAV swarm is used for the RflySim toolchain and the MATLAB/Python-based UAV swarm control interface. Case studies include multi-UAV mission planning based on ant colony algorithms, UAV formation flying, control within virtual tubes, and large-scale control of UAV swarms. These examples help readers understand the principles and implementation methods of UAV swarm control	10. RflySimSwarm

Table 2.3 Internal structure of the example folder

Number	Name	Folder/File
1	Basic interface example folder	0.ApiExps
2	Basic examples folder	1.BasicExps
3	Advanced examples folder	2.AdvExps
4	Lecture introduction file	Intro.pdf
5	Lecture interface documentation file	API.pdf
6	Lecture supplementary slides file	PPT.pdf
7	Lecture all examples retrieval file	Index.pdf

to topics retrieves the desired data from this pool, while publishing topics adds specific data to the pool for use by other modules.

Detailed explanations of this platform's automatic code generation feature are found in: "[*installation directory*]\RflySimAPIs\5.RflySimFlyCtrl\Intro.pdf".

(1) The hardware and software architectures of the CubePilot/Pixhawk autopilot include the CubePilot/Pixhawk hardware (similar to a computer host) and the PX4 autopilot software system (similar to the operating system and applications running on the host).
(2) The entire PX4 autopilot software system is divided into several smaller modules, each running independently (multithreading parallelism), with data transmission and interaction achieved through the subscription and publication functions of the uORB message module.
(3) After deploying code generated by Simulink to the PX4 autopilot software, it does not affect the native PX4 autopilot software operation. However, it adds an independent module named "px4_simulink_app" (running in a separate thread) alongside other modules.
(4) The entire code generation and deployment process, as shown in Fig. 2.15, proceeds as follows:

- The PSP toolbox generates C/C++ code for control algorithms designed in Simulink;
- This code is imported into the source code of the PX4 autopilot, generating a program running independently as "px4_simulink_app";
- The PSP toolbox invokes the compilation tool to compile all code into ".px4" PX4 autopilot software firmware files (similar to system image files);
- The obtained firmware is downloaded and flashed into the CubePilot/Pixhawk autopilot, which executes the PX4 software with the generated algorithm code.

(5) To prevent potential conflicts where native PX4 control algorithms may need to access the same hardware output resources as "px4_simulink_app", a one-click deployment script is provided. This script automatically blocks the PX4 native firmware from accessing the actuators (shown in the last option in Fig. 2.15),

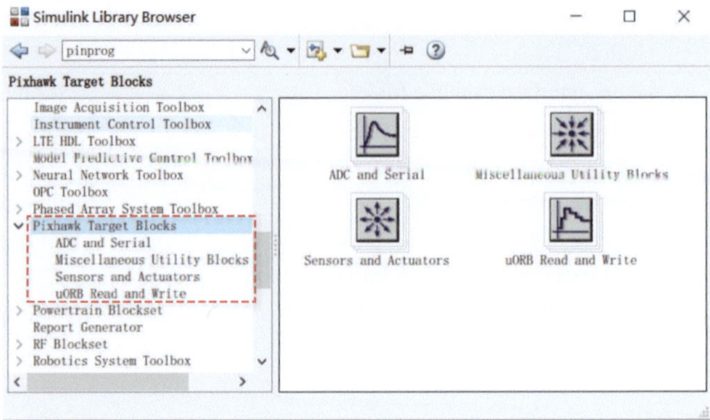

Fig. 2.16 Simulink PSP Toolbox

ensuring that only the "px4_simulink_app" module can output motor control signals.

Once the PSP toolbox has been installed, as demonstrated in Fig. 2.18, the Simulink library will display the "Pixhawk Target Blocks" module. This library offers a range of Simulink interface modules that allow access to Pixhawk hardware resources and PX4 software internal messages.

The interface modules are viewed in the "Pixhawk Target Blocks" toolbox in Simulink, as shown in Fig. 2.16. It consists of four sub-libraries: "ADC and Serial" (Analog to Digital Conversion and Serial Communication library), "Miscellaneous Utility Blocks" (Auxiliary tool library), "Sensors and Actuators" (Sensor and Actuator interface library), and "uORB Read and Write" (uORB message read-write library). Figure 2.17 shows several key input-output modules in the "Sensors and Actuators" library. Through these modules, Pixhawk sensor data or filtered state data is conveniently obtained, and outputs for motors, LED lights, and buzzers are controlled. For detailed descriptions of each module, refer to " [*installation directory*]\RflySimAPIs\5.RflySimFlyCtrl\API.pdf", as shown in Table 2.4.

2.3.3 *Control Experiment Based on Visual Feedback*

Research on UAV perception and avoidance technology primarily focuses on two key aspects: flight environment perception and control of avoidance decisions. To effectively measure the aerial environment, sensors and data links are used to estimate the state and collision threat level of potential targets. Regarding avoidance decisions, the UAV plans paths and controls maneuvers to reduce the probability of collision with potential threats and ensure safe flight. Detailed information about the vision-based control function of this platform is found in: " [*installation directory*]\RflySimAPIs\8.RflySimVision\Intro.pdf".

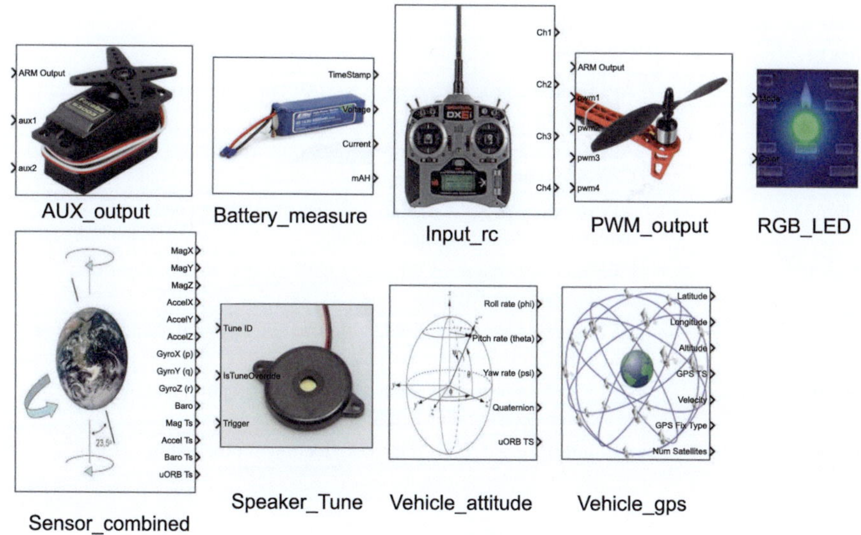

Fig. 2.17 PSP Toolkit sensor and actuator interface library

Table 2.4 Description of each module in the toolbox

ADC and serial	Read ADC channels	Outputting the input of external ADC
	Serial	Serial communication module
Miscellaneous utility blocks	binary_logger	Data logging module
	ExamplePrintFcn	Print function example
	ParamUpdate	Custom storage class parameter update module
Sensors and actuators	Battery_measure	Battery data module
	Input_rc	Remote Control input module
	PWM_output	Motor PWM module
	Speaker_Tune	Buzzer module
	Vehicle attitude	Attitude data module
	Vehicle gps	GPS data module
uORB read Async uORB	uORB Read Function-Call Trigger	uORB message read callback trigger module
	uORB write	uORB message data publication interface module
	uORB write advanced	uORB message data publication interface advanced module
	uORB write advanced_dai–uORB	Message data publication interface advanced module

Table 2.5 RflySimSDK introduction

Folder	Introduction
comm	API interface programs related to UAV communication and swarm networking
ctrl	API interface programs for MavLink protocol control, DLL files, and CopterSim invocation
phm	API interface programs related to security testing and health assessment of the RflySim toolchain
swarm	API interface programs related to UAV swarm control algorithm development for RflySim
ue	API interface programs related to three-dimensional scene modeling and simulation for RflySim
vision	API interface programs related to visual perception and obstacle avoidance decisions for RflySim
word	API interface programs related to document processing

After installing the RflySim toolchain, deploy a Python environment located at "[*installation directory*]/Python38". This environment includes the majority of libraries required for vision-based control experiments. Visual Studio Code is recommended for reading and executing programs. The installation method can be found at: https://code.visualstudio.com/. Detailed configuration processes are available in the files within the RflySim toolchain: "[*installation directory*]\RflySimAPIs\1.RflySimIntro\2.AdvExps\e3_PythonConfig\Readme.pdf".

The relevant interface files for the vision-based control function of this platform are integrated into the RflySimSDK. Table 2.5 provides an overview of the RflySimSDK. If readers prefer to use their own Python development environments such as Pycharm or Anaconda, they can directly run " [*installation directory*]\RflySimAPIs\RflySimSDK \ReLabPath.py" file and install dependencies such as pymavlink and python-opencv as prompted. This allows the global invocation of various interfaces in the SDK.[14]

Figure 2.18 illustrates a quadcopter simulation involving takeoff, navigation through ring-shaped frames, and landing. The program execution displays real-time results of the image captured by the UAV's front camera after a series of image processing. The UAV must continuously align the ring center with the center of the image. Based on the detected error in the image, the UAV is controlled to move toward the ring center and approach the ball's direction until it successfully passes through the ring. The detailed experimental process is shown in the file in the RflySim toolchain located at: "[*installation directory*]\RflySimAPIs\8.RflySimVision\1.BasicExps\1-VisionCtrlDemos\e4_CrossRing\Index.pdf".

Figure 2.20 shows fixed-wing UAVs sending messages to RflySim3D during flight for scene control and requesting images. The image stream is transmitted to the local

[14] For more detailed instructions, please refer to the API page located at: "[*installation directory*]\RflySimAPIs\RflySimSDK\Readme.html".

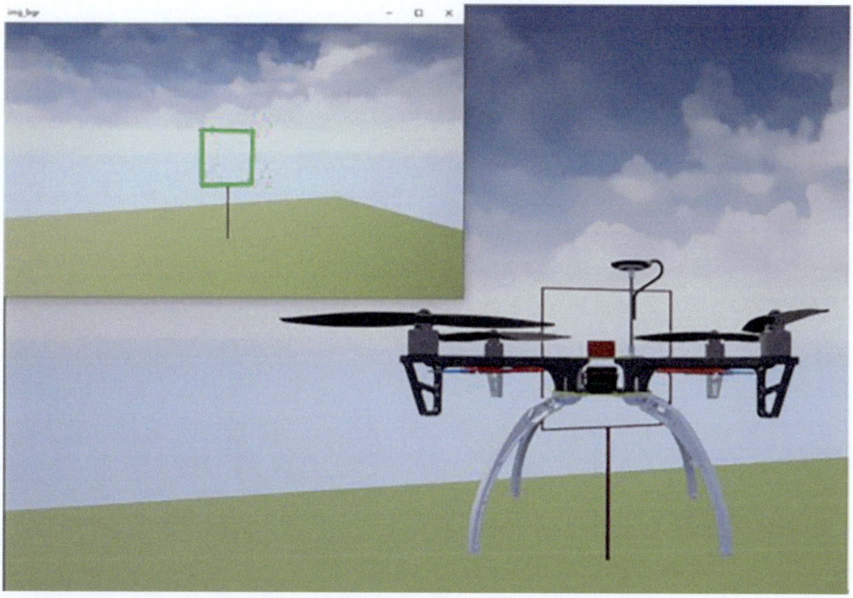

Fig. 2.18 A quadcopter is preparing to pass through a ring-shaped frame

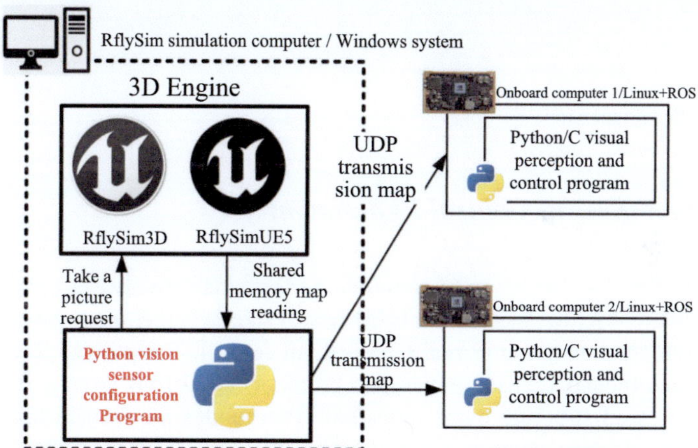

Fig. 2.19 Image transmission mode

vision program through methods such as shared memory and network communication. The framework is illustrated in Fig. 2.19. The configuration can be customized through the Config.json file to obtain the image stream from the RflySim3D software. For detailed experimental procedures, please refer to the files in the RflySim toolchain located at: "[*installation directory*]\RflySimAPIs\5.RflySimFlyCtrl \1.BasicExps\e10-FixedWingCtrl\code-2\e2.1\Readme.pdf".

Fig. 2.20 Viewfinder window

2.3.4 UAV Swarm Control Experiment

Communication bandwidth and computing performance are always important bottlenecks that restrict the increase in UAV swarm size, whether from the perspective of multi-vehicle simulation or real UAV swarm control. To optimize these factors, the RflySim toolchain focuses on two data protocols: MAVLink communication and UDP compressed structure communication. Two modes are formed according to different optimization methods. (1) Full mode, which contains as much data as possible and sends it at the highest frequency to ensure integrity. This mode is suitable for UAV swarm development with up to eight UAVs. (2) Simple mode, which contains only essential data and reduces the sending frequency to ensure real-time and smooth communication in a large-scale UAV swarm. This mode is suitable for UAV swarm development with more than four UAVs. The platform offers five communication modes: UDP_Full, UDP_Simple, MAVLink_Full, MAVLink_Simple, and MAVLink_NoSend, based on the aforementioned data protocols and optimization methods. Additionally, it provides two communication interface programs: a MATLAB/Simulink program and a Python program. The MATLAB/Simulink interface module utilizes Simulink S functions implemented through C++ hybrid compilation. This approach offers high efficiency, low computation, and low latency, making it the main interface program used in this book. For detailed explanations of the UAV swarm control functionality of this platform, please refer to: " [*installation directory*]\RflySimAPIs\10.RflySimSwarm\Intro.pdf".

The RflySim toolchain offers a range of models for UAV swarm control, including PX4 flight control integration based on high-precision dynamic models and custom controller design in the Simulink environment. It also supports the simulation and control of various physical models, such as point mass models, multicopter models, and fixed-wing UAV models. From a model accuracy standpoint, using a high-precision 6DOF model (CopterSim) in conjunction with an actual flight control system (PX4) in a closed-loop SIL/HIL simulation can effectively enhance the credibility of the model, thereby reducing the gap between simulation and real-world experiments. However, the computational complexity of the code in the above architecture restricts the number of UAVs that can run on a single computer (usually less than 20 for SIL and less than 30 for HIL).

As shown in Fig. 2.21, through the Simulink UAV swarm control transmission module provided by the RflySim toolchain, the status information of the UAVs is received, and then the planning of the motion of the four UAVs is carried out, the control information is sent to this module, and then the simulation is carried out. For detailed experimental procedures, please refer to the files in the RflySim toolchain located at:" [*installation directory*]\RflySimAPIs
\10.RflySimSwarm\1.BasicExps\e1_RflyUdpSwarmExp\2.RflyUdpFullFour_Mat\
Readme.pdf".

As shown in Fig. 2.22, it describes a swarm flight for four fixed-wing UAVs capable of executing various flight modes, including taxiing, climbing, level flight, and hovering at a target point. For detailed experimental procedures, please refer to the files in the RflySim toolchain located at:"[*installation directory*]\RflySimAPIs\5.RflySimFlyCtrl\1.BasicExps\e10-FixedWingCtrl\code-2\e2.2\Readme.pdf".

Fig. 2.21 A quadcopter swarm simulation

Fig. 2.22 Fixed-wing swarm control

References

1. X. Dai, C. Ke, Q. Quan, K.-Y. Cai, Rflysim: automatic test platform for UAV autopilot systems with FPGA-based hardware-in-the-loop simulations. Aerosp. Sci. Technol. **114**, 106–727 (2021)
2. X. Dai, C. Ke, Q. Quan, K.-Y. Cai, Simulation credibility assessment methodology with FPGA-based hardware-in-the-loop platform. IEEE Trans. Ind. Electron. **68**(4), 3282–321 (2021)
3. Q. Quan, X. Dai, S. Wang, *Multicopter Design and Control Practice: A Series of Experiments based on MATLAB and Pixhawk* (Springer, Singapore, 2020)
4. Y.C. Paw, Synthesis and validation of flight control for UAV, M.S. thesis, University of Minnesota, Minneapolis, MN, 2011

Chapter 3
Level Flight Performance Evaluation Experiment

Xinquan Chen, Wenhan Gao, and Quan Quan

Level flight is the most common flight state for micro-small fixed-wing Unmanned Aerial Vehicles (UAVs). This chapter introduces the principles for calculating the level flight performance of electrically driven micro-small fixed-wing UAVs and related experiments. First, a kinematic analysis and mathematical modeling of the electric propulsion system for fixed-wing UAVs are conducted. Subsequently, the performance of micro-small fixed-wing UAVs is assessed through a series of experiments—basic, analysis, and design experiments—to enhance the reader's comprehension of flight principles. In the basic experiments, readers will learn how to calculate the maximum level flight speed, cruise speed, and endurance speed, along with their corresponding flight distances. The analysis experiment will examine the relationship between these flight speeds and the parameters of fixed-wing aircraft. Readers will learn to design and select wings and batteries for micro-small fixed-wing UAVs according to flight missions in the design experiment. For more information on this chapter, please refer to the link

https://rflysim.com/doc/en/C/5.FW_Book.html.

3.1 Experimental Principles of Level Flight Performance Evaluation

The most common motion of fixed-wing UAVs is level flight, also known as the *steady level flight*. For micro-small fixed-wing UAVs, takeoff is replaced by catapulting, throwing, or vertical takeoff, rather than the traditional runway takeoff; landing can also be achieved by stalling or vertical descent. These two phases generally have a short flight time and distance. Therefore, the level flight performance of micro-small fixed-wing UAVs with extended flight time and distance is often the most critical aspect of their overall flight performance.

Q. Quan et al., *Micro-Small Fixed-Wing UAV Control*,
https://doi.org/10.1007/978-981-95-1954-5_3

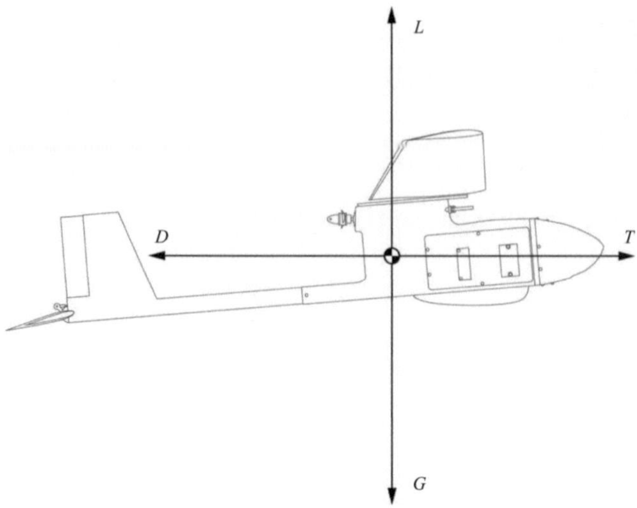

Fig. 3.1 Force analysis in level flight

3.1.1 Level Flight Performance Analysis

1. Thrust Required and Thrust Available

When the aircraft is in steady-level flight, the rate of change of flight speed is zero, i.e., $\frac{\mathrm{d}V}{\mathrm{d}t} = 0$. Therefore, the force balance equation is described as:

$$\begin{aligned} T &= D \\ L &= G \end{aligned} \quad (3.1)$$

where T represents the thrust provided by the current propulsion system of the aircraft; D represents the drag experienced by the aircraft; L represents the lift at the current speed; G is the gravity of the aircraft. At this time, the thrust T (or pull) satisfying the above motion equation is called the *thrust required* for steady level flight, denoted as T_{re}.

(1) **Thrust Required**

The expression for thrust required is obtained by multiplying the upper and lower equations in Eq. (3.1), i.e.,

$$T_{\mathrm{re}} = \frac{D}{L}G = \frac{G}{Q} \quad (3.2)$$

where $Q = L/D$ is the lift-to-drag ratio, which is the ratio of lift to drag when the aircraft is flying at a certain speed and angle of attack. Also, in steady-level flight, the thrust required T_{re} is equal to the drag D, so the variation of thrust required T_{re} is the same as the variation of drag D during flight. Drag is mainly divided into zero-lift drag D_0 and induced drag D_i, and from Eq. (1.19), the formula for calculating drag is

$$T_{re} = D = D_0 + D_i = \frac{1}{2}C_{D_0}\rho V^2 S + \frac{1}{2}AC_L^2\rho V^2 S. \tag{3.3}$$

At the same time, gravity is also balanced, i.e.,

$$G = L = \frac{1}{2}C_L\rho V^2 S. \tag{3.4}$$

This yields the lift coefficient at this time

$$C_L = \frac{2G}{\rho V^2 S}. \tag{3.5}$$

Substituting the lift coefficient into Eq. (3.3), the thrust required is expressed as

$$T_{re} = \frac{1}{2}C_{D_0}\rho V^2 S + \frac{AG^2}{\frac{1}{2}\rho V^2 S}. \tag{3.6}$$

(2) **Thrust Available**

Whether the aircraft can achieve steady level flight also depends on the *thrust available* T_{av} of the propulsion system, i.e., the maximum thrust provided by the aircraft's propulsion system at a specific altitude during level flight at different speeds. The maximum thrust of the propulsion system often decreases with increasing airspeed. The thrust at zero speed is referred to as *static thrust*. Let T be the static thrust and K be the decay coefficient. The decay law of dynamic thrust T_{dyn} (i.e., thrust available T_{av}) at different speeds is approximated by the following linear function:

$$T_{dyn} = T - K\frac{V}{\sqrt{T}}. \tag{3.7}$$

The thrust required, T_{re}, and the thrust available, T_{av}, together determine the steady-level flight performance of micro-small fixed-wing UAVs, as described by the maximum level flight speed, endurance speed, and cruise speed, among others.

2. Maximum Level Flight Speed

The *maximum level flight* speed refers to the highest speed an aircraft can achieve level flight at a given altitude, with its propulsion system operating continuously at full available thrust. In general, the variation range of flight altitude for micro-small fixed-wing UAVs is small, and it is assumed that within their working altitude range, parameters such as air density change little, and the variation of the maximum thrust caused by flight altitude can be ignored. Taking micro-small fixed-wing UAVs using propeller propulsion systems as an example, the maximum speed is satisfied as

$$T - K\frac{V}{\sqrt{T}} = \frac{1}{2}C_{D_0}\rho V^2 S + \frac{AG^2}{\frac{1}{2}\rho V^2 S}. \tag{3.8}$$

The above equation is analytically complex and is typically solved by visually analyzing graphs of thrust required and thrust available versus speed to determine the maximum level flight speed, $V_{\max}$.

3. Endurance Speed

Endurance speed refers to the flight speed at which the aircraft's propulsion system consumes the least energy per unit of time, thereby maximizing the aircraft's total flight time (endurance) in level flight. The energy consumption per unit time of propeller aircraft is proportional to the output power of the propulsion system, i.e.,

$$\frac{\mathrm{d}E}{\mathrm{d}t} = -P. \tag{3.9}$$

In Eq. (3.9), $E > 0$ represents energy, and $P > 0$ represents power. As determined by Eq. (3.9), the power required accounts only for overcoming the aircraft's drag. However, energy consumption also depends on other factors, such as the energy consumed by the propeller's rotation. Therefore, the ratio of the power required to the total power available, known as the *thrust-to-power ratio*, is denoted as η. Consequently, the required power is given by:

$$\begin{aligned} P_{\mathrm{re}} &= \frac{T_{\mathrm{re}}V}{\eta} \\ &= \left(C_{D0} \cdot \frac{1}{2}\rho V^2 S + A \cdot \frac{2G^2}{\rho V^2 S}\right)\frac{V}{\eta} \\ &= \frac{1}{\eta}\left(C_{D0} \cdot \frac{1}{2}\rho V^3 S + A \cdot \frac{2G^2}{\rho V S}\right). \end{aligned} \tag{3.10}$$

Suppose the power remains at its minimum value during level flight. In that case, the energy consumption of the UAV also remains at its minimum, allowing the UAV to achieve the longest flight time. To find the minimum power, Eq. (3.10) is differentiated with respect to velocity and set equal to zero; then, the velocity corresponding to the minimum power is solved. For small UAVs using fixed-pitch propellers, the propeller efficiency is approximately constant. At this time, the endurance speed is obtained as:

$$V_{\mathrm{lt}} = \sqrt{\frac{2G}{\rho S\sqrt{\frac{3C_{D_0}}{A}}}}. \tag{3.11}$$

4. Cruise Speed

Corresponding to endurance speed, the *cruise speed* is another important speed. As the name implies, cruise speed is the flight speed at which the aircraft's propulsion system uses all its energy to achieve the maximum flight distance (range) in level flight. It can also be understood as the speed with the least energy consumption per unit of distance. Considering Eq. (3.9), the range calculation is as follows:

$$\begin{aligned} L &= \int V\mathrm{d}t \\ &= -\int_{E_0}^{0} \frac{V}{P_{\mathrm{re}}}\mathrm{d}E \\ &= \int_{0}^{E_0} \frac{\mathrm{d}E}{(P_{\mathrm{re}}/V)}. \end{aligned} \tag{3.12}$$

When the UAV's required power-to-speed ratio, i.e., the thrust required, remains at its minimum, the UAV can achieve the longest range, so the cruise speed is

$$V_{\mathrm{ld}} = \sqrt{\frac{2G}{\rho S \sqrt{\frac{C_{D_0}}{A}}}}. \tag{3.13}$$

From Eqs. (3.11) and (3.13), it is found that the ratio of cruise speed to endurance speed is

$$\frac{V_{\mathrm{ld}}}{V_{\mathrm{lt}}} = \sqrt[4]{3} \approx 1.32. \tag{3.14}$$

There are many other methods for calculating the performance of fixed-wing aircraft, and readers can also refer to [1, 2].

3.1.2 Propulsion System Modeling

The most common propulsion system used in micro-small fixed-wing UAVs is the motor-propeller system, which consists of four primary components: the propeller, motor, Electronic Speed Controller (ESC), and battery. To ensure the practicality of the modeling method, the inputs of each calculation model are mainly based on the nominal parameters provided by the manufacturers, which can usually be found in the product descriptions, as shown in Table 3.1.

1. Propeller Modeling

Micro-small fixed-wing UAVs typically use fixed-pitch propellers. The descriptions for thrust T (unit: N) and torque M (unit: N·m) of the propeller are as follows:

Table 3.1 Propulsion system parameters

Component	Parameters
Propeller	Θ_p={diameter D_p, pitch H_p, number of blades B_p, propeller weight G_p}
Motor	Θ_m={nominal no-load KV K_{V0}, maximum current I_{mMax}, nominal no-load current I_{m0}, nominal no-load voltage U_{m0}, internal resistance R_m, motor weight G_m}
ESC	Θ_e={maximum ESC current I_{eMax}, internal resistance R_e, ESC weight G_e}
Battery	Θ_b={capacity C_b, internal resistance R_b, total voltage U_b, maximum discharge rate K_b, battery weight G_b}

$$T = C_T\rho\left(\frac{N}{60}\right)^2 D_p^4 \tag{3.15}$$

$$M = C_M\rho\left(\frac{N}{60}\right)^2 D_p^5 \tag{3.16}$$

where N (unit: RPM, revolutions per minute) is the propeller speed; D_p (unit: m) is the propeller diameter; C_T and C_M are dimensionless thrust and torque coefficients, respectively. These equations correspond only to static thrust and do not account for factors such as the incoming flow velocity and angle of attack of the propeller disk, which affect the actual thrust. It is worth noting that ρ (unit: kg/m^3) is the air density, which is a function of flight altitude H (unit: m) and temperature T_t (unit: °C), expressed as follows:

$$\rho = \frac{273P_a}{101325\,(273+T_t)}\rho_0 \tag{3.17}$$

where the standard atmospheric density $\rho_0 = 1.293\text{kg/m}^3$ (environmental conditions at 0 °C, 273 K), and atmospheric pressure P_a (unit: Pa) is further represented as:

$$P_a = 101325\left(1 - 0.0065\frac{H}{273+T_t}\right)^{5.2561}. \tag{3.18}$$

Micro-small fixed-wing UAVs fly at relatively high speeds, and there is significant incoming flow along the axis of the propeller, leading to a decrease in the provided thrust. In Sect. 3.1.1, it is described that the decay of dynamic thrust (thrust available) is linearly related to the level flight speed. Here, for the case where there is incoming flow along the axis of the propeller, the ratio of the axial velocity to the rotational linear velocity of the propeller is defined as the advance ratio J, represented as:

$$J = \frac{V}{60\pi N D_p}. \tag{3.19}$$

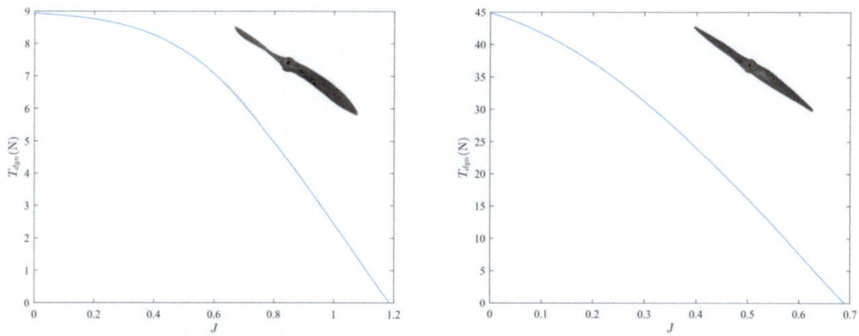

Fig. 3.2 Relationship between propeller trust and forward ratio

Based on the blade element theory analysis, the thrust generated by the propeller decreases as the advance ratio increases. For the convenience of numerical simulation calculations, the relationship between thrust and advance ratio in some intervals is approximately represented as a linear function, expressed as:

$$T_{\text{dyn}} = T - k\frac{V}{60\pi N D_{\text{p}}} \tag{3.20}$$

where $k > 0$ and T represents the thrust when the forward velocity is 0, i.e., the static thrust of the propeller. Figure 3.2 shows the relationship between APC propeller thrust and advance ratio, indicating that propeller thrust is approximately linear with the advance ratio at large advance ratios [3]. It is also known that for a well-designed propeller, there exists a relationship between speed and thrust $N \propto \sqrt{T}$. Hence, the simplified expression is shown in Eq. (3.7). Equation (3.7) is mainly used for steady-state analysis of micro-small fixed-wing aircraft during forward flight, so the propeller thrust will not reach zero.

2. Motor Modeling

Currently, the motors used in fixed-wing aircraft are brushless DC motors (simply called brushless motors), which are equivalent to a permanent magnet DC motor model. The equivalent circuit of the motor is shown in Fig. 3.3, where U_{m} (unit: V) is the equivalent voltage of the motor; I_{m} (unit: A) is the equivalent current of the motor; E_{a} (unit: V) is the back electromotive force of the armature; R_{m} (unit: Ω) is the internal resistance of the armature; L_{m} (unit: H) is the armature inductance; $\hat{I}_0$ (unit: A) is the no-load current. The no-load current compensates for mechanical losses from motor friction and air resistance, as well as iron losses from eddy currents and hysteresis, and is assumed constant during operation. After subtracting the no-load current, the remaining current $I_{\text{a}} = I_{\text{m}} - \hat{I}_0$ is used to generate electromagnetic torque to drive the rotor.

The primary purpose of motor modeling is to use the motor load torque M (equal to the propeller torque), motor speed N (equal to the propeller speed), and motor

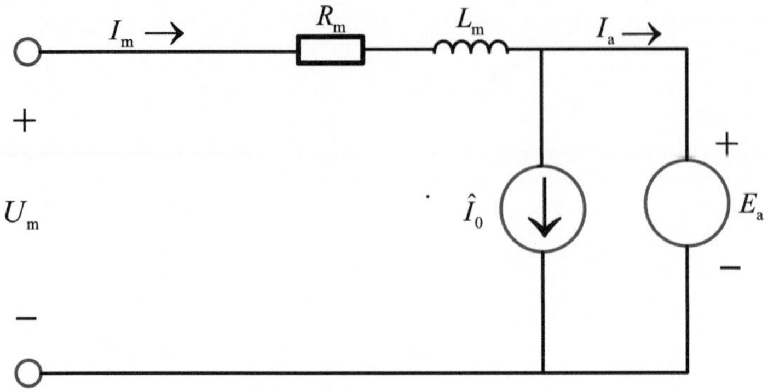

Fig. 3.3 Equivalent circuit of a brushless DC motor

basic parameter set Θ_m as inputs to solve for the equivalent voltage U_m (unit: V) and equivalent current I_m (unit: A) of the motor. The solution equations are abstractly represented as follows:

$$\begin{aligned} U_m &= f_{U_m}(\Theta_m, M, N) \\ I_m &= f_{I_m}(\Theta_m, M, N) \end{aligned} \tag{3.21}$$

where M and N are determined based on Eqs. (3.15) and (3.16), and the specific definition of Θ_m is observed in Table 3.3. The final results are as follows:

$$\begin{cases} U_m = f_{U_m}(\Theta_m, M, N) \triangleq \left(\dfrac{M K_{V0} U_{m0}}{9.55 (U_{m0} - I_{m0} R_m)} + I_{m0} \right) R_m + \dfrac{U_{m0} - I_{m0} R_m}{K_{V0} U_{m0}} N \\ I_m = f_{I_m}(\Theta_m, M, N) \triangleq \dfrac{M K_{V0} U_{m0}}{9.55 (U_{m0} - I_{m0} R_m)} + I_{m0}. \end{cases} \tag{3.22}$$

3. ESC Modeling

In brushless motors, the ESC modulates the battery voltage. Then, the resulting current goes into the motor's three-phase drive circuit, generating an alternating magnetic field in the armature that drives rotor rotation. The equivalent circuit of the ESC modulation brushless motor is shown in Fig. 3.4.

Fig. 3.4 Equivalent circuit of an ESC

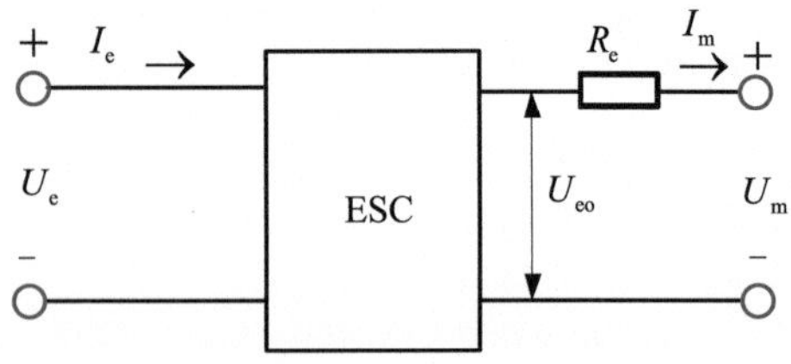

ESC modeling primarily uses the motor-derived U_{m} and I_{m}, along with the ESC parameter set Θ_{e} and battery parameter set Θ_{b} (see Table 3.3), to calculate the ESC throttle command σ (dimensionless, ranging from 0 to 1), input current I_{e} (A), and input voltage U_{e} (V). The following abstract equations are established:

$$\begin{cases} \sigma = f_{\sigma}\left(\Theta_{\mathrm{m}}, U_{\mathrm{m}}, I_{\mathrm{m}}, U_{\mathrm{b}}\right) \\ I_{\mathrm{e}} = f_{I_{\mathrm{e}}}\left(\sigma, I_{\mathrm{m}}\right) \\ U_{\mathrm{e}} = f_{U_{\mathrm{e}}}\left(\Theta_{\mathrm{b}}, I_{\mathrm{b}}\right). \end{cases} \tag{3.23}$$

As shown in Fig. 3.4, U_{eo} is the equivalent DC voltage after the ESC modulation, which is expressed as:

$$U_{\mathrm{eo}} = U_{\mathrm{m}} + I_{\mathrm{e}} R_{\mathrm{e}}. \tag{3.24}$$

Based on Eq. (3.24), the ESC throttle command σ satisfies (assuming that the ESC input voltage U_{e} is approximately equal to the battery voltage U_{b})

$$\sigma = \frac{U_{\mathrm{eo}}}{U_{\mathrm{e}}} \approx \frac{U_{\mathrm{eo}}}{U_{\mathrm{b}}}. \tag{3.25}$$

Considering that the input power of the ESC is equal to the output power, the input current I_{e} of the ESC satisfies

$$I_{\mathrm{e}} = \sigma I_{\mathrm{m}}. \tag{3.26}$$

The size of the ESC input current is limited as follows

$$I_{\mathrm{e}} \leqslant I_{\mathrm{eMax}} \tag{3.27}$$

where I_{eMax} (unit: A, from Θ_{e}) is the maximum continuous current of the ESC. Considering the voltage drop due to the internal resistance, the relationship between the ESC input voltage U_{e} and the battery voltage U_{b} (unit: V, from Θ_{b}) is as follows

$$U_{\mathrm{e}} = U_{\mathrm{b}} - I_{\mathrm{b}} R_{\mathrm{b}} \tag{3.28}$$

where I_{b} (unit: A) represents the battery current, R_{b} (unit: Ω, from Θ_{b}) is the internal resistance of the battery. For a fixed-wing UAV, the number of ESCs equals the number of motors, so

$$I_{\mathrm{b}} = n_{\mathrm{r}} I_{\mathrm{e}} + I_{\mathrm{other}} \tag{3.29}$$

where I_{other} (unit: A) includes the current of other onboard equipment (such as autopilot and camera) and some other current losses. Usually, it is simply assumed that $I_{\mathrm{other}} \approx 1\mathrm{A}$. Thus, the following relationship holds:

$$\sigma = f_\sigma(\Theta_e, U_m, I_m, U_b) \triangleq \frac{U_m + I_m R_e}{U_b} \tag{3.30}$$

$$I_e = f_{I_e}(\sigma, I_m) \triangleq \sigma I_m \tag{3.31}$$

$$U_e = f_{U_e}(\Theta_b, I_b) \triangleq U_b - I_b R_b. \tag{3.32}$$

4. Battery Modeling

The battery model mainly uses the battery parameters Θ_b (see Table 3.3) and the battery current I_b (see Eq. (3.29)) to solve for the endurance time T_b of the fixed-wing UAV, which is abstractly represented as follows:

$$T_b = f_{T_b}(\Theta_b, I_b). \tag{3.33}$$

The battery modeling simplifies the actual discharge process of the battery, assuming that the voltage remains constant during the discharge process and the battery's remaining capacity decreases linearly. Considering the minimum remaining capacity C_{min} (unit: mAh) set by the battery discharge protection, the specific modeling is as follows:

$$T_b = f_{T_b}(\Theta_b, I_b) = \frac{C_b - C_{min}}{I_b} \times \frac{60}{1000}. \tag{3.34}$$

The percentage of the minimum capacity C_{min} to the total capacity C_b (unit: mAh, from Θ_b) is set arbitrarily, generally between $0.15\ C_b$ and $0.2\ C_b$. The total battery current I_b should not exceed its maximum discharge current, which depends on the maximum discharge rate K_b (dimensionless, from Θ_b) of the battery, satisfying

$$I_b \leqslant \frac{C_b K_b}{1000}. \tag{3.35}$$

Ignoring the energy consumption during takeoff and landing phases, the distance that the aircraft can fly in steady-level flight is estimated as

$$L = V T_b. \tag{3.36}$$

3.1.3 Flight Envelope

When considering the operational characteristics of fixed-wing UAVs, altitude and speed are key factors. The speed range and maximum altitude at which a UAV can maintain stable flight are influenced by various factors, including the propulsion system and aircraft structure. Typically, speed is plotted on the horizontal axis, and altitude on the vertical axis. The upper and lower speed limits at each altitude are plotted to form a boundary line, known as the *flight envelope*. The aircraft can only operate within this envelope. However, the thrust and lift limits do not solely

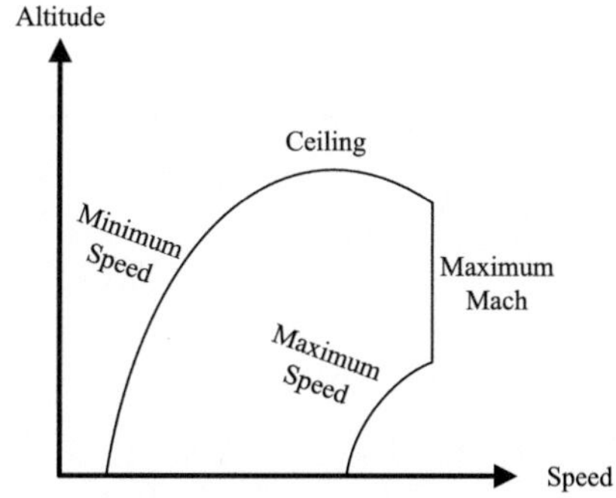

Fig. 3.5 Flight envelope

determine the flight envelope. Low speeds may cause the plane to lose longitudinal stability at certain altitudes, while high speeds can lead to structural flutter.

The *minimum speed*[1] in the flight envelope refers to the minimum speed at which a fixed-wing UAV can maintain stable-level flight. According to the lift formula (1.13), it is observed that the lift of the UAV is proportional to the velocity. Therefore, the minimum speed corresponds to the speed at which the lift balances the weight at the maximum lift coefficient. Lift is also proportional to air density; as altitude increases, air density decreases, causing the minimum speed to increase. Thus, in the velocity versus altitude coordinate system, the curve in Fig. 3.5 gradually bends. When a UAV flies at low speeds, the angle of attack needs to be increased to increase the lift coefficient. However, due to the limitation of the maximum angle of attack, it is impossible to increase the lift coefficient infinitely to prevent the aircraft from stalling. Therefore, during flight, the speed of the fixed-wing UAV must be greater than the minimum speed.

The *maximum speed* in the flight envelope is influenced by several factors, mainly related to the aircraft's maximum thrust and structural strength. When the maximum lift available to the aircraft cannot exceed the current weight of the plane, the aircraft cannot climb further, and the obtained altitude is the ceiling of the UAV, which is the *maximum altitude* in the flight envelope.

[1] The constraints on the minimum speed are divided into two parts: lift constraints during low-altitude flight and thrust constraints during high-altitude flight. Since micro-small fixed-wing UAVs typically operate at low altitudes, they are only subject to lift constraints.

3.2 Basic Experiment

3.2.1 Experimental Objective

1. Things to prepare
Parameters of a micro-small fixed-wing UAV are given as shown in Tables 3.2 and 3.3.

2. Objective
Use MATLAB to plot the thrust required curve and the available curve of the fixed-wing UAV, analyze and calculate its maximum level flight speed, endurance speed, and cruise speed, and calculate the corresponding flight time and distance at each speed.

Table 3.2 Parameters and environmental parameters of a micro-small fixed-wing UAV

Parameter	Value
Total mass	$m = 8.165$ (unit: kg)
Wing area	$S = 0.982$ (unit: m^2)
Span	$b = 2.795$ (unit: m)
Chord	$c = 0.351$ (unit: m)
Zero lift drag coefficient	$C_{D_0} = 0.022$
Lift-induced drag factor	$A = 0.057$
Gravity acceleration	$g = 9.81$ (unit: m/s^2)
Air density	$\rho = 1.225$ (unit: kg/m^3)

Table 3.3 Parameters of the propulsion system of a micro-small fixed-wing UAV

Component	Parameters and values
Propeller	Diameter $D_p = 15$ inch, Pitch $H_p = 8$ inch, Number of blades $B_p = 2$
APC1580	Thrust coefficient $C_T = 0.0984$, Torque coefficient $C_M = 0.0068$, Decay coefficient $K = 12$
Motor	KV value $K_{V0} = 465$ RPM/V, Maximum power W_{mMax}=1700 W, Maximum voltage U_{mMax}=22.2 V, Nominal no-load current $I_{m0} = 2.4$ A,
SUNNYSKY X4120-8	Nominal no-load voltage $U_{m0} = 10$ V, Internal resistance R_m=0.016 Ω
ESC	Maximum ESC current $I_{eMax} = 80$ A
Hobbywing Xrotor Pro HV	Internal resistance $R_e = 0.008$ Ω
Battery	Capacity $C_b = 5000$ mAh, Internal resistance R_b=0.0084 Ω, Total voltage
ACE 5000 mAh	$U_b = 22.2$ V, Maximum discharge rate $K_b = 65$ C

3.2.2 Experimental Procedure

The objective of the basic experiment is to access the operational performance of a micro-small fixed-wing UAV, focusing on its maximum level flight speed, endurance speed, and cruise speed. First, in Step 1, MATLAB plots the thrust required curve and the available curve to obtain data support for the maximum level flight speed. Next, in Step 2, based on Eqs. (3.11) and (3.13), MATLAB programs are written to calculate the endurance speed and cruise speed. These speed values are crucial for evaluating the aircraft's performance on long-distance and long-duration flights. After obtaining the maximum level flight speed, endurance speed, and cruise speed, in Step 3, the endurance time is calculated based on the parameters of the micro-small fixed-wing UAV's propulsion system to determine the flight distance at different speeds.

1. Step 1: Plotting thrust required and thrust available curves in MATLAB
(1) Write the "DrawP.m" file to first initialize the parameters of the fixed-wing UAV. Initialize the parameters of the micro-small fixed-wing UAV and environment according to Tables 1 and 2. For example, wing span $b = 2.795$ m, chord length $c = 0.351$ m, wing area $S = 0.982\,\mathrm{m}^2$, total mass $m = 8.165$ kg, zero-lift drag coefficient $C_{D_0} = 0.022$, induced drag factor $A = 0.057$, gravitational acceleration $g = 9.81$ m/s^2, air density $\rho = 1.225$ kg/m^3, thrust coefficient $C_\mathrm{T} = 0.0984$, torque coefficient $C_\mathrm{M} = 0.0068$, propeller diameter $D_\mathrm{p} = 0.381$ m, motor nominal no-load KV value $K_\mathrm{V0} = 465$ RPM/V, maximum motor voltage $U_{m\mathrm{Max}} = 22.2$ V, propeller dynamic thrust decay coefficient $K = 12$. The example code is shown in Code 3.1.

Code 3.1 Parameter setting, from "e1/DrawP.m"

```
uavspan = 2.795;
uavchord = 0.351;
uavS = 0.982;
uavmass = 8.165;
uavCD0 = 0.022;
uavApolar = 0.057;
envg = 9.81;
envrho0 = 1.225;
Ct = 0.0984;
Cm = 0.0068;
Dp = 0.381;
Kv = 465;
Umax = 22.2;
Kdecay = 12;
```

(2) As shown in Code 3.2, use the example code to obtain the relationship among thrust required, thrust available, and velocity. The example code constructs functions relating thrust required T_re and velocity V according to Equation (3.6), and thrust available T_av and velocity V according to Equation (3.7). Run the "DrawP.m" file to plot the thrust required curve and thrust available curve, as shown in Fig. 3.6.

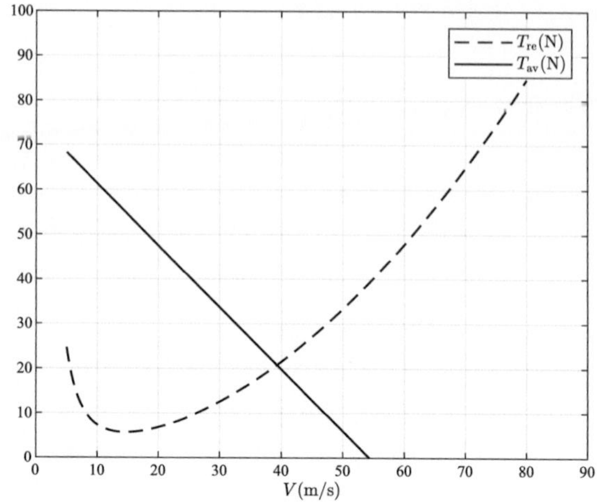

Fig. 3.6 Thrust required and thrust available curves

Code 3.2 Constructor relationship, from "e1/DrawP.m"

```
V = 5:0.1:80;
% Calculate the thrust required for steady-level flight
P_re = @(V) uavCD0 * 0.5 * envrho0 * uavS * V.^2 + uavApolar *
    ( uavmass * envg )^2 ./ (0.5 * envrho0 * uavS * V.^2 );
% Calculate the available thrust of the engine with a given
    parameter
N = Kv * Umax;
P_av = Ct * envrho0 * (N / 60)^2 * Dp^4;
P_av_V = @(V)P_av - Kdecay * V /sqrt(P_av);
```

2. Step 2: Calculating the maximum speed, endurance speed, and cruise speed

(1) Maximum Level Flight Speed

When the UAV is at the maximum level flight speed, the thrust available T_{av} equals the thrust required T_{re}. From Fig. 3.6, the maximum level flight speed is approximately 40 m/s. In the "DrawP.m" file, use the example code in Code 3.3 to solve for the maximum level flight speed. The example code constructs an auxiliary function regarding velocity V. Use the built-in MATLAB library function "fsolve()" to solve for the root of the auxiliary function, corresponding to the maximum level flight speed V_{max}.

Code 3.3 Solving the maximum level flight speed, from "e1/DrawP.m"

```
F = @(V) P_av_V(V) - P_re(V);
% Maximum level flight speed
Vmax = fsolve(F,[0.1,50]);
Vmax = Vmax(2)
```

(2) Endurance Speed
Substituting the parameters of the fixed-wing UAV into Eq. (3.11) yields the endurance speed:

$$V_{\mathrm{lt}} = \sqrt{\frac{2g}{\rho S \sqrt{\frac{3C_{D0}}{A}}}} = \sqrt{\frac{2 \times 9.81 \times 8.165}{1.225 \times 0.982 \times \sqrt{\frac{3\times 0.022}{0.057}}}} = 11.1247.$$

Thus, the endurance speed V_{lt} is 11.1247 m/s.

(3) Cruise Speed
Substituting the parameters of the fixed-wing UAV into Eq. (3.13) yields the cruise speed:

$$V_{\mathrm{ld}} = \sqrt{\frac{2g}{\rho S \sqrt{\frac{C_{D0}}{A}}}} = \sqrt{\frac{2 \times 9.81 \times 8.165}{1.225 \times 0.982 \times \sqrt{\frac{0.022}{0.057}}}} = 14.6409.$$

Thus, the cruise speed V_{ld} is 14.6409 m/s.

3. Step 3: Writing a MATLAB program to solve for flight time and distance at the maximum speed, endurance speed, and cruise speed.
(1) Write the "PropulsionSystemCal.m" file to first initialize all parameters and set the speeds to be solved. Initialize the parameters of the micro-small fixed-wing UAV and environment according to Tables 1 and 2. For example, wing span $b = 2.795$ m, chord length $c = 0.351$ m, wing area $S = 0.982\ \mathrm{m}^2$, total mass $m = 8.165$ kg, zero-lift drag coefficient $C_{D_0} = 0.022$, induced drag factor $A = 0.057$, gravitational acceleration $g = 9.81\ \mathrm{m/s}^2$, air density $\rho = 1.225\ \mathrm{kg/m}^3$, thrust coefficient $C_{\mathrm{T}} = 0.0984$, torque coefficient $C_{\mathrm{M}} = 0.0068$, propeller diameter $D_{\mathrm{p}} = 0.381$ m, motor nominal no-load KV value $K_{\mathrm{V0}} = 465$ RPM/V, maximum motor voltage $U_{m\mathrm{Max}} = 22.2$ V, propeller dynamic thrust decay coefficient $K = 12$, motor no-load voltage $U_{\mathrm{m0}} = 10$ V, motor no-load current $I_{\mathrm{m0}} = 2.4$ A, motor internal resistance $R_m = 0.016$ ohm, electric modulation maximum current $I_{\mathrm{eMax}} = 80$ A, electric modulation internal resistance $R_{\mathrm{e}} = 0.008$ ohm, battery capacity $C_{\mathrm{b}} = 5000$ mAh, minimum battery capacity C_{min}, current loss $I_{\mathrm{other}} = 0.5$ A, battery voltage $U_{\mathrm{b}} = 22.2$ V. The example code is shown in Code 3.4.

Code 3.4 Parameter setting, from "e1/PropulsionSystemCal.m"

```
uavspan = 2.795;
uavchord = 0.351;
uavS = 0.982;
uavmass = 8.165;
uavCD0 = 0.022;
uavApolar = 0.057;
% Environmental parameter
envg = 9.81;
envrho0 = 1.225;
% Propeller parameter
Ct = 0.0984;
Cm = 0.0068;
Dp = 0.381;
% Motor parameter
Kv = 465;
Umax = 22.2;
Kdecay = 12;
Um0 = 10;
Im0 = 2.4;
Rm = 0.016;
% Electric modulation parameter
Re =0.008;
%Battery parameter
Cb = 5000;
Cmin = Cb * 0.15;
Rb = 0.0084;
Iother = 0.5;
Ub= 22.2;
```

(2) In the "PropulsionSystemCal.m" file, the example code shown in Code 3.5 is used to calculate the flight distance of a fixed-wing UAV at different flight speeds. In the example code, the first line sets the speed to be solved for, while the second line calculates the thrust required at the current speed according to Eq. (3.6). Lines 3–5 calculate the static thrust generated by the propeller. Lines 7–9 determine the motor speed and propeller torque based on the obtained static thrust, using Eqs. (3.15) and (3.16) from the propeller model. Lines 11–12 calculate the motor's equivalent current and voltage based on the obtained motor speed and propeller torque, using Equation (3.22) from the motor model. Lines 14–17 determine the ESC throttle command, ESC equivalent current, and ESC equivalent voltage based on the obtained motor's equivalent current and voltage, using Eqs. (3.30), (3.31), and (3.32) from the ESC model. Finally, lines 19–22 calculate the flight distance of the micro-small fixed-wing UAV at the given speed based on the ESC throttle command, ESC equivalent current, and ESC equivalent voltage, using Eqs. (3.34) and (3.36) from the battery model.

Code 3.5 Flight distance solution, from "e1/PropulsionSystemCal.m"

```
V = 39.2865
P_re = uavCD0 * 0.5 * envrho0 * uavS * V.^2 + uavApolar * (
    uavmass * envg )^2 ./ (0.5 * envrho0 * uavS * V.^2 )
equation = @(x) x - Kdecay*V/sqrt(x) - P_re;
initialGuess = P_re;
x_solution = fsolve(equation, initialGuess)
% Calculate the speed of the motor
N = 60 * sqrt(x_solution / (envrho0 * Dp^4 * Ct));
% Calculate the torque of the propeller
M = Cm * envrho0 * (N / 60)^2 * Dp^5;
% Calculate the equivalent current and voltage of the motor
Im = M * Kv * Um0 / 9.55 / (Um0 - Im0 * Rm) + Im0;
Um = Im * Rm + (Um0 - Im0 * Rm) * N / (Kv * Um0);
% Calculate the input throttle, input current, and input
    voltage of the electric modulation
sigma = (Um + Im * Re) / Ub;
Ie = Im * sigma;
Ib = Ie + Iother;
Ue = Ub - Ib * Rb;
% Calculate the battery discharge duration
Tb_s = (Cb - Cmin) / Ib;
Tb_mins = Tb_s * 60 /1000
% Calculating flight distance
L = V * Tb_s*3600/1000
```

Running the "PropulsionSystemCal.m" file with different speed inputs allows obtaining the thrust required, flight time, and flight distance corresponding to the maximum speed, endurance speed, and cruise speed, as shown in Table 3.4. These results indicate that when the UAV flies at its maximum level flight speed, both flight time and range are relatively short. Flying at cruise speed can achieve the maximum range, but it reduces the flight time. Flying at endurance speed can achieve the maximum flight time, but the range is reduced. Therefore, when utilizing a fixed-wing UAV to perform tasks, it is essential to consider various factors and select the optimal flight plan.

Table 3.4 Time and flight distance corresponding to different speeds

Item (Unit)	Speed (m/s)	Thrust required (N)	Flight time (min)	Distance (m)
Maximum level flight speed	39.2865	20.8173	2.3201	5467
Cruise speed	14.6409	5.6729	7.1707	6299
Endurance speed	11.1247	6.5505	8.7868	5865

3.3 Analysis Experiment

3.3.1 Experimental Objective

1. Things to prepare
Refer to Tables 3.2 and 3.3 for parameters of the micro-small fixed-wing UAV.

2. Objective
Analyze the effects of weight, wing area, aspect ratio, and zero-lift drag on the maximum level flight speed and cruise speed (Note: the impact of related variables on endurance speed and cruise speed are the same).

3.3.2 Experimental Procedure

The objective of the analysis experiment is to use simulation to verify how the maximum level flight speed and cruise speed of the micro-small fixed-wing UAVs vary with different parameter settings. To ensure the comparability of the experiment, when studying the impact of a single parameter on aircraft performance, other parameters will be kept constant for comparative analysis to understand the degree of influence that parameter has. In Step 1, modifications will be made to the parameter values to be analyzed. MATLAB will be used to draw the required and available thrust curves to obtain the maximum level flight speed of the micro-small fixed-wing UAVs. In Step 2, MATLAB programs will be written based on Eqs. (3.11) and (3.13) to calculate the cruise speed. In Step 3, the effects of parameter changes on the maximum level flight speed and cruise speed will be analyzed. The purpose of the analysis experiment is to systematically evaluate the influence of each parameter on the performance of the micro-small fixed-wing UAVs. By gradually adjusting parameters and observing changes in speed indicators, a deeper understanding of the impact of each parameter on UAV performance is gained, providing valuable insights for optimized design.

1. Step 1: Analyzing the effects of wing area variation on maximum level flight speed $V_{\max}$ and Cruise Speed V_{ld}
(1) Write "DrawVel.m" file, and initialize parameters first. According to Tables 3.2 and 3.3, initialize parameters of the micro-small fixed-wing UAVs and environment. It is known that the wingspan $b = 2.795$ m, chord length $c = 0.351$ m, total mass $m = 8.165$ kg, zero-lift drag coefficient $C_{D_0} = 0.022$, lift-induced drag factor $A = 0.057$, gravitational acceleration $g = 9.81$ m/s^2, air density $\rho = 1.225$ kg/m^3, thrust coefficient $C_{\mathrm{T}} = 0.0984$, torque coefficient $C_{\mathrm{M}} = 0.0068$, propeller diameter $D_{\mathrm{p}} = 0.381$ m, motor nominal no-load KV value $K_{\mathrm{V0}} = 465$ RPM/V, maximum motor voltage $U_{\mathrm{mMax}} = 22.2$ V, propeller dynamic thrust decay coefficient $K = 12$. The wing area is the variable to be analyzed, and its variation range is set to 0.2–3 m^2, with a step size of 0.1 m^2. The example code is shown in Code 3.6.

Code 3.6 Parameter settings, from "e1/DrawVel.m"

```
uavspan = 2.795;
uavchord = 0.351;
uavS = 0.2:0.1:3;
uavmass = 8.165;
uavCD0 = 0.022;
uavApolar = 0.057;
envg = 9.81;
envrho0 = 1.225;
Ct = 0.0984;
Cm = 0.0068;
Dp = 0.381;
Kv = 465;
Umax = 22.2;
Kdecay = 12;
```

(2) The example code in Code 3.7 is used to solve for the maximum level flight speed of a fixed-wing UAV under different wing areas. The first line of the example code initializes the maximum speed list, and lines 3–15 set up a loop to calculate the corresponding maximum level flight speeds for different wing areas. Line 6 constructs a function relationship between the thrust required T_{re} and speed V based on Eq. (3.6). Lines 8–10 construct a function relationship between the thrust available T_{av} and speed V based on Eq. (3.7). Line 11 constructs an auxiliary function by subtracting the thrust required T_{re} from the thrust available T_{av}. Lines 13–14 solve for the zeros of the auxiliary function using MATLAB's "fsolve()" function, which corresponds to the maximum level flight speed for the current wing area.

Code 3.7 Solving for the maximum level flight speed with different wing areas, from "e1/DrawVel.m"

```
1  % Maximum level flight speed
2  Vmax_list = [];
3  for uavSi = uavS
4  V = 5:0.1:80;
5  % Calculate P_px
6  P_re = @(V) uavCD0 * 0.5 * envrho0 * uavSi * V.^2 +
       uavApolar * ( uavmass * envg )^2 ./ (0.5 * envrho0 *
       uavSi * V.^2 );
7  % Calculate P_ky_V
8  N = Kv * Umax;
9  P_av = Ct * envrho0 * (N / 60)^2 * Dp^4;
10 P_av_V = @(V)P_av - Kdecay * V /sqrt(P_av);
11 F = @(V) P_av_V(V) - P_re(V);
12 % Maximum level flight speed
13 Vmax = fsolve(F,[0.1,50]);
14 Vmax_list(end+1) = Vmax(2);
15 end
```

(3) The example code in Code 3.8 is used to solve for the cruise speed of fixed-wing UAVs under different wing areas. Line 2 calculates the gravity of the UAV based on its mass, and line 3 constructs a function for the cruise speed in terms of the wing area based on Eq. (3.13).

Code 3.8 Solving for cruise speed under different wing areas, from "e1/DrawVel.m"

```
1 % calculate V_ld
2 G = uavmass*envg;
3 V_ld =
      @(uavS)sqrt(2*G./(envrho0*uavS*sqrt(uavCD0/3/uavApolar)));
```

(4) Drawing the velocity change curve, the curves for the two speeds can be obtained as shown in Fig. 3.7.

From Fig. 3.7, it is observed that as the wing area of the fixed-wing UAV increases, both the maximum level flight speed V_{max} and the cruise speed V_{ld} gradually decrease. Compared to the maximum level flight speed, the cruise speed is more affected and decreases faster. This is because, with the increase in wing area, the fixed-wing UAV experiences greater air resistance, requiring more thrust to overcome it. However, the available thrust for fixed-wing aircraft remains constant, decreasing the maximum level flight speed. According to Eq. (3.13), the cruise speed is inversely proportional to the wing area, so as the wing area increases, the cruise speed decreases.

2. Step 2: Analyzing the effect of aircraft mass variation on the maximum level flight speed V_{max} and cruise speed V_{ld}

Taking the mass of the fixed-wing UAV as a variable, ranging from 5–20 kg, repeating Step 1 can obtain the curves for the two speeds as the mass of the fixed-wing UAV changes, as shown in Fig. 3.8. As the weight of the fixed-wing UAV increases, the maximum level flight speed V_{max} exhibits a gradual decline; however, this decrease is minimal and can be considered negligible. In contrast, the cruise speed V_{ld} signif-

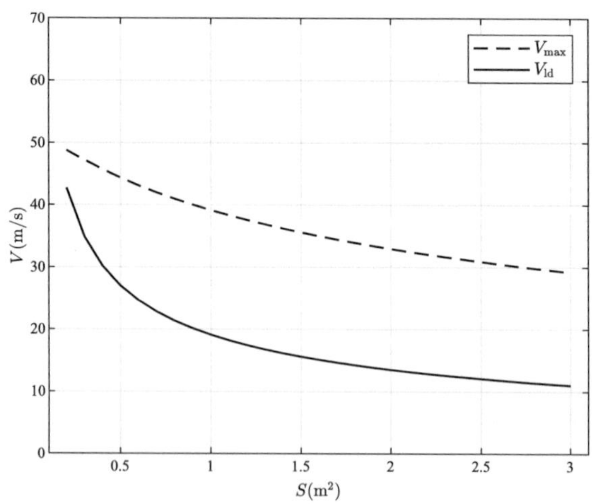

Fig. 3.7 Maximum level flight speed and cruise speed changing with wing area

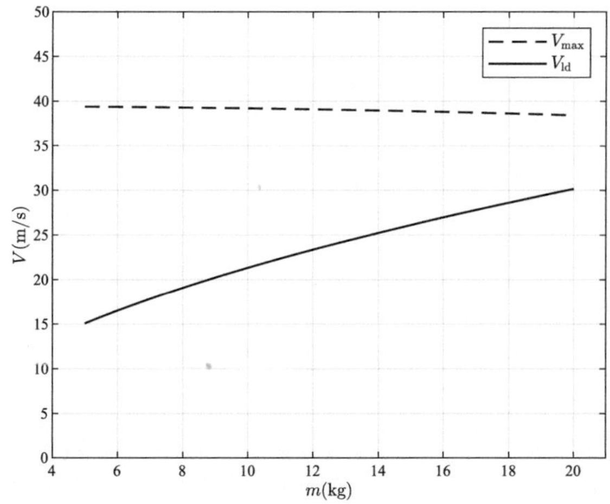

Fig. 3.8 Maximum level flight speed and cruise speed changing with mass

icantly increases. This phenomenon occurs for two reasons. First, according to Eq. (3.6), the increase in gravitational force has a negligible effect on the thrust required. Secondly, based on Eq. (3.7), it does not influence the available thrust. Consequently, the change in the maximum level flight speed remains insignificant. According to Eq. (3.13), the cruise speed is directly proportional to the UAV's weight, so as the weight of the aircraft increases, the cruise speed also increases.

3. Step 3: Analyzing the effect of aspect ratio variation on the maximum level flight speed V_{max} and cruise speed V_{ld}

By varying the aspect ratio of the fixed-wing UAV from 1 to 16 and repeating Step 1, the resulting curves for the two speeds as the aspect ratio changes are illustrated in Fig. 3.9. As the aspect ratio increases, the maximum level flight speed, V_{max}, shows a slight increase, although this increase is minimal and can be disregarded. Conversely, the cruise speed V_{ld} gradually decreases, with the rate of decrease slowing over time. This occurs because, according to Eq. (3.6), an increased aspect ratio reduces the required thrust, while the corresponding increase in speed slightly diminishes the available thrust. Since these changes offset each other, the variation in the maximum level flight speed remains negligible. According to Eq. (3.13), cruising speed is positively correlated with the lift-induced drag factor of the UAV. In contrast, by definition, the aspect ratio is inversely correlated with the lift-induced drag factor. Therefore, as the aspect ratio increases, the cruising speed decreases.

4. Step 4: Analyzing the effect of zero-lift drag coefficient variation on the maximum level flight speed V_{max} and cruise speed V_{ld}

By varying the zero-lift drag coefficient of the fixed-wing UAV from 0.005 to 0.1 and repeating Step 1, the resulting speed curves as the zero-lift drag coefficient changes are shown in Fig. 3.10. As the zero-lift drag coefficient increases, both the

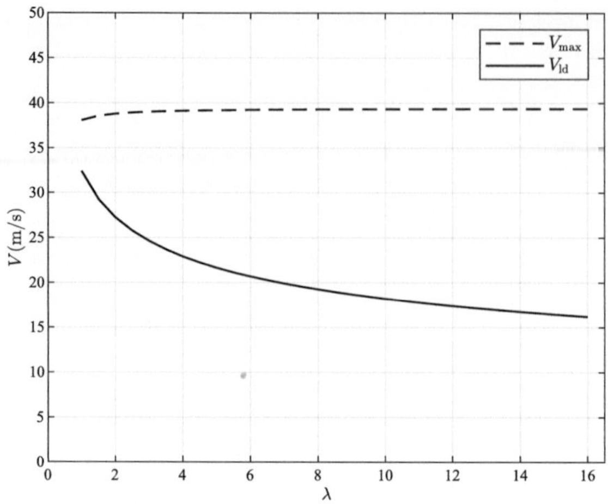

Fig. 3.9 Maximum level flight speed and cruise speed changing with aspect ratio

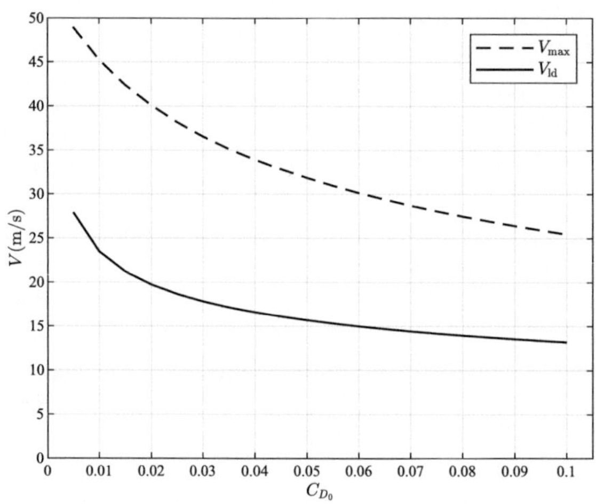

Fig. 3.10 Maximum level flight speed and cruise speed changing with zero-lift drag coefficient

maximum level flight speed V_{max} and cruise speed V_{ld} progressively decrease. This occurs because an increase in the zero-lift drag coefficient results in higher air resistance, which requires more thrust to overcome. However, since the available thrust remains constant, the maximum level flight speed declines. Additionally, according to Eq. (3.13), the cruise speed is inversely proportional to the zero-lift drag coefficient, resulting in a decrease in cruise speed as the drag coefficient increases.

3.4 Design Experiment

3.4.1 Experimental Objective

1. Things to prepare

The parameters of a micro-small fixed-wing UAV other than the wing and battery are described in Sect. 3.2.

2. Objective

With the rapid development of the economy, the demand for logistics is gradually increasing, and UAV transportation is now a new research direction. Currently, assuming a micro-small fixed-wing UAV with a 15 kg payload, it needs to transport goods to a destination 8 km away at a cruise speed of 22 m/s. The goal is to design suitable wing parameters and select the appropriate battery so the UAV can complete the transportation task. The available battery models and their corresponding parameters are listed in Table 3.5.

3.4.2 Experimental Design

The first step in designing the experiment is to determine a suitable wingspan and chord length, which is a critical step based on the mission objectives. Next, in Step 2, the range of the micro-small fixed-wing UAV at the given cruising speed will be calculated based on the energy situation under different battery signals. This range calculation will help assess whether the aircraft can complete the mission. After obtaining the range data, the most suitable battery will be evaluated and selected. This selection will consider the aircraft's energy requirements, battery performance, and the optimal balance between performance and cost to meet the mission's objectives.

1. Step 1: Solving for a suitable wingspan, chord length, wing area, and choosing the right battery model.

As stated in Sect. 3.1.1, the formula for estimating cruise speed is:

$$V_{\mathrm{ld}} = \sqrt{\frac{2G}{\rho S\sqrt{\frac{C_{D_0}}{A}}}}$$

Table 3.5 Available battery models and their parameters

Battery model	TATTU 4500 mAh	ACE 5000 mAh	TATTU 8000 mAh	TATTU 10000 mAh
Battery capacity	4500 mAh	5000 mAh	8000 mAh	10000 mAh
Battery weight	650 g	800 g	1200 g	1400 g

where the wing area S and lift-to-drag factor A are both related to the wingspan b and chord length c, being coupled variables, represented as:

$$S = bc$$
$$A = \frac{1}{\pi \lambda e} = \frac{c}{\pi be}.$$

Substituting the relevant formulas and specific data into the cruise speed formula (3.13) yields:

$$\begin{aligned} V_{\mathrm{ld}} &= \sqrt{\frac{2G}{\rho S \sqrt{\frac{C_{D_0}}{A}}}} \\ &= \sqrt{\frac{2G}{1.225 \times \sqrt{0.022 \times 3.14 \times 0.7} \times \sqrt{b^3} \times \sqrt{c}}} \\ &= 22. \end{aligned} \tag{3.37}$$

The current cruising speed requirement exceeds the value provided in Sect. 3.2. Analysis in Sect. 3.3 indicates that reducing wing area, increasing aircraft mass, and decreasing aspect ratio can all contribute to an increase in cruising speed. Here, the aspect ratio will remain constant, and a comprehensive approach will be taken to adjust aircraft mass and wing area to achieve the mission objectives. Maintaining the aspect ratio from the basic experiment in Sect. 3.2 results in:

$$\frac{b}{c} = \frac{2.795}{0.351}. \tag{3.38}$$

Solving Eqs. (3.37) and (3.38) gets:

$$b = 0.1239\sqrt{2.8217G}$$
$$c = 0.0156\sqrt{2.8217G}.$$

As selected in Table 3.3, the battery model is ACE 5000 mAh. After obtaining the battery weight, the weight of the fixed-wing UAV is 7.365 kg. Thus, the total weight of the UAV, excluding the battery, is 22.365 kg. Therefore, one has:

$$G = (22.365 + m_{\mathrm{b}})g$$

where m_{b} is the battery mass. When calculating, it is assumed that changing the wingspan parameters b and chord length c will not affect the UAV's weight, and only the variation in battery mass is considered. In contrast, other parameters are assumed to remain unchanged.

Table 3.6 UAV parameters and flight performance corresponding to different battery models

Battery model (Units)	Wingspan b (m)	Chord length c (m)	Duration (min)	Range (m)
TATTU 4500 mAh	3.1257	0.3935	3.5898	4739
ACE 5000 mAh	3.1359	0.3948	3.9799	5254
TATTU 8000 mAh	3.1628	0.3982	6.3306	8356
TATTU 10000 mAh	3.1762	0.3999	7.8902	10415

2. Step 2: Solving flight performance
Based on the above process, solve for the wingspan b and chord length c corresponding to each battery, then apply the obtained wing area to the "PropulsionSystemCal.m" file written in the basic experiment for solving. The results are shown in Table 3.6.

According to the mission requirements, the range of the UAV at a cruising speed of 22 m/s needs to be greater than 8 km. Therefore, only the last two types of UAVs meet the requirements. Considering the high power consumption during takeoff and landing, it is necessary to have some backup power available. Therefore, a TATTU 10000mAh battery can be selected.

3.5 Exercise

1. During the takeoff process of a micro-small fixed-wing UAV, acceleration on the runway is necessary to generate sufficient lift to overcome gravity and complete takeoff. In consideration of the design parameters of the micro-small fixed-wing UAV (as illustrated in Table 3.2) and the distinctive attributes of its propulsion system (refer to Table 3.3), with a payload weight of 1 kg, determine the requisite runway length for takeoff. In this scenario, it is assumed that the fixed-wing UAV maintains a constant angle of attack during takeoff, corresponding to a lift coefficient of $C_L = 0.3587$. The coefficient of friction between the UAV and the ground is set at $\mu = 0.01$, with the propeller speed fixed at $N = 7000$ RPM. Based on the parameters above, calculate the requisite runway length for takeoff. As illustrated in Fig. 3.11, the aircraft carrier's bow is typically curved upward, which can reduce the necessary runway length. Make reasonable assumptions about the data and then calculate and analyze them. This calculation considers multiple factors, including ground friction, lift generation, and payload weight, which are critical parameters in evaluating the UAV's takeoff performance.
2. Given the specified parameters for the micro-small fixed-wing UAV (see Table 3.2) and the characteristics of the propulsion system (see Table 3.3), with a payload

Fig. 3.11 Bow of an aircraft carrier

weight of 1 kg, the micro-small fixed-wing UAV successfully transports the payload to the designated destination, releases the payload, and returns to the starting point. Determine the maximum distance at which the UAV can successfully deliver the payload. To determine the maximum distance the UAV can deliver the payload, several factors must be considered, including the aircraft's flight performance and the payload's mass. This distance is of great consequence in evaluating the UAV's capacity to transport goods, as it reflects pivotal parameters of its range and performance.

3. In consideration of the specified parameters for the micro-small fixed-wing UAV (see Table 3.2), it can be assumed that the UAV employs an oil-powered system as the primary source of power. During the flight process of an oil-powered UAV, power is derived from fuel consumption, resulting in a gradual decrease in the aircraft's mass. In this problem, it is assumed that the UAV carries 2 kg of fuel and that the fuel consumption rate is 0.03 kg/N·h (aircraft fuel consumption (kg) = fuel consumption rate × thrust × time). Calculate the maximum flight distance the UAV can achieve under this oil-powered mode. This calculation will consider factors such as fuel consumption, changes in aircraft mass over time, and power consumption during flight to determine the longest distance an oil-powered UAV can fly with the given fuel mass.
4. By the specified parameters of the micro-small fixed-wing UAV (as detailed in Table 3.2) and the characteristics of the propulsion system (outlined in Table 3.3), the UAV ascends at a consistent velocity throughout its flight trajectory, exhibiting a climbing angle that ranges from 0° to 10°. In this problem, the objective is to calculate the maximum altitude a fixed-wing UAV can reach while climbing at a constant speed. It is essential to acknowledge that variations in the aircraft's

altitude result in corresponding changes in environmental pressure and air density, which significantly impact the aircraft's performance and lift characteristics. This calculation will take into account various factors, including the aircraft's climbing speed, climbing angle, and the environmental pressure and air density under which the aircraft operates. This calculation of the maximum altitude reached is essential for evaluating the aircraft's performance, high-altitude flight capabilities, and environmental influences.

5. In consideration of the specified parameters for the micro-small fixed-wing UAV (see Table 3.2) and the attributes of the propulsion system (see Table 3.3), the UAV is currently operating at an endurance speed at an altitude of 500 m. Assuming that during the acceleration process, the lift coefficient of the UAV is $C_L = 0.3587$, and the UAV operates in a no-wind environment. The propulsion system suddenly ceases to function, necessitating a powerless glide for the UAV. A powerless glide is typically regarded as a uniform, straight descent. The objective is to calculate the current equilibrium state of the UAV after losing power and determine the distance the UAV can glide without power. This calculation requires knowledge of the flight state and characteristics after losing power, including the UAV's glide ratio, descent speed, and glide distance. Given the UAV's altitude and stall conditions, this parameter is a key factor in evaluating the UAV's flight performance and emergency response after losing power.
6. In consideration of the parameters of the micro-small fixed-wing UAV (see Table 3.2) and the characteristics of the propulsion system (see Table 3.3), it is requested that the flight envelope of the UAV be calculated and approximately plotted, assuming that only the aerodynamic and power effects are considered, without accounting for structural issues.

References

1. Y. Zhang, W. Guo, D. Ma, J. Xiang, Calculation of range and endurance of jet aircraft at constant altitude. J. Beijing Univ. Aeronaut. Astronaut. **29**(7), 565–569 (2003). in Chinese
2. M. Zhang, H. Chen, Estimated model of range and endurance of small electric UAVs. Acta Aeronautica et Astronautica Sinica **42**(3), 625–085 1 (2021). in Chinese
3. Performance data—APC propellers. [Online]. Available: https://www.apcprop.com/technical-information/performance-data

Chapter 4
Controlled Dynamic System Modeling Experiment

Lin Xu, Wenhan Gao, Hao Sha, Xunhua Dai, and Quan Quan

Establishing the controlled dynamic model of a micro-small fixed-wing Unmanned Aerial Vehicle (UAV) is fundamental. In order to describe the attitude and position of the micro-small fixed-wing UAV, appropriate coordinate systems must be established. The attitude of the fixed-wing UAV in space is mainly described by the rotation relationship among the ground coordinate system, the body coordinate system, the wind coordinate system, and the flight path coordinate system. Based on these coordinate systems and attitude descriptions, considering the forces acting on the micro-small fixed-wing UAV, a controlled dynamic model is established for system identification and flight control. Readers can deepen their understanding of the controlled dynamic model of micro-small fixed-wing UAVs through a basic experiment, an analysis experiment, and a design experiment. In the basic experiment, readers will understand the influence of actuator inputs on flight states and the relationship between aerodynamic coefficients and the deflection of ailerons/rudders of micro-small fixed-wing UAVs. In the analysis experiment, readers will understand the trimming methods of micro-small fixed-wing UAVs, then obtain linear dynamic equations, and finally compare the nonlinear dynamic model and the linear mathematical model of micro-small fixed-wing UAVs. In the design experiment, readers will learn to build a dynamic model of the micro-small fixed-wing UAV using Simulink, observe the UAV's flight status in FlightGear, learn code generation, and finally perform a Hardware-In-the-Loop (HIL) simulation using the RflySim toolchain. For more information on this chapter, please refer to the link

https://rflysim.com/doc/en/C/5.FW_Book.html.

Q. Quan et al., *Micro-Small Fixed-Wing UAV Control*,
https://doi.org/10.1007/978-981-95-1954-5_4

4.1 Preliminary

This section describes the coordinate system employed in modeling a fixed-wing UAV, as well as the equations of kinematics and dynamics. It then analyzes in detail the forces and moments to which the UAV is subjected, followed by small perturbation linearization and further analysis of the UAV's lateral motion and mode.

4.1.1 Coordinate System

1. Ground coordinate system
The ground coordinate system $o_e x_e y_e z_e$ is a coordinate system fixed to the Earth, with its origin o_e located at a point on the ground. As shown in Fig. 4.1 on the left, $o_e x_e$ axis points to the north, $o_e y_e$ axis points to the east, and $o_e z_e$ axis points vertically downward. Therefore, the ground coordinate system is also commonly referred to as the North-East-Down (NED) coordinate system.

2. Body coordinate system
The body coordinate system $o_b x_b y_b z_b$ is fixed to the body of the fixed-wing UAV, with its origin o_b located at the center of mass of the UAV. The $o_b x_b$ axis lies within the UAV's symmetry plane, aligned with the aircraft's longitudinal axis, and points toward the nose. The $o_b z_b$ axis, also in the symmetry plane, is perpendicular to the $o_b x_b$ axis and points downward. The $o_b y_b$ axis is perpendicular to the symmetry plane, pointing to the right. The relationship between the body coordinate system of the fixed-wing UAV and the ground coordinate system is shown in Fig. 4.1.

The rotation between the ground coordinate system and the body coordinate system consists of three basic rotations, namely, rotations about $o_b z_b$, $o_b y'_b$, and $o_b x''_b$ axes, respectively, corresponding to the Euler angles ψ, θ, and ϕ as shown in Fig. 4.2.

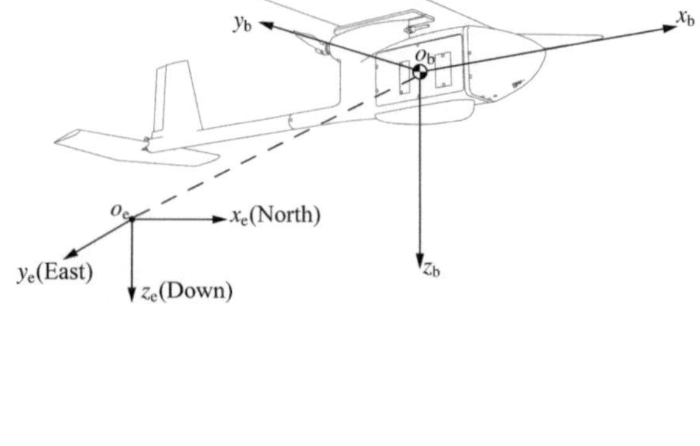

Fig. 4.1 Ground coordinate system and body coordinate system

Fig. 4.2 Coordinate system rotation and Euler angles

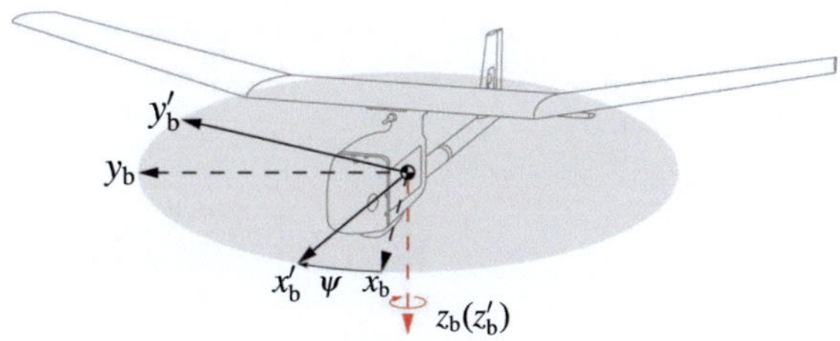

(a) Yaw angle

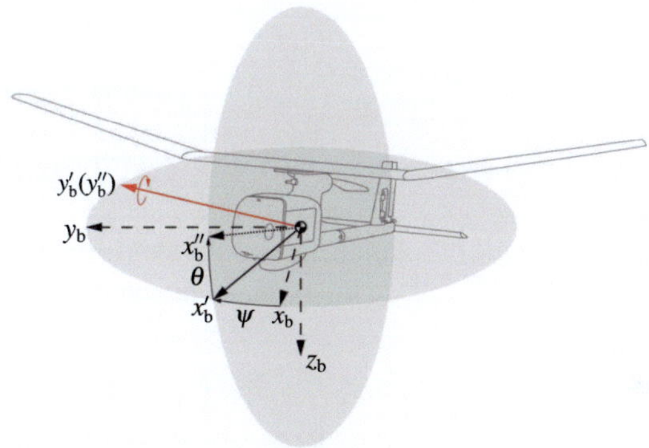

(b) Pitch angle

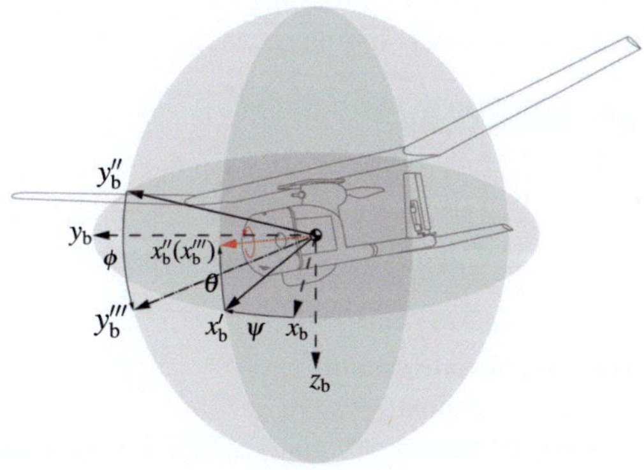

(c) Roll angle

(1) The angle of rotation about $o_b z_b$ axis is the yaw angle ψ. It is defined as positive when the aircraft's nose turns to the right, as shown in Fig. 4.2a. The rotation matrix is given by

$$\mathbf{R}(\psi) = \begin{bmatrix} \cos\psi & \sin\psi & 0 \\ -\sin\psi & \cos\psi & 0 \\ 0 & 0 & 1 \end{bmatrix}. \tag{4.1}$$

(2) The angle of rotation about $o_b y_b'$ axis is the pitch angle θ. It is defined as positive when the aircraft pitches up, as shown in Fig. 4.2b. The rotation matrix is given by

$$\mathbf{R}(\theta) = \begin{bmatrix} \cos\theta & 0 & -\sin\theta \\ 0 & 1 & 0 \\ \sin\theta & 0 & \cos\theta \end{bmatrix}. \tag{4.2}$$

(3) The angle of rotation about $o_b x_b''$ axis is the roll angle ϕ. It is defined as positive when the aircraft rolls to the right, as shown in Fig. 4.2c. The rotation matrix is given by

$$\mathbf{R}(\phi) = \begin{bmatrix} 1 & 0 & 0 \\ 0 & \cos\phi & \sin\phi \\ 0 & -\sin\phi & \cos\phi \end{bmatrix}. \tag{4.3}$$

Thus, by sequentially rotating the Euler angles ψ, θ, and ϕ, the rotation matrix $\mathbf{R}_e^b$ from the ground coordinate system to the body coordinate system is obtained as

$$\begin{aligned} \mathbf{R}_e^b &= \mathbf{R}(\phi)\mathbf{R}(\theta)\mathbf{R}(\psi) \\ &= \begin{bmatrix} \cos\theta\cos\psi & \cos\theta\sin\psi & -\sin\theta \\ \sin\phi\sin\theta\cos\psi - \cos\phi\sin\psi & \sin\phi\sin\theta\sin\psi + \cos\theta\cos\psi & \sin\phi\cos\theta \\ \cos\phi\sin\theta\cos\psi + \sin\phi\sin\psi & \cos\phi\sin\theta\sin\psi - \sin\phi\cos\psi & \cos\phi\cos\theta \end{bmatrix}. \end{aligned} \tag{4.4}$$

3. Wind coordinate system

(1) Stability coordinate system

The velocity of the aircraft relative to the surrounding air is called the *airspeed vector* $\mathbf{v}_a$, and its magnitude V_a is commonly referred to as the *airspeed*. The angle between the line of airspeed direction and the projection of the aircraft's longitudinal axis on the plane of symmetry, known as the *attack angle* α, is positive, as shown in Fig. 4.3. The attack angle α is the angle between the body coordinate system $o_b x_b$ axis and the projection $\mathbf{v}_{ax}$ of the airspeed vector $\mathbf{v}_a$ in the plane of the body coordinate system $o_b x_b z_b$. A new coordinate system, denoted as $o_s x_s y_s z_s$, is defined with its origin o_s coinciding with o_b. The $o_s x_s$ axis aligns with the direction of the vector $\mathbf{v}_{ax}$, $o_s y_s$ axis coincides with $o_b y_b$ axis, and $o_s z_s$ axis is determined by the right-hand rule. This coordinate system $o_s x_s y_s z_s$ is called the *stability coordinate system*. The rotation

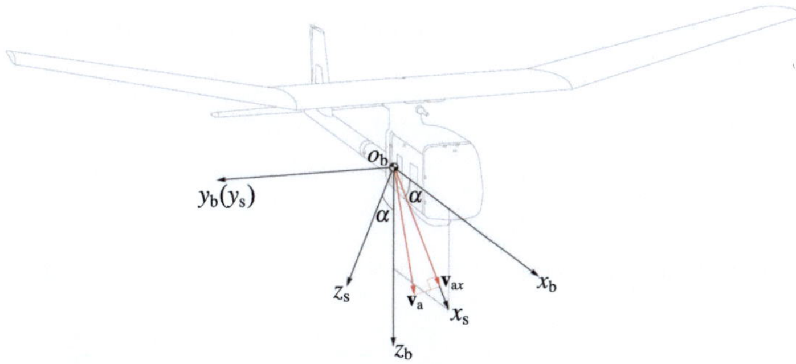

Fig. 4.3 Stability coordinate system and attack angle

matrix from the body coordinate system to the stability coordinate system is given by

$$\mathbf{R}(\alpha) = \begin{bmatrix} \cos\alpha & 0 & \sin\alpha \\ 0 & 1 & 0 \\ -\sin\alpha & 0 & \cos\alpha \end{bmatrix}. \tag{4.5}$$

(2) Wind coordinate system
The angle between the airspeed vector $\mathbf{v}_a$ and the plane of the body coordinate system $o_b x_b z_b$ is called the *sideslip angle* β. As shown in Fig. 4.4, based on the stability coordinate system $o_s x_s y_s z_s$, a new coordinate system $o_w x_w y_w z_w$ is defined, where $o_w x_w$ coincides with the direction of the airspeed vector $\mathbf{v}_a$, $o_w z_w$ axis coincides with $o_s z_s$ axis, and $o_w y_w$ axis is determined by the right-hand rule. Since $o_w x_w$ is aligned with the airspeed direction, this coordinate system $o_w x_w y_w z_w$ is called the

Fig. 4.4 Wind coordinate system and sliding angle

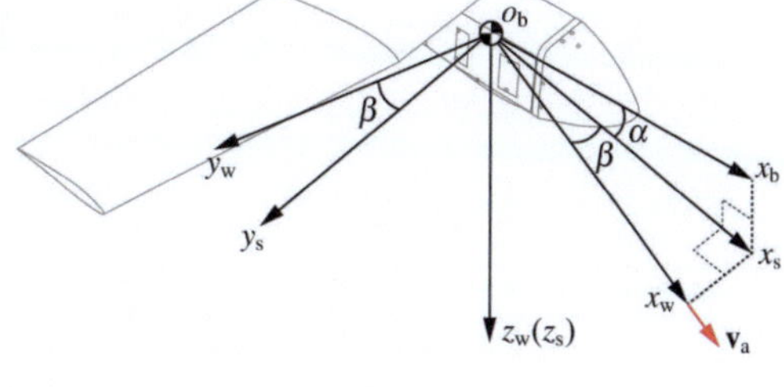

wind coordinate system. The rotation matrix from the stability coordinate system to the wind coordinate system is given by

$$\mathbf{R}(\beta) = \begin{bmatrix} \cos\beta & \sin\beta & 0 \\ -\sin\beta & \cos\beta & 0 \\ 0 & 0 & 1 \end{bmatrix}. \tag{4.6}$$

Therefore, the rotation matrix from the body coordinate system to the wind coordinate system is

$$\mathbf{R}_{\mathrm{b}}^{\mathrm{w}} = \mathbf{R}(\beta)\mathbf{R}(\alpha) = \begin{bmatrix} \cos\beta\cos\alpha & \sin\beta & \cos\beta\sin\alpha \\ -\sin\beta\cos\alpha & \cos\beta & -\sin\beta\cos\alpha \\ -\sin\alpha & 0 & \cos\alpha \end{bmatrix}. \tag{4.7}$$

4. Flightpath coordinate system

The velocity of the fixed-wing UAV relative to the Earth (ground coordinate system) is called the *ground speed*, represented as $\mathbf{v}_{\mathrm{g}}$.[1] The origin of the flightpath coordinate system $o_{\mathrm{k}}x_{\mathrm{k}}y_{\mathrm{k}}z_{\mathrm{k}}$ is located at the center of mass of the fixed-wing UAV. The $o_{\mathrm{k}}x_{\mathrm{k}}$ axis is aligned with the direction of the ground speed, $o_{\mathrm{k}}z_{\mathrm{k}}$ axis lies in the vertical plane containing $o_{\mathrm{k}}x_{\mathrm{k}}$ and points downward relative to the aircraft, and $o_{\mathrm{k}}y_{\mathrm{k}}$ axis is determined by the right-hand rule. The angle between the ground speed $\mathbf{v}_{\mathrm{g}}$ and the plane of the ground coordinate system $o_{\mathrm{e}}x_{\mathrm{e}}y_{\mathrm{e}}$ is called the *flight path angle* γ, with positive values indicating an upward trajectory, as shown in Fig. 4.5. The angle between the projection of the ground speed $\mathbf{v}_{\mathrm{g}}$ onto $o_{\mathrm{e}}x_{\mathrm{e}}y_{\mathrm{e}}$ plane and $o_{\mathrm{e}}x_{\mathrm{e}}$ axis is called the *course angle* χ, with positive values indicating a rightward direction.

5. Airspeed, Windspeed, and Ground Speed

The velocity of a fixed-wing UAV relative to the airflow is represented as the *airspeed vector* $\mathbf{v}_{\mathrm{a}}$. The velocity of the airflow relative to the ground coordinate system is denoted as $\mathbf{v}_{\mathrm{w}}$. Their relationship with the ground speed is given by

$$\mathbf{v}_{\mathrm{a}} = \mathbf{v}_{\mathrm{g}} - \mathbf{v}_{\mathrm{w}}. \tag{4.8}$$

The components of the ground speed vector in the body coordinate system are described as

$${}^{\mathrm{b}}\mathbf{v}_{\mathrm{g}} = \begin{bmatrix} v_{\mathrm{g}x_{\mathrm{b}}} \\ v_{\mathrm{g}y_{\mathrm{b}}} \\ v_{\mathrm{g}z_{\mathrm{b}}} \end{bmatrix}. \tag{4.9}$$

[1] Under windless conditions, the ground speed is consistent with the airspeed.

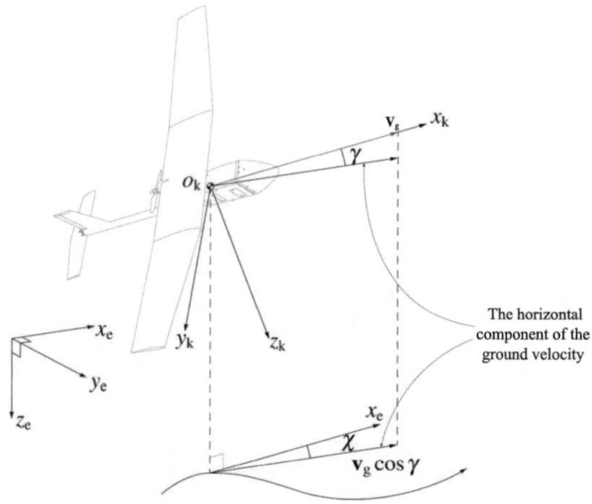

Fig. 4.5 Flightpath coordinate system

Similarly, the windspeed vector in the body coordinate system is represented as

$$^{\mathrm{b}}\mathbf{v}_{\mathrm{w}} = \begin{bmatrix} v_{\mathrm{w}x_{\mathrm{b}}} \\ v_{\mathrm{w}y_{\mathrm{b}}} \\ v_{\mathrm{w}z_{\mathrm{b}}} \end{bmatrix} = \mathbf{R}_{\mathrm{e}}^{\mathrm{b}} \mathbf{v}_{\mathrm{w}}. \tag{4.10}$$

The airspeed vector in the body coordinate system is obtained as

$$^{\mathrm{b}}\mathbf{v}_{\mathrm{a}} = \begin{bmatrix} v_{\mathrm{a}x_{\mathrm{b}}} \\ v_{\mathrm{a}y_{\mathrm{b}}} \\ v_{\mathrm{a}z_{\mathrm{b}}} \end{bmatrix} = \begin{bmatrix} v_{\mathrm{g}x_{\mathrm{b}}} - v_{\mathrm{w}x_{\mathrm{b}}} \\ v_{\mathrm{g}y_{\mathrm{b}}} - v_{\mathrm{w}y_{\mathrm{b}}} \\ v_{\mathrm{g}z_{\mathrm{b}}} - v_{\mathrm{w}z_{\mathrm{b}}} \end{bmatrix}. \tag{4.11}$$

The airspeed vector in the wind coordinate system is represented as

$$^{\mathrm{w}}\mathbf{v}_{\mathrm{a}} = \begin{bmatrix} V_{\mathrm{a}} \\ 0 \\ 0 \end{bmatrix}. \tag{4.12}$$

In modeling and simulation, the windspeed vector in the ground coordinate system is provided by a wind model, and the aircraft's velocity relative to the ground coordinate system is also obtained from kinematic equations. Thus, combining the above expressions yields

$$^{\mathrm{b}}\mathbf{v}_{\mathrm{a}} = \begin{bmatrix} v_{\mathrm{a}x_{\mathrm{b}}} \\ v_{\mathrm{a}y_{\mathrm{b}}} \\ v_{\mathrm{a}z_{\mathrm{b}}} \end{bmatrix} = \mathbf{R}_{\mathrm{w}}^{\mathrm{b}} \begin{bmatrix} V_{\mathrm{a}} \\ 0 \\ 0 \end{bmatrix}. \tag{4.13}$$

Inverting the rotation matrix in Eq. (4.7) and substituting it into Eq. (4.13) result in

$$
{}^{\mathrm{b}}\mathbf{v}_{\mathrm{a}} = \begin{bmatrix} v_{\mathrm{a}x_{\mathrm{b}}} \\ v_{\mathrm{a}y_{\mathrm{b}}} \\ v_{\mathrm{a}z_{\mathrm{b}}} \end{bmatrix} = V_{\mathrm{a}} \begin{bmatrix} \cos\alpha\cos\beta \\ \sin\beta \\ \sin\alpha\cos\beta \end{bmatrix}. \tag{4.14}
$$

Through the above analysis, the following relationships are derived as

$$
\begin{aligned}
V_{\mathrm{a}} &= \sqrt{v_{\mathrm{a}x_{\mathrm{b}}}^2 + v_{\mathrm{a}y_{\mathrm{b}}}^2 + v_{\mathrm{a}z_{\mathrm{b}}}^2} \\
\alpha &= \arctan\left(\frac{v_{\mathrm{a}z_{\mathrm{b}}}}{v_{\mathrm{a}x_{\mathrm{b}}}}\right) \\
\beta &= \arcsin\left(\frac{v_{\mathrm{a}y_{\mathrm{b}}}}{\sqrt{v_{\mathrm{a}x_{\mathrm{b}}}^2 + v_{\mathrm{a}y_{\mathrm{b}}}^2 + v_{\mathrm{a}z_{\mathrm{b}}}^2}}\right).
\end{aligned} \tag{4.15}
$$

4.1.2 *Kinematics and Dynamics*

1. Assumptions for rigid body aircraft equation derivation

The following assumptions are made to simplify the calculation process when deriving the equations of motion for an aircraft's rigid body.

(1) The surface of the UAV does not deform, and its mass remains constant.
(2) The Earth's rotation rate is ignored, and the inertial coordinate system is chosen as the ground coordinate system.
(3) The ground is treated as an ideal two-dimensional plane.
(4) The acceleration due to gravity remains constant.
(5) The structure of the UAV is symmetric; its plane of symmetry is the plane of the body coordinate system $o_{\mathrm{b}}x_{\mathrm{b}}z_{\mathrm{b}}$. It has a symmetric mass distribution, with the inertia products satisfying $J_{xy} = J_{zy} = 0$, and the UAV's moments of inertia about the body axes remain constant.

2. Kinematic Equations

Let the position vector of the center of gravity of the micro-small fixed-wing UAV be ${}^{\mathrm{e}}\mathbf{p} \in \mathbb{R}^3$. Then, the motion equation is given by

$$
{}^{\mathrm{e}}\dot{\mathbf{p}} = {}^{\mathrm{e}}\mathbf{v} \tag{4.16}
$$

where ${}^{\mathrm{e}}\mathbf{v} \in \mathbb{R}^3$ represents the velocity of the fixed-wing UAV in the ground coordinate system.

For the attitude kinematic equation, let the body angular velocity be ${}^{\mathrm{b}}\omega = \begin{bmatrix} \omega_{x_{\mathrm{b}}} & \omega_{y_{\mathrm{b}}} & \omega_{z_{\mathrm{b}}} \end{bmatrix}^{\mathrm{T}} \in \mathbb{R}^3$. Then, the relation between the Euler angle attitude rate and the body rotation angular velocity is given by

$$\begin{bmatrix} \omega_{x_b} \\ \omega_{y_b} \\ \omega_{z_b} \end{bmatrix} = \begin{bmatrix} 1 & 0 & -\sin\theta \\ 0 & \cos\phi & \cos\theta\sin\phi \\ 0 & -\sin\phi & \cos\theta\cos\phi \end{bmatrix} \begin{bmatrix} \dot{\phi} \\ \dot{\theta} \\ \dot{\psi} \end{bmatrix}. \tag{4.17}$$

Furthermore, one has

$$^{\mathrm{b}}\omega = \mathbf{W} \cdot \dot{\boldsymbol{\Theta}} \tag{4.18}$$

where

$$\dot{\boldsymbol{\Theta}} \triangleq \begin{bmatrix} \dot{\phi} \\ \dot{\theta} \\ \dot{\psi} \end{bmatrix} \quad \mathbf{W} \triangleq \begin{bmatrix} 1 & 0 & -\sin\theta \\ 0 & \cos\phi & \cos\theta\sin\phi \\ 0 & -\sin\phi & \cos\theta\cos\phi \end{bmatrix}.$$

2. Dynamic Equations

(1) Position

According to Newton's second law, one has

$$^{\mathrm{e}}\dot{\mathbf{v}} = \frac{^{\mathrm{e}}\mathbf{f}}{m} \tag{4.19}$$

where m represents the total mass of the fixed-wing UAV, and the total external force $^{\mathrm{e}}\mathbf{f}$ includes gravity $\mathbf{g}$, thrust (pull) from the propulsion system $^{\mathrm{b}}\mathbf{t}$, and aerodynamic force $^{\mathrm{b}}\mathbf{f}_{\mathrm{a}}$, expressed as

$$^{\mathrm{e}}\mathbf{f} = \mathbf{g} + \mathbf{R}_{\mathrm{b}}^{\mathrm{e}}(^{\mathrm{b}}\mathbf{t} +^{\mathrm{b}} \mathbf{f}_{\mathrm{a}}). \tag{4.20}$$

Furthermore, due to

$$^{\mathrm{e}}\mathbf{v} = \mathbf{R}_{\mathrm{b}}^{\mathrm{e}} \cdot {}^{\mathrm{b}}\mathbf{v}, \tag{4.21}$$

combining Eqs. (4.19), (4.20), and (4.21) yields

$$\begin{aligned} ^{\mathrm{b}}\dot{\mathbf{v}} &= -[^{\mathrm{b}}\omega]_{\times}\, {}^{\mathrm{b}}\mathbf{v} + \frac{^{\mathrm{b}}\mathbf{f}}{m} \\ ^{\mathrm{b}}\mathbf{f} &= (\mathbf{R}_{\mathrm{b}}^{\mathrm{e}})^{-1}\mathbf{g} +^{\mathrm{b}} \mathbf{t} +^{\mathrm{b}} \mathbf{f}_{\mathrm{a}}. \end{aligned} \tag{4.22}$$

(2) Attitude

The attitude dynamic equation in the body coordinate system is established as

$$\begin{aligned} \mathbf{J} \cdot^{\mathrm{b}} \dot{\omega} &= -^{\mathrm{b}}\omega \times (\mathbf{J} \cdot^{\mathrm{b}} \omega) +^{\mathrm{b}} \mathbf{m} \\ ^{\mathrm{b}}\mathbf{m} &= \mathbf{g}_{\mathrm{y}} +^{\mathrm{b}} \mathbf{m}_{\mathrm{a}} +^{\mathrm{b}} \mathbf{m}_{\mathrm{p}} \end{aligned} \tag{4.23}$$

where $\mathbf{J} \in \mathbb{R}^{3\times3}$ is the rotational inertia of the fixed-wing UAV; $\mathbf{g}_{\mathrm{y}} \in \mathbb{R}^3$ is the gyroscopic moment; $^{\mathrm{b}}\mathbf{m}_{\mathrm{a}} \in \mathbb{R}^3$ is the aerodynamic moment experienced by the fixed-wing

UAV in the body coordinate system; ${}^{\mathrm{b}}\mathbf{m}_{\mathrm{p}}$ represents the moment generated by the propellers, which is ignored in subsequent modeling; ${}^{\mathrm{b}}\mathbf{m} \in \mathbb{R}^3$ is the total moment experienced by the fixed-wing UAV in the body coordinate system. The rotational inertia $\mathbf{J}$ is expressed as

$$\mathbf{J} = \begin{bmatrix} J_{xx} & 0 & -J_{xz} \\ 0 & J_{yy} & 0 \\ -J_{zx} & 0 & J_{zz} \end{bmatrix}$$

where $J_{xx}, J_{yy}, J_{zz} \in \mathbb{R}_+$ are the principal moments of inertia, and $J_{xz} = J_{zx} \in \mathbb{R}_+$ is the moment of inertia product.

4.1.3 Force and Moment

1. Gravity

The force of gravity acts on the center of mass of the fixed-wing UAV and is represented in component form in the ground coordinate system as

$$\mathbf{g} = \begin{bmatrix} 0 \\ 0 \\ mg \end{bmatrix}. \tag{4.24}$$

In practical modeling and simulation, it needs to be represented in the body coordinate system, i.e.,

$${}^{\mathrm{b}}\mathbf{g} = \mathbf{R}_{\mathrm{e}}^{\mathrm{b}} \begin{bmatrix} 0 \\ 0 \\ mg \end{bmatrix} = \begin{bmatrix} -mg\sin\theta \\ mg\cos\theta\sin\phi \\ mg\cos\theta\cos\phi \end{bmatrix}. \tag{4.25}$$

2. Aerodynamic forces and moments

(1) Longitudinal aerodynamics

The longitudinal aerodynamic forces and moments include lift, drag, and pitch moment, causing the aircraft to move within $o_{\mathrm{e}}x_{\mathrm{e}}z_{\mathrm{e}}$ plane, also known as the *pitch plane*, as shown in Fig. 4.6. These forces and moments are mainly influenced by the attack angle α, the pitch rate $\omega_{y_{\mathrm{b}}}$, and the elevator deflection δ_{e}. Therefore, the formulas for lift f_{az}, drag f_{ax}, and pitch moment m_{ay} are written as [1]

$$\begin{aligned} f_{\mathrm{az}} &\approx \frac{1}{2}\rho V_{\mathrm{a}}^2 S C_L(\alpha, \dot{\alpha}, \omega_{y_{\mathrm{b}}}, \delta_{\mathrm{e}}) \\ f_{\mathrm{ax}} &\approx \frac{1}{2}\rho V_{\mathrm{a}}^2 S C_D(\alpha, \dot{\alpha}, \omega_{y_{\mathrm{b}}}, \delta_{\mathrm{e}}) \\ m_{\mathrm{ay}} &\approx \frac{1}{2}\rho V_{\mathrm{a}}^2 S c_{\mathrm{A}} C_m(\alpha, \dot{\alpha}, \omega_{y_{\mathrm{b}}}, \delta_{\mathrm{e}}). \end{aligned} \tag{4.26}$$

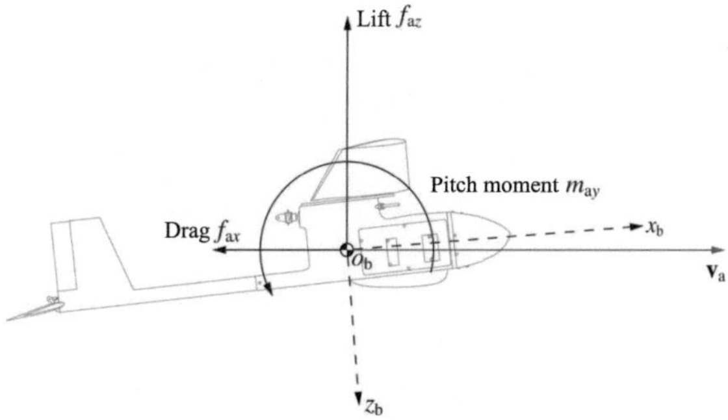

Fig. 4.6 Longitudinal aerodynamics

These force and moment equations are nonlinear functions of the attack angle α, pitch rate ω_{y_b}, and elevator deflection δ_e. However, when the attack angle is small, the airflow around the wing is relatively stable. In this case, lift, drag, and pitch moment can be used to represent linear models. Taking the lift as an example, a first-order Taylor expansion is performed on the lift equation as

$$C_L = C_{L_0} + \frac{\partial C_L}{\partial \alpha}\alpha + \frac{\partial C_L}{\partial \dot{\alpha}}\dot{\alpha} + \frac{\partial C_L}{\partial q}\omega_{y_b} + \frac{\partial C_L}{\partial \delta_e}\delta_e \tag{4.27}$$

where C_{L_0} is the value of C_L when $\alpha = \omega_{x_b} = 0$. It is common to nondimensionalize the partial derivatives of this linear approximation. Since C_L and α, δ_e are dimensionless, while the units for the angular velocities ω_{x_b}, $\dot{\alpha}$ are rad/s, so the non-dimensionalization of $\partial C_L/\partial q$ and $\partial C_L/\partial \dot{\alpha}$ is necessary. Introducing the standard factor $c_A/(2V_a)$, Eq. (4.27) above is rewritten as

$$C_L = C_{L_0} + C_{L_\alpha}\alpha + C_{L_{\dot{\alpha}}}\frac{c_A}{2V_a}\dot{\alpha} + C_{L_q}\frac{c_A}{2V_a}\omega_{y_b} + C_{L_{\delta_e}}\delta_e \tag{4.28}$$

where C_{L_0}, $C_{L_\alpha} = \partial C_L/\partial \alpha$, $C_{L_{\dot{\alpha}}} = (\partial C_L/\partial \dot{\alpha})/(c_A/2V_a)$, $C_{L_q} = (\partial C_L/\partial q)/(c_A/2V_a)$, $C_{L_{\delta_e}} = \partial C_L/\partial \delta_e$ Orthographic are all non-dimensional values.

The choice of the standard factor is related to the specific derivation of aerodynamic forces, as detailed in [2]; C_{L_α}, $C_{L_{\dot{\alpha}}}$, and C_{L_q} are commonly referred to as the *aerodynamic derivatives*, and $C_{L_{\delta_e}}$ is referred to as the *control derivative*[2]

[2] Aerodynamic derivatives and aerodynamic coefficients are two concepts commonly used in aircraft design and aerodynamics, and they have some differences. Aerodynamic coefficient: a dimensionless coefficient that describes the interaction between a vehicle and the air in a particular flight state. Aerodynamic derivative: refers to the rate of change of a vehicle in flight concerning a particular parameter (e.g., angle of attack, sideslip angle, etc.). For example, C_L in Eq. (4.28) is the lift coefficient, while C_{L_α} is the aerodynamic derivative. The aerodynamic and control derivatives are sometimes collectively referred to as coefficients for descriptive convenience.

Similarly, longitudinal aerodynamics drag and pitching moment equations are also linearized through the first-order Taylor expansion. Eventually, the longitudinal aerodynamic forces and moments are represented as:

$$\begin{aligned} f_{\mathrm{az}} &\approx \frac{1}{2}\rho V_{\mathrm{a}}^2 S\left(C_{L_0} + C_{L_\alpha}\alpha + C_{L_{\dot\alpha}}\frac{c_{\mathrm{A}}}{2V_{\mathrm{a}}}\dot\alpha + C_{L_q}\frac{c_{\mathrm{A}}}{2V_{\mathrm{a}}}\omega_{y_{\mathrm{b}}} + C_{L_{\delta_{\mathrm{e}}}}\delta_{\mathrm{e}}\right) \\ f_{\mathrm{ax}} &\approx \frac{1}{2}\rho V_{\mathrm{a}}^2 S\left(C_{D_0} + C_{D_\alpha}\alpha + C_{D_{\dot\alpha}}\frac{c_{\mathrm{A}}}{2V_{\mathrm{a}}}\dot\alpha + C_{D_q}\frac{c_{\mathrm{A}}}{2V_{\mathrm{a}}}\omega_{y_{\mathrm{b}}} + C_{D_{\delta_{\mathrm{e}}}}\delta_{\mathrm{e}}\right) \\ m_{\mathrm{ay}} &\approx \frac{1}{2}\rho V_{\mathrm{a}}^2 S c_{\mathrm{A}}\left(C_{m_0} + C_{m_\alpha}\alpha + C_{m_{\dot\alpha}}\frac{c_{\mathrm{A}}}{2V_{\mathrm{a}}}\dot\alpha + C_{m_q}\frac{c_{\mathrm{A}}}{2V_{\mathrm{a}}}\omega_{y_{\mathrm{b}}} + C_{m_{\delta_{\mathrm{e}}}}\delta_{\mathrm{e}}\right). \end{aligned} \tag{4.29}$$

(2) Lateral aerodynamics

The lateral aerodynamic forces and moments cause the fixed-wing UAV to move along $o_{\mathrm{b}}y_{\mathrm{b}}$ axis and also induce roll and yaw motions, as depicted in Fig. 4.7. Lateral aerodynamic forces are primarily influenced by the sideslip angle β, as well as the roll rate, yaw rate, aileron deflection, and rudder deflection, represented as:

$$\begin{aligned} f_{\mathrm{ay}} &\approx \frac{1}{2}\rho V_{\mathrm{a}}^2 S C_Y(\beta, \omega_{x_{\mathrm{b}}}, \omega_{z_{\mathrm{b}}}, \delta_{\mathrm{a}}, \delta_{\mathrm{r}}) \\ m_{\mathrm{ax}} &\approx \frac{1}{2}\rho V_{\mathrm{a}}^2 S b C_l(\beta, \omega_{x_{\mathrm{b}}}, \omega_{z_{\mathrm{b}}}, \delta_{\mathrm{a}}, \delta_{\mathrm{r}}) \\ m_{\mathrm{az}} &\approx \frac{1}{2}\rho V_{\mathrm{a}}^2 S b C_n(\beta, \omega_{x_{\mathrm{b}}}, \omega_{z_{\mathrm{b}}}, \delta_{\mathrm{a}}, \delta_{\mathrm{r}}). \end{aligned} \tag{4.30}$$

Equations (4.30) are linearized using a first-order Taylor expansion, resulting in the linear representations of lateral aerodynamic forces and moments:

$$\begin{aligned} f_{\mathrm{ay}} &\approx \frac{1}{2}\rho V_{\mathrm{a}}^2 S\left(C_{Y_0} + C_{Y_\beta}\beta + C_{Y_p}\frac{b}{2V_{\mathrm{a}}}\omega_{x_{\mathrm{b}}} + C_{Y_r}\frac{b}{2V_{\mathrm{a}}}\omega_{z_{\mathrm{b}}} + C_{Y\delta_{\mathrm{a}}}\delta_{\mathrm{a}} + C_{Y\delta_{\mathrm{r}}}\delta_{\mathrm{r}}\right) \\ m_{\mathrm{ax}} &\approx \frac{1}{2}\rho V_{\mathrm{a}}^2 S b\left(C_{l_0} + C_{l_\beta}\beta + C_{l_p}\frac{b}{2V_{\mathrm{a}}}\omega_{x_{\mathrm{b}}} + C_{l_r}\frac{b}{2V_{\mathrm{a}}}\omega_{z_{\mathrm{b}}} + C_{l\delta_{\mathrm{a}}}\delta_{\mathrm{a}} + C_{l\delta_{\mathrm{r}}}\delta_{\mathrm{r}}\right) \\ m_{\mathrm{az}} &\approx \frac{1}{2}\rho V_{\mathrm{a}}^2 S b\left(C_{n_0} + C_{n_\beta}\beta + C_{n_p}\frac{b}{2V_{\mathrm{a}}}\omega_{x_{\mathrm{b}}} + C_{n_r}\frac{b}{2V_{\mathrm{a}}}\omega_{z_{\mathrm{b}}} + C_{n\delta_{\mathrm{a}}}\delta_{\mathrm{a}} + C_{n\delta_{\mathrm{r}}}\delta_{\mathrm{r}}\right). \end{aligned} \tag{4.31}$$

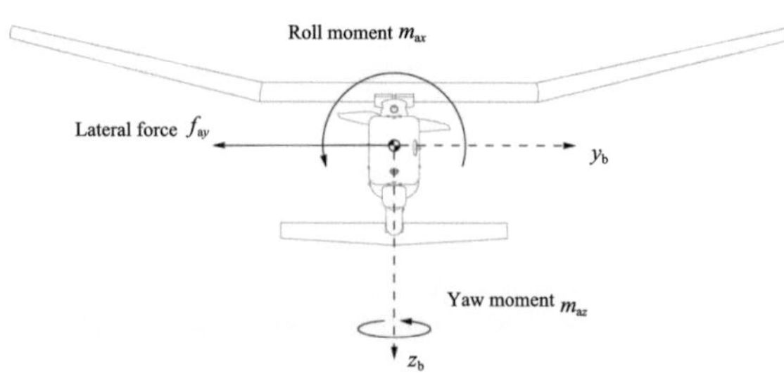

Fig. 4.7 Lateral aerodynamics

3. Thrust by propulsion system

In the process of modeling, it is assumed that the propulsion system of the fixed-wing UAV is installed along $o_b x_b$ axis of the body coordinate system. In the analysis of the longitudinal performance in Sect. 3.1.1, the thrust generated by the propeller propulsion system with zero airspeed is

$$T = C_T \rho \left(\frac{N}{60}\right)^2 D_p^4. \tag{4.32}$$

Therefore, when the airspeed is V_a, according to Eq. (3.7), the thrust by the propulsion system is represented in the body coordinate system as

$$^{b}\mathbf{t} = \begin{bmatrix} T - K\dfrac{V_a}{\sqrt{T}} \\ 0 \\ 0 \end{bmatrix}. \tag{4.33}$$

For Eq. (4.33), it is suitable for steady-state flight of fixed-wing UAVs. However, the thrust change needs to be considered in establishing a controlled model, including $T = 0$. To avoid division by zero, the thrust model is rewritten as follows:

$$T_{dyn} = T - K_0(K_1 + K_2 J + K_3 J^2)\frac{V_a}{\sqrt{T + \epsilon}} \tag{4.34}$$

where $K_0, K_1, K_2, K_3 \in \mathbb{R}$ are the decay coefficients, $\epsilon > 0$ is a very small value, introduced to prevent division by zero, and J represents the advance ratio. Taking $K_0 = 4.034$, $K_1 = 0.2641$, $K_2 = 0.4361$, $K_3 = 0.001263$, $\epsilon = 0.00001$ as an example, the thrust curve with respect to airspeed is shown in Fig. 4.8.

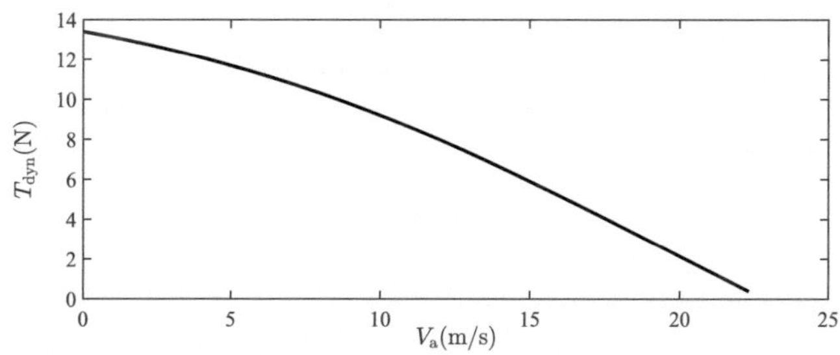

Fig. 4.8 Propeller thrust changing with airspeed

4.1.4 Trimming and Linearization

1. Linearization method based on small perturbation theory
Consider a general nonlinear dynamics

$$\dot{\mathbf{x}} = \mathbf{f}(\mathbf{x}, \mathbf{u}) \tag{4.35}$$

where $\mathbf{x} \in \mathbb{R}^n$ is the n-dimensional state vector, $\mathbf{u} \in \mathbb{R}^m$ is the m-dimensional control input vector, and $\mathbf{f}(\cdot)$ is a vector function composed of n nonlinear functions. Furthermore, Eq. (4.35) can be written in an implicit form as

$$\mathbf{f}(\dot{\mathbf{x}}, \mathbf{x}, \mathbf{u}) = 0. \tag{4.36}$$

The solution $(\mathbf{x}^*, \mathbf{u}^*)$ satisfying the condition $\dot{\mathbf{x}} \equiv 0$, $\mathbf{u} \equiv 0$ or $\mathbf{u}$ is constant and $\mathbf{f}(\dot{\mathbf{x}}, \mathbf{x}, \mathbf{u}) = 0$ is called the *equilibrium point*. The motion under the equilibrium condition is referred to as the *reference motion*. If the aircraft's deviation from the reference motion is small, the motion is called the *small perturbation motion*, with small perturbations denoted as $\Delta\mathbf{x}$.

Let $\Delta\mathbf{x} \triangleq \mathbf{x} - \mathbf{x}^*$, $\Delta\mathbf{u} \triangleq \mathbf{u} - \mathbf{u}^*$, one has

$$\begin{aligned} \Delta\dot{\mathbf{x}} &= \dot{\mathbf{x}} - \dot{\mathbf{x}}^* \\ &= \mathbf{f}(\mathbf{x}, \mathbf{u}) - \mathbf{f}(\mathbf{x}^*, \mathbf{u}^*) \\ &= \mathbf{f}(\mathbf{x} + \mathbf{x}^* - \mathbf{x}^*, \mathbf{u} + \mathbf{u}^* - \mathbf{u}^*) - \mathbf{f}(\mathbf{x}^*, \mathbf{u}^*) \\ &= \mathbf{f}(\mathbf{x}^* + \Delta\mathbf{x}, \mathbf{u}^* + \Delta\mathbf{u}) - \mathbf{f}(\mathbf{x}^*, \mathbf{u}^*). \end{aligned} \tag{4.37}$$

Using the first term of the Taylor series expansion as the equilibrium state, one further has

$$\begin{aligned} \Delta\dot{\mathbf{x}} &= \mathbf{f}(\mathbf{x}^*, \mathbf{u}^*) + \left.\frac{\partial \mathbf{f}(\mathbf{x}, \mathbf{u})}{\partial \mathbf{x}}\right|_{\mathbf{x}=\mathbf{x}^*, \mathbf{u}=\mathbf{u}^*} \Delta\mathbf{x} + \left.\frac{\partial \mathbf{f}(\mathbf{x}, \mathbf{u})}{\partial \mathbf{u}}\right|_{\mathbf{x}=\mathbf{x}^*, \mathbf{u}=\mathbf{u}^*} \Delta\mathbf{u} \quad + o(\Delta\mathbf{x}, \Delta\mathbf{u}) - \mathbf{f}(\mathbf{x}^*, \mathbf{u}^*) \\ &\approx \left.\frac{\partial \mathbf{f}(\mathbf{x}, \mathbf{u})}{\partial \mathbf{x}}\right|_{\mathbf{x}=\mathbf{x}^*, \mathbf{u}=\mathbf{u}^*} \Delta\mathbf{x} + \left.\frac{\partial \mathbf{f}(\mathbf{x}, \mathbf{u})}{\partial \mathbf{u}}\right|_{\mathbf{x}=\mathbf{x}^*, \mathbf{u}=\mathbf{u}^*} \Delta\mathbf{u}. \end{aligned} \tag{4.38}$$

Thus, under the equilibrium conditions, linearized dynamic equations are determined by calculating $\partial\mathbf{f}/\partial\mathbf{x}$ and $\partial\mathbf{f}/\partial\mathbf{u}$. Then, Eq. (4.36) is expanded into scalar form:

$$\begin{aligned} &\text{fun1}(\dot{\mathbf{x}}, \mathbf{x}, \mathbf{u}) = 0 \\ &\qquad\vdots \\ &\text{fun}n(\dot{\mathbf{x}}, \mathbf{x}, \mathbf{u}) = 0. \end{aligned} \tag{4.39}$$

Taking the first-order term of the Taylor series expansion at equilibrium conditions results in

$$\begin{gathered}(\nabla_{\dot{\mathbf{x}}}\text{fun}1)\Delta\dot{\mathbf{x}} + (\nabla_{\mathbf{x}}\text{fun}1)\Delta\mathbf{x} + (\nabla_{\mathbf{u}}\text{fun}1)\Delta\mathbf{u} = 0 \\ \vdots \\ (\nabla_{\dot{\mathbf{x}}}\text{fun}n)\Delta\dot{\mathbf{x}} + (\nabla_{\mathbf{x}}\text{fun}n)\Delta\mathbf{x} + (\nabla_{\mathbf{u}}\text{fun}n)\Delta\mathbf{u} = 0\end{gathered} \tag{4.40}$$

where ∇ is the row vector of the differential operator concerning the first order. For example

$$\nabla_{\mathbf{x}}\text{fun}1 = [\frac{\partial \text{fun}1}{\partial x_1} \quad \frac{\partial \text{fun}1}{\partial x_2} \quad \cdots \quad \frac{\partial \text{fun}1}{\partial x_n}]. \tag{4.41}$$

Equation (4.40) is rewritten in the form of linear state equations, namely

$$\mathbf{E}\dot{\mathbf{x}} = \mathbf{A}\mathbf{x} + \mathbf{B}\mathbf{u} \tag{4.42}$$

where

$$\mathbf{E} = \begin{bmatrix}\nabla_{\dot{\mathbf{x}}}\text{fun}1 \\ \vdots \\ \nabla_{\dot{\mathbf{x}}}\text{fun}n\end{bmatrix}_{\substack{\mathbf{x}=\mathbf{x}^*,\\ \mathbf{u}=\mathbf{u}^*}}, \mathbf{A} = -\begin{bmatrix}\nabla_{\mathbf{x}}\text{fun}1 \\ \vdots \\ \nabla_{\mathbf{x}}\text{fun}n\end{bmatrix}_{\substack{\mathbf{x}=\mathbf{x}^*,\\ \mathbf{u}=\mathbf{u}^*}}, \mathbf{B} = -\begin{bmatrix}\nabla_{\mathbf{u}}\text{fun}1 \\ \vdots \\ \nabla_{\mathbf{u}}\text{fun}n\end{bmatrix}_{\substack{\mathbf{x}=\mathbf{x}^*,\\ \mathbf{u}=\mathbf{u}^*}}.$$

2. Longitudinal channel trim

For a symmetric fixed-wing UAV, lateral balance is assumed. In the condition of level flight with no sideslip or roll, the roll angle $\phi = 0$, the roll rate $\omega_{x_b} = 0$, and the sideslip angle $\beta = 0$. Without a sideslip, the stability coordinate system coincides with the wind coordinate system. Also, under no-wind conditions, the airspeed equals the ground speed. The forces acting on the aircraft in the trimmed condition are shown in Fig. 4.9, wherein the wind coordinate system, the total forces along $o_w x_w$ and $o_w z_w$ axes are zero. The pitch moment is also zero, simultaneously satisfying:

$$\begin{cases} T\cos\alpha - f_{ax} + mg\sin\gamma = 0 \\ T\sin\alpha + f_{az} - mg\cos\gamma = 0 \\ m_{ay} = 0. \end{cases} \tag{4.43}$$

Substituting Eq. (4.29) into Eq. (4.43), the trim equations are obtained:

$$\begin{cases} T\cos\alpha - \frac{1}{2}\rho V_a^2 S(C_{D_0} + C_{D_\alpha}\alpha + C_{D_{\dot{\alpha}}}\frac{c_A}{2V_a}\dot{\alpha} + C_{D_q}\frac{c_A}{2V_a}\omega_{y_b} + C_{D_{\delta_e}}\delta_e) - mg\sin\gamma = 0 \\ T\sin\alpha + \frac{1}{2}\rho V_a^2 S(C_{L_0} + C_{L_\alpha}\alpha + C_{L_{\dot{\alpha}}}\frac{c_A}{2V_a}\dot{\alpha} + C_{L_q}\frac{c_A}{2V_a}\omega_{y_b} + C_{L_{\delta_e}}\delta_e) - mg\cos\gamma = 0 \\ \frac{1}{2}\rho V_a^2 S c_A(C_{m_0} + C_{m_\alpha}\alpha + C_{m_{\dot{\alpha}}}\frac{c_A}{2V_a}\dot{\alpha} + C_{m_q}\frac{c_A}{2V_a}\omega_{y_b} + C_{m_{\delta_e}}\delta_e) = 0. \end{cases} \tag{4.44}$$

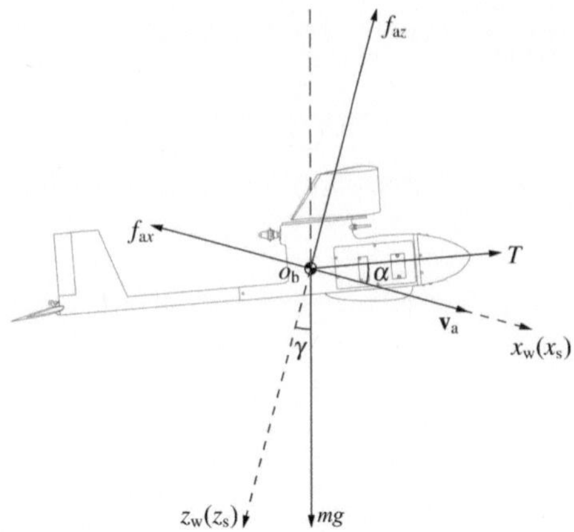

Fig. 4.9 Force analysis under trim conditions

By substituting the airspeed and altitude of the fixed-wing stable flight into the trim equation, the trimmed thrust T^*, elevator δ_e^*, and angle of attack α^* are obtained.

3. Model linearization

(1) Longitudinal attitude motion equation linearization

The motion of a fixed-wing UAV at the trim point is considered as trim motion, and small perturbation motion deviates only slightly from the trim motion. At this point, the motion of the UAV is regarded as the superposition of the trim motion $\mathbf{x}^*$ and the small perturbation $\Delta\mathbf{x}$.

Differentiating Eq. (4.14) yields:

$$\begin{cases} \dot{v}_{\mathrm{a}x_\mathrm{b}} = \dot{V}_\mathrm{a}\cos\alpha\cos\beta - \dot{\alpha}V_\mathrm{a}\sin\alpha\cos\beta - \dot{\beta}V_\mathrm{a}\cos\alpha\sin\beta \\ \dot{v}_{\mathrm{a}y_\mathrm{b}} = \dot{V}_\mathrm{a}\sin\beta + \dot{\beta}V_\mathrm{a}\cos\beta \\ \dot{v}_{\mathrm{a}z_\mathrm{b}} = \dot{V}_\mathrm{a}\sin\alpha\cos\beta + \dot{\alpha}V_\mathrm{a}\cos\alpha\cos\beta - \dot{\beta}V_\mathrm{a}\sin\alpha\sin\beta. \end{cases} \tag{4.45}$$

The forces acting on each axis in the body coordinate system are given by:

$$\begin{bmatrix} f_{x_\mathrm{b}} \\ f_{y_\mathrm{b}} \\ f_{z_\mathrm{b}} \end{bmatrix} = \begin{bmatrix} T - mg\sin\theta - f_{\mathrm{a}x}\cos\alpha\cos\beta + f_{\mathrm{a}z}\sin\alpha - f_{\mathrm{a}y}\cos\alpha\sin\beta \\ f_{\mathrm{a}y}\cos\beta + mg\sin\phi\cos\theta - f_{\mathrm{a}x}\sin\beta \\ mg\cos\phi\cos\theta - f_{\mathrm{a}x}\sin\alpha\cos\beta - f_{\mathrm{a}z}\cos\alpha - f_{\mathrm{a}y}\sin\alpha\sin\beta \end{bmatrix}. \tag{4.46}$$

Utilizing Eqs. (4.45) and (4.46), the first vector equation in Eq. (4.22) is rewritten as follows:

$$\begin{cases} m\dot{V}_\mathrm{a} = T\cos\alpha\cos\beta - f_{\mathrm{a}x} + G_{x\mathrm{w}} \\ mV_\mathrm{a}\dot{\beta} = -T\cos\alpha\sin\beta + f_{\mathrm{a}y} - mV_\mathrm{a}(-\omega_{x_\mathrm{b}}\sin\alpha + \omega_{z_\mathrm{b}}\cos\alpha) + G_{y\mathrm{w}} \\ mV_\mathrm{a}\cos\beta\dot{\alpha} = -T\sin\alpha - f_{\mathrm{a}z} + mV_\mathrm{a}(-\omega_{x_\mathrm{b}}\cos\alpha\sin\beta + \omega_{y_\mathrm{b}}\cos\beta - \omega_{z_\mathrm{b}}\sin\alpha\sin\beta) + G_{z\mathrm{w}} \end{cases} \tag{4.47}$$

where

$$\begin{bmatrix} G_{x_\mathrm{w}} \\ G_{y_\mathrm{w}} \\ G_{z_\mathrm{w}} \end{bmatrix} = \begin{bmatrix} mg(-\cos\alpha\cos\beta\sin\theta + \sin\beta\sin\phi\cos\theta + \sin\alpha\cos\beta\cos\theta) \\ mg(\cos\alpha\sin\beta\sin\theta + \cos\beta\sin\phi\cos\theta - \sin\alpha\sin\beta\cos\phi\cos\theta) \\ mg(\sin\alpha\sin\theta + \cos\alpha\cos\phi\cos\theta) \end{bmatrix}.$$

According to Eqs. (4.47), (4.17) and (4.23), the kinematics and dynamics equations for the longitudinal channel of a fixed-wing UAV are given by:

$$\begin{cases} m\dot{V}_\mathrm{a} = T\cos\alpha\cos\beta - f_{\mathrm{a}x} + mg(-\cos\alpha\cos\beta\sin\theta + \sin\beta\sin\phi\cos\theta + \sin\alpha\cos\beta\cos\theta) \\ mV_\mathrm{a}\cos\beta\dot{\alpha} = -T\sin\alpha - f_{\mathrm{a}z} + mV_\mathrm{a}(-\omega_{x_\mathrm{b}}\cos\alpha\sin\beta + \omega_{y_\mathrm{b}}\cos\beta - \omega_{z_\mathrm{b}}\sin\alpha\sin\beta) \\ \qquad +mg(\sin\alpha\sin\theta + \cos\alpha\cos\phi\cos\theta) \\ \dot{\theta} = \omega_{y_\mathrm{b}}\cos\phi - \omega_{z_\mathrm{b}}\sin\phi \\ \dot{\omega}_{y_\mathrm{b}} = \dfrac{1}{J_{yy}}\left(\omega_{x_\mathrm{b}}\omega_{z_\mathrm{b}}(J_{zz} - J_{xx}) + J_{xz}(\omega_{z_\mathrm{b}}^2 - \omega_{x_\mathrm{b}}^2) + m_{\mathrm{a}y}\right). \end{cases} \tag{4.48}$$

The state variables for the longitudinal channel are $\mathbf{x}_\mathrm{lon} = \left[\Delta V_\mathrm{a}\ \Delta\alpha\ \Delta\theta\ \Delta\omega_{y_\mathrm{b}}\right]^\mathrm{T}$, and $\mathbf{x}^*$ is the steady-state value of the reference motion, obtained through trimming, with input $\mathbf{u}_\mathrm{lon} = [\Delta T\ \Delta\delta_\mathrm{e}]^\mathrm{T}$. Equation (4.48) is linearized by the small perturbation linearization principle; the linearized expression for the longitudinal channel of a fixed-wing UAV is as follows:

$$\mathbf{E}_\mathrm{lon}\dot{\mathbf{x}}_\mathrm{lon} = \mathbf{A}_\mathrm{lon}\mathbf{x}_\mathrm{lon} + \mathbf{B}_\mathrm{lon}\mathbf{u}_\mathrm{lon} \tag{4.49}$$

where

$$\mathbf{E}_\mathrm{lon} = \begin{bmatrix} m & \left(\dfrac{\partial f_{\mathrm{a}x}}{\partial\dot{\alpha}}\right)^* & 0\ 0 \\ 0 & mV_\mathrm{a}^* + \left(\dfrac{\partial f_{\mathrm{a}z}}{\partial\dot{\alpha}}\right)^* & 0\ 0 \\ 0 & 0 & 1\ 0 \\ 0 & -\dfrac{1}{J_{yy}}\left(\dfrac{\partial m_{\mathrm{a}y}}{\partial\dot{\alpha}}\right)^* & 0\ 1 \end{bmatrix} = \begin{bmatrix} m & X_{\dot{\alpha}} & 0\ 0 \\ 0 & mV_\mathrm{a}^* + Z_{\dot{\alpha}} & 0\ 0 \\ 0 & 0 & 1\ 0 \\ 0 & -\dfrac{1}{J_{yy}}M_{y\dot{\alpha}} & 0\ 1 \end{bmatrix}$$

$$\mathbf{A}_\mathrm{lon} = \begin{bmatrix} \left(\frac{\partial T}{\partial V_\mathrm{a}}\right)^*\cos\alpha^* - \left(\frac{\partial f_{\mathrm{a}x}}{\partial V_\mathrm{a}}\right)^* & -T^*\sin\alpha^* - \left(\frac{\partial f_{\mathrm{a}x}}{\partial\alpha}\right)^* + mg\cos\gamma^* & -mg\cos\gamma^* & \left(-\frac{\partial f_{\mathrm{a}x}}{\partial\omega_{y_\mathrm{b}}}\right)^* \\ -\left(\frac{\partial T}{\partial V_\mathrm{a}}\right)^*\sin\alpha^* - \left(\frac{\partial f_{\mathrm{a}z}}{\partial V_\mathrm{a}}\right)^* & -T^*\cos\alpha^* - \left(\frac{\partial f_{\mathrm{a}z}}{\partial\alpha}\right)^* + mg\sin\gamma^* & -mg\sin\gamma^* & -\left(\frac{\partial f_{\mathrm{a}z}}{\partial\omega_{y_\mathrm{b}}}\right)^* + mV_\mathrm{a}^* \\ 0 & 0 & 0 & 1 \\ \frac{1}{J_{yy}}\left(\frac{\partial m_{\mathrm{a}y}}{\partial V_\mathrm{a}}\right)^* & \frac{1}{J_{yy}}\left(\frac{\partial m_{\mathrm{a}y}}{\partial\alpha}\right)^* & 0 & \frac{1}{J_{yy}}\left(\frac{\partial m_{\mathrm{a}y}}{\partial\omega_{y_\mathrm{b}}}\right)^* \end{bmatrix}$$

$$= \begin{bmatrix} T_V\cos\alpha^* - X_V & -T^*\sin\alpha^* - X_\alpha + mg\cos\gamma^* & -mg\cos\gamma^* & -X_{\omega y} \\ -T_V\sin\alpha^* - Z_V & -T^*\cos\alpha^* - Z_\alpha + mg\sin\gamma^* & -mg\sin\gamma^* & -Z_{\omega_y} + mV_\mathrm{a}^* \\ 0 & 0 & 0 & 1 \\ \frac{1}{J_{yy}}M_{yV} & \frac{1}{J_{yy}}M_{y\alpha} & 0 & \frac{1}{J_{yy}}M_{y\omega_y} \end{bmatrix}$$

$$\mathbf{B}_{\mathrm{lon}}=\begin{bmatrix}\cos\alpha^* & -\left(\dfrac{\partial f_{\mathrm{ax}}}{\partial\delta_{\mathrm{e}}}\right)^* \\ -\sin\alpha^* & -\left(\dfrac{\partial f_{\mathrm{az}}}{\partial\delta_{\mathrm{e}}}\right)^* \\ 0 & 0 \\ 0 & \dfrac{1}{J_{yy}}\left(\dfrac{\partial m_{\mathrm{ay}}}{\partial\delta_{\mathrm{e}}}\right)^*\end{bmatrix}=\begin{bmatrix}\cos\alpha^* & -X_{\delta_{\mathrm{e}}} \\ -\sin\alpha^* & -Z_{\delta_{\mathrm{e}}} \\ 0 & 0 \\ 0 & \dfrac{1}{J_{yy}}M_{y\delta_{\mathrm{e}}}\end{bmatrix}.$$

Note: $\gamma=\theta-\alpha$ only exists in the state of equilibrium.

(2) Lateral attitude motion equation linearization
According to Eqs. (4.47), (4.17), and (4.23), the kinematic and dynamic equations of the lateral channel of a fixed-wing UAV are obtained as follows:

$$\begin{cases} mV_{\mathrm{a}}\dot{\beta}=-T\cos\alpha\sin\beta+f_{\mathrm{ay}}-mV_{\mathrm{a}}(-\omega_{x_{\mathrm{b}}}\sin\alpha+\omega_{z_{\mathrm{b}}}\cos\alpha) \\ \qquad +mg(\cos\alpha\sin\beta\sin\theta+\cos\beta\sin\phi\cos\theta-\sin\alpha\sin\beta\cos\phi\cos\theta) \\ \dot{\phi}=\omega_{x_{\mathrm{b}}}+\tan\theta(\omega_{y_{\mathrm{b}}}\sin\phi+\omega_{z_{\mathrm{b}}}\cos\phi) \\ \dot{\omega}_{x_{\mathrm{b}}}=\dfrac{1}{J_{xx}J_{zz}-J_{xz}^2}(J_{zz}m_{\mathrm{ax}}+J_{xz}m_{\mathrm{az}}+\omega_{y_{\mathrm{b}}}\omega_{z_{\mathrm{b}}}(J_{zz}J_{yy}-J_{zz}^2-J_{xz}^2)-\omega_{x_{\mathrm{b}}}\omega_{y_{\mathrm{b}}}(J_{yy}J_{xz}-J_{zz}J_{xz}-J_{xx}J_{xz})) \\ \dot{\omega}_{z_{\mathrm{b}}}=\dfrac{1}{J_{xx}J_{zz}-J_{xz}^2}(J_{xx}m_{\mathrm{az}}+J_{xz}m_{\mathrm{ax}}+\omega_{x_{\mathrm{b}}}\omega_{y_{\mathrm{b}}}(J_{xx}^2+J_{xz}^2-J_{xx}J_{yy})+\omega_{y_{\mathrm{b}}}\omega_{z_{\mathrm{b}}}(J_{yy}J_{xz}-J_{zz}J_{xz}-J_{xx}J_{xz})). \end{cases} \tag{4.50}$$

The state variables of the lateral channel are $\mathbf{x}_{\mathrm{lat}}=\left[\Delta\beta\ \Delta\phi\ \Delta\omega_{x_{\mathrm{b}}}\ \Delta\omega_{z_{\mathrm{b}}}\right]^{\mathrm{T}}$, and the inputs are $\mathbf{u}_{\mathrm{lat}}=[\Delta\delta_{\mathrm{a}}\ \Delta\delta_{\mathrm{r}}]^{\mathrm{T}}$. Linearizing Eq. (4.50) yields

$$\mathbf{E}_{\mathrm{lat}}\dot{\mathbf{x}}_{\mathrm{lat}}=\mathbf{A}_{\mathrm{lat}}\mathbf{x}_{\mathrm{lat}}+\mathbf{B}_{\mathrm{lat}}\mathbf{u}_{\mathrm{lat}} \tag{4.51}$$

where

$$\mathbf{E}_{\mathrm{lat}}=\begin{bmatrix} mV_{\mathrm{a}}^* & 0 & 0 & 0 \\ 0 & 1 & 0 & 0 \\ 0 & 0 & 1 & 0 \\ 0 & 0 & 0 & 1\end{bmatrix}$$

$$\mathbf{A}_{\mathrm{lat}}=\begin{bmatrix} \left(\dfrac{\partial f_{\mathrm{ay}}}{\partial\beta}\right)^*-T^*\cos\alpha^*+mg\sin\gamma^* & mg\cos\gamma^* & \left(\dfrac{\partial f_{\mathrm{ay}}}{\partial\omega_{x_{\mathrm{b}}}}\right)^*+mV_{\mathrm{a}}^*\sin\alpha^* & \left(\dfrac{\partial f_{\mathrm{ay}}}{\partial\omega_{z_{\mathrm{b}}}}\right)^*-mV_{\mathrm{a}}^*\cos\alpha^* \\ 0 & 0 & \dfrac{\cos\gamma^*}{\cos\theta^*} & \dfrac{\sin\gamma^*}{\cos\theta^*} \\ \Gamma_1\left(\dfrac{\partial m_{\mathrm{ax}}}{\partial\beta}\right)^*+\Gamma_2\left(\dfrac{\partial m_{\mathrm{az}}}{\partial\beta}\right)^* & 0 & \Gamma_1\left(\dfrac{\partial m_{\mathrm{ax}}}{\partial\omega_{x_{\mathrm{b}}}}\right)^*+\Gamma_2\left(\dfrac{\partial m_{\mathrm{az}}}{\partial\omega_{x_{\mathrm{b}}}}\right)^* & \Gamma_1\left(\dfrac{\partial m_{\mathrm{ax}}}{\partial\omega_{z_{\mathrm{b}}}}\right)^*+\Gamma_2\left(\dfrac{\partial m_{\mathrm{az}}}{\partial\omega_{z_{\mathrm{b}}}}\right)^* \\ \Gamma_3\left(\dfrac{\partial m_{\mathrm{az}}}{\partial\beta}\right)^*+\Gamma_2\left(\dfrac{\partial m_{\mathrm{ax}}}{\partial\beta}\right)^* & 0 & \Gamma_3\left(\dfrac{\partial m_{\mathrm{az}}}{\partial\omega_{x_{\mathrm{b}}}}\right)^*+\Gamma_2\left(\dfrac{\partial m_{\mathrm{ax}}}{\partial\omega_{x_{\mathrm{b}}}}\right)^* & \Gamma_3\left(\dfrac{\partial m_{\mathrm{az}}}{\partial\omega_{z_{\mathrm{b}}}}\right)^*+\Gamma_2\left(\dfrac{\partial m_{\mathrm{ax}}}{\partial\omega_{z_{\mathrm{b}}}}\right)^* \end{bmatrix}$$

$$=\begin{bmatrix} Y_\beta-T^*\cos\alpha^*+mg\sin\gamma^* & mg\cos\gamma^* & Y_{\omega_x}+mV_{\mathrm{a}}^*\sin\alpha^* & Y_{\omega_z}-mV_{\mathrm{a}}^*\cos\alpha^* \\ 0 & 0 & \dfrac{\cos\gamma^*}{\cos\theta^*} & \dfrac{\sin\gamma^*}{\cos\theta^*} \\ \Gamma_1 M_{x\beta}+\Gamma_2 M_{z\beta} & 0 & \Gamma_1 M_{x\omega_x}+\Gamma_2 M_{z\omega_x} & \Gamma_1 M_{x\omega_z}+\Gamma_2 M_{z\omega_z} \\ \Gamma_3 M_{z\beta}+\Gamma_2 M_{x\beta} & 0 & \Gamma_3 M_{z\omega_x}+\Gamma_2 M_{x\omega_x} & \Gamma_3 M_{z\omega_z}+\Gamma_2 M_{x\omega_z} \end{bmatrix}$$

$$\mathbf{B}_{\text{lat}} = \begin{bmatrix} \left(\frac{\partial f_{\text{ay}}}{\partial \delta_{\text{a}}}\right)^* & \left(\frac{\partial f_{\text{ay}}}{\partial \delta_{\text{r}}}\right)^* \\ 0 & 0 \\ \Gamma_1\left(\frac{\partial m_{\text{ax}}}{\partial \delta_{\text{a}}}\right)^* + \Gamma_2\left(\frac{\partial m_{\text{az}}}{\partial \delta_{\text{a}}}\right)^* & \Gamma_1\left(\frac{\partial m_{\text{ax}}}{\partial \delta_{\text{r}}}\right)^* + \Gamma_2\left(\frac{\partial m_{\text{az}}}{\partial \delta_{\text{r}}}\right)^* \\ \Gamma_3\left(\frac{\partial m_{\text{az}}}{\partial \delta_{\text{a}}}\right)^* + \Gamma_2\left(\frac{\partial m_{\text{ax}}}{\partial \delta_{\text{a}}}\right)^* & \Gamma_3\left(\frac{\partial m_{\text{az}}}{\partial \delta_{\text{r}}}\right)^* + \Gamma_2\left(\frac{\partial m_{\text{az}}}{\partial \delta_{\text{r}}}\right)^* \end{bmatrix} = \begin{bmatrix} Y_{\delta_{\text{a}}} & Y_{\delta_{\text{r}}} \\ 0 & 0 \\ \Gamma_1 M_{x\delta_{\text{a}}} + \Gamma_2 M_{z\delta_{\text{a}}} & \Gamma_1 M_{x\delta_{\text{r}}} + \Gamma_2 M_{z\delta_{\text{r}}} \\ \Gamma_3 M_{z\delta_{\text{a}}} + \Gamma_2 M_{x\delta_{\text{a}}} & \Gamma_3 M_{z\delta_{\text{r}}} + \Gamma_2 M_{x\delta_{\text{r}}} \end{bmatrix}$$

$$\Gamma_1 = \frac{J_{zz}}{J_{xx}J_{zz} - J_{xz}^2}, \quad \Gamma_2 = \frac{J_{xz}}{J_{xx}J_{zz} - J_{xz}^2}, \quad \Gamma_3 = \frac{J_{xx}}{J_{xx}J_{zz} - J_{xz}^2}.$$

4. Transfer function model

(1) Transfer function of the longitudinal channel

The variables related to the longitudinal channel are pitch angle θ, pitch rate $\omega_{y_{\text{b}}}$, altitude $H(p_{z_{\text{e}}})$, and airspeed V_{a}. The control commands affecting longitudinal dynamics are elevator deflection δ_{e} and thrust provided by throttle T. For convenience, the transfer function from the elevator deflection to the pitch angle and altitude is analyzed in a constant-speed state.

(a) Pitch angle

The transfer function from elevator deflection to pitch angle is derived as follows, assuming that the pitch angle is primarily influenced by elevator deflection, with other variables having minimal impact. Extracting the state variables and input related to the pitch angle from Eq. (4.49) yields

$$\begin{bmatrix} \Delta\dot{\theta} \\ \Delta\dot{\omega}_{y_{\text{b}}} \end{bmatrix} = \mathbf{A}'\begin{bmatrix} \Delta\theta \\ \Delta\omega_{y_{\text{b}}} \end{bmatrix} + \mathbf{B}'\Delta\delta_{\text{e}} \tag{4.52}$$

where

$$\mathbf{A}' = \begin{bmatrix} 0 & 1 \\ 0 & \frac{1}{J_{yy}}M_{y\omega_y} \end{bmatrix}, \mathbf{B}' = \begin{bmatrix} 0 \\ \frac{1}{J_{yy}}M_{y\delta_{\text{e}}} \end{bmatrix}.$$

Regarding elevator deflection $\Delta\delta_{\text{e}}$ as the control input and pitch angle $\Delta\theta$ as the output, the transfer function is expressed as the ratio of two determinants. The denominator is the determinant of the matrix $s\mathbf{I}_2 - \mathbf{A}'$. The numerator is the determinant obtained by replacing the column corresponding to $\Delta\theta$ in matrix $\mathbf{B}'$ with the input column of the matrix $s\mathbf{I}_2 - \mathbf{A}'$. Thus, the transfer function is given by:

$$\begin{aligned} \frac{\Delta\theta(s)}{\Delta\delta_{\text{e}}(s)} &= \frac{1}{|s\mathbf{I}_2 - \mathbf{A}'|}\begin{vmatrix} 0 & -1 \\ \frac{1}{J_{yy}}M_{y\delta_{\text{e}}} & s - \frac{1}{J_{yy}}M_{y\omega_y} \end{vmatrix} \\ &= \frac{M_{y\delta_{\text{e}}}}{J_{yy}s(s - M_{y\omega_y})}. \end{aligned} \tag{4.53}$$

(b) Altitude

For a fixed-wing UAV flying at a constant airspeed, the pitch angle directly affects the aircraft's altitude. Therefore, the transfer function is directly derived from the pitch angle to the altitude. Based on the translational motion Eq. (4.16) and the rotation matrix $\mathbf{R}_{\mathrm{e}}^{\mathrm{b}}$ from Eq. (4.4), the motion equation for the altitude in the ground coordinate system is obtained as:

$$\dot{H} = v_{\mathrm{g}x_{\mathrm{b}}} \sin\theta - v_{\mathrm{g}y_{\mathrm{b}}} \sin\phi\cos\theta - v_{\mathrm{g}z_{\mathrm{b}}} \cos\phi\cos\theta. \tag{4.54}$$

During a straight and level flight, $v_{\mathrm{g}y_{\mathrm{b}}} \approx 0$, $v_{\mathrm{g}z_{\mathrm{b}}} \approx 0$, $v_{\mathrm{g}x_{\mathrm{b}}} \approx V_{\mathrm{a}}$ hold, and $\phi \approx 0$. If θ is small, then $\sin\theta \approx \theta$. Equation (4.54) is simplified to:

$$\dot{H} = V_{\mathrm{a}}\theta. \tag{4.55}$$

Assuming $V_{\mathrm{a}} = V_{\mathrm{a}}^{*}$, taking the Laplace transform of Eq. (4.55) yields

$$H(s) = \frac{V_{\mathrm{a}}^{*}}{s}\theta(s). \tag{4.56}$$

Then, according to the small perturbation theory, one has

$$\Delta H(s) = \frac{V_{\mathrm{a}}^{*}}{s}\Delta\theta(s). \tag{4.57}$$

So far, the transfer function from the altitude to the pitch angle has been obtained. Combined with the transfer function from the pitch angle to the elevator deflection, the longitudinal transfer function is obtained from the elevator deflection to the altitude.

(2) Transfer function of lateral channel

The variables related to lateral dynamics are roll angle ϕ, roll rate $\omega_{x_{\mathrm{b}}}$, yaw angle ψ, and yaw rate $\omega_{z_{\mathrm{b}}}$. The aileron δ_{a} directly controls the roll angle, while the rudder deflection δ_{r} directly controls the yaw angle.

(a) Roll angle

According to Cramer's rule [3], the transfer function is expressed as the ratio of two determinants. The denominator is the determinant of the matrix $s\mathbf{E}_{\mathrm{lat}} - \mathbf{A}_{\mathrm{lat}}$, and the numerator is obtained by replacing the column corresponding to the output variable in matrix $\mathbf{B}_{\mathrm{lat}}$ with the input column of the matrix $s\mathbf{E}_{\mathrm{lat}} - \mathbf{A}_{\mathrm{lat}}$. Thus, the transfer function is given by:

$$s\mathbf{E}_{\mathrm{lat}} - \mathbf{A}_{\mathrm{lat}} = \begin{bmatrix} mV_{\mathrm{a}}^{*}s - Y_{\beta}^{*} & -mg\cos\gamma^{*} & -Y_{\omega_x}^{*} & -Y_{\omega_z}^{*} \\ 0 & s & -\dfrac{\cos\gamma^{*}}{\cos\theta^{*}} & -\dfrac{\sin\gamma^{*}}{\cos\theta^{*}} \\ -(\Gamma_1 M_{x\beta} + \Gamma_2 M_{z\beta}) & 0 & s - (\Gamma_1 M_{x\omega_x} + \Gamma_2 M_{z\omega_x}) & -(\Gamma_1 M_{x\omega_z} + \Gamma_2 M_{z\omega_z}) \\ -(\Gamma_3 M_{z\beta} + \Gamma_2 M_{x\beta}) & 0 & -(\Gamma_3 M_{z\omega_x} + \Gamma_2 M_{x\omega_x}) & s - (\Gamma_3 M_{z\omega_z} + \Gamma_2 M_{x\omega_z}) \end{bmatrix} \tag{4.58}$$

where $Y_\beta^* = Y_\beta - T^* \cos\alpha^* + mg \sin\gamma^*$, $Y_{\omega_x}^* = Y_{\omega_x} + mV_\mathrm{a}^* \sin\alpha^*$, $Y_{\omega_z}^* = Y_{\omega_z} - mV_\mathrm{a}^* \cos\alpha^*$. First, the transfer function with aileron deflection $\Delta\delta_\mathrm{a}$ as control input and roll perturbation $\Delta\phi$ as state variable is studied. The transfer function can be written as a ratio of two determinants, where the denominator is the determinant of the matrix $s\mathbf{E}_\mathrm{lat} - \mathbf{A}_\mathrm{lat}$ denoted by $|s\mathbf{E}_\mathrm{lat} - \mathbf{A}_\mathrm{lat}|$. The numerator is a determinant formed by replacing the column corresponding to the aileron deflection perturbation $\Delta\delta_\mathrm{a}$ (the first column) in matrix $\mathbf{B}_\mathrm{lat}$ with the column corresponding to the roll perturbation $\Delta\phi$ (the second column) in determinant $|s\mathbf{E}_\mathrm{lat} - \mathbf{A}_\mathrm{lat}|$, i.e.,

$$\frac{\Delta\phi(s)}{\Delta\delta_\mathrm{a}(s)} = \frac{1}{|s\mathbf{E}_\mathrm{lat} - \mathbf{A}_\mathrm{lat}|}\begin{vmatrix} mV_\mathrm{a}^* s - Y_\beta^* & Y_{\delta_\mathrm{a}} & -Y_{\omega_x}^* & -Y_{\omega_z}^* \\ 0 & 0 & -\dfrac{\cos\gamma^*}{\cos\theta^*} & -\dfrac{\sin\gamma^*}{\cos\theta^*} \\ -(\Gamma_1 M_{x\beta} + \Gamma_2 M_{z\beta}) & \Gamma_1 M_{x\delta_\mathrm{a}} + \Gamma_2 M_{z\delta_\mathrm{a}} & s - (\Gamma_1 M_{x\omega_x} + \Gamma_2 M_{z\omega_x}) & -(\Gamma_1 M_{x\omega_z} + \Gamma_2 M_{z\omega_z}) \\ -(\Gamma_3 M_{z\beta} + \Gamma_2 M_{x\beta}) & \Gamma_3 M_{z\delta_\mathrm{a}} + \Gamma_2 M_{x\delta_\mathrm{a}} & -(\Gamma_3 M_{z\omega_x} + \Gamma_2 M_{x\omega_x}) & s - (\Gamma_3 M_{z\omega_z} + \Gamma_2 M_{x\omega_z}) \end{vmatrix}. \tag{4.59}$$

Similarly, the transfer function from the rudder deflection perturbation to the roll perturbation is obtained as follows:

$$\frac{\Delta\phi(s)}{\Delta\delta_\mathrm{r}(s)} = \frac{1}{|s\mathbf{E}_\mathrm{lat} - \mathbf{A}_\mathrm{lat}|}\begin{vmatrix} mV_\mathrm{a}^* s - Y_\beta^* & Y_{\delta_\mathrm{r}} & -Y_{\omega_x}^* & -Y_{\omega_z}^* \\ 0 & 0 & -\dfrac{\cos\gamma^*}{\cos\theta^*} & -\dfrac{\sin\gamma^*}{\cos\theta^*} \\ -(\Gamma_1 M_{x\beta} + \Gamma_2 M_{z\beta}) & \Gamma_1 M_{x\delta_\mathrm{r}} + \Gamma_2 M_{z\delta_\mathrm{r}} & s - (\Gamma_1 M_{x\omega_x} + \Gamma_2 M_{z\omega_x}) & -(\Gamma_1 M_{x\omega_z} + \Gamma_2 M_{z\omega_z}) \\ -(\Gamma_3 M_{z\beta} + \Gamma_2 M_{x\beta}) & \Gamma_3 M_{z\delta_\mathrm{r}} + \Gamma_2 M_{x\delta_\mathrm{r}} & -(\Gamma_3 M_{z\omega_x} + \Gamma_2 M_{x\omega_x}) & s - (\Gamma_3 M_{z\omega_z} + \Gamma_2 M_{x\omega_z}) \end{vmatrix}. \tag{4.60}$$

(b) Coordinated turn

The conditions for a coordinated turn are as follows: the roll angle ϕ is constant; the yaw rate ω_{z_b} is constant; the vertical speed is zero (horizontal turn); and the sideslip angle is zero ($\beta = 0$). In a coordinated turn, the forces acting on the UAV are balanced: the horizontal component of the lift equals the centrifugal force, and the vertical component of the lift equals the gravity, as shown in Fig. 4.10. Therefore, the force relationship is given by:

$$\begin{cases} mg = f_\mathrm{a}z\cos\phi \\ f_\mathrm{c} = f_{\mathrm{a}z}\sin\phi. \end{cases} \tag{4.61}$$

In the absence of wind, $V_\mathrm{g} = V_\mathrm{a}$, and the centrifugal force is given by

$$\begin{aligned} f_\mathrm{c} &= m\frac{{V_\mathrm{a}}^2}{R} \\ &= mV_\mathrm{a}\omega \\ &= mV_\mathrm{a}\dot{\psi}. \end{aligned} \tag{4.62}$$

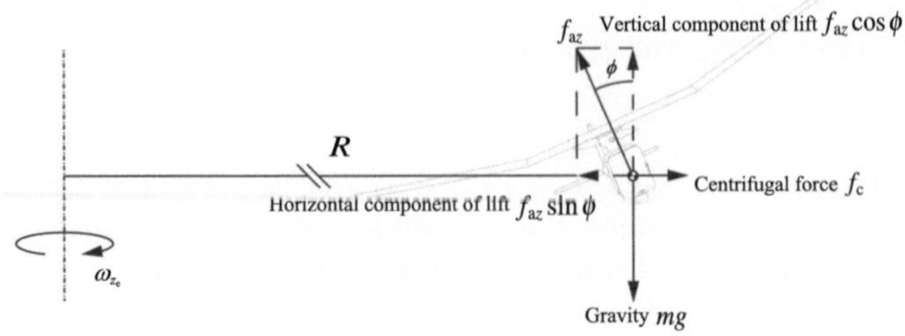

Fig. 4.10 Force analysis under coordinated turn

By rearranging Eqs. (4.61) and (4.62), the relationship between the yaw rate $\dot{\psi}$ and the roll angle ϕ is obtained as

$$\dot{\psi} = \frac{g}{V_a}\tan\phi. \tag{4.63}$$

If the vertical speed of the fixed-wing UAV is not zero (climbing in a spiral), there will be a pitch angle. In this case, Eq. (4.61) becomes:

$$\begin{cases} mg = f_{az}\cos(\theta - \alpha)\cos\phi \\ f_c = f_{az}\sin\phi. \end{cases} \tag{4.64}$$

From this, the relationship between the yaw rate $\dot{\psi}$ and the roll angle ϕ is derived as

$$\dot{\psi} = \frac{g}{V_a}\tan\phi\cos(\theta - \alpha). \tag{4.65}$$

(c) Yaw angle
For a coordinated turn without wind, the roll angle directly influences the yaw angle. Therefore, the transfer function is directly derived from the roll angle to the yaw angle. If ϕ is small, then $\tan\phi \approx \phi$. According to the coordinated turn relationship (4.63), one has

$$\dot{\psi} = \frac{g}{V_a}\phi. \tag{4.66}$$

Assuming $V_a = V_a^*$, taking the Laplace transform of Eq. (4.66) yields

$$\psi(s) = \frac{g}{V_a^* s}\phi(s). \tag{4.67}$$

Then, based on the small perturbation theory, it is obtained

$$\Delta\psi(s) = \frac{g}{V_a^* s}\Delta\phi(s). \tag{4.68}$$

This yields the transfer function between the yaw angle and roll angle, which is used to derive the transfer functions between the yaw angle and aileron, as well as the yaw angle and rudder deflection.

4.1.5 Analysis of Fixed-Wing UAV Lateral Motion

The coupling between yaw and roll motions in fixed-wing UAVs is significant. When the UAV rolls, the yaw angle changes synchronously due to aerodynamic interactions, primarily involving the wing aerodynamics, vertical tail, and changes in sideslip angle. The kinematic equation for the yaw rate is obtained from the attitude kinematic Eq. (4.17) as follows:

$$\dot{\psi} = \frac{1}{\cos\theta}\left(\omega_{z_{\mathrm{b}}}\cos\phi + \omega_{y_{\mathrm{b}}}\sin\phi\right). \tag{4.69}$$

In horizontal turns without wind, the body pitch angle of the UAV is small, and the pitch rate is zero. From Eq. (4.69), it is inferred that the body yaw rate $\dot{\psi}$ is mainly influenced by the rotation rate $\omega_{z_{\mathrm{b}}}$ about the body axis z_{b} and the roll angle ϕ. From the kinematic and dynamic equations of the lateral channel of fixed-wing UAVs (4.50), the dynamic equation for the angular rate about the body axis z_{b} is:

$$\begin{aligned}\dot{\omega}_{z_{\mathrm{b}}} = &\frac{1}{J_{xx}J_{zz} - J_{xz}^2}\big(J_{xx}m_{\mathrm{az}} + J_{xz}m_{\mathrm{ax}} + \omega_{x_{\mathrm{b}}}\omega_{y_{\mathrm{b}}}(J_{xx}^2 + J_{xz}^2 - J_{xx}J_{yy})\\ &+ \omega_{y_{\mathrm{b}}}\omega_{z_{\mathrm{b}}}(J_{yy}J_{xz} - J_{zz}J_{xz} - J_{xx}J_{xz})\big).\end{aligned} \tag{4.70}$$

The change in yaw angle is mainly influenced by the yaw moment m_{az}. Therefore, the influence of roll motion on the yaw moment is analyzed. From Eq. (4.31), one has

$$m_{\mathrm{az}} = \frac{1}{2}\rho V_{\mathrm{a}}^2 Sb(C_{n_0} + C_{n_\beta}\beta + C_{n_p}\frac{b}{2V_{\mathrm{a}}}\omega_{x_{\mathrm{b}}} + C_{n_r}\frac{b}{2V_{\mathrm{a}}}\omega_{z_{\mathrm{b}}} + C_{n\delta_{\mathrm{a}}}\delta_{\mathrm{a}} + C_{n\delta_{\mathrm{r}}}\delta_{\mathrm{r}}). \tag{4.71}$$

Here, the yaw moment caused by the rolling motion of the fixed-wing UAV itself is considered without considering the variation of the control signal. It is observed that both sideslip angle β and roll rate $\omega_{x_{\mathrm{b}}}$ affect the yaw moment, with sideslip angle being the primary factor.

1. Influence of sideslip angle

Considering a fixed-wing UAV with a roll angle ϕ, the force analysis is shown in Fig. 4.11. At this time, the horizontal component of wing lift $f_{az}\sin\phi$ will cause the UAV to generate a rightward velocity component in the body axis y ($o_{\mathrm{b}}y_{\mathrm{b}}$) direction, resulting in a right sideslip, $\beta > 0$. The fuselage and vertical tail mainly generate the aerodynamic yaw moment caused by the sideslip angle. Generally, the aerodynamic yaw moment generated by the fuselage is smaller than that of the vertical

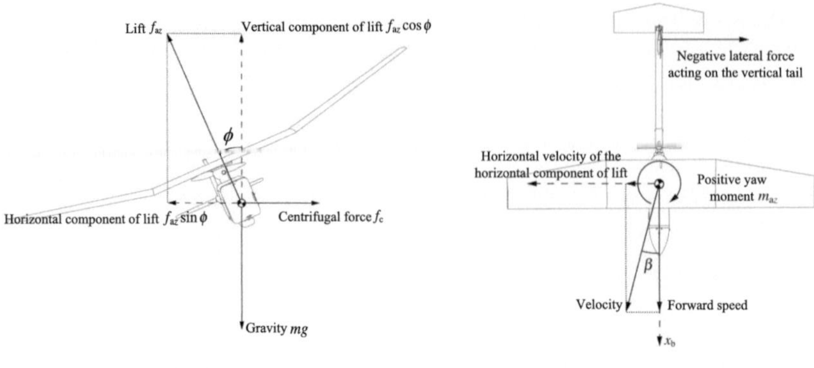

(a) Front view (b) Top view

Fig. 4.11 Influence of sideslip angle

tail. Therefore, the following analysis uses the vertical tail as an example to examine the mechanism of the aerodynamic yaw moment induced by the sideslip angle β. When the fixed-wing UAV slides to the right, it generates a negative lateral force (in the body axis system) on the vertical tail. Since the vertical tail is located aft of the UAV's center of gravity, the lateral force generates a positive yaw moment, causing the nose to turn and reducing the sideslip angle, thereby providing yaw stability to the UAV.

2. Influence of roll rate

Considering that when a fixed-wing UAV rolls to the right while flying forward horizontally at an airspeed Va, the roll rate is greater than zero. At this point, the yaw moment caused by this angular rate is mainly composed of two parts: the wing and the vertical tail, and the wing plays a major role.

(1) For the vertical tail, the UAV rolls to the right, the vertical tail also has a rightward velocity component, equivalent to generating a local sideslip angle $\beta > 0$ at the vertical tail, resulting in a negative side force and a positive yaw moment.

(2) The effect of the wings is more complex, and the following discussion is about the principle behind it. The UAV has an angle of attack α, and when it rolls to the right, the roll rate $\omega_{x_b} > 0$. The right-wing of the UAV moves downward, with a downward velocity increment ΔV, and the resultant velocity is V'_a. Thus, there is a relative angle of attack increment $\Delta\alpha$, with $\Delta\alpha \ll \alpha$. Therefore, the lift $f_{az,\text{right}}$ and drag $f_{ax,\text{right}}$ of the right-wing both increase and similarly, the lift $f_{az,\text{left}}$ and drag $f_{ax,\text{left}}$ of the left-wing both decrease. The lift and drag are respectively parallel and perpendicular to the new airspeed direction, so the horizontal components of the lift and drag for the right and left wings are $f_{ax,\text{right}}\cos(\Delta\alpha) - f_{az,\text{right}}\sin(\Delta\alpha)$ and $f_{ax,\text{left}}\cos(\Delta\alpha) + f_{az,\text{left}}\sin(\Delta\alpha)$, respectively. Using a small angle approximation, they are $f_{ax,\text{right}} - f_{az,\text{right}}\Delta\alpha$ and $f_{ax,\text{left}} + f_{az,\text{left}}\Delta\alpha$. Their magnitudes depend on the angle of attack of the UAV, which is difficult to determine directly [1]. In certain situations, the horizontal resultant force generated by the left wing is greater than

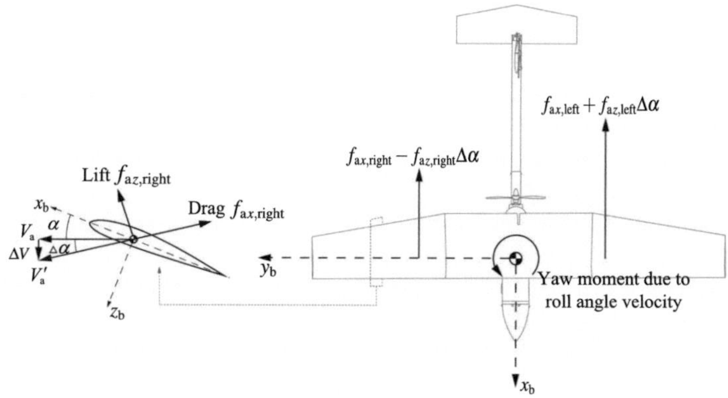

Fig. 4.12 Influence of roll rate

that generated by the right wing, as shown in Fig. 4.12. Therefore, it is easy to deduce that it generates a negative yaw moment.

In conclusion, fixed-wing UAVs' roll and yaw motions are closely related. The impact of the roll motion on the yaw angle, caused by the configuration of the UAV's vertical tail, wings, fuselage, and other parts, is briefly covered in the study above. Based on the analysis, it is clear that the fuselage's structural design plays a significant role in the UAV's yaw stability and should be considered throughout the design phase.

4.1.6 Fixed-Wing UAV Modal Analysis

The linearized expressions for the longitudinal and lateral motions of the fixed-wing UAV are obtained from Eqs. (4.49) and (4.51) as follows:

$$\begin{aligned}\mathbf{E}_{\mathrm{lon}}\dot{\mathbf{x}}_{\mathrm{lon}} &= \mathbf{A}_{\mathrm{lon}}\mathbf{x}_{\mathrm{lon}} + \mathbf{B}_{\mathrm{lon}}\mathbf{u}_{\mathrm{lon}} \\ \mathbf{E}_{\mathrm{lat}}\dot{\mathbf{x}}_{\mathrm{lat}} &= \mathbf{A}_{\mathrm{lat}}\mathbf{x}_{\mathrm{lat}} + \mathbf{B}_{\mathrm{lat}}\mathbf{u}_{\mathrm{lat}}.\end{aligned} \tag{4.72}$$

In Sect. 4.1.4, Cramer's rule is introduced to solve the transfer functions between the state and input variables. This section will provide a brief analysis and introduction to the characteristic equations for lateral motion.

1. Longitudinal motion

The determinant $\Delta_{\mathrm{lon}} = |s\mathbf{E}_{\mathrm{lon}} - \mathbf{A}_{\mathrm{lon}}|$ is commonly referred to as the *characteristic determinant* of longitudinal motion, while $\Delta_{\mathrm{lon}} = 0$ represents the characteristic equation of longitudinal motion. The characteristic equation is written in the following form:

$$\Delta_{\mathrm{lon}} = s^4 + a_1 s^3 + a_2 s^2 + a_3 s + a_4 = 0. \tag{4.73}$$

The characteristic equation of the fixed-wing UAV describes the inherent stability of the UAV itself. The fundamental characteristics of the perturbed motion of the UAV can be understood by analyzing the roots of the characteristic equation. The time-domain solutions obtained by the inverse Laplace transform are referred to as modes of perturbed motion, and the total perturbed motion is obtained by linearly superposing various modes.

Generally, the characteristic equation of the UAV's longitudinal motion is factored to represent the phugoid and short-period modes, respectively. Typically, the phugoid mode corresponds to a pair of small, complex-conjugate roots with a long oscillation period and slow convergence. In contrast, the short-period mode corresponds to a pair of large complex conjugate roots with a short oscillation period and fast convergence.

The general longitudinal motion characteristic equation of a UAV can be written as the product of two quadratic factors, representing the phugoid mode and the short-period mode, respectively. Generally, the phugoid mode corresponds to a pair of smaller, complex-conjugate roots, characterized by a long oscillation period and slow convergence. The short-period mode corresponds to a pair of larger complex conjugate roots, characterized by a short oscillation period and fast convergence. In Step 4 of Analysis Experiment Sect. 4.3, an example of modal analysis is given: with zero input and a given initial positive angle of attack disturbance, the self-perturbation motion of each state variable in the pitch channel is observed as shown in Fig. 4.29. It is not difficult to find that: in the short-period mode, the changes in ΔV_{a} and ΔH are very small; in the phugoid mode, the changes in $\Delta\alpha$ and $\omega_{y_{\mathrm{b}}}$ are almost ignored; and $\Delta\theta$ obviously changes in the both short-period and phugoid mode, but mainly exhibits the characteristics in the phugoid mode. When disturbances or control inputs occur, the short-period mode typically dominates the initial period of the UAV's longitudinal motion, and its transient time is often very brief. Meanwhile, during this period, the changes in airspeed and pitch angle corresponding to the phugoid mode are insignificant. After the short-period motion ends, the UAV enters the phugoid mode, where the phugoid mode mainly reflects the trajectory of the UAV's center of gravity. Utilizing the linearized model of the fixed-wing UAV provided in this book, the roots of the characteristic equation of the longitudinal channel are given as:

$$\lambda_1 = -40.4623; \quad \lambda_2 = -6.3114; \quad \lambda_{3,4} = -0.2713 \pm 0.5968i$$

where the two negative real roots with smaller absolute values correspond to the phugoid mode, characterized by slow convergence and long adjustment time. The pair of conjugate complex roots with larger negative real parts corresponds to the short-period mode, characterized by a short oscillation period and rapid convergence.

2. Lateral motion

Similarly, the characteristic determinant $\Delta_{\mathrm{lat}} = |s\mathbf{E}_{\mathrm{lat}} - \mathbf{A}_{\mathrm{lat}}|$ is calculated. Cramer's rule obtains the transfer functions between various state and input variables. The lateral motion includes both rolling and yawing motions, and the coupling between the two motions is significant. Unlike the longitudinal channel, where each state variable corresponds to a predominant mode, in the lateral channel, three motion modes correspond to the roots of the characteristic equation: a pair of complex conjugate roots and one large root. The complex conjugate roots correspond to the Dutch roll mode, the large negative root indicates the rapid convergence rolling mode, and the small root (positive or negative) represents the slow spiral divergence mode. Utilizing the linearized model of the fixed-wing UAV provided in this book, the roots of the characteristic equation of the lateral channel are given as:

$$\lambda_1 = -6.2228; \quad \lambda_2 = -0.02258; \quad \lambda_{3,4} = -2.9813 \pm 2.8964i.$$

4.2 Basic Experiment

4.2.1 Experimental Objective

1. Things to prepare

(1) Software: MATLAB R2022b or above.
(2) Program: Experimental package "e2/e2-1". The "e2-1" folder contains "InitData.m" file for initializing the parameters of the fixed-wing UAV, and "SmallFixedWingUAVnoctrl.slx" file is the Simulink simulation model of the fixed-wing UAV.

The micro-small fixed-wing UAV in this book is based on the UltraStick-25E manufactured by the University of Minnesota, as shown in Fig. 2.1. This UAV is a remote-controlled commercial fixed-wing UAV with a symmetrical airfoil wing. The UltraStick-25E features horizontal and vertical tail control surfaces, including elevator and rudder control surfaces, as well as ailerons. The flaps are not actuated during the simulation; therefore, flap control is not considered in this analysis. The main physical parameters of this model[3] are mass $m = 1.9$ kg, span $b = 1.2$ m, chord $c = 0.3$ m and wing area $S = 0.32\,\text{m}^2$. The aerodynamic coefficients related to aerodynamic forces and moments [4] are shown in Table 4.1. The simulation structure of the fixed-wing UAV is based on the open-source fixed-wing UAV source code developed by MathWorks [5], with the parameters in the "InitData.m" file primarily sourced from this code.

2. Objective

(1) Observe and record the effects of elevator, aileron, rudder, and propeller thrust adjustments on the fixed-wing UAV's flight states, and plot the corresponding response curves.
(2) Observe and record the changes in aerodynamic coefficients resulting from aileron and rudder deflections in the fixed-wing UAV and analyze the results of roll and yaw angle changes in the lateral channel.

---

[3] The fundamental physical parameters of the UAV under consideration are drawn from [4] and are essentially analogous to those referenced in Chap. 2.

Table 4.1 Aerodynamic coefficients of UltraStick-25E fixed-wing UAV

Coefficient	Value	Coefficient	Value	Coefficient	Value
C_{L0}	0.23	$C_{Y_{\delta_r}}$	0.191	C_{n_r}	–0.411
C_{L_α}	4.58	C_{m0}	0.135	C_{l_β}	–0.04
$C_{L_{\delta_e}}$	0.124	C_{m_α}	–1.50	$C_{l_{\delta_a}}$	0.0677
$C_{L_{\dot{\alpha}}}$	1.97	$C_{m_{\delta_e}}$	–1.13	$C_{l_{\delta_r}}$	0.0168
C_{L_q}	7.95	$C_{m_{\dot{\alpha}}}$	–10.4	C_{l_p}	–0.414
$C_{L_{\min}}$	0.23	C_{m_q}	–50.8	C_{l_r}	0.399
C_{D0}	0.0434	C_{n_β}	0.0344	J_{xx}	0.0894
$C_{D_{\delta_e}}$	0.0135	$C_{n_{\delta_a}}$	–0.012	J_{yy}	0.144
$C_{D_{\delta_r}}$	0.0303	$C_{n_{\delta_r}}$	–0.0345	J_{zz}	0.162
C_{Y_β}	–0.83	C_{n_p}	-0.075	J_{xz}	0.00013

4.2.2 Experimental Procedure

The basic experiment aims to understand the direct performance of each control input on the UAV's motion through simulation testing in Simulink. In Steps 1–5, changes in elevator, aileron, rudder, and propeller throttle under step input are observed and recorded, along with the changes in relevant flight states and aerodynamic coefficients of the fixed-wing UAV. In Step 6, an analysis based on principles is conducted.

1. Step 1: Observing and recording the responses of pitch and altitude by elevator deflection.

(1) First, open and run the initialization file "e2-1/InitData.m". Then open "SmallFixedWingUAVnoctrl.slx" as shown in Fig. 4.13. This model consists of eight modules: dashed box① - "Command", command module for inputting commands to the fixed-wing UAV; dashed box②—"Actuator Model", actuator module, which introduces delay to input commands; dashed box③—"Motor Model", propulsion system

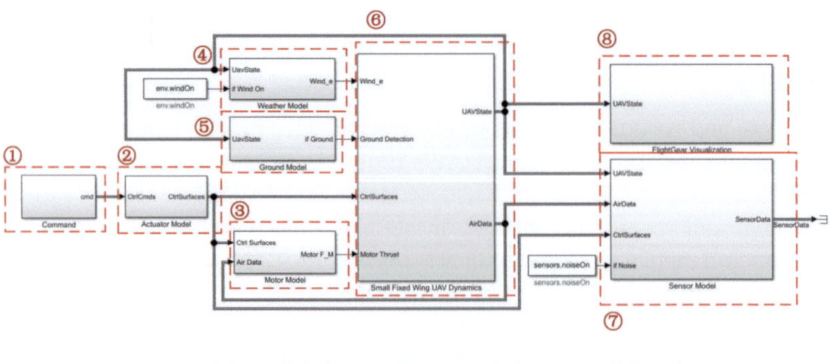

Fig. 4.13 Simulink model of the fixed-wing UAV, from "e2/e2-1/SmallFixedWingUAVnoctrl.slx"

module, which converts throttle commands into propeller thrust; dashed box④—"Weather Model", atmospheric model, which simulates atmospheric data such as air density, pressure, and wind speed; dashed box⑤—"Ground Model", ground model, which provides ground support force and moment; dashed box⑥—"Small Fixed Wing UAV Dynamics", fixed-wing UAV dynamics module, which calculates the attitude, position, and environmental information of the fixed-wing UAV; dashed box⑦—"Sensor Model", sensor module, which simulates sensor data to obtain the attitude, position, and environmental information of the fixed-wing UAV; dashed box⑧—"FlightGear Visualization", FlightGear visualization module, which visualizes the simulation using FlightGear software.

(2) Double-click to open the leftmost "command" module in the model, as shown in Fig. 4.14a. Here, each control command is set to its default neutral position of 0, with the throttle at 37.24%. The step signal to the right of each control command corresponds to the respective input. For instance, the signal change in the "elevator" module, depicted in Fig. 4.15, shows a negative deflection of 0.04 rad at 35 s. To use it, disconnect the left step signal output from the command module and connect the adder to the corresponding step signal under "change actuators" on the right side, as illustrated in Fig. 4.14b.

(3) Run program "SmallFixedWingUAVnoctrl.slx" by clicking "Run" in the toolbar above Simulink, as shown in Fig. 4.16. After the run, click "Data Inspector" as

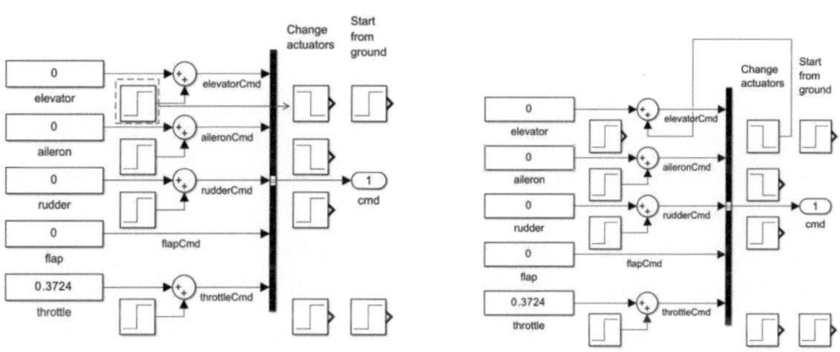

(a) Command module default connection

(b) Command module elevator output connection

Fig. 4.14 Input command module, from "e2/e2-1/SmallFixedWingUAVnoctrl.slx"

Fig. 4.15 Signal changes in "elevator"

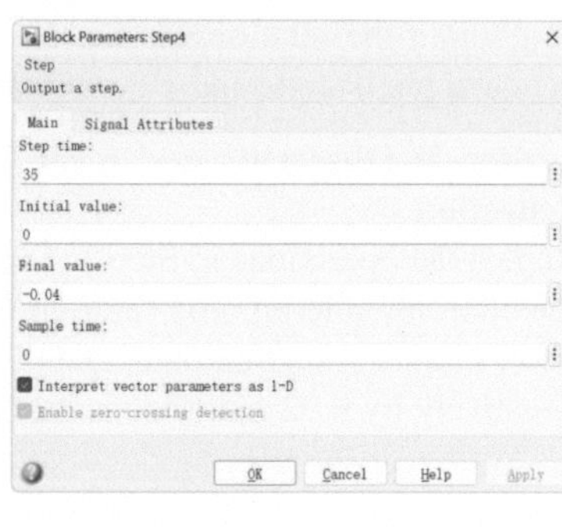

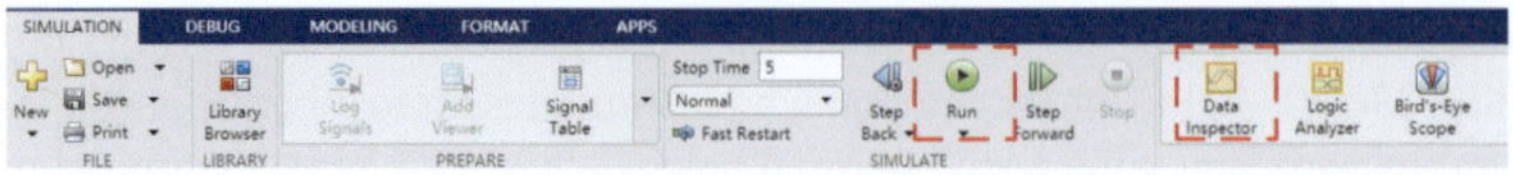

Fig. 4.16 Toolbar

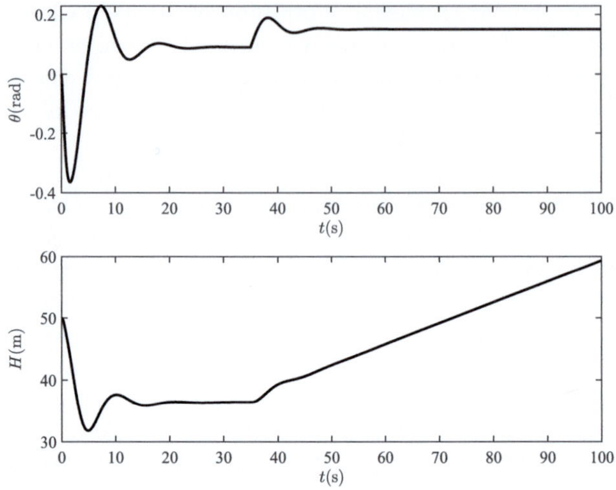

Fig. 4.17 Flight state response with elevator deflection

shown in Fig. 4.16. Relevant state signals are found on the left side of the "Data Inspector" interface. Expand "Euler" to see the Euler angles, in order of roll, pitch, and yaw; expand "SensorData" to see the sensor data, where "airspeed" is speed, and "gps_h" is altitude.

(4) Observe the simulation data and output the response curve. The UAV reaches the level flight state within 20 s after the simulation begins, and the elevator deflection response is observed. At 35 s of negative elevator deflection of 0.04 rad, the UAV's pitch angle increases positively, i.e., the nose is raised. As the pitch angle increases, the UAV altitude increases, and the flight state changes, as shown in Fig. 4.17. The relationship between pitch and elevator deflection is obtained through Eqs. (4.29) and (4.23). The relationship between altitude and elevator deflection is obtained by combining the relationship between altitude and pitch angle with the relationship between pitch angle and elevator deflection in series.

2. Step 2: Observing and recording the responses of roll and yaw to aileron deflection

Repeat the operations in Step 1. Connect the step signal generator corresponding to the aileron ("aileron") on the right side of the "command" module with the adder, observe the simulation data, and output response curves. When the aileron deflects negatively by 0.02 rad, the aircraft rolls negatively, and the aileron also affects the

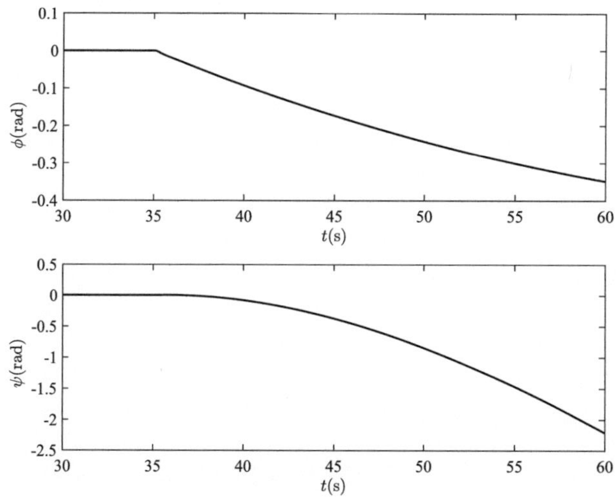

Fig. 4.18 Flight state response with aileron deflection

aircraft's yaw angle. The yaw angle decreases. Figure 4.18 shows the flight state changes.

3. Step 3: Observing and recording the response of roll and yaw to rudder deflection

Repeat the operations in Step 1. Connect the step signal generator corresponding to the rudder ("rudder") on the right side of the "command" module with the adder, observe the simulation data, and output responses. When the rudder deflects positively by 0.02 rad, the aircraft undergoes a negative yaw motion, and the rudder also affects the aircraft's roll angle. The roll angle decreases. The flight state changes are shown in Fig. 4.19.

4. Step 4: Observing and recording the response of altitude and airspeed to propeller thrust

Repeat the operations in Step 1. Connect the step signal generator for the propeller's throttle ("throttle") on the right side of the "command" module to the adder, observe the data, and output the images. When the throttle is increased to 50%, the aircraft's speed rises rapidly. However, a subsequent decrease in speed may occur due to speed saturation in the absence of control. The increased speed generates lift exceeding the aircraft's weight, resulting in a rise in altitude. The flight state changes are shown in Fig. 4.20.

5. Step 5: Setting aileron and rudder deflections separately and observing and recording changes in the relevant aerodynamic coefficients in the lateral channel.

(1) First, open and run the initialization file "e2-1/InitData.m". Then open file "SmallFixedWingUAVnoctrl.slx", and double-click to enter the module "Small Fixed Wing

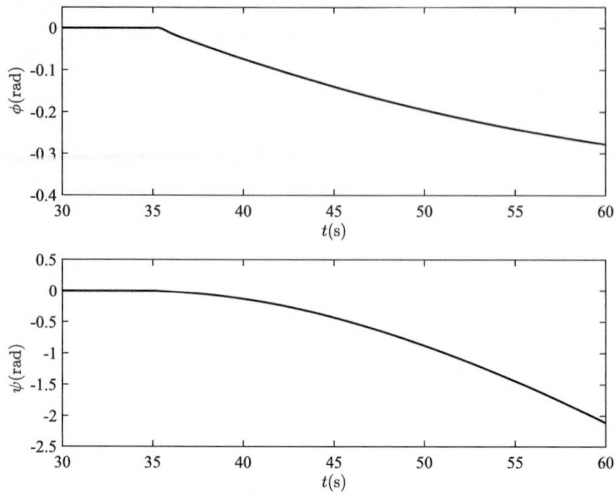

Fig. 4.19 Flight state response with rudder deflection

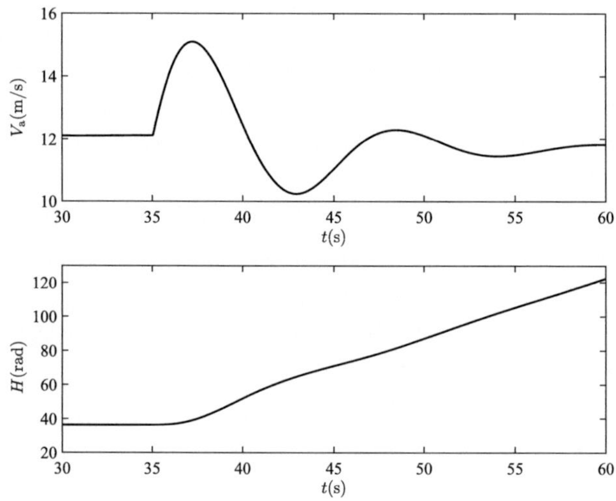

Fig. 4.20 Flight state response with propeller thrust

UAV Dynamics" → "Aerodynamic Forces and Moments" → "Compute Body Frame Forces and Moments" → "Wind frame F & M" → "Lateral Channel". In "Lateral Channel", there are "Cl" and "Cn" modules, which are the modules for the roll moment coefficient and yaw moment coefficient, respectively. Taking the "Cl" module as an example, click on the relevant coefficient signal line, and select "Log Selected Signal", as shown in Fig. 4.21.

(2) Increase the simulation time to 100s. Following the same steps as in Sect. 4.2, set a step signal input for the aileron and rudder separately, run "SmallFixed-

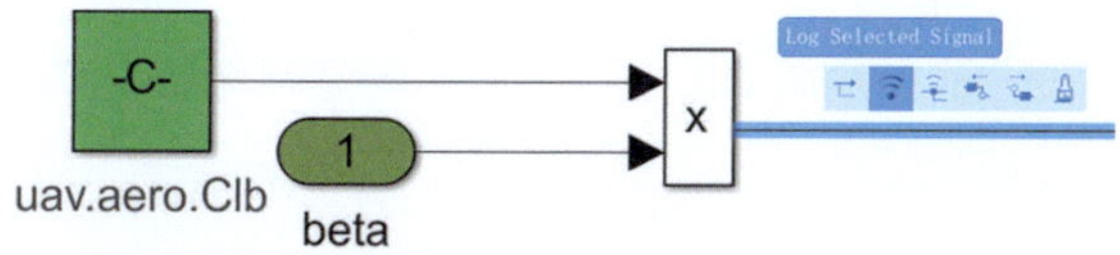

Fig. 4.21 Data logging for relevant coefficients

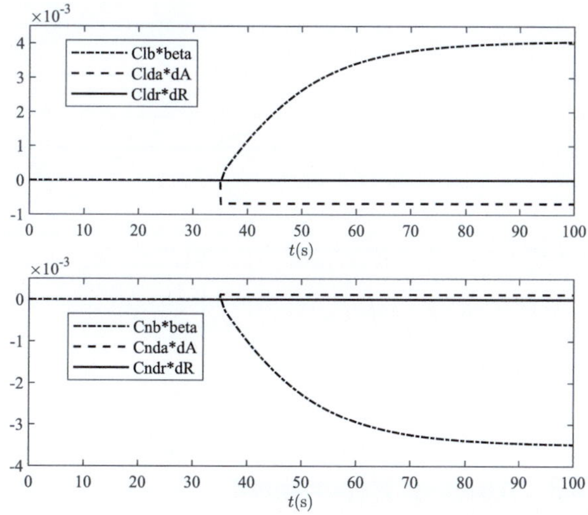

Fig. 4.22 Aerodynamic coefficient response with aileron deflection

WingUAVnoctrl.slx", and observe the recorded change curves in "Data Inspector". The aerodynamic coefficient changing when the aileron is deflected negatively is shown in Fig. 4.22, and the aerodynamic coefficient changing when the rudder is deflected positively is shown in Fig. 4.23.

6. **Step 6: Analyzing the trends of roll and yaw changes.**
When the ailerons deflect negatively, $C_{l_{\delta_a}}\delta_a$ and $C_{l_\beta}\beta$ in Eq. (4.31) (corresponding to "Clda ∗ dA" and "Clb ∗ beta" in Fig. 4.22a) play a significant role, with $C_{l_\beta}\beta$ having a relatively large absolute value. Similarly, $C_{n_{\delta_a}}\delta_a$ and $C_{n_\beta}\beta$ in Equation (4.31) (corresponding to "Cnda ∗ dA" and "Cnb ∗ beta" in Fig. 4.22b) play a significant role, with $C_{n_\beta}\beta$ having a relatively large absolute value. The terms $C_{l_{\delta_a}}\delta_a$ and $C_{n_{\delta_a}}\delta_a$ are caused by aileron deflection, while $C_{l_\beta}\beta$ and $C_{n_\beta}\beta$ are caused by the sideslip angle. The sideslip angle has a predominant influence on the aerodynamic moments, with the coefficients associated with ailerons playing a secondary role compared to those related to the sideslip angle.

When the rudder deflects positively, $C_{l_{\delta_r}}\delta_r$ and $C_{l_\beta}\beta$ in Eq. (4.31) (corresponding to "Cldr ∗ dR" and "Clb ∗ beta" in Fig. 4.23a) play a significant role, with $C_{l_\beta}\beta$ having a relatively large absolute value. Similarly, $C_{n_\beta}\beta$ and $C_{n_{\delta_r}}\delta_r$ in Eq. (4.31) (corresponding to "Cnb ∗ beta" and "Cndr ∗ dR" in Fig. 4.23(b)) play a significant role, with $C_{n_\beta}\beta$ having a relatively large absolute value. The terms $C_{l_{\delta_r}}\delta_r$ and $C_{n_{\delta_r}}\delta_r$ are caused by rudder deflection. In contrast, $C_{l_\beta}\beta$ and $C_{n_\beta}\beta$ are influenced by the sideslip angle, which has a dominant effect on the aerodynamic moments.

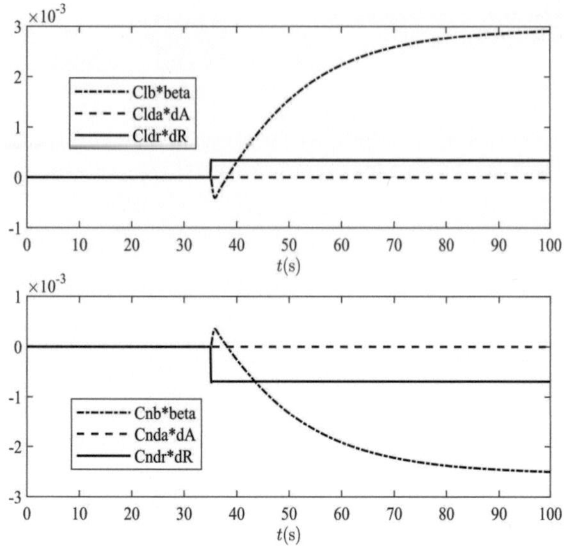

Fig. 4.23 Aerodynamic coefficient response with rudder deflection

4.3 Analysis Experiment

4.3.1 Experimental Objective

1. Things to prepare

(1) Software: MATLAB R2022b or above.
(2) Program: Experimental package "e2/e2-2". The "e2-2" folder includes the trim file "balancing.m", the state matrix derivation file "derivation.m", the longitudinal transfer function file "transfer_lon.m", the lateral transfer function file "transfer_lat.m", the initial state file for fixed-wing UAVs "InitData.m", and the controlled model "SmallFixedWingUAVnoctrl.slx".

2. Objective

(1) Obtain the fixed-wing trim motion flight state through longitudinal trim, and analyze the linear motion equation under the trim motion.
(2) For the same control input, observe the state response of various channels of the fixed-wing nonlinear and linear mathematical models, then perform a comparative analysis.

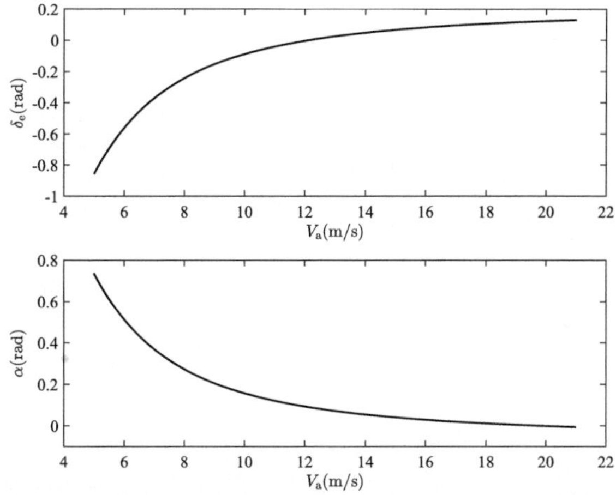

Fig. 4.24 Trim results at different airspeeds

4.3.2 *Experimental Procedure*

Taking the longitudinal trim motion of fixed-wing UAVs as an example, the simulation results of the linear model and nonlinear model are compared in three steps. In Step 1, the equilibrium state of the fixed-wing UAV in a specific speed range is obtained through longitudinal trim, and the trim motion state is selected. In Step 2, the longitudinal and lateral motion equations at the trim motion state are linearized. In Step 3, the responses of the linearized model and the nonlinear model under the same input are compared and analyzed.

1. Step 1: Longitudinal trim of fixed-wing UAV

(1) In the steady straight flight state of the UAV, the flight path angle $\gamma = 0$. According to Table 3.4 in Chap. 3, the flight speed of the fixed-wing UAV is in the range of 5–20 m/s. Open and run file "e2-2/balancing.m", and obtain the relationship between airspeed and trim elevator deflection and the relationship between airspeed and trim angle of attack within the range of 5–20 m/s, as shown in Fig. 4.24. The figure shows that as airspeed increases, the trim elevator declination also increases. With the increase in airspeed, the trim angle of attack decreases. After 19.7 m/s, a negative angle of attack will occur. An excessive angle of attack can lead to stalling, while a negative angle of attack creates a negative lift, which is detrimental to the flight of a UAV. As a result, the airspeed for trimming is selected in the range of 9–19 m/s.

(2) Finally, a constant airspeed of $V_a = 11.4\text{m/s}$ and a height of 50m are selected. Substituting them into Eq. (4.44), the trim deflection angle of elevator $\delta_e^* = -0.02482$ rad, thrust $T^* = 2.4509$ N, angle of attack $\alpha^* = 0.1087$ rad, and pitch angle $\theta^* = 0.1087$ rad are obtained, which are taken as the trim state.

2. Step 2: Linearization of fixed-wing flight model

(1) Open file "derivation.m", run the "Longitudinal function" code block, and enter "E1", "A1", "B1" in the command window to obtain the state matrices $\mathbf{E}_{\text{lon}}$, $\mathbf{A}_{\text{lon}}$,

and $\mathbf{B}_{\text{lon}}$ in the longitudinal linear model. The linearized equation is obtained by substituting the trim state into Eq. (4.49), as follows:

$$\begin{bmatrix} 1.9 & 0.09622 & 0 & 0 \\ 0 & 22.3170 & 0 & 0 \\ 0 & 0 & 1 & 0 \\ 0 & 7.2262 & 0 & 1 \end{bmatrix} \begin{bmatrix} \Delta\dot{V}_{\text{a}} \\ \Delta\dot{\alpha} \\ \Delta\dot{\theta} \\ \Delta\dot{\omega}_{y_{\text{b}}} \end{bmatrix} = \begin{bmatrix} -0.8269 & 1.3656 & -18.6326 & -0.3883 \\ -3.1787 & -118.5267 & 0 & 19.0085 \\ 0 & 0 & 0 & 1 \\ 0 & -79.2106 & 0 & -35.2974 \end{bmatrix} \begin{bmatrix} \Delta V_{\text{a}} \\ \Delta\alpha \\ \Delta\theta \\ \Delta\omega_{y_{\text{b}}} \end{bmatrix} + \begin{bmatrix} 1.3179 & -0.8248 \\ -0.1438 & -3.2952 \\ 0 & 0 \\ 0 & -59.6720 \end{bmatrix} \begin{bmatrix} \Delta T \\ \Delta\delta_{\text{e}} \end{bmatrix}. \tag{4.74}$$

Running the “Lateral function” part and then entering “e1”, “a1”, “b1” in the command window, the state matrices $\mathbf{E}_{\text{lat}}$, $\mathbf{A}_{\text{lat}}$, and $\mathbf{B}_{\text{lat}}$ in the lateral linear model are obtained. Thus, the lateral linearization equation is given by

$$\begin{bmatrix} 21.6600 & 0 & 0 & 0 \\ 0 & 1 & 0 & 0 \\ 0 & 0 & 1 & 0 \\ 0 & 0 & 0 & 1 \end{bmatrix} \begin{bmatrix} \Delta\dot{\beta} \\ \Delta\dot{\phi} \\ \Delta\dot{\omega}_{x_{\text{b}}} \\ \Delta\dot{\omega}_{z_{\text{b}}} \end{bmatrix} = \begin{bmatrix} -26.8661 & 18.5227 & 2.3498 & -21.5322 \\ 0 & 0 & 1 & 0.1091 \\ -12.7707 & 0 & -7.6329 & 6.5982 \\ 5.3553 & 0 & -1.4008 & -3.4913 \end{bmatrix} \begin{bmatrix} \Delta\beta \\ \Delta\phi \\ \Delta\omega_{x_{\text{b}}} \\ \Delta\omega_{z_{\text{b}}} \end{bmatrix} + \begin{bmatrix} 0 & 4.8414 \\ 0 & 0 \\ 22.9921 & 4.7660 \\ -0.2661 & -6.0658 \end{bmatrix} \begin{bmatrix} \Delta\delta_{\text{a}} \\ \Delta\delta_{\text{r}} \end{bmatrix}. \tag{4.75}$$

If the matrices $\mathbf{E}_{\text{lon}}$ and $\mathbf{E}_{\text{lat}}$ are nonsingular, Eqs. (4.74) and (4.75) are written in the standard form of linear state equations through $\bar{\mathbf{A}} = \mathbf{E}^{-1}\mathbf{A}$ and $\bar{\mathbf{B}} = \mathbf{E}^{-1}\mathbf{B}$:

$$\dot{\mathbf{X}} = \bar{\mathbf{A}}\mathbf{X} + \bar{\mathbf{B}}\mathbf{X}. \tag{4.76}$$

(2) Open “transfer_lon.m” file, run Code 4.1, and enter “f2” in the command window to substitute the parameters in Eqs. (4.74) into (4.53). Then

$$\frac{\Delta\theta(s)}{\Delta\delta_{\text{e}}(s)} = \frac{-59.672}{s^2 - 0.5915s}. \tag{4.77}$$

Furthermore, using Eq. (4.57), the transfer function from ΔH to $\Delta\delta_{\text{e}}$ is

$$\frac{\Delta H(s)}{\Delta\delta_{\text{e}}(s)} = \frac{-680.2608}{s^3 - 0.5915s^2}. \tag{4.78}$$

Code 4.1 Transfer function from elevator deflection to pitch angle, from "e2/e2-2/transfer_lon.m"

```
%% Simplified transfer function
syms s
I2=eye(2); % 2*2 Identity matrix
A2=[0 1;0 -35.2974];
a2=det(s*I2-A2); % determinant
B2=[0 -1; -59.6720 s+35.2974];
b2=det(B2);
f2=b2/a2;
```

(3) Open "transfer_lat.m" file and run it (see Code 4.2). Then, enter "f1" in the command window to get

$$\frac{\Delta\phi(s)}{\Delta\delta_a(s)} = \frac{22.9870s^2 + 102.7759s + 203.6605}{s^4 + 12.208s^3 + 54.6572s^2 + 109.1560s + 6.9516}. \tag{4.79}$$

Furthermore, using Eq. (4.68), the transfer function from the input $\Delta\delta_a$ to the yaw angle $\Delta\psi$ is

$$\frac{\Delta\psi(s)}{\Delta\delta_a(s)} = \frac{19.7742s^2 + 88.4112s + 175.1954}{s(s^4 + 12.2080s^3 + 54.6572s^2 + 109.1560s + 6.9206)}. \tag{4.80}$$

Code 4.2 Transfer functions of lateral channel, from "e2/e2-2/transfer_lat.m"

```
%%sI-A
I=eye(4);
A=[-1.0838   0.8552   0.1085  -0.9941;
0     0     1     0.0191;
-12.7707   0   -7.6329   6.5982;
5.3553     0   -1.4008  -3.4913];
[V,G]=eig(A);
syms s
a=det(s*I-A); % determinant
B=[s+1.0838  0    -0.1085  0.9941;
0     0     -1     -0.0191;
12.7707       22.9921 s+7.6329   -6.5982;
-5.3553    -0.2661  1.4008   s+3.4913];
b=det(B); % % determinant
C=[s+1.0838  0.2235    -0.1085  0.9941;
0           0           -1     -0.0191;
12.7707       4.7660 s+7.6329   -6.5982;
-5.3553    -6.0658  1.4008   s+3.4913];
c=det(C);
f1=b/a; %transfer function between roll and aileron
f2=c/a; %transfer function between roll and rudder
```

(4) Open "transfer_lat.m" file and run it, then enter "f2" in the command window to get

$$\frac{\Delta\phi(s)}{\Delta\delta_{\mathrm{t}}(s)} = \frac{-4.6501s^2 - 22.1872s - 79.9991}{s^4 + 12.2080s^3 + 54.6572s^2 + 109.1560s + 6.9516}. \tag{4.81}$$

3. Step 3: Comparison of linearized and nonlinear model responses

(1) First, open file "e2-2/InitData.m" and run the initialization file. Set the initial Euler angle to $[0 \quad 0.1087 \quad 0]^{\mathrm{T}}$ rad, and choose an initial airspeed of 11.4m/s. Then open file "e2-2/SmallFixedWingUAVnoctrl.slx", which includes the nonlinear mathematical model of a fixed-wing UAV (built in Sect. 4.2.2 of the basic experiment) and the linear mathematical model constructed based on Eqs. (4.74) and (4.75). To ensure level flight, set the throttle command in the "throttle" module to 0.3724. Connect the step signal generator on the right side of the elevator "elevator" in "command" instruction module to the adder. That is, when the elevator is deflected negatively by 0.04 rad, observe the data output of "Small Fixed Wing UAV Linear Dynamics" module and compare them with the output of the nonlinear model "Small Fixed Wing UAV Dynamics" module, specifically "pitch" and "altitude" channels in "Compare" module, as shown in Fig. 4.25.

According to the open-loop transfer function (4.77), since the numerator is negative, the pitch angle increases positively when the elevator deflection increment is negative. It has two real open-loop poles, so there is no overshoot phenomenon. A pole being zero acts as the dominant pole, and then, under a step input, the output shows ramp growth. The relationship between altitude and elevator given by Eq. (4.78) has three real poles, with two poles at 0 playing a dominant role. Under the step input, the output shows quadratic growth. In the nonlinear model, the altitude does not remain strictly at 50 m, resulting in a deviation from the linear model.

Fig. 4.25 Comparison of flight state response with elevator deflection

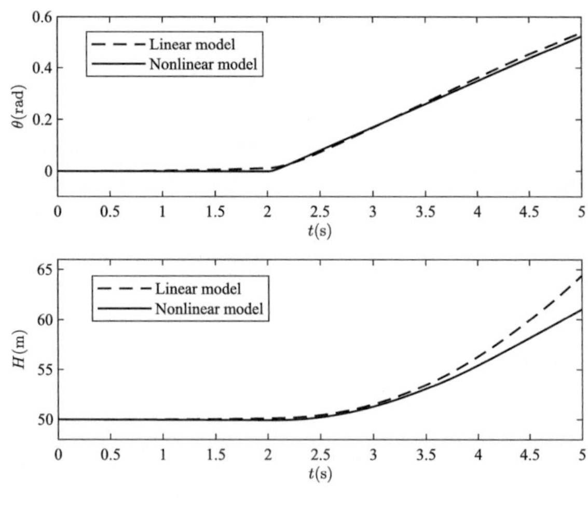

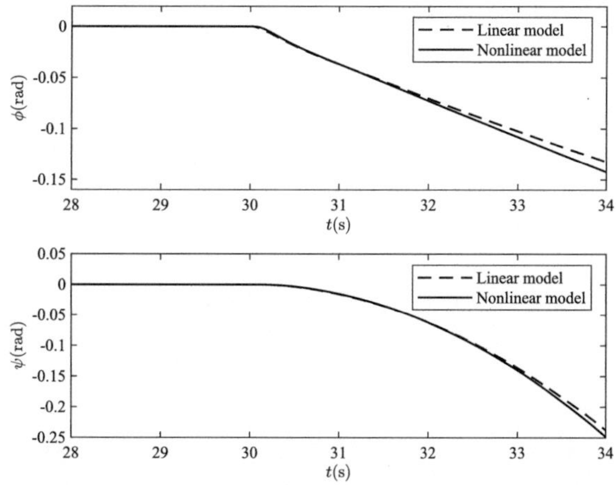

Fig. 4.26 Comparison of flight state response with aileron deflection

(2) When studying the lateral channel, it is assumed to be independent of the longitudinal model, so the longitudinal motion is kept as the trim motion, with an initial airspeed of 11.4 m/s and an initial pitch angle of 0.1087 rad. Set "uav.ic.Vb_0" in "InitData.m" file to [11.4; 0; 0], and set "uav.ic.Euler_0" to [0; 0.1087; 0]. Run the "InitData.m" file. Next, set the throttle in the "throttle" module to 0.3724, and connect the step signal generator on the right side of the "aileron" module to the adder. Run file "SmallFixedWingUAVnoctrl.slx". When the aileron is deflected negatively by 0.002 rad at 30 s, observe the data output images of "Small Fixed Wing UAV Linear Dynamics" module and compare them with the output images of the nonlinear model "Small Fixed Wing UAV Dynamics" module, specifically "roll" and "yaw" channels in "Compare" module, as shown in Fig. 4.26.

The transfer function (4.79) has two open-loop zeros $z_{1,2} = -2.2355 \pm 1.9653i$ and four poles $p_1 = -6.2058,\ p_2 = -6.5823,\ p_{3,4} = -2.9682 \pm 2.8650i$. Since the conjugate poles and zeros are close, their performances on the output can cancel each other, resulting in no overshoot but slight oscillations. Two real poles mainly influence the output. Since pole p_1 is far from the imaginary axis, its performance on the output is small, while the negative real pole is close to the origin, resulting in a ramp output with slight oscillations for a step input. The transfer function from the aileron deflection perturbation to the yaw angle perturbation has an additional zero compared to the transfer function from the aileron deflection perturbation to the roll angle perturbation. This causes the yaw angle output to exhibit quadratic growth in response to a step input in aileron deflection.

(3) Repeat the operation in step (2), run the "InitData.m" file, reset other channels in the "command" instruction module, and only connect the step signal generator on the right side of the rudder to the adder. Run file "SmallFixedWingUAVnoctrl.slx". When the aileron deflects by 0.002 rad, observe the data output of "Small Fixed Wing UAV Linear Dynamics" module and compare them with the output of the nonlinear model "Small Fixed Wing UAV Dynamics" module, specifically "roll" and "yaw" channels in "Compare" module, as shown in Fig. 4.27.

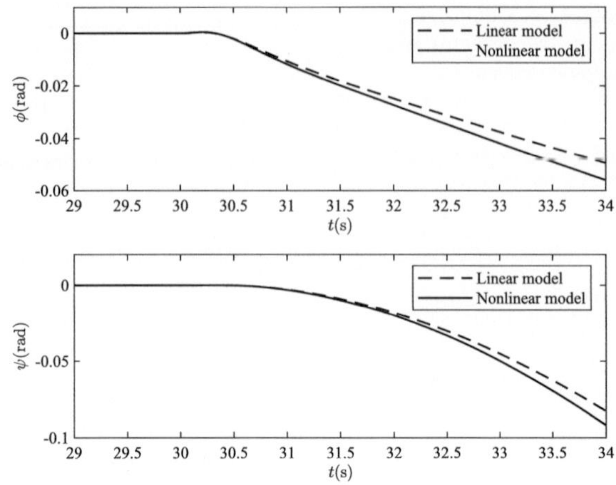

Fig. 4.27 Comparison of flight state response with rudder deflection

The transfer function (4.81) has two open-loop zeros $z_{1,2} = -2.3857 \pm 3.3930i$, and four poles $p_1 = -6.2058,\ p_2 = -6.5823,\ p_{3,4} = -2.9682 \pm 2.8650i$. Due to the presence of conjugate poles, overshooting appears in the output. During linearization, it is assumed that the lateral linear model is independent of the longitudinal model and that the airspeed is constant. However, in the nonlinear model, the throttle's airspeed change is not constant, affecting the lateral aerodynamic moments. Therefore, an error occurs between the lateral and nonlinear models.

(4) Repeat the operation in step (1), run the "InitData.m" file, connect the step signal generator on the right side of the throttle "throttle" in the "command" instruction module to the adder, and increase the throttle to 50%. Observe the simulation output of the "Small Fixed Wing UAV Linear Dynamics" module and compare it with the output of the nonlinear model "Small Fixed Wing UAV Dynamics" module, specifically the "airspeed" channel in the "Compare" module, as shown in Fig. 4.28.

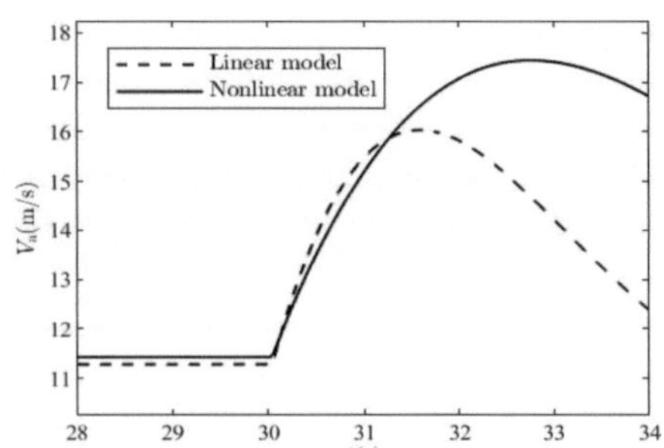

Fig. 4.28 Comparison of airspeed response with propeller thrust changes

4. Step 4: Modal analysis

(1) First, compare the phugoid and short-period modes of the fixed-wing UAV's longitudinal model. Open the file "modalanalysis.m" and execute the "Longitudinal channel" code block. Figure 4.29 illustrates the temporal evolution of airspeed, angle of attack, pitch angle, and pitch rate. The airspeed and pitch angle exhibit slow convergence with extended settling times, while the angle of attack and pitch rate demonstrate rapid oscillations with short periods, converging fast.

(2) Run code in the "Lateral channel" block in the "modalanalysis.m" file. Figure 4.30 shows the changing of sideslip angle, roll angle, roll rate, and yaw rate over time. The roll and yaw rates exhibit obvious oscillations corresponding to the Dutch roll mode. At the same time, the sideslip angle quickly converges, corresponding to the rolling mode, and the roll angle corresponds to the spiral divergence mode.

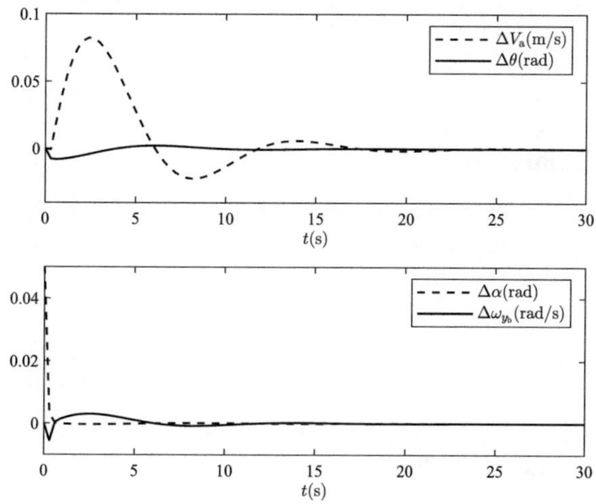

Fig. 4.29 State response in the longitudinal channel

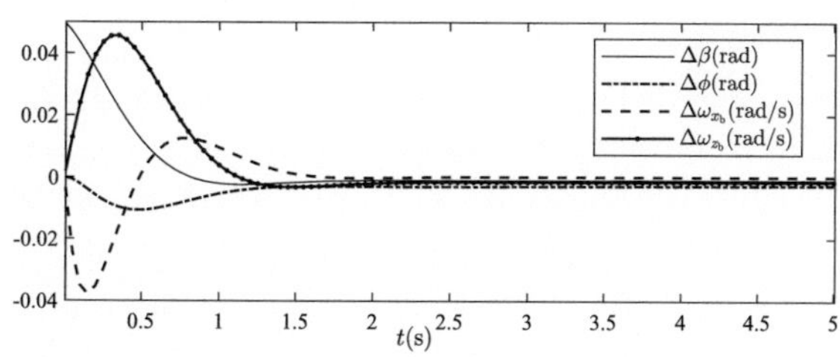

Fig. 4.30 State response in the lateral channel

4.4 Design Experiment

4.4.1 Experimental Objective

1. Things to prepare

(1) Software: MATLAB R2022b or above, RflySim toolchain. The RflySim toolchain provides a HIL simulation environment for micro-small fixed-wing UAVs.
(2) Program: Experimental package "e2/e2-3" and "e2/e2-4". The "e2-3" folder contains the initial state file "InitData.m" for the fixed-wing UAV, the controlled model "SmallFixedWingUAVnoctrl.slx", and the Software-in-the-Loop (SIL) simulation script "runfg.bat". The "e2-4" folder contains the initial state file "Init.m" for the fixed-wing UAV, the HIL simulation model "SmallFixedWingUAVnoctrlHIL.slx" for the fixed-wing UAV, the model dynamic library generation file "GenerateModelDLLFile.p", parameters of the micro-small fixed-wing UAV and environmental parameters are initialized in the "InitData.m" file.
(3) Hardware: a CubePilot/Pixhawk autopilot, a remote controller, and a receiver.

2. Objective

(1) Establish a complete dynamic system model of the fixed-wing UAV in MATLAB/Simulink.
(2) Perform SIL simulation in MATLAB/Simulink with visual display using FlightGear.
(3) Generate the model code for the RflySim toolchain, use PX4 official fixed-wing flight control firmware for HIL simulation, and visualize in 3D using RflySim3D.

4.4.2 Experimental Design

The controlled dynamic model of the fixed-wing UAV mainly includes the actuator delay model, the motor-propeller propulsion system (see Sect. 3.1.2), the rigid body dynamics and kinematics model (see Sect. 4.1.2), the aerodynamic force and moment model (see Sect. 4.1.2), and environmental model, as shown in Fig. 4.31.

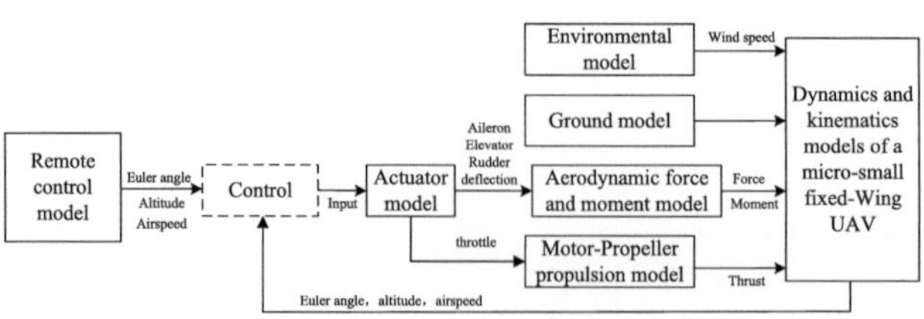

Fig. 4.31 Model of a controlled dynamic system of a micro-small fixed-wing UAV

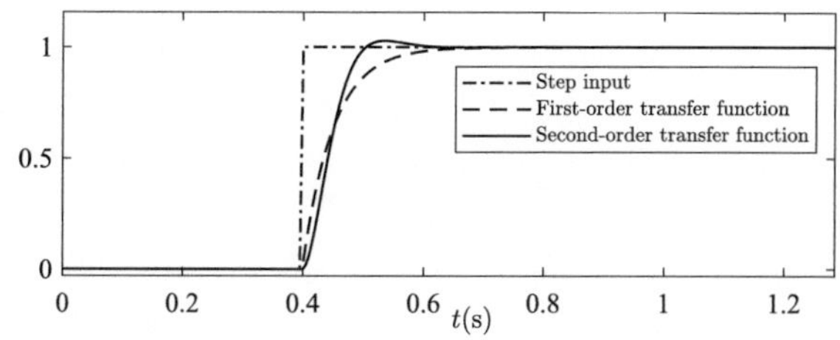

Fig. 4.32 Performance of the first-order and second-order transfer functions by step inputs

In the actual flight process of a fixed-wing UAV, there is a delay in the action of the actuators. Generally, a first-order or second-order transfer function is used to illustrate this delay [1, 6]. Its delayed performance is shown in Fig. 4.32.

(1) Motor-Propeller Propulsion Model: This model is derived from the equations governing propeller thrust at zero airspeed (4.32) and airspeed V_a (4.33), respectively.

(2) Rigid Body Dynamics and Kinematics Model: The inputs are the forces and moments in the body coordinate system. The velocities and angular velocities are obtained based on Eqs. (4.19) and (4.23). Positions and Euler angles are obtained based on Eqs. (4.16) and (4.18).

(3) Aerodynamic Force and Moment Model: Inputs include aerodynamic data (angle of attack, sideslip angle, air density, and airspeed), as well as angular velocities and control surface deflection angles. Aerodynamic forces and moments are obtained based on the longitudinal and lateral aerodynamic forces, namely Eq. (4.29) and moment, namely Eq. (4.31).

(4) Environmental Model: This module mainly includes the atmospheric model, which obtains wind speed, air density, and pressure information. The ground model determines altitude, while the sensor model accounts for sensor noise and measurement errors.

4.4.3 Experimental Procedure

The experimental design aims to establish a comprehensive design for each module of the UAV model and to conduct both SIL and HIL simulations. Step 1 involves constructing a complete dynamic system model of the UAV in MATLAB/Simulink, utilizing the parameters provided by a micro-small fixed-wing UAV. Step 2 entails executing the SIL simulation, while Step 3 focuses on completing the HIL simulation.

1. Step 1: Developing a complete dynamic model of the micro-small fixed-wing UAV in MATLAB/Simulink.

(1) Actuator module design

The actuator unit is divided into two parts: the control module, referred to as the "Actuator Model", and the propulsion system module, referred to as the "Motor Model". Since there is a delay in the action of actuators during actual flight, the control module "Actuator Model" models the servos as second-order systems. MATLAB

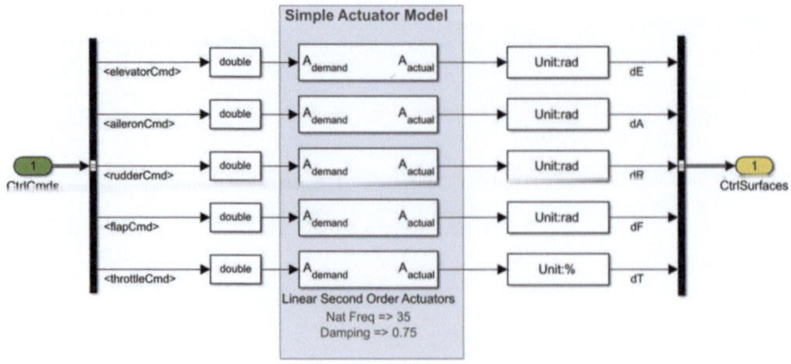

Fig. 4.33 Actuator module, from "e2/e2-3/SmallFixedWingUAVnoctrl.slx"

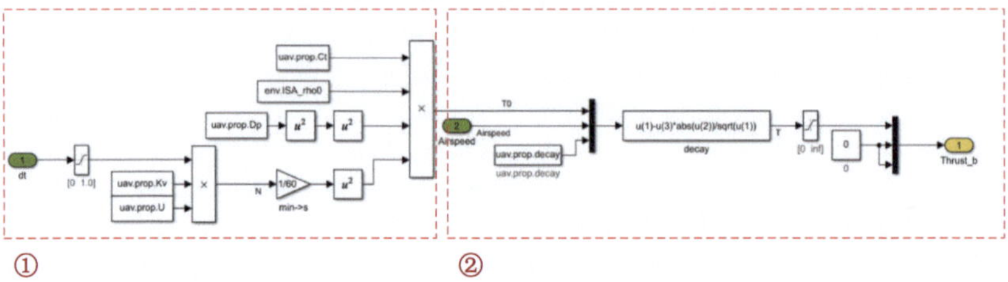

Fig. 4.34 Propulsion system module, from "e2/e2-3/SmallFixedWingUAVnoctrl.slx"

provides the "Linear Second Order Actuators" module to simulate the actual response of the control module, as shown in Fig. 4.33.

The propulsion system module models the motor-propeller propulsion system. Since the throttle command is directly proportional to the motor speed, the model is simplified here by taking the throttle command percentage as equivalent to the motor speed percentage. The modeling is based on Eqs. (3.7) and (3.15) in Sect. 3.1.1, as shown in Fig. 4.34.

(2) Rigid body dynamics and kinematics module design

Rigid body dynamics and kinematics use MATLAB's "6DOF (Euler Angles)" module. The internal design is shown in Fig. 4.35, where modules① and② correspond to Eqs. (4.22) and (4.23), respectively. Forces and moments are fed into the six-degree-of-freedom module as shown in Fig. 4.36.

(3) Aerodynamic force and moment module design

The aerodynamic force and moment module is the most important part of modeling a fixed-wing UAV. The main design includes the aerodynamic data calculation module "Compute Air Data" and the aerodynamic force and moment calculation module "Compute Body Frame Forces and Moments". The aerodynamic data calculation module is shown in Fig. 4.37, used to calculate airspeed, angle of attack, sideslip angle, and dynamic pressure. Dashed box① corresponds to Eq. (4.15) in the experimental principles, and dashed box② is used to calculate dynamic pressure $q = 1/2\rho V_{\mathrm{a}}^2$.

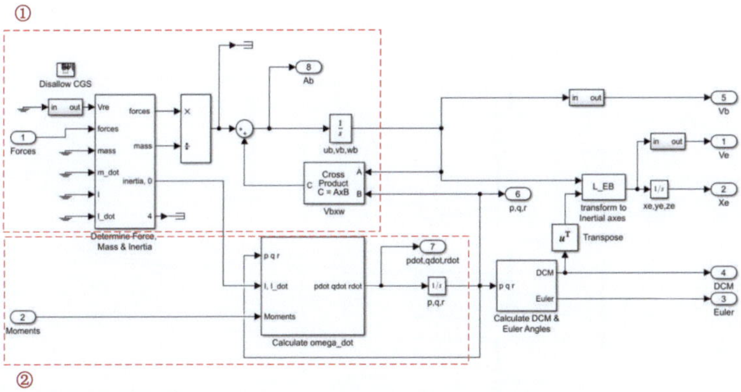

Fig. 4.35 6DOF (Euler Angles), from "e2/e2-3/SmallFixedWingUAVnoctrl.slx"

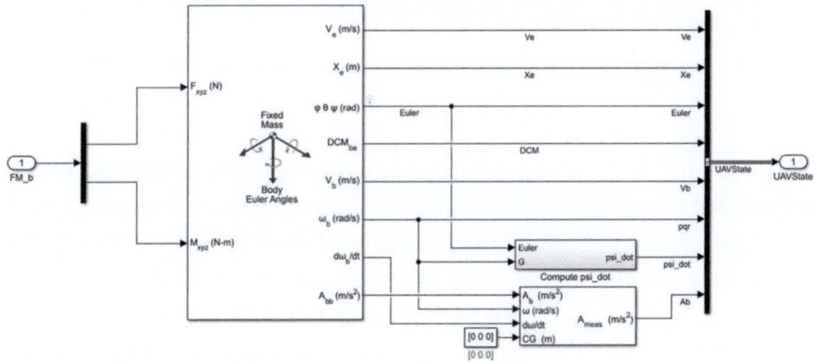

Fig. 4.36 Rigid body dynamics and kinematics module, from "e2/e2-3/SmallFixedWing-UAVnoctrl.slx"

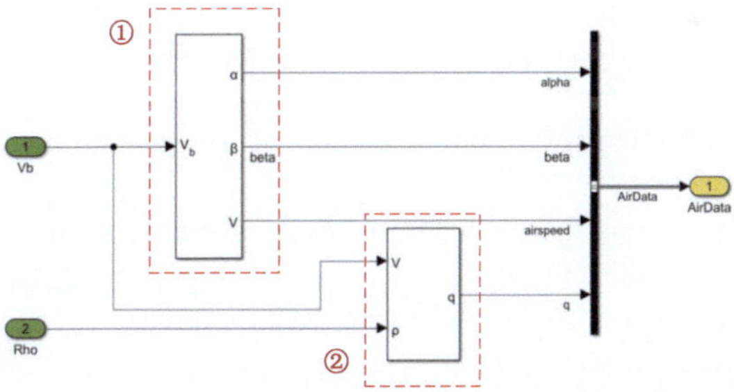

Fig. 4.37 Aerodynamic data calculation module, from "e2/e2-3/SmallFixedWingUAVnoctrl.slx"

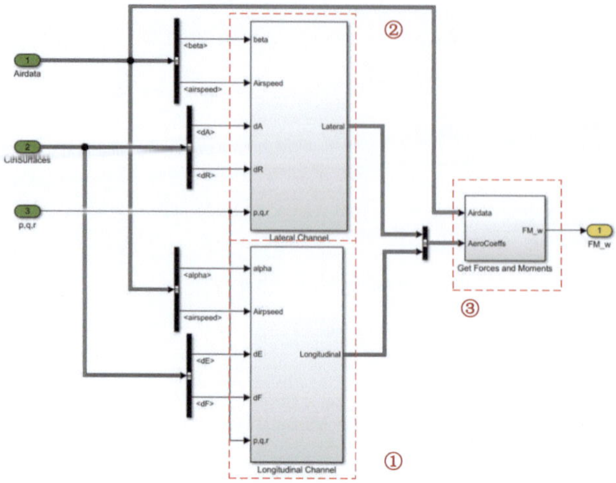

Fig. 4.38 Lateral and longitudinal channels, from "e2/e2-3/SmallFixedWingUAVnoctrl.slx"

The aerodynamic force and moment calculation module is divided into a lateral and a longitudinal channel. The lateral channel is established based on Eq. (4.31), and the longitudinal channel is established based on Eq. (4.29). As shown in Fig. 4.38, the design approach is first to calculate the sum of various aerodynamic coefficients within each channel. Dashed box① calculates the pitch moment aerodynamic coefficients, and dashed box② calculates the roll moment and yaw moment aerodynamic coefficients.

Taking the calculation of the roll moment aerodynamic coefficients in the lateral channel as an example, as shown in Fig. 4.39, it includes aerodynamic coefficients generated by sideslip angle "$C_{l_\beta}\beta$", aerodynamic coefficients generated by aileron deflection "$C_{l_{\delta_a}}\delta_a$", aerodynamic coefficients generated by rudder deflection "$C_{l_{\delta_r}}\delta_r$" and aerodynamic coefficients generated by angular rates "$C_{l_p}p$" and "$C_{l_r}r$". After summing these coefficients, the total roll moment aerodynamic coefficient "C_l" is obtained.

After summarizing coefficients within various channels, dashed box③ multiplies each coefficient with the current aerodynamic data according to Eqs. (4.29) and (4.31) to obtain the final aerodynamic forces and moments.

(4) Environmental model-design

The environmental model includes the atmospheric, ground, and sensor models. The atmospheric model "Weather Model" can determine air density, pressure, wind speed, and other relevant data. The ground model "Ground Model" provides support forces and moments on the ground for the fixed-wing UAV. The sensor model "Sensor Model" incorporates errors and noise into sensor measurements. These parts are not elaborated in this section and are considered known information when building the simulation model.

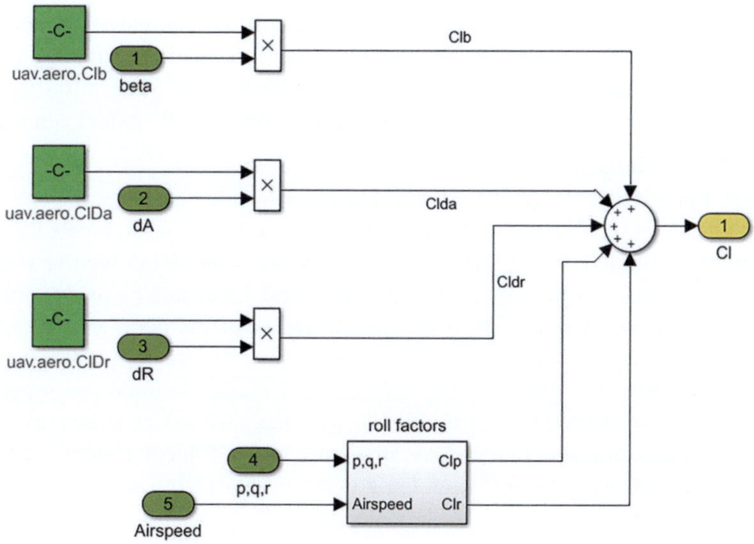

Fig. 4.39 Roll moment coefficient calculation, from "e2/e2-3/SmallFixedWingUAVnoctrl.slx"

(5) Model building

The overall connection of the model is shown in Fig. 4.40. The control module "Actuator Model" is connected to the instruction module "Command" used in the basic experiment in Sect. 4.3. After obtaining the relevant actuator's actions, the throttle command is fed into the propulsion system module "Motor Model" to calculate the current thrust. The deflection of control surfaces is fed into the fixed-wing UAV dynamics module "Small Fixed Wing UAV Dynamics", which includes the aerodynamic data calculation module "Compute Air Data" and the aerodynamic force

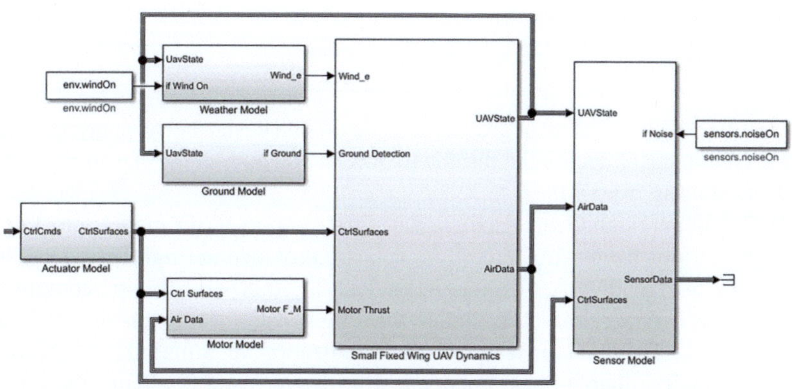

Fig. 4.40 Overall model connection, from "e2/e2-3/SmallFixedWingUAVnoctrl.slx"

and moment calculation module "Compute Body Frame Forces and Moments". The fixed-wing UAV dynamics module outputs all state information of the fixed-wing UAV based on the current forces and moments. Finally, the state information is connected to the environmental-related modules to complete the model connection.

2. Step 2: SIL simulation

(1) First, initialize the UAV state. In the initialization file "InitData.m", configure the initial state of the fixed-wing UAV, as shown in Code 4.3. Then, create a parameter structure "uav.ic", where "uav.ic.Pos_0" represents the initial relative position in xyz order, "uav.ic.Euler_0" represents the initial Euler angles in the order of roll, pitch, and yaw, "uav.ic.PQR_0" represents the initial angular rates corresponding to the Euler angles, "uav.ic.Vb_0" represents the initial velocity corresponding to the initial position, "uav.ic.gsLL" represents the current latitude and longitude, "uav.ic.gsH" represents the current altitude, with negative values indicating above sea level. Finally, different initial states are configured to achieve either "Start from ground" with stationary takeoff from the ground or "Start from air" with an initial airborne flight status.

Code 4.3 UAV parameter settings, from "e2/e2-3/InitData"

```
%% Initial Condition
% Start from ground
uav.ic = struct;
uav.ic.Pos_0 = [0; 0; 0];
uav.ic.Euler_0 = [0; 0; 0];
uav.ic.PQR_0 = [0; 0; 0];
uav.ic.Vb_0 = [0.01; 0; 0];
uav.ic.gsLL = [45 120];
uav.ic.gsH = -1;
% Start from air
% uav.ic = struct;
% uav.ic.Pos_0 = [0; 0; 50];
% uav.ic.Euler_0 = [0; 0.01906; 0];
% uav.ic.PQR_0 = [0; 0; 0];
% uav.ic.Vb_0 = [11.4; 0; 0];
% uav.ic.gsLL = [45 120];
% uav.ic.gsH = -1;
```

(2) Utilize the initial state of ground taxi takeoff and design corresponding control surface commands. Increase the throttle to 50% at 30 s and set the elevator's pitch at –0.01 rad starting from 40 s.

(3) Configure FlightGear visualization. First, import a 3D model into FlightGear. The various fixed-wing aircraft models are provided in FlightGear. Here, use UltraStick-25E for 3D visualization. Move "e2-3/UltraStick-25E.zip" compressed file to "/data/Aircraft" directory under the FlightGear installation directory and decompress it. Then, set up FlightGear visualization in Simulink, as shown in Fig. 4.41. Finally, double-click "Generate Run Script" and configure the aircraft model and FlightGear installation path as shown in Fig. 4.42.

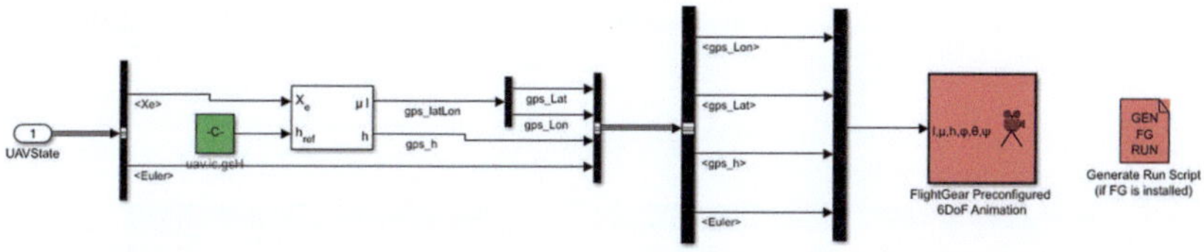

Fig. 4.41 FlightGear visualization module, from "e2/e2-3/SmallFixedWingUAVnoctrl.slx"

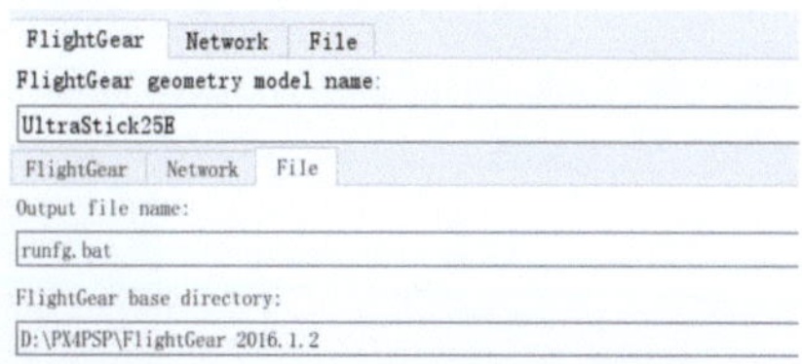

Fig. 4.42 FlightGear generation setting

Fig. 4.43 FlightGear display interface

(4) Open "e2-3/runfg.bat", wait for FlightGear initialization, and run the model "SmallFixedWingUAVnoctrl.slx". FlightGear visualization is shown in Fig. 4.43.

3. Step 3: HIL Simulation

(1) First, open "e2-4" file in MATLAB and run the initialization file "Init.m" to import parameters into the workspace. Then, open the HIL simulation model "SmallFixedWingUAVnoctrlHIL.slx". In the toolbar "APPS", find "Embedded Coder" as shown in Fig. 4.44a. Finally, click to reveal the "C++ CODE" tab, and click "Generate Code" as shown in Fig. 4.44b.

(2) Wait for the program to finish compiling. Right-click in MATLAB and run "GenerateModelDLLFile.p" in the current folder to generate the model dynamic library file "SmallFixedWingUAVnoctrlHIL.dll".

(3) Place the generated model dynamic library file "SmallFixedWingUAVnoctrlHIL.dll" into "CopterSim/external/model" directory under the RflySim installation directory.

(a) Embedded Coder

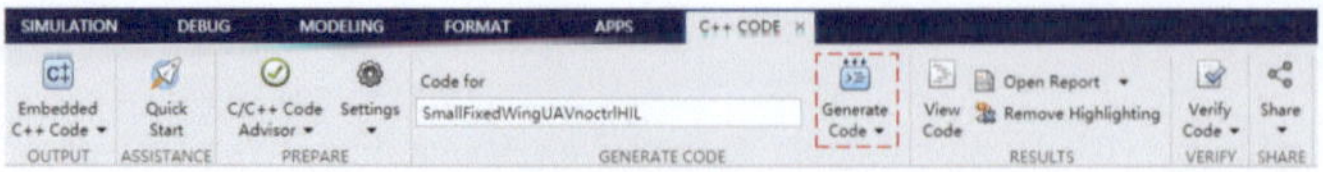

(b) Generate Code

Fig. 4.44 Code compilation and generation

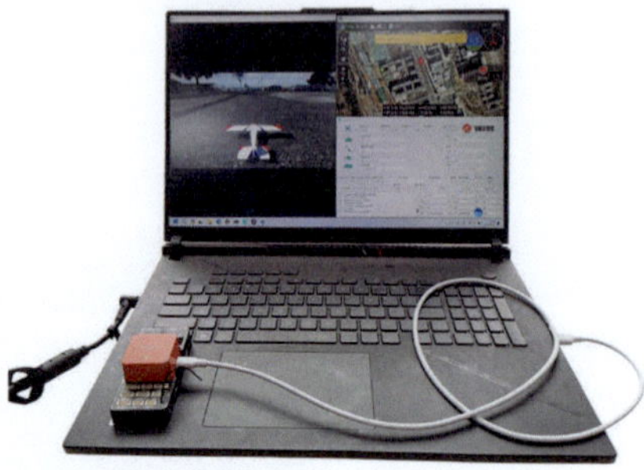

Fig. 4.45 Hardware layout of HIL simulation

(4) Connect the CubePilot/Pixhawk autopilot to the computer via USB and open the ground station, which is shown in Fig. 4.45. The ground station will automatically detect the hardware in the firmware flashing interface. A firmware configuration window will appear on the right side of the interface. Check "PX4 Pro" and click "OK". QGroundControl (QGC) will automatically download and install the latest PX4 firmware into the CubePilot/Pixhawk autopilot hardware.

(5) Open CopterSim and RflySim3D software (or LowGPU version), and in "Use DLL Model File", select "SmallFixedWingUAVnoctrlHIL.dll" as shown in Fig. 4.46.

(6) Click "Start Simulation", wait for initialization to finish, unlock the UAV through the ground station, and control the fixed-wing UAV for HIL simulation using the remote controller. The simulation performance is shown in Fig. 4.47.

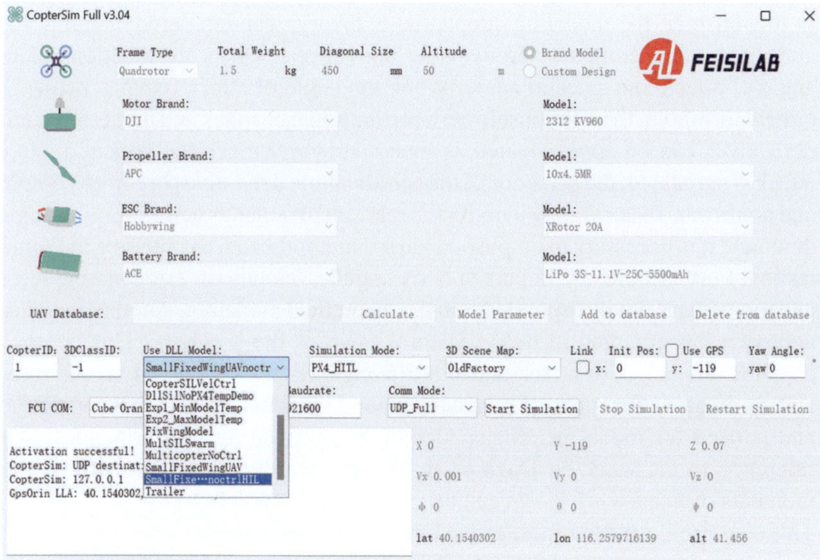

Fig. 4.46 Configuring CopterSim

Fig. 4.47 Simulated view of a fixed-wing UAV flying in the HIL simulation

4.5 Exercise

1. It is of paramount importance that accurate modeling be employed in order to ensure the reliable flight of UAVs. During flight, several faults may occur, necessitating the implementation of active fault-tolerant control measures to ensure the safe and reliable operation of the system. The foundation of active fault-tolerant control is modeling the flight after the occurrence of a fault. To illustrate, if the left aileron of a fixed-wing UAV becomes stuck at 23°, the problem can be simplified. The left aileron is assumed to be fixed at a deflection angle of $\delta_{\mathrm{a}} = 23\pi/180$ rad. Using this deflection as a disturbance, please revise the aerodynamic force and moment model established based on Eqs. (4.29) and (4.31).

2. In this chapter, the modeling adopts the π theory, wherein aerodynamic forces and moments encompass aerodynamic derivatives, which are functions of the angle of attack and sideslip angle. When the angle of attack remains below the critical threshold, the relationship between the angle of attack and the aerodynamic derivatives can be approximated as linear. However, once the critical angle of attack is surpassed, the behavior of the aerodynamic derivatives becomes complex and nonlinear. To accurately identify the relevant parameters across varying angles of attack, it is necessary to employ advanced methods such as precise wind tunnel testing. Furthermore, the impact of body angular velocity on aerodynamic forces and moments is not incorporated into the theoretical modeling. Another modeling approach has emerged in recent years, known as the *ϕ theory*. This approach abandons the angle of attack and sideslip angle, expressing aerodynamic forces and moments as functions of body and angular velocity. It has global consistency and is used for model parameter identification based on first principles. With reference to the work by Daga et al. [7], the aerodynamics and aerodynamic moments of fixed-wing UAVs can be analyzed and modeled.
3. This exercise is adapted from an example in [8]. For a single-input-single-output linear transfer function, if the zero of the numerator is in the right half-plane, the transfer function is considered a non-minimum phase system. Analyze the pitch control dynamics of a micro-small fixed-wing UAV, beginning with an analysis of the moment balance in the longitudinal channel. Considering a simplified situation where the pitch moment causes the fixed-wing UAV to change pitch, the linearized expression for the longitudinal channel can be derived from Eq. (4.49) as follows:

$$J_{yy}\ddot{\theta} = M_{\alpha}\alpha + M_{\omega_y}\omega_y + M_{\delta_e}\delta_e. \tag{4.82}$$

To analyze the non-minimum phase characteristics of the fixed-wing UAV, force analysis of the altitude channel, as shown in Fig. 4.48, is conducted. The forces are simplified to the lift affected by the angle of attack and the force affected

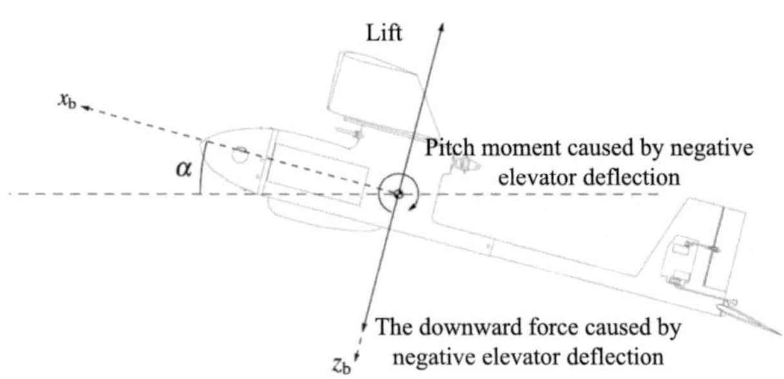

Fig. 4.48 Force analysis of altitude channel

by the deflection of the elevator (providing downward force but also a pitch-up moment to increase lift). By Newton's second law, one has

$$m\ddot{H} = Z_{\delta_e}\delta_e + Z_\alpha\alpha. \tag{4.83}$$

Considering $\dot{\theta} = \omega_y$ and treating pitch angle and attack angle as equal, Eqs. (4.82) and (4.83) are simplified as:

$$J_{yy}\ddot{\theta} = M_\alpha\theta + M_{\omega_y}\dot{\theta} + M_{\delta_e}\delta_e \tag{4.84}$$

$$m\ddot{H} = Z_{\delta_e}\delta_e + Z_\alpha\theta. \tag{4.85}$$

Setting parameters $J_{yy} = 1$, $M_\alpha = -4$, $M_{\omega_y} = -4$, $M_{\delta_e} = -3, m = 1, Z_{\delta_e} = 8$, $Z_\alpha = 12$. Try to find the transfer function from the elevator to the altitude of the fixed-wing UAV and analyze whether the altitude model controlled by the elevator for a small fixed-wing UAV is a non-minimum phase system. Given a negative pulse deflection of the elevator, plot the response of the fixed-wing UAV's altitude and observe its response characteristics. Analyze how this characteristic affects fixed-wing UAVs, and analyze the relationship between this characteristic and system stability. Furthermore, explore the relationship between the system's transfer function and its response characteristic.

4. Develop a model for the hovering UAV referenced in the Exercises of Chap. 1, and analyze how the fixed-wing UAV depicted in Fig. 4.49 executes attitude and position control. (Note: The propeller's downwash does not encompass all control surfaces; for instance, it only partially affects the aileron, leading to variations in the airspeed experienced by the affected control surfaces.)
5. The coordinated turn equation in this chapter is established under no wind. In practice, with wind, the airspeed is calculated by combining the ground speed and the wind speed. According to the reference [9], derive the coordinated turn equation through force analysis under windy conditions.

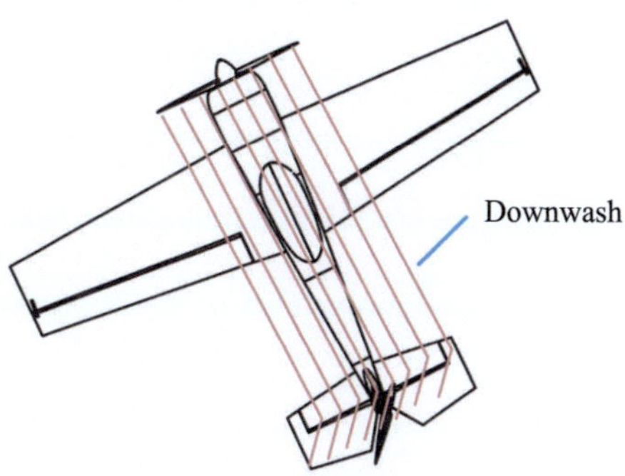

Fig. 4.49 A hovering fixed-wing UAV

References

1. R.Y. Thomas, L.M. Steven, E.B. David, F.H. Wayne, *Introduction to Aircraft Flight Mechanics: Performance, Static Stability, Dynamic Stability, and Classical Feedback Control* (American Institute of Aeronautics and Astronautics, Reston, 2014)
2. J.D. Anderson, *Fundamentals of Aerodynamics*, 6th edn. (McGraw-Hill Education, New York, 2016)
3. G. Strang, *Introduction to Linear Algebra*, 5th edn. (Springer Science Business Media, LLC, 2016)
4. Y.C. Paw, Synthesis and validation of flight control for UAV, M.S. thesis, University of Minnesota, Minneapolis, MN, 2011
5. R. Pablo, L. Mariano, K. John, Simulink drone reference application. Accessed 22 June 2023 [Online]. Available: https://github.com/mathworks/simulinkDroneReferenceApp
6. F.F. Gene, J.D. Powell, E.-N. Abbas, *Feedback Control of Dynamic Systems* (Pearson Education, Texas, 1986)
7. R.L. Leandro, D. Franois, J.-M. Moschetta, Global singularity-free aerodynamic model for algorithmic flight control of tail sitters. J. Guidance Control Dyn. **42**(2), 303–316 (2019)
8. J.-J. Slotine, W.P. Li, *Applied Nonlinear Control* (Prentice-Hall, New Jersey, 1991)
9. J. Osborne, R. Rysdyky, *Waypoint Guidance for Small UAVs in Wind, in AIAA Guidance, Navigation, and Control Conference* (Arlington, Virginia, 2005)

Chapter 5
System Identification Experiment

Runxiao Liu, Xunhua Dai, and Quan Quan

System identification is a technique used to generate mathematical models of dynamic systems based on observation data. This chapter explains how to use system identification methods to obtain guidance models and aerodynamic parameters of micro-small fixed-wing Unmanned Aerial Vehicles (UAVs) through four progressive experiments: a basic experiment, an analysis experiment, a design experiment, and a Hardware-In-the-Loop (HIL) experiment. The objective is to deepen the understanding of fixed-wing UAV identification. This chapter describes the introduction of the MATLAB System Identification Toolbox in the basic experiment for identifying guidance models using flight log data. The analysis experiment involves identifying aerodynamic derivatives and examining the impact of various input signals on the identification results. The design experiment involves conducting experiments to identify the parameters of three attitude moment channels. Finally, the guidance model and aerodynamic parameters will be identified based on HIL simulation data. For more information on this chapter, please refer to the link

https://rflysim.com/doc/en/C/5.FW_Book.html.

5.1 Preliminary

System identification involves stimulating the system with well-designed input signals and collecting the corresponding input-output signals to obtain the system's model and parameters. The high costs associated with wind tunnel experiments, which are necessary for developing micro-small fixed-wing UAVs, make them an impractical choice. Therefore, a more cost-effective and relatively accurate method is necessary. Aircraft can be flown either manually or automatically, and aerodynamic parameters can be obtained through parameter identification. Furthermore,

Q. Quan et al., *Micro-Small Fixed-Wing UAV Control*,
https://doi.org/10.1007/978-981-95-1954-5_5

it is also important to obtain the guidance model of micro-small fixed-wing UAVs for the development of swarm control algorithms and other purposes (for example, Chaps. 7, 8, 9, and 10).

5.1.1 Steps of System Identification

The steps of system identification are shown in Fig. 5.1, mainly including the following six steps: prior knowledge, experimental design, data acquisition, model selection, model computation, and model validation [1].

(1) **Prior knowledge**: This includes existing knowledge about system characteristics, data acquisition methods, and other aspects of the system to be identified. This knowledge plays a crucial role in selecting candidate models, designing experiments, determining computational methods, and evaluating criteria. Due to different identification purposes, even for the same system, the prior knowledge applied may vary significantly.
(2) **Experimental design**: The purpose is to obtain input-output data of the system and make it show the system's performance as much as possible under known conditions. In the process of collecting input/output data, special experimental designs may be required for the identification process. Users may need to determine which signals to measure and when to measure these signals.
(3) **Data acquisition**: Obtain input/output data through well-designed experiments.
(4) **Model selection**: Choose a set of candidate models and determine the most suitable model through subsequent validation. A model with unknown parameters

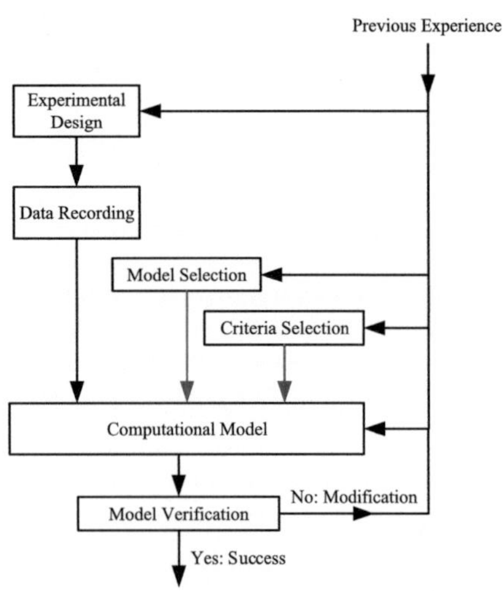

Fig. 5.1 Steps of system identification

can be obtained through mathematical modeling, and then the unknown parameters in the model can be computed through parameter identification methods.

(5) **Model computation**: Use suitable optimization methods to compute the unknown parameters of candidate models.

(6) **Model validation**: It is necessary to establish a standard to enable the verification of whether candidate models and computed parameters meet the design requirements. The determination of this standard is typically based on observed data, prior knowledge, and the intended use of the identification model. If the model and its parameters are validated, it can be considered the final identified model; otherwise, the process must be repeated until the model is validated.

5.1.2 Guidance Model Identification for Micro-Small Fixed-Wing UAVs

The complexity of the six-degree-of-freedom full-state equations for micro-small fixed-wing UAVs presents a challenge when designing swarm control algorithms for these UAVs equipped with autopilots. In such cases, it is often sufficient to consider a simplified yet controllable model that captures the system's basic characteristics, namely the guidance model. The guidance model may vary based on different assumptions and autopilot control modes [2]. Here, one commonly used guidance model is presented. Assuming the autopilot controls airspeed, altitude, and course angle, the guidance model is represented by Eq. (5.1):

$$\begin{cases} \dot{p}_{x_e} = V_a \cos\psi + v_{wx} \\ \dot{p}_{y_e} = V_a \sin\psi + v_{wy} \\ \psi = \chi + f\left(v_{wx}, v_{wy}\right) \\ V_a(s) = G_{V_a}(s)\, V_{ad}(s) \\ \chi(s) = G_\chi(s)\, \chi_d(s) \\ H(s) = G_H(s)\, H_d(s)\,. \end{cases} \tag{5.1}$$

In Eq. (5.1), p_{x_e} represents the northward position of the micro-small fixed-wing UAV in the North-East-Down coordinate system, p_{y_e} represents the eastward position of the micro-small fixed-wing UAV in the North-East-Down coordinate system, V_a is the airspeed(scalar), v_{wx} and v_{wy} are the northward and eastward wind speeds, respectively; ψ is the yaw angle of the micro-small fixed-wing UAV, and χ is the course angle. When there is no wind or the wind speed is low, v_{wx}, v_{wy}, and $f\left(v_{wx}, v_{wy}\right)$ can be ignored. H is the actual altitude of the micro-small fixed-wing UAV relative to the ground, H_d is the desired altitude, V_{ad} is the desired airspeed (scalar), and χ_d is the desired course angle.

In Eq. (5.1), the altitude and course angle channels can be considered as second-order systems, while the velocity channel can be considered as a first-order system.

For the identification of such linear systems, MATLAB System Identification Toolbox or CIFER (Comprehensive Identification From Frequency Response) Toolbox [3] can be used.

MATLAB System Identification Toolbox: The toolbox provides MATLAB functions, Simulink function blocks, and an application that constructs dynamic systems using measured input-output data. Users can obtain dynamic models of systems that are difficult to model using physical laws and identify continuous or discrete transfer functions, process models, and state-space models from time or frequency domain input-output data. Additionally, the toolbox offers various identification methods, including the maximum likelihood method, the prediction error minimization method, and the subspace identification method. For the identification of nonlinear systems, models such as the Hammerstein-Weiner model, the nonlinear ARX model with wavelet network, and other nonlinear models can be considered. The toolbox also supports "grey-box" system identification, where custom models are used for parameter estimation. The identified model can then be used to model and predict system responses.

CIFER (Comprehensive Identification From Frequency Response) Toolbox: CIFER is an integrated toolbox for system identification based on comprehensive frequency response methods. It is suitable for solving complex system identification problems, especially those related to aircraft modeling and controller design. The basic principle of the CIFER method is to extract a complete set of high-quality, non-parametric, multi-input, multi-output frequency response data. Without prior assumptions, the coupling characteristics of the system can be directly extracted from the response data. CIFER employs advanced linear chirp Z-transform and combined optimization window techniques as its theoretical foundation. In comparison to the standard Fast Fourier Transform, there is a notable enhancement in the quality of the frequency response. With a known complete frequency response data set, various complex nonlinear search algorithms can be utilized to obtain state-space models. Consequently, this toolbox is extensively utilized in the system identification of diverse real aircraft.

5.1.3 Aerodynamic Parameter Identification for Micro-Small Fixed-Wing UAVs

The dynamic equations of micro-small fixed-wing UAVs encompass numerous aerodynamic parameters that play a crucial role in modeling and controller design for these vehicles. Currently, standard methods for acquiring aerodynamic parameters include wind tunnel experiments, Computational Fluid Dynamics (CFD) simulations, and parameter identification.

1. Dynamic Equations of Micro-small Fixed-wing UAVs

Summarizing the dynamic equations of micro-small fixed-wing UAVs mentioned in Chap. 4 of this book, the following force and moment equations are derived:

$$\begin{aligned} f_{\mathrm{ax}} &= \frac{1}{2}\rho V_{\mathrm{a}}^2 S\left(C_{D_0} + C_{D_\alpha}\alpha + C_{D_{\dot\alpha}}\frac{c_{\mathrm{A}}}{2V_{\mathrm{a}}}\dot\alpha + C_{D_q}\frac{c_{\mathrm{A}}}{2V_{\mathrm{a}}}\omega_{y_{\mathrm{b}}} + C_{D_{\delta_{\mathrm{e}}}}\delta_{\mathrm{e}}\right) \\ f_{\mathrm{ay}} &= \frac{1}{2}\rho V_{\mathrm{a}}^2 S\left(C_{Y_0} + C_{Y_\beta}\beta + C_{Y_p}\frac{b}{2V_{\mathrm{a}}}\omega_{x_{\mathrm{b}}} + C_{Y_r}\frac{b}{2V_{\mathrm{a}}}\omega_{z_{\mathrm{b}}} + C_{Y\delta_{\mathrm{a}}}\delta_{\mathrm{a}} + C_{Y\delta_{\mathrm{r}}}\delta_{\mathrm{r}}\right) \\ f_{\mathrm{az}} &= \frac{1}{2}\rho V_{\mathrm{a}}^2 S\left(C_{L_0} + C_{L_\alpha}\alpha + C_{L_{\dot\alpha}}\frac{c_{\mathrm{A}}}{2V_{\mathrm{a}}}\dot\alpha + C_{L_q}\frac{c_{\mathrm{A}}}{2V_{\mathrm{a}}}\omega_{y_{\mathrm{b}}} + C_{L_{\delta_{\mathrm{e}}}}\delta_{\mathrm{e}}\right). \end{aligned} \tag{5.2}$$

$$\begin{aligned} m_{\mathrm{ax}} &= \frac{1}{2}\rho V_{\mathrm{a}}^2 Sb\left(C_{l_0} + C_{l_\beta}\beta + C_{l_p}\frac{b}{2V_{\mathrm{a}}}\omega_{x_{\mathrm{b}}} + C_{l_r}\frac{b}{2V_{\mathrm{a}}}\omega_{z_{\mathrm{b}}} + C_{l\delta_{\mathrm{a}}}\delta_{\mathrm{a}} + C_{l\delta_{\mathrm{r}}}\delta_{\mathrm{r}}\right) \\ m_{\mathrm{ay}} &= \frac{1}{2}\rho V_{a}^2 Sc_{\mathrm{A}}\left(C_{m_0} + C_{m_\alpha}\alpha + C_{m_{\dot\alpha}}\frac{c_{\mathrm{A}}}{2V_{\mathrm{a}}}\dot\alpha + C_{m_q}\frac{c_{\mathrm{A}}}{2V_{\mathrm{a}}}\omega_{y_{\mathrm{b}}} + C_{m_{\delta_{\mathrm{e}}}}\delta_{\mathrm{e}}\right) \\ m_{\mathrm{az}} &= \frac{1}{2}\rho V_{\mathrm{a}}^2 Sb\left(C_{n_0} + C_{n_\beta}\beta + C_{n_p}\frac{b}{2V_{\mathrm{a}}}\omega_{x_{\mathrm{b}}} + C_{n_r}\frac{b}{2V_{\mathrm{a}}}\omega_{z_{\mathrm{b}}} + C_{n\delta_{\mathrm{a}}}\delta_{\mathrm{a}} + C_{n\delta_{\mathrm{r}}}\delta_{\mathrm{r}}\right). \end{aligned} \tag{5.3}$$

It can be observed that the equations contain numerous aerodynamic parameters, which are the focus of identification in this chapter.

2. Wind Tunnel Experiments for Aerodynamic Parameter Identification

A wind tunnel is a primary facility for conducting aerodynamic experiments, and almost all aerodynamic experiments take place in various types of wind tunnels. The principle of a wind tunnel involves using a power device to drive a controllable airflow through a specially designed duct, passing over a stationary model in the test section to simulate the motion of the object in still air, as shown in Fig. 5.2. Measurements of the aerodynamic forces on the model and observations of the flow phenomena around and on the model's surface are conducted. Furthermore, the icing phenomenon can be replicated in a wind tunnel, as illustrated in Fig. 5.3.

Wind tunnel experiments offer several advantages for obtaining aerodynamic parameters. First, they permit the accurate control of experimental conditions, such as airflow velocity, pressure, and temperature. Secondly, indoor experiments are less affected by weather conditions and time constraints, making the installation, operation, and use of models and testing instruments more convenient. Thirdly, they permit

Fig. 5.2 The inside of a wind tunnel

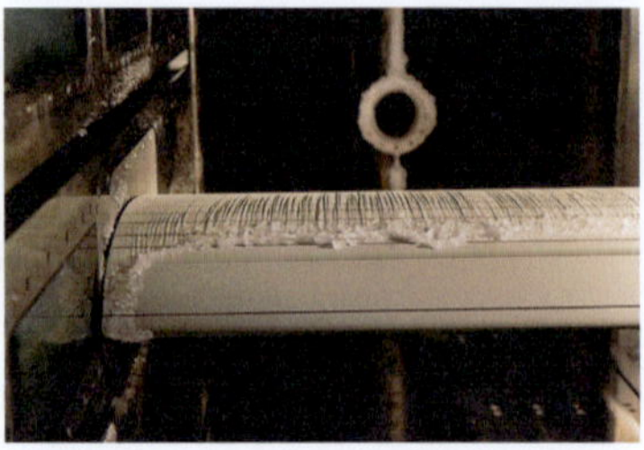

Fig. 5.3 Icing captured in wind tunnel tests

the execution of diverse experimental projects and the generation of high-precision results. Fourthly, they are relatively safe and efficient.

However, wind tunnel experiments also have certain drawbacks:

(1) In real flight, the stationary atmosphere is unbounded, while in a wind tunnel, the airflow has boundaries, leading to boundary effects that affect experimental accuracy.
(2) Support structures are required to hold the model in place during wind tunnel experiments, which can introduce disturbances to the model's flow field and potentially cause experimental errors.
(3) Wind tunnel experiments are expensive, especially for the development of micro-small fixed-wing UAVs.

3. CFD for Aerodynamic Parameter Identification

CFD is an interdisciplinary field that draws upon the principles of applied mathematics, fluid dynamics, and computing technology [4]. The deployment of CFD software for identifying aerodynamic parameters offers several distinctive advantages, including a relatively low cost, high efficiency of execution when steady-state conditions are considered, a comprehensive data set, and the capacity to simulate a wide range of operational conditions. The most commonly utilized software for aerodynamic calculations is Ansys Fluent® [5] and XFLR5 [6]. Fluent®, a widely utilized commercial CFD software, has extensive applications in fluid dynamics, heat transfer, and chemical reactions. It is often employed to obtain the aerodynamic parameters of aircraft. XFLR5, developed by the Massachusetts Institute of Technology (MIT), is an open-source software that integrates the XFOIL [7] solver for assessing the aerodynamics and calculating the aerodynamic derivatives for micro- and small fixed-wing aircraft at lower Reynolds numbers.

To maintain computational accuracy, the pertinent software typically employs finite-volume methodologies based on multi-resolution, unstructured meshes [8]. Due to the considerable computational burden involved, the flow equations are typically solved to obtain a steady-state solution. Figure 5.4 provides an example of a CFD software program that employs the finite-volume method on an unstructured mesh. However, complex simulations often necessitate the consideration of unsteady flow fields, such as the aerodynamic interaction between the rotor and fuselage, which represents a significant challenge in the application of Vertical Take-Off and Landing (VTOL) aircraft. The existing software solutions are inadequate to meet these

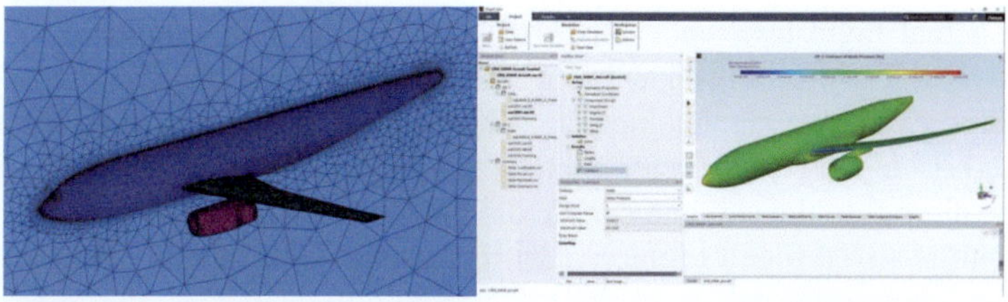

Fig. 5.4 Meshing and aerodynamic simulation from traditional CFD software

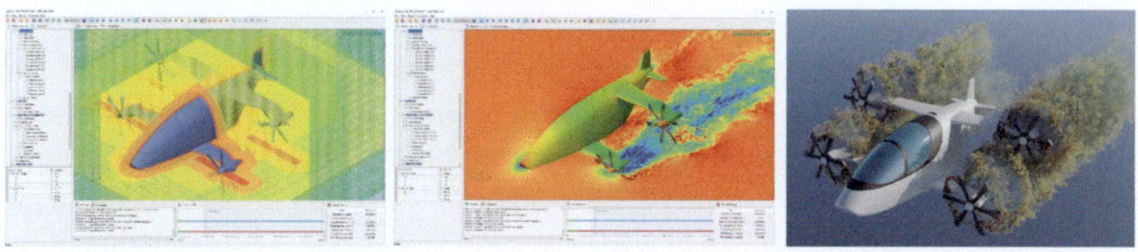

Fig. 5.5 Rapid aerodynamic simulation and visualization of composite wing vehicles with Aerocae System®

demands; however, the low-dissipation lattice Boltzmann solver [9] offers an alternative. This solver is suitable for low-speed, weakly compressible aerodynamics and aeroacoustics simulations and can be applied to the design of micro and small aircraft with complex structures. The most commonly used software systems are Dassault PowerFLOW® [10] and XFlow® [11]. While the aforementioned software is designed to operate on large-scale CPU clusters, the Aerocae System®, developed by a Chinese domestic team, is capable of running directly on GPUs. This capacity endows the software with the flexibility to perform complex analyses in small workstations with significantly enhanced performance. Figure 5.5 illustrates an example of simulation results from the Aerocae System®.

4. Obtaining Aerodynamic Parameters Based on Parameter Identification
When operating a fixed-wing UAV under closed-loop control of PX4 open-source flight control, it is recommended to use "Offboard Mode". This mode enables an external computer to directly control the aircraft's attitude and thrust through a communication interface. The specific flight modes available include "Manual Mode", "Altitude Mode", "Position Mode", "Auto Mode", and "Return Mode". In "Offboard Mode", the aircraft is controlled entirely by an external computer, without any input from the remote controller. This mode allows for advanced autonomous flight and mission execution, including path planning, target tracking, and autonomous landing. "Offboard mode" enables users to implement their own algorithms and control logic to perform more complex tasks and applications. To use "Offboard mode", communication with the autopilot is required via MAVLink to send control commands and receive status information. This mode enables control of the UAV's waypoints, speeds, or attitudes, facilitating the UAV's ability to remain in a relatively stable state, which is beneficial for aerodynamic parameter identification. It is recommended to

conduct actual flight tests under no wind or very low wind conditions (less than Beaufort scale 2). Installing an airspeed sensor can improve identification accuracy.

Standard excitation signals include step input, pulse input, square wave input, and "3-2-1-1" input. However, for micro-small fixed-wing UAVs, the step input can cause the aircraft to deviate in a certain direction, which can decrease flight safcty. Pulse input is also difficult to implement in practical applications, and excessively fast pulse inputs often fail to excite the aircraft's flight modes due to servo rate limitations. The short-period mode of the UAV cannot be excited by the "3-2-1-1" input signal because its time span far exceeds the short-period time span of the UAV. A square wave signal, with a mathematical model provided in Eq. (5.4), would be a more appropriate input signal. This signal has a broad frequency spectrum and contains energy in the high-frequency region as well. Adjusting Δt appropriately can ensure that the signal contains sufficient energy in the frequency band of interest and overcome the usage limitations of other signals. It is a simple and effective input signal. Other excitation signals can also be used to ensure adequate excitation of the corresponding channels of the fixed-wing UAV, thereby improving their actuation and increasing identification accuracy.

$$\delta(t) = \begin{cases} 0, & \text{for } t \in [0, t_1), t \in (t_1 + 2\Delta t, +\infty), \\ a, & \text{for } t \in [t_1, t_1 + \Delta t), \\ -a, & \text{for } t \in [t_1 + \Delta t, t_1 + 2\Delta t). \end{cases} \tag{5.4}$$

Following the completion of the flight experiment and data exportation, aerodynamic parameters can be identified through data preprocessing using the least squares method and stepwise regression. This is a fundamental estimation method in the field of system identification. The least squares estimation algorithm can be briefly described mathematically as follows.

Assume a known linear relationship between the dependent variable y and independent variables $x_1, x_2, \ldots, x_n$, i.e.:

$$y = \lambda_1 x_1 + \cdots + \lambda_n x_n \tag{5.5}$$

where $\lambda_1, \ldots, \lambda_n$ are unknown coefficients. To determine these coefficients, it is usually necessary to collect m sets of observed data $(x_1^{(k)}, \ldots, x_n^{(k)}, y^{(k)})$, $k = 1, \ldots, m$. Define

$$\mathbf{A} = \begin{bmatrix} x_1^{(1)} & x_2^{(1)} & \cdots & x_n^{(1)} \\ x_1^{(2)} & x_2^{(2)} & & \\ \vdots & & \ddots & \\ x_1^{(m)} & & & x_n^{(m)} \end{bmatrix}, \quad \mathbf{b} = \begin{bmatrix} y^{(1)} \\ y^{(2)} \\ \vdots \\ y^{(m)} \end{bmatrix}, \quad \boldsymbol{\lambda} = \begin{bmatrix} \lambda_1 \\ \lambda_2 \\ \vdots \\ \lambda_n \end{bmatrix}, \quad \hat{\boldsymbol{\lambda}} = \begin{bmatrix} \hat{\lambda}_1 \\ \hat{\lambda}_2 \\ \vdots \\ \hat{\lambda}_n \end{bmatrix}, \tag{5.6}$$

where $\hat{\boldsymbol{\lambda}}$ is the optimal estimate of $\boldsymbol{\lambda}$, satisfying

$$\hat{\boldsymbol{\lambda}} = \arg\min_{\boldsymbol{\lambda}} \|\mathbf{A}\boldsymbol{\lambda} - \mathbf{b}\|^2 . \tag{5.7}$$

When $\mathbf{A}^{\mathrm{T}}\mathbf{A}$ is nonsingular, the optimal estimate of $\boldsymbol{\lambda}$ is

$$\hat{\boldsymbol{\lambda}} = (\mathbf{A}^{\mathrm{T}}\mathbf{A})^{-1}\mathbf{A}^{\mathrm{T}}\mathbf{b}. \tag{5.8}$$

Based on the least squares method, stepwise regression is a regression analysis method more suitable for parameter identification because it not only identifies parameters corresponding to variables but also determines whether to introduce a variable based on the significance of its partial regression square sum test result. The mathematical concept of stepwise regression involves introducing variables into the regression equation one at a time. When introducing a variable, judge whether the test result of its partial regression sum of squares is significant. If significant, accept it; otherwise, exclude it. Moreover, each time a new variable is introduced, the previously introduced variables must be tested, and the nonsignificant ones are removed. After traversing all variables, the optimal variable subset can be obtained, where each variable is significant. Specifically, suppose there are n independent variables $x_j (j = 1, 2, \ldots, n)$ and m observed points for the dependent variable y. The estimated value of y is:

$$\hat{y} = \lambda_0 + \lambda_1 x_1 + \lambda_2 x_2 + \cdots + \lambda_l x_l, \tag{5.9}$$

where $l < n$, $x_i (i = 1, 2, \ldots, l)$ are factors selected from n independent variables based on statistical tests.

In MATLAB, there are corresponding functions for stepwise regression. This book uses the "regress" function and the "stepwise" function for data processing in parameter identification. Besides providing identified parameters, the "regress" function and "stepwise" function also provide parameters for evaluating whether the model parameters are accurate—regression goodness (coefficient of determination), denoted as R^2.

Given a series of true values y_i and corresponding predicted values $\hat{y}_i$, R^2 is defined as

$$R^2 = 1 - \frac{\sum_i \left(\hat{y}_i - y_i\right)^2}{\sum_i \left(\hat{y}_i - \bar{y}\right)^2}, \tag{5.10}$$

where $\bar{y}$ is the average value of all true values y_i:

$$\bar{y} = \frac{\sum_i y_i}{m}. \tag{5.11}$$

The meaning of the regression goodness R^2 is the proportion of the variance of the variable explained by the predicted value y_i, measuring the degree of fit of the predicted value to the actual value. In simple terms, assuming the variance of y_i is one unit, R^2 represents "how much the variance of the residuals of y_i has been reduced" after using the model. For example, if R^2 equals 0.8, the variance of the residuals after using the model is 20% of the variance of the original y_i values.

The criteria for R^2 are basically as follows:

- $R^2 = 1$: Ideal situation, all predicted values are equal to true values;
- $R^2 = 0$: One possibility is "simple prediction that all y values are equal to the average of y", i.e., all $\hat{y}_i$ are equal to $\bar{y}$ (the average of the true y values), but other possibilities exist;
- $R^2 < 0$: Poor predictive ability of the model, the effect of "simple prediction that all y values are equal to the average of y" is poor. This suggests that the incorrect model may have been used or that the model's assumptions are unreasonable.

The minimum value of the indicator R^2 has no lower limit because predictions can be arbitrarily bad. Therefore, the range of R^2 is $(-\infty, 1]$. Note: R^2 is not the square of some number so that it can be negative.

5.2 Basic Experiment

5.2.1 Experimental Objective

1. Things to prepare

(1) Software: MATLAB R2022b or above.
(2) Program: Experimental package "e3/e3-1" includes a micro-small fixed-wing UAV guidance model system identification example, including speed channel "SpeedChannel", course angle channel "HeadingChannel", and altitude channel "HeightChannel". For example, in the altitude channel, "log_trajectory_setpoint_0.csv" and "log_vehicle_local _position_0.csv", respectively, record the altitude command signal and actual signal in the flight log. "HeightChannel_Input.mat" and "HeightChannel_Output.mat" are the exported data from the log files, "HeightChannel_DataPreprocess.mlx" is the data preprocessing script, and "HeightChannel_Identification.sid" stores the identification results.

2. Objective

(1) Understand the guidance model of the micro-small fixed-wing UAV, master the data time registration preprocessing method (including start-stop time alignment and uniform sampling frequency), and learn how to use the MATLAB system identification toolbox.

(2) Use the collected input-output data from the speed, course angle, and altitude channels, and complete the identification of the micro-small fixed-wing guidance model (5.1) using the MATLAB system identification toolbox.

5.2.2 *Experimental Procedure*

The experimental procedure includes the following steps. First, in Step 1, the micro-small fixed-wing flight log files are imported and preprocessed. Then, in Step 2, system identification is performed using the MATLAB System Identification Toolbox. Finally, in Step 3, the identification results are validated.

Step 1. Importing the Fixed-Wing Flight Log Files and Preprocess
The experimental procedures are illustrated using altitude channel identification as an example. The purpose of the altitude channel guidance model identification is to identify $G_H(s)$ in Eq. (5.1). The built-in "Data Import" function of MATLAB can conveniently import the "log_trajectory_setpoint_0.csv" file into the MATLAB workspace and save it as the "HeightChannel_Input.mat" file. It should be noted that in the data import interface, as shown in Fig. 5.6, select "Numeric matrix" as "Output type", set "Exclude rows with the following items" for "Unable to import cells", then select "timestamp" representing the timestamp and "z" column representing the altitude setpoint (input), and click "Import Selected Content" to import the data into the workspace, then appropriately name and save it as ".mat" format. The processing of the "log_vehicle_local_position_0.csv" file is similar to the above process.

Following the above steps, export "HeightChannel_Input.mat" and "HeightChannel_Output.mat" from "log_trajectory_setpoint_0.csv" and "log_vehicle_local_position_0.csv" files, respectively, corresponding to $H_{\mathrm{d}}(s)$ and $H(s)$ in Eq. (5.1). Next, input-output data processing is necessary for time alignment. The time alignment here needs to address two problems: one is aligning the start and end times, and the other is aligning the sampling frequency. To align the start and end times, examine the input-output data and select the appropriate data segment for identification. For example, in identifying the altitude channel, the phase of the altitude

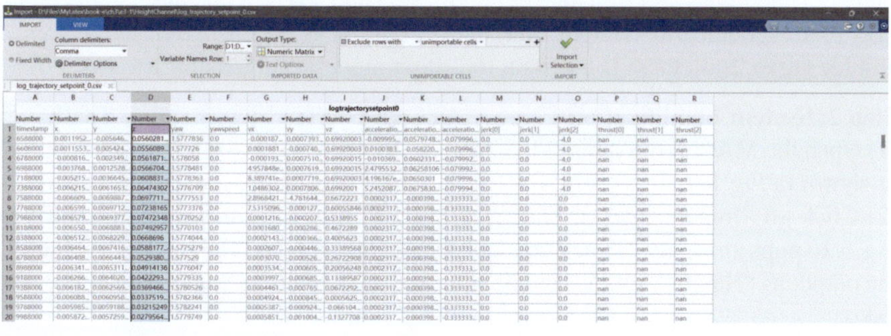

Fig. 5.6 UAV flight log files

channel step change should be selected after the UAV completes takeoff and enters waypoint flight mode, according to the preset waypoints. This determines the start and end times and then truncates the corresponding data from the input-output data. For the alignment of sampling frequency, since the timestamp of the flight control log is in microseconds, after observing the timestamp of the input-output data, it can be known that the sampling frequency of the input data 5 Hz (the sampling frequency may be less 5 Hz in the initialization of the flight control hardware, but will stabilize 5 Hz after initialization), and the sampling frequency of the output data 10 Hz. Therefore, it is necessary to align the timestamps of the input-output data by resampling, i.e., increasing the sampling frequency of the input data 10 Hz. Since the input data maintains the value from the previous moment until the next update, it is only necessary to ensure that the values of the added input data sampling points are consistent with the previous moment to obtain the required "Height" (actual altitude) and "HeightDesResample" (resampled desired altitude) for identification.

The key code in "HeightChannel_DatePreprocess.mlx" is as follows in Code 5.1. Line 1 clears the workspace. Lines 2–3 import data. Lines 4–5 truncate data and convert it to matrix format. Lines 6–10 perform resampling.

Code 5.1 Preprocessing code for height channel data, from "e3-1/HeightChannel_DatePreprocess.mlx"

```
1   clear
2   HeightData1=importdata( "HeightChannel_Input.mat");
        %Import data
3   HeightData2=importdata( "HeightChannel_Output.mat");
4   HeightDes=table2array(HeightData1(320:671,2));   %Transform
        table to array
5   Height=table2array(HeightData2(800:1501,2));
6   HeightDesResample=[];                           %Resample data
7   for i=1:2:size(Height)-1
8       HeightDesResample(i,1)=HeightDes((i+1)/2,1);
9       HeightDesResample(i+1,1)=HeightDes((i+1)/2,1);
10  end
```

Using the "plot" command to plot "HeightDesResample" and "Height" variables, the input-output data curves for identification can be obtained, as shown in Fig. 5.7. From the figure, the height expectation curve "H_d" (corresponding to "HeightDesResample") undergoes a step change between –82.5 and –87.5, while the actual altitude "H" (corresponding to "Height" variable) curve tracks the change of the reference altitude curve. Still, there is a certain transition process and tracking error.

Step 2: System Identification Using MATLAB System Identification Toolbox

(1) Open the MATLAB System Identification Toolbox "APP-System Identification" as shown in Fig. 5.8.

(2) Click on "Import data" as shown in Fig. 5.9, then after the window shown in Fig. 5.10 pops up, select "Time domain data", set the input as "HeightDesResample", the output as "Height", set the sample time as 0.1 s, and click "Import" to import the data.

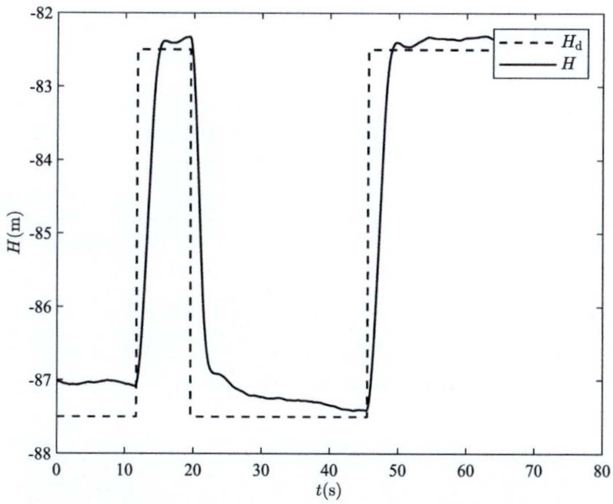

Fig. 5.7 Altitude channel response

Fig. 5.8 "System Identification" toolbox

Fig. 5.9 Toolbox interface

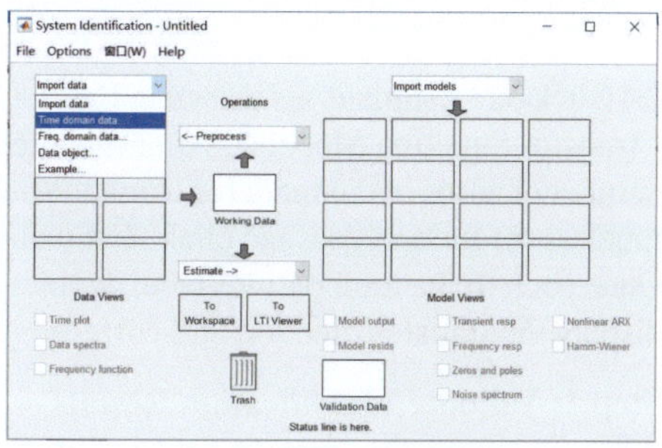

Fig. 5.10 Time domain setting interface

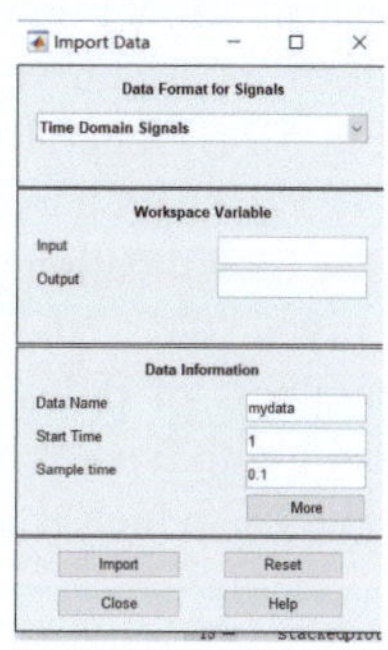

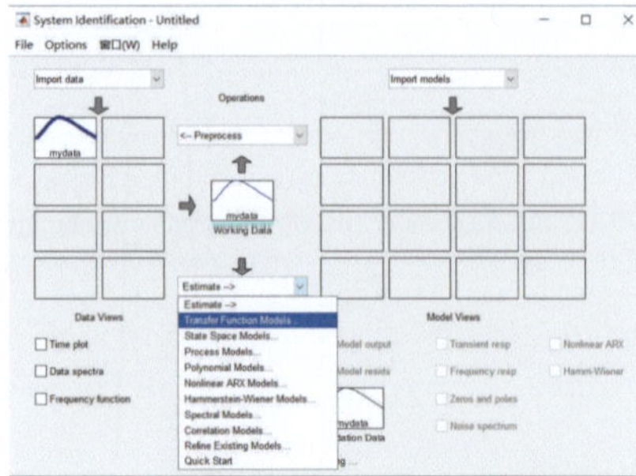

Fig. 5.11 Establishing transfer function

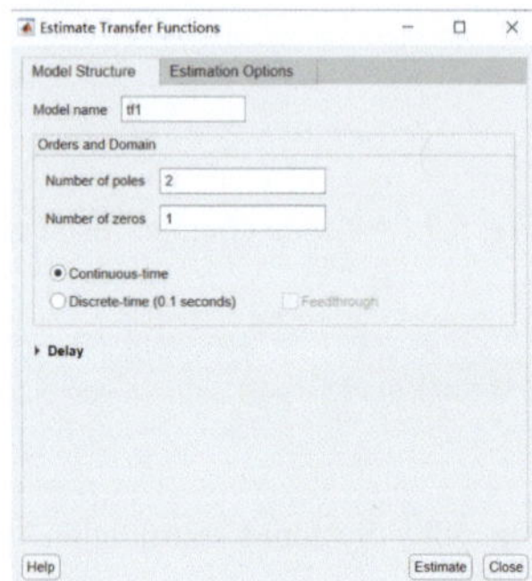

Fig. 5.12 Setting transfer function

(3) Click on "Estimate" as shown in Fig. 5.11, then select the transfer function model "Transfer Function Models". After the interface shown in Fig. 5.12 pops up, set the number of zeros of the model to 0 and the number of poles to 1, choose "Continuous-time" as the model type, and click "Estimate" to identify the transfer function model. Once back to the toolbox interface, double-click on the established system model to display the transfer function and corresponding fits,[1] as shown in Fig. 5.13.

Step 3: Validate Identification Results

After obtaining the transfer function in Step 2, select the established model and check "Model output" in the bottom right corner of the toolbox interface to obtain the response curve and agreement of the model under the input signal, as shown in Fig. 5.14. Generally, an agreement above 70% is considered good identification.

(1) Velocity Channel

The identification results for the velocity channel are shown in Fig. 5.15, where the curves "tf1", "tf2", and "tf3" represent the results obtained by selecting different numbers of poles and zeros. Comparing the results, the transfer function model built with 2 poles and 0 zeros, labeled as "tf2", has the highest agreement, reaching 83.37%, which is as expected.

[1] See MATLAB System Identification Toolbox official documentation.

Fig. 5.13 Transfer function

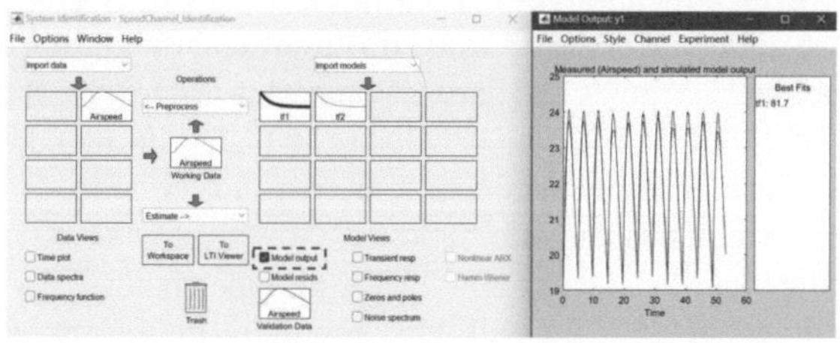

Fig. 5.14 Model output

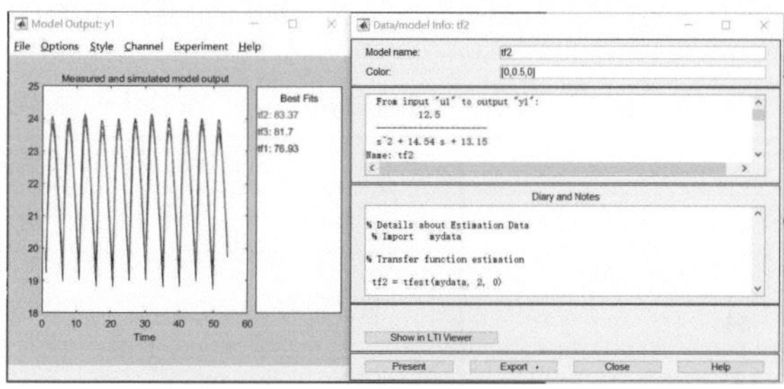

Fig. 5.15 Velocity channel identification

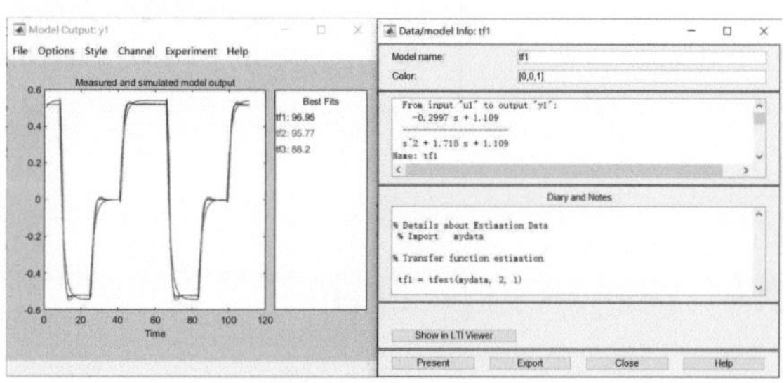

Fig. 5.16 Course angle channel identification

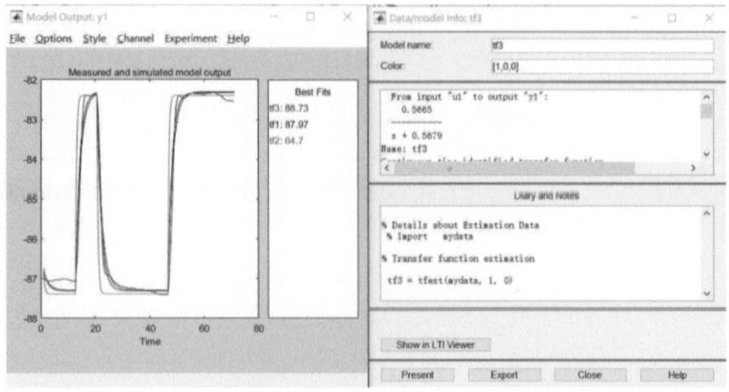

Fig. 5.17 Altitude channel identification

(2) Course angle Channel

The identification results for the course angle channel are shown in Fig. 5.16, where the curves "tf1", "tf2", and "tf3" represent the results obtained by selecting different numbers of poles and zeros. Comparing the results, the transfer function model built with 2 poles and 1 zero, labeled as "tf1", has the highest agreement, reaching 96.95%, which is as expected.

(3) Altitude Channel

The identification results for the altitude channel are shown in Fig. 5.17, where the curves "tf1", "tf2", and "tf3" represent the results obtained by selecting different numbers of V_a, poles, and zeros. Comparing the results, the transfer function model built with 1 pole and 0 zeros, labeled as "tf3", has the highest agreement, reaching 88.73%, which is as expected.

In summary, the identified guidance model is represented as:

$$\begin{cases} \dot{p}_{x_e} = V_a \cos\psi \\ \dot{p}_{y_e} = V_a \sin\psi \\ \psi = \chi \\ V_a(s) = \dfrac{12.5}{s^2 + 14.54s + 13.15} V_{ad}(s) \\ \chi(s) = \dfrac{-0.2997s + 1.109}{s^2 + 1.715s + 1.109} \chi_d(s) \\ H(s) = \dfrac{0.5665}{s + 0.5679} H_d(s). \end{cases} \tag{5.12}$$

5.3 Analysis Experiment

5.3.1 Experimental Objective

1. Things to prepare

(1) Software: MATLAB R2022b or above.

(2) Program: Experimental package "e3/e3-2", including: "model" folder contained with a micro-small fixed-wing model file "SmallFixedWingUAVupper.slx" and the initialisation file "InitDatactrl.m". Raw data required for identification and obtained by running the model "Cm_Data.mat", data processing script "Cm_DataPreprocess.mlx", data directly used for identification "Cm_DataFiltered.mat" and the script "Cm_Identification.mlx" designed for identifying the aerodynamic parameters of the pitch moment channel.

2. Objective

Taking the pitch moment channel as an example, identify the parameters $C_{m0}, C_{m\alpha}, C_{mq}, C_{m\delta_e}$ in Eq. (5.3). Understand the process and concept of aerodynamic parameter identification. Compare and analyze the identification results using dipole square waves with different amplitudes or durations, or sweep signals with different amplitudes or frequency ranges as excitation signals.

5.3.2 Experimental Procedure

The experimental procedure includes the following steps. First, in Step 1, dipole square wave signals with varying amplitudes and durations are set. Then, in Step 2, software simulation is run, required state variables are recorded, and filtering is performed. Following that, in Step 3, aerodynamic parameters are identified using the least squares method and the stepwise regression method. Finally, in Step 4, the identification results are validated.

Step 1. Setting Dipole Square Wave Signals with Different Amplitudes and Durations

In "SmallFixedWingUAVupper.slx", under "Command" path, "Signal Builder (PitchChannel)" module can be employed to inject excitation into the pitch channel. For identifying the pitch channel, dipole square wave signals of varying amplitudes and durations can be used as an excitation source. The excitation signal depicted in Fig. 5.18 is a dipole square wave with an amplitude of 20° and a duration of 4s. To prevent other channels from affecting pitch identification, input only the excitation signal to the pitch channel, keeping the other channels unchanged.

Step 2. Running Software Simulation, Recording Required State Variables, and Performing Filtering

(1) Open "model" folder, first run the initialisation file "InitDatactrl.m", then run "SmallFixedWingUAVupper.slx" program. After running the program, the data required for pitch moment channel identification can be automatically exported and saved as a raw data file "Cm_Data.mat". The data rows in "Cm_Data.mat" file are: angle of attack α ("alpha"), elevator deflection angle δ_e ("dE"), pitch rate ω_{y_b} ("wby"), airspeed V_a ("Airspeed"), pitch angular acceleration $\dot{\omega}_{y_b}$ ("wby_dot").

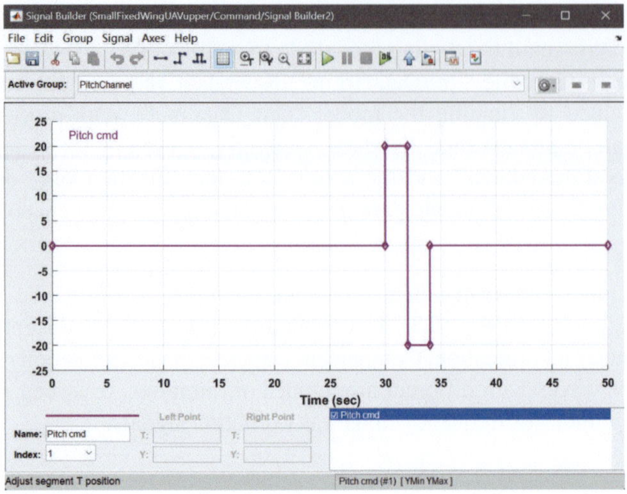

Fig. 5.18 Changing pitch moment channel excitation signal in Signal Builder

Fig. 5.19 Filter performance

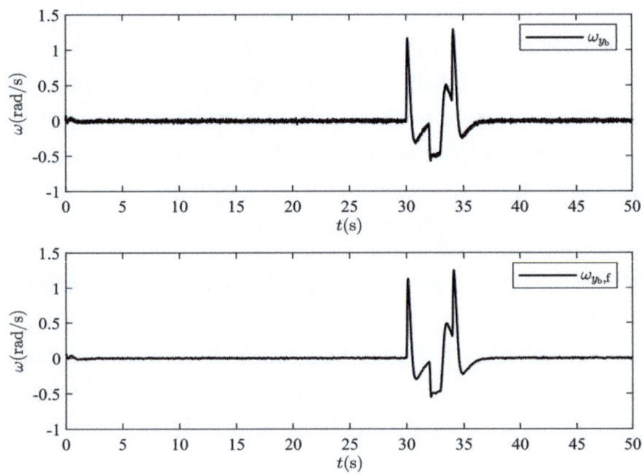

(2) Since the collected state signals contain noise, the data needs to be filtered before performing aerodynamic parameter identification. In this experiment, use the built-in MATLAB moving average filter for processing. This filter is capable of passing low frequencies and blocking high frequencies, thereby eliminating the influence of high-frequency noise. The filter parameters can be selected by observing the frequency spectrum characteristics of each signal using the MATLAB Signal Analyzer. To ensure that the time delay caused by the filter is the same for all signals and to consider the filtering effect, here, choose a moving average filter window length of 15. The effect is shown in Fig. 5.19 (using pitch rate as an example), where ω_{y_b} is the original data and $\omega_{y_b,f}$ is the filtered data. The comparison shows that the filtering performance is obvious.

The data processing code is in "SpeedChannel_DatePreprocess.mlx", as shown in the following Code 5.2. Line 1 clears the workspace, line 2 reads the data from ".mat" file and stores it as "data", lines 4–9 split "data" into elevator deflection angle (before correction), pitch rate, angle of attack, airspeed, dynamic pressure, and pitch angular acceleration data. Lines 11–17 perform moving average filtering on each state's data separately, with a window length of 15. The "movmean" function is provided by MATLAB, and its code is "M = movmean (A, k)", which returns an array composed of the mean of the local "k" data points, with each mean calculated based on a sliding window of length "k" over the adjacent elements of "A". When "k" is odd, the window is centered on the current position's element. When "k" is even, the window is centered on the current element and its previous element. When there are not enough elements to fill the window, the window will be truncated at the endpoints. When the window is truncated, the mean is calculated only based on the elements within the window.

Code 5.2 Pitch moment channel data preprocessing code, from "e3/e3-1/Cm_DatePreprocess.mlx"

```
1  clear;clc;
2  data = load("Cm_Data.mat");
3  % Split data
4  alpha = data.data(2,:);
5  dE = data.data(3,:);
6  q = data.data(4,:);
7  Airspeed = data.data(5,:);
8  alpha_dot = data.data(6,:);
9  wby_dot = data.data(7,:);
10 % Moving average filter
11 window = 15;
12 data_filtered.alpha=movmean(alpha,window);
13 data_filtered.dE=movmean(dE,window);
14 data_filtered.wby=movmean(q,window);
15 data_filtered.Airspeed=movmean(Airspeed,window);
16 data_filtered.alpha_dot=movmean(alpha_dot,window);
17 data_filtered.wby_dot=movmean(wby_dot,window);
```

Step 3: Identifying Aerodynamic Parameters Using Least Squares and Stepwise Regression

(1) Writing code to identify the aerodynamic parameters based on the micro-small fixed-wing attitude dynamics shown in Eq. (4.23) using the least squares method and stepwise regression after selecting an aircraft for steady flight. The actual pitch moment coefficient is computed, and MATLAB's built-in functions "regress" and "stepwise" are used for least squares estimation and stepwise regression, respectively, to obtain the pitch channel aerodynamic parameters.

The code is located in "Cm_Identification.mlx", as shown in Code 5.3. The first line clears the workspace; the second line reads data from the ".mat" file and stores it as "data". Lines 3–8 split "data" into data corresponding to pitch rate, pitch angular acceleration, airspeed, elevator deflection, angle of attack, and rate of angle of attack. Lines 10–14 input the aircraft's moment of inertia and geometric parameters required for this channel identification. Lines 15–24 calculate the pitch moment coefficient and correct the elevator deflection based on the rigid-body rotation equation. Line 26 generates the zero-lift moment term, line 27 merges the state variable data used for identification into a matrix, and line 28 calls MATLAB's "regress" function for identification. Here, "regress" function is a multivariate linear regression function, with code "[b, bint, r, rint, stats] = regress(y, X)", which returns: the vector "b" as the coefficient estimates of the multivariate linear regression of the response in "y" on the predictor variables in matrix "X"; "bint" as the matrix of 95% confidence intervals for the coefficient estimates; "r" as the residual vector; "rint" as the interval that can be used to diagnose outliers, where outliers are any data points that differ significantly from other observations in the set; "stats" containing the R^2 statistic and other statistical indicators (not used in this experiment). Note that to compute coefficient estimates for a model with a constant term (intercept), a column of ones should be included in the matrix "X", which corresponds to the zero-lift moment term here. Line 30 performs the identification using the "stepwise" function. The "stepwise" function is essentially the same as the "regress" function, but it provides a visualization that allows incremental addition of identified data and significance testing, and it can display various results obtained from the "regress" function. Line 31 calculates the pitch moment coefficient using the identified aerodynamic parameters.

Code 5.3 Pitch moment channel identification code, from "e3-2/Cm_Identification.mlx"

```
1  clear;
2  load('Cm_DataFiltered.mat'); %Load data
3  wby=data_filtered.wby(5500:7500)'; %Split data
4  wby_dot=data_filtered.wby_dot(5500:7500)';
5  Va=data_filtered.Airspeed(5500:7500)';
6  deltaE=data_filtered.dE(5500:7500)';
7  alpha=data_filtered.alpha(5500:7500)';
8  dalpha=data_filtered.alpha_dot(5500:7500)';
9  % Aircraft geometric parameters
10 Iy=0.144;
11 c=0.3;
12 b=1.2;
13 S=0.32;
14 rho = 1.225;
15 qbar=zeros(2001,1);
16 Cm=zeros(2001,1);
17 dalpha_after=zeros(2001,1);
18 q_after=zeros(2001,1);
19 for i=1:2001 %Calculate qbar Cm dalpha_after q_after
20     qbar(i)=0.5*rho*Va(i)^2;
21     Cm(i)=(Iy*wby_dot(i))/(qbar(i)*c*S);
22     dalpha_after(i)=dalpha(i)*c/(2*Va(i));
23     q_after(i)=wby(i)*c/(2*Va(i));
24 end
25 % Regress
26 Cm0=ones(2001,1);
27 XCm=[Cm0,alpha,dalpha_after,q_after,deltaE];
28 [bCm,bintCm,rCm,rintCm,statsCm]=regress(Cm,XCm);
```

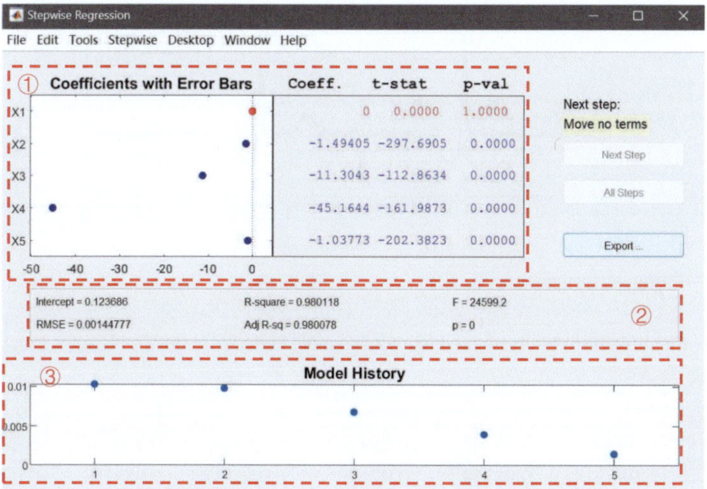

Fig. 5.20 Stepwise regression

```
% Stepwise
stepwise(XCm,Cm);
Cm_identify=XCm*bCm;
```

(2) Identification Result: Using a dipole square wave signal with an amplitude of 20° and a duration of 4s as an example to illustrate.
Figure 5.20 shows the interface of "stepwise" function, which mainly consists of three parts. The left graph of the dashed box① displays the identification results of regression coefficients with error bars. The y-axis from "X1" to "X4" represents the independent variables involved in stepwise regression, while the x-axis represents the corresponding error values. The right part corresponds to each row on the left, with three columns, where only the "Coeff." column needs to be focused on. For example, "–1.4940", "–11.3043", "–45.1644" and "–1.03773" are the parameters identified in this example. The other columns are other statistical quantities. The dashed box② is the final regression result, including "intercep"; the coefficient of determination R^2, the closer R^2 is to 1, the better the result is; "F-statistic": the larger the indicator is, the better the result of the fit is; "RMSE": refers to the root mean square error, the smaller the indicator, the better; "Adjusted R-square": This is not used here and does not need to be concerned; "p" represents the statistical p-value, which is the probability of error after accepting the fitted equation. The dashed box③ shows the change in "RMSE" as independent variables are sequentially introduced.

In this example, "X2" to "X5" represent the aerodynamic parameters C_{m_α}, $C_{m_{\dot{\alpha}}}$, C_{m_q}, $C_{m_{\delta_e}}$, while the intercept corresponds to C_{m_0}. The parameter of determination R^2 is 0.98, indicating relatively accurate regression results. In this chapter's experiment, only the identified intercept, regression coefficients, and the coefficient of determination R^2 need to be focused on. Figure 5.21 shows the fitting performance

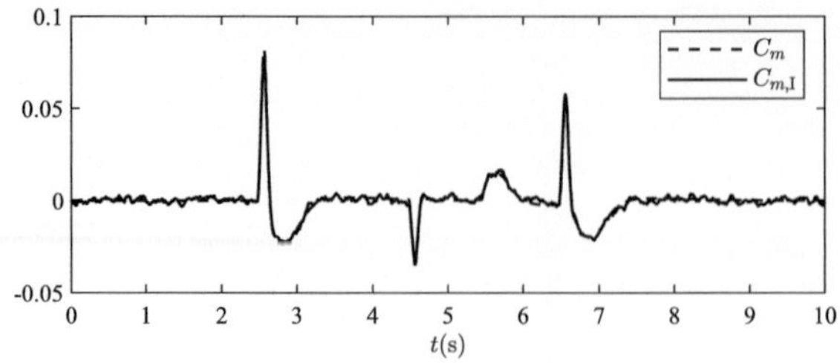

Fig. 5.21 Fitting performance

Table 5.1 Identification results of pitch moment channel under different dipole square wave signals

Amplitude	20°			5°		
Length	4s	6s	8s	4s	6s	8s
Intercept	0.28986	0.287052	0.289222	0.284535	0.278196	0.284088
R^2	0.980118	0.98354	0.97199	0.956337	0.958191	0.930923

of the aerodynamic parameter identification results, where C_m curve is the true value of the pitching moment coefficient and $C_{m,\mathrm{I}}$ curve is the pitching moment coefficient fitted according to the identified aerodynamic parameters. The comparison of the two shows that the identified aerodynamic parameters are accurate.

Step 4: Validating Identification Results and Comparing Accuracy of Different Input Signals.

(1) Identification Results: Six comparative experiments are conducted by replacing dipole square waves with different amplitudes and durations, obtaining the data shown in Table 5.1.
(2) Comparing the Identification Accuracy by Different Input Signals: By comparison, it can be observed that when the length of the dipole square wave input is 6s, R^2 is larger than that when it is 4s and 8s. With the increase of the dipole square wave amplitude, R^2 becomes closer to 1, indicating better fitting results.
(3) Analysis: First, analyze the influence of the input signal amplitude on the identification results. To ensure the stability of the UAV, a larger input signal amplitude means a larger actuation amplitude of the UAV. On the one hand, this is conducive to fully exciting the motion modes of the corresponding channels of the UAV, thereby improving the identification accuracy. On the other hand, the generated attitude angles, angular velocities, and other data will vary more, making it easier to separate them from noise data and thereby improving the accuracy of identification. Secondly, the influence of the length of the bipolar square-wave signal on the identification results is analyzed. The length of the bipolar square wave signal will affect its energy distribution on the spectrum. An appropriate frequency energy distribution can better excite the response in the corresponding frequency band of the UAV, thereby improving identification accuracy.

5.4 Design Experiment

5.4.1 Experimental Objective

1. Things to prepare

(1) Software: MATLAB R2022b or above.
(2) Program: Experimental package "e3/e3-3", including: "model" folder contained with a micro-small fixed-wing model file "SmallFixedWingUAVupper.slx" and the initialization file "InitDatactrl.m". Raw data required for identification and obtained by running the model "Cl_Data.mat", data processing script "Cm_DataPreprocess.mlx", data directly used for identification "Cl_DataFiltered.mat" and the script "Cl_Identification.mlx" designed for identifying the aerodynamic parameters of the roll moment channel.

2. Objective
In the micro-small fixed-wing UAV Simulink model, design appropriate input signals to excite the corresponding channels, collect and process data, identify the aerodynamic derivatives of the roll moment channel, i.e., C_{l_0}, C_{l_β}, C_{l_p}, C_{l_r}, $C_{l_{\delta_a}}$, $C_{l_{\delta_r}}$ in Eq. (5.3), and verify the fit of the obtained nonlinear model.

5.4.2 Experimental Design

The following steps need to be performed to identify the aerodynamic derivatives of the roll moment channel. First, in Step 1, appropriate excitation signals are designed. Then, in Step 2, the moment balance equation is analyzed to obtain the relationship between the moment and angular velocity. Afterward, in Step 3, the aerodynamic equation is analyzed to derive the relationship between the moment and aerodynamic parameters. Subsequently, in Step 4, the acquisition of aerodynamic angles is completed. Finally, in Step 5, the parameter identification method is determined(consistent with the analysis experiment).

Step 1. Designing Excitation Signals
Based on the results of the basic experiment and the analysis experiment, a square wave with a duration of 2 s and an amplitude of 10°, or a square wave with a duration of 2 s and an amplitude of 5°, is used as the excitation signal for the roll moment channel. Next, using a square wave with a duration of 2 s and an amplitude of 5° is illustrated.

Step 2. Analyzing the Moment Balance Equation
To identify the roll moment channel, it is necessary to calculate the roll moment m_{ax} in Eq. (5.3) through the moment balance equation given by:

$$m_{\mathrm{ax}} \approx J_{xx}\dot{\omega}_{x_{\mathrm{b}}}. \tag{5.13}$$

According to this equation, the roll moment m_{ax} can be calculated using the estimated roll angular acceleration $\dot{\omega}_{x_{\mathrm{b}}}$ and the known moment of inertia J_{xx}.

Step 3. Analysis of Aerodynamic Equations
According to Eq. (5.3), one has

$$m_{\mathrm{ax}} = \frac{1}{2}\rho V_{\mathrm{a}}^2 Sb(C_{l_0} + C_{l_\beta}\beta + C_{l_p}\frac{b}{2V_{\mathrm{a}}}\omega_{x_{\mathrm{b}}} + C_{l_r}\frac{b}{2V_{\mathrm{a}}}\omega_{z_{\mathrm{b}}} + C_{l\delta_{\mathrm{a}}}\delta_{\mathrm{a}} + C_{l\delta_{\mathrm{r}}}\delta_{\mathrm{r}}). \tag{5.14}$$

Once the state variables ρ, V_{a}, β, $\omega_{x_{\mathrm{b}}}$, $\omega_{z_{\mathrm{b}}}$, δ_{a}, δ_{r} and geometric parameters S, b are obtained, the required identification parameters C_{l_0}, C_{l_β}, C_{l_p}, C_{l_r}, $C_{l_{\delta_{\mathrm{a}}}}$, $C_{l_{\delta_{\mathrm{r}}}}$ can be obtained through the least squares method.

Step 4. Obtaining Aerodynamic Angles
In MATLAB simulation, it is assumed that using an aerodynamic angle sensor can measure the angle of attack α and sideslip angle β. For micro-small fixed-wing UAVs without an installed aerodynamic angle sensor, estimation is used to obtain α and β. The estimation method is given as follows.

In general, the autopilot provides quaternion and velocity in the ground coordinate system. Attitude angles of the UAV, namely(ϕ, θ, ψ), can be calculated from quaternions using Eq. (5.15) as follows:

$$\begin{aligned}
\phi &= \arctan \frac{2\left(q_0 q_1 + q_2 q_3\right)}{1 - 2\left(q_1^2 + q_2^2\right)} \\
\theta &= \arcsin\left(2\left(q_0 q_2 - q_1 q_3\right)\right). \\
\psi &= \arctan \frac{2\left(q_0 q_3 + q_1 q_2\right)}{1 - 2\left(q_2^2 + q_3^2\right)}
\end{aligned} \tag{5.15}$$

Combined with the known velocity in the ground coordinate system $\mathbf{v}_{\mathrm{g}}$, the velocity in the body coordinate system is obtained as:

$${}^{\mathrm{b}}\mathbf{v}_{\mathrm{g}} = \begin{bmatrix} v_{\mathrm{g}x_{\mathrm{b}}} \\ v_{\mathrm{g}y_{\mathrm{b}}} \\ v_{\mathrm{g}z_{\mathrm{b}}} \end{bmatrix} = \begin{bmatrix} \cos\theta\cos\psi & \cos\theta\sin\psi & -\sin\theta \\ \sin\theta\cos\psi\sin\phi - \sin\psi\cos\phi & \sin\theta\sin\psi\sin\phi + \cos\psi\cos\phi & \cos\theta\sin\phi \\ \sin\theta\cos\psi\cos\phi + \sin\psi\sin\phi & \sin\theta\sin\psi\cos\phi - \cos\psi\sin\phi & \cos\theta\cos\phi \end{bmatrix} \mathbf{v}_{\mathrm{g}}. \tag{5.16}$$

Ignoring the influence of wind $\mathbf{v}_{\mathrm{w}}$, one has ${}^{\mathrm{b}}\mathbf{v}_{\mathrm{g}} \approx {}^{\mathrm{b}}\mathbf{v}_{\mathrm{a}}$. Further, according to Eq. (4.15), the following holds

$$\begin{aligned}
\alpha &= \arctan\left(\frac{v_{\mathrm{a}z_{\mathrm{b}}}}{v_{\mathrm{a}x_{\mathrm{b}}}}\right) \\
\beta &= \arcsin\left(\frac{v_{\mathrm{a}y_{\mathrm{b}}}}{\sqrt{v_{\mathrm{a}x_{\mathrm{b}}}^2 + v_{\mathrm{a}y_{\mathrm{b}}}^2 + v_{\mathrm{a}z_{\mathrm{b}}}^2}}\right).
\end{aligned} \tag{5.17}$$

This provides the angle of attack α and sideslip angle β.

Step 5. Parameter Identification Methods

After obtaining the roll moment m_{ax} on the left side of Eq. (5.13) and the state variables ρ, V_a, β, ω_{x_b}, ω_{z_b}, δ_a, δ_r, as well as geometric parameters S and b on the right side, the parameter identification can be carried out using "regress" function and "stepwise" function in MATLAB. The former employs the least squares method, while the latter uses stepwise regression. In general, the results of parameter identification using these two methods do not differ significantly.

5.4.3 *Experimental Procedure*

The experimental procedure includes the following steps. First, in Step 1, the excitation signal is designed. Then, in Step 2, Simulink software simulation is run, the required actuation for identification is performed, and the corresponding data is recorded. Next, in Step 3, the data for each state variable is preprocessed, such as through filtering. Afterward, in Step 4, the aerodynamic parameters are identified using the least squares method and stepwise regression. Finally, in Step 5, the identification results are validated.

Step 1: Designing Excitation Signal

For the excitation signal design, it is ensured that the aircraft primarily undergoes motion in the channel being identified, thereby aiming to improve the identification accuracy. In "SmallFixedWingUAVupper.slx", under "Command" path, "Signal Builder (RollChannel)" module can be employed to inject excitation into the roll channel. To prevent other channels from affecting the roll channel identification, only the excitation signal is fed to the roll channel, keeping other channels unchanged, as shown in Fig. 5.22.

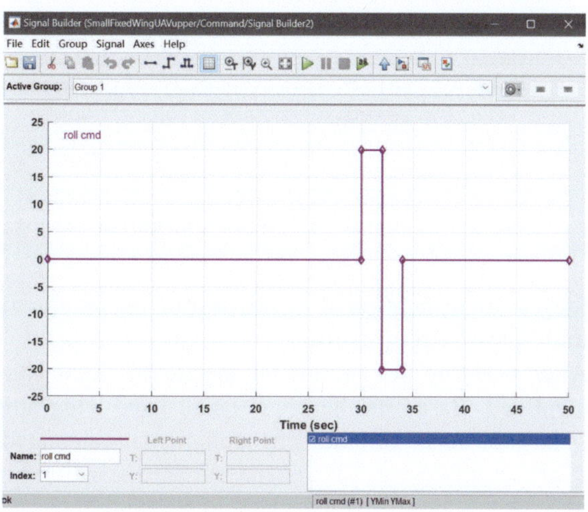

Fig. 5.22 Changing roll moment channel excitation signal in Signal Builder

Step 2: Running Simulink Software Simulation and Recording Required Actuation Data
Open the "model" folder, first run the initialization file "InitDatactrl.m", then run the "SmallFixedWingUAVupper.slx" program. After running the program, the data required for pitch moment channel identification can be automatically exported and saved as a raw data file "Cl_Data.mat". For the roll moment channel, the required data is: sideslip angle β ("beta"), aileron deflection δ_{a} ("dA"), rudder deflection,δ_{r} ("dR"), airspeed V_{a} ("Airspeed"), roll rate $\omega_{x_{\mathrm{b}}}$ ("wbx"), yaw rate $\omega_{z_{\mathrm{b}}}$ ("wbz") and roll acceleration $\dot{\omega}_{x_{\mathrm{b}}}$ ("wby_dot").
Step 3: Preprocessing of State Variable Data with Filtering
Through the analysis experiment, it has been verified that using a sliding average filter with a window length of 15 provides good filtering and noise reduction effects on the data. The data processing code is located in "Cl_DataPreprocess.mlx", as shown in the following Code 5.4. In the design experiment, the same method is applied to preprocess the data and obtain denoised data. The first two lines and lines 12–19 are consistent with the corresponding code in the analysis experiment. It should be noted that lines 4–10 split "data" into the sideslip angle, aileron deflection, rudder deflection, airspeed, roll rate, yaw rate, and roll acceleration data required for identification.

Code 5.4 Roll moment channel data preprocessing code, from "e3/e3-3/Cl_DataPreprocess.mlx"

```
1   clear;clc;
2   data = load("Cl_Data.mat");
3   % Split data
4   beta = data.data(2,:);
5   dA = data.data(3,:);
6   dR = data.data(4,:);
7   Airspeed = data.data(5,:);
8   wbx = data.data(6,:);
9   wbz = data.data(7,:);
10  wbx_dot = data.data(8,:);
11  % Average sliding filter
12  window = 15;
13  data_filtered.beta=movmean(beta,window);
14  data_filtered.dA=movmean(dA,window);
15  data_filtered.dR=movmean(dR,window);
16  data_filtered.Airspeed=movmean(Airspeed,window);
17  data_filtered.wbx=movmean(wbx,window);
18  data_filtered.wbz=movmean(wbz,window);
19  data_filtered.wbx_dot=movmean(wbx_dot,window);
```

Step 4: Identification of Aerodynamic Parameters using Least Squares Method and Stepwise Regression
After selecting the aircraft for the steady flight phase, the corresponding data is identified. Based on the micro-small fixed-wing attitude dynamics shown in Eq. (4.23), the actual roll moment coefficient is calculated. The least squares estimation

and stepwise regression are performed using MATLAB's built-in functions "regress" and "stepwise" to obtain the aerodynamic parameters of the roll channel.

The roll channel code is located in "Cl_Identification.mlx", as shown in the following Code 5.5. The first line clears the workspace; the second line reads data from the ".mat" file and stores it in the variable "data". Lines 4–10 split "data" into roll rate, roll acceleration, airspeed, aileron deflection, rudder deflection, sideslip angle, and yaw rate. Lines 12–16 input the aircraft's moment of inertia and geometric parameters required for channel identification. Lines 17–28 calculate the dynamic pressure and the roll moment coefficient based on the rigid body rotation Eq. (4.23), as well as correct the roll rate and the yaw rate. Line 30 generates the zero lift moment term, line 31 merges the state variable data used for identification into a matrix, and line 32 calls the "regress" function provided by MATLAB for identification. Line 34 performs identification using the "stepwise" function. Explanations for "regress" and "stepwise" functions were provided in the analysis experiment and are not repeated here. Line 35 calculates the roll moment coefficient using the identified aerodynamic parameters.

Code 5.5 Roll moment channel identification code, from "e3-3/Cl_Identification.mlx"

```
1   clear;
2   load('Cl_DataFiltered.mat'); %Load data
3   % Split data
4   wbx=data_filtered.wbx(5500:7500)';
5   wbx_dot=data_filtered.wbx_dot(5500:7500)';
6   Va=data_filtered.Airspeed(5500:7500)';
7   dA=data_filtered.dA(5500:7500)';
8   dR=data_filtered.dR(5500:7500)';
9   beta=data_filtered.beta(5500:7500)';
10  wbz=data_filtered.wbz(5500:7500)';
11  % Aircraft geometric parameters
12  Ix=0.0894;
13  c=0.3;
14  b=1.2;
15  S=0.32;
16  rho = 1.225;
17  qbar=zeros(2001,1); %Calculate qbar Cl wbx_compensation
        wbz_compensation
18  Cl=zeros(2001,1);
19  dalpha_after=zeros(2001,1);
20  q_after=zeros(2001,1);
21  wbx_compensation=zeros(2001,1);
22  wbz_compensation=zeros(2001,1);
23  for i=1:2001
24      qbar(i)=0.5*rho*Va(i)^2;
25      Cl(i)=(Ix*wbx_dot(i))/(qbar(i)*b*S);
26      wbx_compensation(i)=0.5*wbx(i)*b/(Va(i));
27      wbz_compensation(i)=0.5*wbz(i)*b/(Va(i));
28  end
```

```
% Regress
Cl0=ones(2001,1);
XCl=[Cl0,beta,wbx_compensation,wbz_compensation,dA,dR];
[bCl,bintCl,rCl,rintCl,statsCl]=regress(Cl,XCl);
% Stepwise
stepwise(XCl,Cl);
Cl_identify=XCl*bCl;
```

Step 5: Validation of Identification Results

The identification results of the roll moment channel are shown in Fig. 5.23: R^2 reaches 0.998, indicating a good regression.

As shown in Table 5.2, the identified aerodynamic parameters are compared with the ground truth, and the identified results are in good agreement.

As shown in Fig. 5.24, the actual roll moment derivative is compared with the fitted roll moment derivative from the identification results, and they almost completely overlap, demonstrating the high accuracy of the identification.

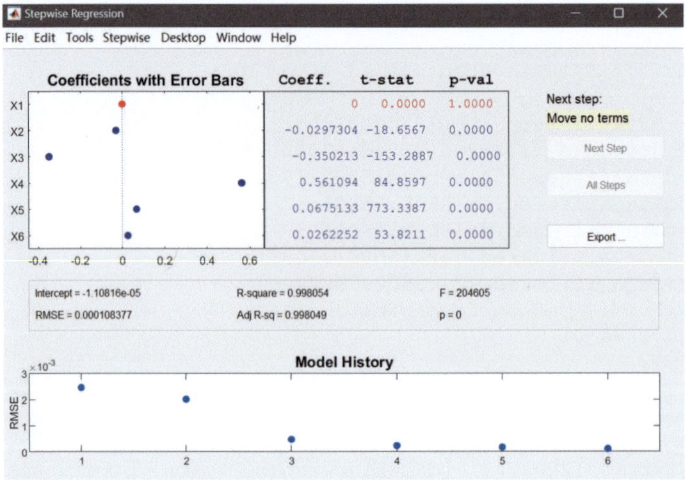

Fig. 5.23 Stepwise regression of roll moment channel

Table 5.2 Comparison of identified results and ground truth for the roll moment channel

Aerodynamic Parameter	C_{l_0}	C_{l_β}	C_{l_p}	C_{l_r}	$C_{l_{\delta_a}}$	$C_{l_{\delta_r}}$
Ground truth	0	–0.04	–0.414	0.399	0.0677	0.0168
Identified result	0.00001	–0.0297304	–0.350213	0.561094	0.0675133	0.0262252

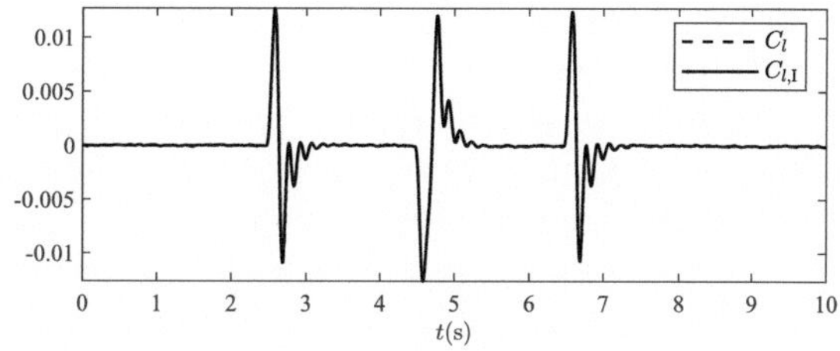

Fig. 5.24 Fitting performance of roll moment channel

5.5 HIL Experiment

5.5.1 Guidance Model Identification HIL Simulation Experiment

5.5.1.1 Experimental Objective

1. Things to prepare

(1) Software: MATLAB R2022b or above, RflySim toolchain. The RflySim toolchain provides a HIL simulation environment for small fixed-wing UAVs, which can be used to send commands to the micro-small fixed-wing UAV in the "Offboard" mode via MAVLINK through a Python terminal. The commands support two modes: "waypoint + airspeed" and "attitude + throttle".
(2) Hardware: a CubePilot/Pixhawk autopilot.
(3) Program: Experimental package "e3-4" with "GuidanceModel_Identification" experimental package. "GuidanceModel_Identification" package contains the altitude channel experiment code package "HeightChannel" folder, which includes altitude channel identification, trajectory control script "HeightChannel_OffboardController.py", "log.ulg" exported from the flight controller hardware, "log_vehicle_local_position_0.csv" and "log_vehicle_local_position_setpoint_0.csv" obtained through "pyulog" package, "HeightChannel_Input.mat" and "HeightChannel_Output.mat" exported from "csv" files, "HeightChannel_DataPreprocess.mlx" for preprocessing input-output data, and "HeightChannel_Identification.sid" for completing identification.

2. Objective

In the HIL simulation environment, perform system identification of the guidance model of a micro-small fixed-wing UAV (linear model as in Eq. (5.1)). This experiment is similar to the basic experiment, except that it requires using the CubePilot/Pixhawk autopilot to obtain the necessary data.

5.5.1.2 Experimental Procedure

The experimental procedure consists of two steps: first, designing the flight trajectory of the fixed-wing aircraft in Step 1; and then running the HIL simulation in Step 2 to complete the required actuation for identification. Subsequently, exporting flight logs from the CubePilot autopilot in Step 3 for data preprocessing, and finally, using MATLAB System Identification Toolbox to complete the guidance model identification in Step 4.

Step 1. Designing the Flight Trajectory of the Fixed-Wing Aircraft
To illustrate, consider the following example of a HIL experiment for the altitude channel. Controlling a micro-small fixed-wing UAV using a "waypoint + airspeed" command interface, the reference speed for path points can be set to 40m/s. The micro-small fixed-wing UAV maintains straight flight in the lateral direction and performs dipole square wave motion in the altitude channel. As shown in Code 5.6, the first line sets the reference speed for path points; lines 2–13 construct a loop to generate a queue of waypoints. Specifically, line 3 sets the northward waypoint; line 4 sets the eastward waypoint; lines 5–6 set the height of the 0–5th waypoints to 100 m, lines 7–8 set the height of the 6–21st waypoints to 120m, lines 9–10 set the height of the 22–41st waypoints to 80 m, lines 11–12 set the height of the remaining waypoints to 100 m, and line 13 appends the three-dimensional waypoint row vectors in each loop iteration using "append" function, resulting in a list of 22 rows and 3 columns of waypoints.

Code 5.6 Altitude channel trajectory design, from "e3-4/HeightChannel_OffboardController.py"

```
1  v = 40 # Set reference airspeed for mission points
2  for i in range(41): # Mission point set
3      x = -v*i
4      y = 0
5      if i <= 5:
6          z = -100
7      elif 5 < i <= 21:
8          z = -120
9      elif 21 < i <= 41:
10         z = -80
11     else:
12         z = -100
13     missionPoints.append([x, y, z])
```

Step 2. Running HIL Simulation to Complete Required Actuation for Identification
As shown in Fig. 5.25, connect a CubePilot/Pixhawk autopilot to the computer, open the "HeightChannel" folder in the experimental package, run the "AircraftMathworksMavlinkHITLRun.bat" file, and then a prompt window will appear. Enter the appropriate serial port number in the terminal window and press "Enter" to start the simulation platform (Figs. 5.26, 5.27, 5.28 and 5.29).

Fig. 5.25 Hardware layout of HIL simulation

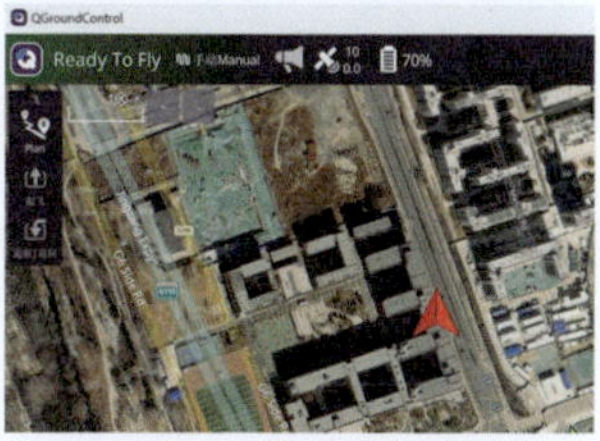

Fig. 5.26 Ready to fly

Fig. 5.27 Simulated view

Step 3: Extracting Flight Logs from Autopilot for Data Preprocessing
After the UAV has landed steadily, click on the icon at the top left corner of the QGC platform and select "Analyse Tools". In the "Log Download" interface, choose the most recent flight data for download, as shown in Fig. 5.30.

The downloaded data is in "ulog" format, which cannot be directly processed using Python or MATLAB. Therefore, the "pyulog" package is required for processing. Install the "pyulog" package in the Python environment using the "pip install" command. Then, use the "ulog2csv" command to convert the "ulog" file into a "csv" format that can be imported into MATLAB. The usage of this command is illustrated in Fig. 5.31.

After exporting the CSV file, the subsequent data preprocessing steps are similar to the basic experiment. The difference lies in observing the height input-output data, where the sampling frequencies are the same. Therefore, there is no need to adjust the sampling frequency; only the start and end times need to be adjusted, and the

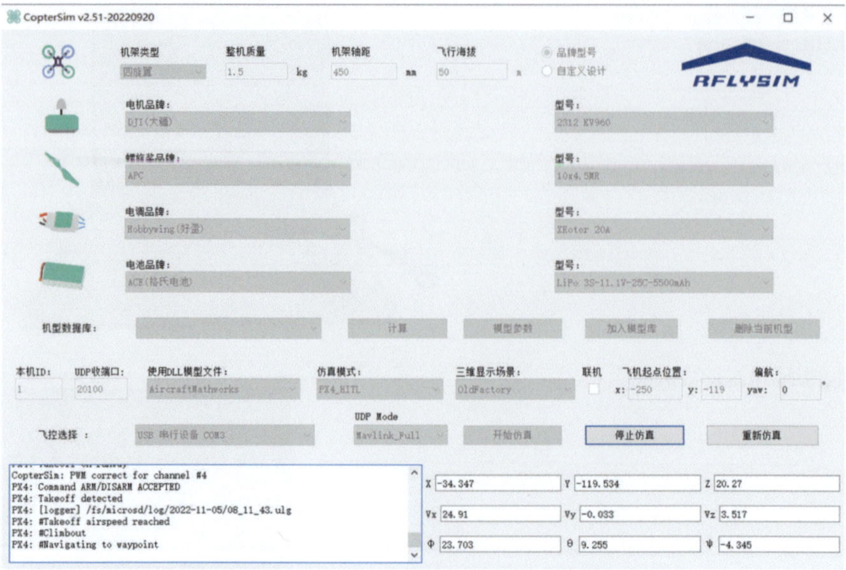

Fig. 5.28 Flight status

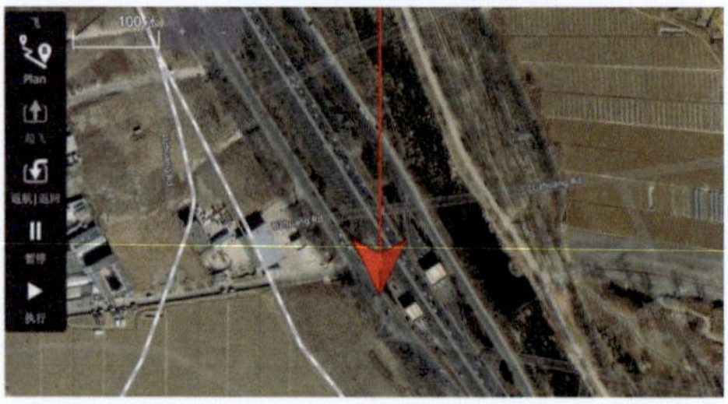

Fig. 5.29 Return to base

Fig. 5.30 Data download

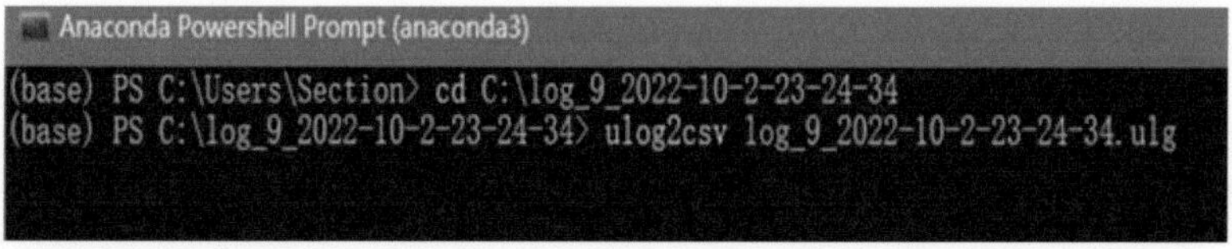

Fig. 5.31 Usage of pyulog

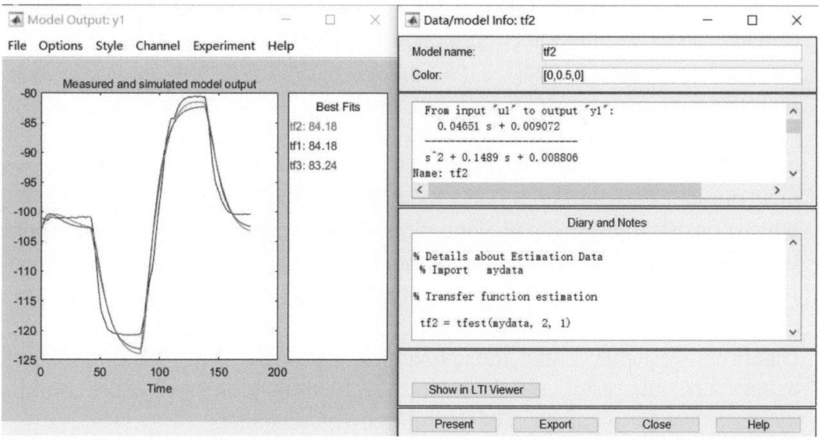

Fig. 5.32 Altitude channel identification

corresponding data segment should be trimmed. The procedure is similar to the basic experiment and will not be repeated.

Step 4: Identifying Guidance Models Using MATLAB System Identification Toolbox

Similar to the basic experiment, the MATLAB System Identification Toolbox is used to identify the guidance model for the altitude channel. The identification results, as shown in Fig. 5.32, achieve an agreement of 84.18%, meeting the expectations.

The guidance model for the altitude channel is given by:

$$G_H(s) = \frac{H(s)}{H_{\mathrm{d}}(s)} = \frac{0.4651\, s + 0.009072}{s^2 + 0.1489s + 0.008806}. \tag{5.18}$$

The identification process for other channels is analogous to that of the altitude channel, with the exception of the designed trajectory and selected input-output data. For details on the practical implementation, please refer to the code provided in the experimental package.

5.5.2 HIL Simulation for Aerodynamic Parameter Identification

5.5.2.1 Experimental Objective

1. Things to prepare

(1) Software: MATLAB R2022b or above, RflySim toolchain. The RflySim toolchain provides a HIL simulation environment for small fixed-wing UAVs.
(2) Program: Experimental package "e3/e3-4/AerodynamicParameter_Identification" experimental package, including the script for HIL experimental control of fixed-wing UAV "AircraftMathworksAttCtrl.py", fixed-wing UAV model DLL file "SmallFixedWingUAVIdentificationHIL.dll", and HIL experiment one-click startup script "UAVIdentification.bat", flight log data folder "log" which consists of flight log file "log.ulog" and corresponding ".csv" file exported from the autopilot hardware; the automation script "load_manipulate_data.m" for importing and aligning data from the CSV file and the data file "Data.mat" exported by this script, the input-output data preprocessing scripts "Cl_DataPreprocess.mlx" and the processed data file "Cm_DataFiltered.mat" exported by this script, and the script "Cl_Identification.mlx" which identifies the aerodynamic parameters of the roll moment channel.
(3) Hardware: a CubePilot/Pixhawk autopilot.

2. Objective
In the HIL simulation environment, aerodynamic parameter identification is performed for the roll moment channel of a fixed-wing UAV. Compared with the analysis and design experiments, the HIL experiment uses a CubePilot/Pixhawk autopilot to conduct experiments and obtain the required data.

5.5.2.2 Experimental Procedure

The difference between aerodynamic parameter identification in HIL simulations and analysis/design experiments lies in the use of CubePilot/Pixhawk autopilot hardware. The fixed-wing UAV HIL experiment control script "AircraftMathworksAttCtrl.py" is used to control the UAV and obtain the flight logs in the HIL experiment. The UAV state data must be derived from the flight logs or estimated using the flight log data, while the subsequent steps of filtering, identification, and validation are the same as in the analysis and design experiments.

Step 1. Designing the Fixed-Wing Aircraft Flight Trajectory
In contrast to the HIL experiments for identifying guidance models, the fixed-wing UAV is controlled via an "attitude + throttle" command interface. This allows the fixed-wing UAV to have a fixed throttle of 0.8 and no movement in the vertical direction, as well as a dipole square wave-type movement in the roll channel with an amplitude of 30° and a duration of 20 s.

The corresponding code is in "AircraftMathworksAttCtrl.py", and the key code is shown in Code 5.7. Lines 1–2 set the fixed-wing UAV's roll angle to 30° and throttle to 0.8 for 10s; lines 3–4 set the fixed-wing UAV's roll angle to –30° and throttle to 0.8 for 10s; line 5 prints a flag indicating the end of the cycle.

Code 5.7 Roll moment channel trajectory design, from "e3/e3-4/AerodynamicParameter_Identification/AircraftMathworksAttCtrl.py"

```
1    mav1.SendAttPX4([30,0,0],0.8)
2    time.sleep(10)
3    mav1.SendAttPX4([-30,0,0],0.8)
4    time.sleep(10)
5    print("arrive")
```

Step 2. Running HIL Simulation to Complete Required Actuation
Similar to the guidance model identification HIL experiment, run the HIL simulation to obtain the corresponding flight log file. First, the fixed-wing UAV model file, namely "SmallFixedWingUAVIdentificationHIL.dll", should be copied to the path "PX4PSP/CopterSim/external/ model"; subsequently, the CubePilot/Pixhawk autopilot must be connected to the computer; once this has been done, the HIL experiment one-click start script, designated as "UAVIdentification.bat", must be run and the appropriate port selected to commence the simulation. Switch to the QGC interface and run the control script "AircraftMathworksAttCtrl.py" when the UAV is in "Ready to fly" state. Note that barometer-related error messages may appear while waiting to fly and can be ignored. Under the control script, the fixed-wing UAV will perform the set attitude excitation. After completing three to five excitation actions, the script should be manually closed, at which point the UAV will enter the hovering state. Afterward, the return procedure of the QGC platform should be executed to facilitate the return of the UAV (for further details, please refer to the HIL experiment on guidance model identification).

Step 3. Exporting Flight Logs from Autopilot for Data Preprocessing
Similar to the HIL experiments for guidance model identification, download the latest flight log from QGC and store it in the folder "log". Use the "ulog2csv" command from the "pyulog" toolkit under the folder "log" to convert the log file to a CSV file.

Extract the following data from specific CSV files: "vehicle_local_position_0.csv" for the UAV's position, velocity, and acceleration in the ground coordinate system; "vehicle_attitude_0.csv" for the quaternion representation of the UAV's attitude; "vehicle_angular_velocity_0.csv" for the UAV's angular velocity; "airspeed_0.csv" for the UAV's airspeed; "vehicle_air_data_0.csv" for the dynamic pressure and air density; "actuator_controls_0_0.csv" for the UAV's actuator control signals.

The manual importation of the aforementioned data and the subsequent alignment of these datasets are processes that are both time-consuming and labor-intensive. It is therefore recommended that the "load_manipulate_data" script be employed. The script facilitates the automated importation of data. The script will automatically import the designated CSV file from the specified "log" folder and perform the necessary time alignment process. It is essential to ensure that the file name and file path are correctly configured and that the data in "log_vehicle_land_detected_0.csv" is accurate in order to utilize this automated script effectively. In particular, it is necessary to ascertain whether the first and last rows of the "landed" column in the aforementioned file contain the value of 1. If this is not the case, it is necessary to manually update the values of the first and last rows to 1 and then save the file. Subsequently, the script "load_manipulate_data.m" should be executed in order to obtain the aligned state data and the corresponding data file "Data.mat".

After time alignment processing, the known state data (velocity and attitude in the ground coordinate system) is used to estimate the required angle of attack and sideslip angle data, following the description in the experimental design for the HIL experiment. Specifically, the UAV's attitude angles (ϕ, θ, ψ) can be easily calculated from quaternions using Eq. (5.15). Combining the known ground velocity $\mathbf{v}_g$, the body coordinate system velocity can be obtained using Eq. (5.16). In the HIL experiment, ignoring the influence of wind, the angle of attack and sideslip angle can be estimated using Equation (4.15). The corresponding code is in "Cl_DataPreprocess.mlx", and the key code is shown in Code 5.8.

Code 5.8 Estimation of angle of attack and sideslip angle, from "e3/e3-4/AerodynamicParameter_Identification/Cl_DataPreprocess.mlx"

```
alpha(i) = atan(vzb(i)/vxb(i)); % Calculate alpha
beta(i) = asin(vyb(i)/sqrt(vxb(i)^2+vyb(i)^2+vzb(i)^2)); % Calculate beta
```

Step 4. Filtering Various State Variables

Similar to the analysis and design experiments, use MATLAB's "movmean" function with an appropriate window length to remove high-frequency noise (the use of hardware in the experimental process brings a reduction of noise, thereby enabling the selection of shorter window lengths). To obtain the filtered data "Cm_DataFiltered.mat", it is sufficient to run the program "Cl_DataPreprocess.mlx".

Step 5. Identification of Aerodynamic Parameters Using Least Squares and Stepwise Regression

Similar to the analysis and design experiments, the roll channel identification Code 5.9 ("Cl_Identification.mlx") is as follows. The first line clears the workspace, the second line reads data from the ".mat" file, and the third to fourth lines set the start and the end of the data to be extracted. Lines 6–12 split "data" into data for roll rate, roll acceleration, airspeed, aileron deflection, rudder deflection, sideslip angle, and yaw rate. Lines 14–18 input the aircraft's rotational inertia and geometric parameters required for the channel identification. Lines 19–28 calculate the dynamic pressure and the roll moment coefficient based on the rigid body rotation equation, and correct the roll rate and the yaw rate. Line 30 generates the zero lift moment term, line 31 combines the state variable data used for identification into a matrix, and line 32 calls the MATLAB-provided "regress" function for identification. Line 34 uses the "stepwise" function for identification, and line 35 calculates the roll moment coefficient using the identified aerodynamic parameters.

Code 5.9 Roll moment channel identification, from "e3/e3-4/AerodynamicParameter_Identification/Cl_Identification.mlx"

```
clear;
load('Cl_DataFiltered.mat');% Load data
startpoint = 7000;% Set intercept point
endpoint = 21000;
% Split data
wbx=data_filtered.wbx(startpoint:endpoint);
wbx_dot=data_filtered.wbx_dot(startpoint:endpoint);
Va=data_filtered.Airspeed(startpoint:endpoint);
dA=data_filtered.dA(startpoint:endpoint);
dR=data_filtered.dR(startpoint:endpoint);
beta=data_filtered.beta(startpoint:endpoint);
wbz=data_filtered.wbz(startpoint:endpoint);
% Aircraft geometric parameters
Ix=0.0894;
c=0.3;
b=1.2;
S=0.32;
rho = 1.15;
qbar=zeros(length(Va),1);
Cl=zeros(length(Va),1);
wbx_compensation=zeros(length(Va),1);
wbz_compensation=zeros(length(Va),1);
for i=1:length(Va)
    qbar(i)=0.5*rho*Va(i)^2;
    Cl(i)=(Ix*wbx_dot(i))/(qbar(i)*b*S);
    wbx_compensation(i)=0.5*wbx(i)*b/(Va(i));
    wbz_compensation(i)=0.5*wbz(i)*b/(Va(i));
end
% Regress
Cl0=ones(length(Va),1);
XCl=[Cl0,beta',wbx_compensation,wbz_compensation,dA',dR'];
[bCl,bintCl,rCl,rintCl,statsCl]=regress(Cl,XCl);
% Stepwise
stepwise(XCl,Cl);
Cl_identify=XCl*bCl;
```

The identification results are as shown in Fig. 5.33, with R^2 reaching 0.55, which is broadly in line with expectations. The fitting results are depicted in Fig. 5.34.

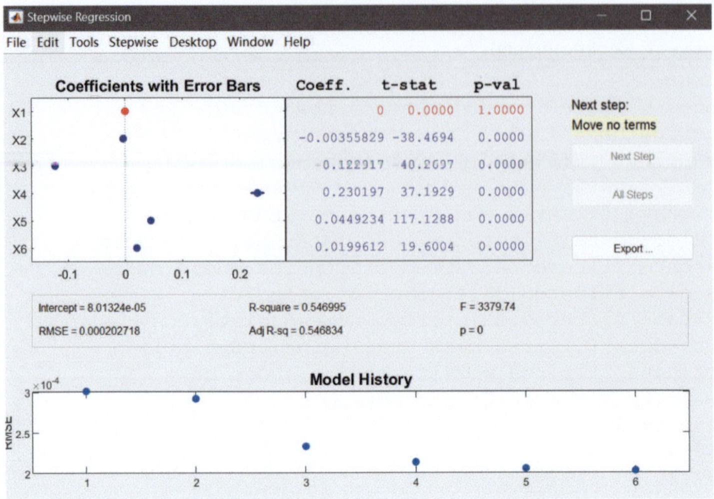

Fig. 5.33 Stepwise regression identified results of roll moment channel HIL

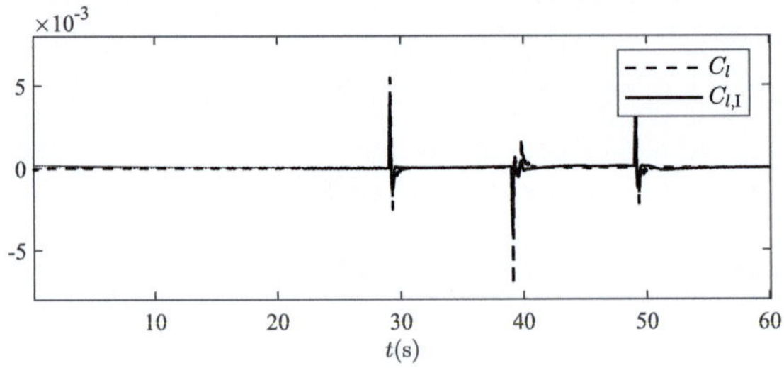

Fig. 5.34 Fitting performance of roll moment channel HIL

Table 5.3 Comparison of identified results for roll moment channel with the ground truth

Aerodynamic parameter	C_{l_0}	C_{l_β}	C_{l_p}	C_{l_r}	$C_{l_{\delta_a}}$	$C_{l_{\delta_r}}$
Ground truth	0	–0.04	–0.414	0.399	0.0677	0.0168
Identified result	0.00008	–0.0356	–0.122917	0.230197	0.0449234	0.0199612

5.5.2.3 Discussion

1. The comparison between identified results and ground truth is shown in Table 5.3. It can be observed that, except for the aerodynamic derivative C_{l_β}, the identified results are in good agreement with the ground truth. It has been verified in both analysis and design experiments that when the angle of attack and sideslip angle can be measured, the identified results are more consistent with the ground truth.

2. The error between the identification result of C_{l_β} in the HIL experiment and the truth can be attributed to the fact that sensors do not directly measure the sideslip angle. Instead, it is estimated through a two-step process. Initially, the velocity in the ground coordinate system and the attitude in the body coordinate system are used to calculate the angle of attack and sideslip angle, respectively. Subsequently, these values are used to estimate the velocity and airspeed in the body coordinate system. The two estimation steps introduce errors, resulting in significant inaccuracies in the identification of C_{l_β} in the HIL experiment. Nevertheless, as the product of the pertinent state variables and aerodynamic derivatives is of relatively small magnitude, its impact on the model is deemed insignificant within the confines of a limited maneuvering range.
3. Overall, from the perspective of R^2 and the fitting performance of the roll moment coefficient, the identified results are generally accurate and can well describe the characteristics of the roll moment channel. This also means that the transfer function with the identified parameters captures the characteristics of the roll moment channel, meeting the requirements of subsequent controller design.

5.6 Exercise

1. Referring to the analysis and design experiments in this chapter, identify the aerodynamic derivatives of the yaw moment channel for the micro-small fixed-wing UAV Simulink model and verify the degree of agreement between the identified results and the ground truth. Note that micro-small fixed-wing aircraft generally do not control yaw directly through the rudder, but use coordinated turns to control the course angle by controlling the roll. In this process, both the yaw moment channel and the roll moment channel are excited, so you can consider using the same excitation as the roll moment channel identification for the yaw moment channel identification.
2. If there are some "bad points" in the identification data, how do you deal with this situation? Please write down the relevant algorithms and verify them yourself. Hint: You can try using MATLAB's "filloutliers" function for verification.
3. If there is always a constant wind disturbance in the guidance model identification process, how do you identify it? Please provide specific identification methods.
4. If there is no stable model before the aerodynamic parameter identification and the flight can only rely on the pilot's control, how do you proceed with the model identification?
5. Researchers have utilized neural network models for establishing dynamic control models [12]. How do we collect data and establish and learn models for micro-small fixed-wing UAVs? For example, using current "actuator outputs + attitude + velocity" as model input, and the next "attitude + velocity" as model output.

References

1. L. Ljung, *System Identification* (Springer, 1998)
2. R.W. Beard, T.W. McLain, *Small Unmanned Aircraft: Theory and Practice* (Princeton University Press, 2012)
3. M.B. Tischler, R.K. Remple, *Aircraft and Rotorcraft System Identification* (American Institute of Aeronautics and Astronautics Reston, VA, 2012)
4. T. Chung, *Computational Fluid Dynamics* (Cambridge University Press, 2002)
5. A. Fluent et al., *Ansys Fluent Theory Guide*, vol. 15317 (Ansys Inc., USA, 2011), pp.724–746
6. XFLR5, Xflr5 v6.02 guidelines, Accessed: 3 Oct 2024, 2 Oct 2014 [Online]. Available: https://sourceforge.net/projects/xflr5/files/Guidelines.pdf/download01/01/2016
7. M. Drela, Xfoil: An analysis and design system for low reynolds number airfoils, in *Low Reynolds Number Aerodynamics: Proceedings of the Conference Notre Dame, Indiana, USA, 5–7 1989* (Springer, 1989), pp. 1–12
8. J. Zhu, C.-W. Shu, A new type of third-order finite volume multi-resolution WENO schemes on tetrahedral meshes. J. Comput. Phys. **406**, 109–212 (2020)
9. T. Krger, H. Kusumaatmaja, A. Kuzmin, O. Shardt, G. Silva, E.M. Viggen, *The Lattice Boltzmann Method* (Springer, 2017)
10. D. Systmes, Powerflow 2024 user's guide, Accessed 3 Oct 2024, 6 Sept 2024 [Online]. Available: https://www.3ds.com/support/documentation/powerflow-2024-product-documentation-and-installation-guides
11. D. Systmes, Xflow 2017x tutorial guide, Accessed 3 Oct 2024, 2017. [Online]. Available: https://www.scribd.com/document/485614627/XFlow-TutorialGuide-pdf
12. T.A. Tutunji, Parametric system identification using neural networks. Appl. Soft Comput. **47**, 251–261 (2016)

Chapter 6
Low-Level Flight Control Experiment

Lin Xu, Wenhan Gao, Dongyuan Zhang, Xunhua Dai, and Quan Quan

The low-level flight control of the micro-small fixed-wing Unmanned Aerial Vehicle (UAV) is the foundation for ensuring the stable flight of UAVs. This chapter provides a detailed overview of the low-level flight control logic, principles, and Proportional Integral Derivative (PID) control methods, including parameter tuning mechanisms and their application in attitude, altitude, and airspeed control of fixed-wing UAVs. Additionally, the PID control in open-source autopilots is introduced for attitude control. Total Energy Control (TEC) is introduced for altitude and airspeed control. Readers will gradually deepen their understanding of the low-level flight control of micro-small fixed-wing UAVs through a basic experiment, an analysis experiment, and a design experiment. In the basic experiment, readers will learn to build simulation models of micro-small fixed-wing UAVs and controllers, use Simulink simulation models, and observe experimental data. In the analysis experiment, readers will explore various PID parameter tuning methods in the time domain and stability analysis in the frequency domain. In the design experiment, readers will apply the frequency correction to achieve the desired performance. Finally, altitude and airspeed control experiments will be implemented in Hardware-In-the-Loop (HIL) simulation. For more information on this chapter, please refer to the link

https://rflysim.com/doc/en/C/5.FW_Book.html.

6.1 Preliminary

The low-level flight control of micro-small fixed-wing UAVs is mainly divided into two loops, which include attitude control and altitude-airspeed control. The closed-loop control framework is shown in Fig. 6.1. The fixed-wing UAV system has three input signals (three desired commands for altitude H_{d}, airspeed V_{ad}, and yaw angle ψ_{d}), six output signals (position $\mathbf{p} \in \mathbb{R}^3$ and attitude $\boldsymbol{\Theta} \in \mathbb{R}^3$), and four control

Q. Quan et al., *Micro-Small Fixed-Wing UAV Control*,
https://doi.org/10.1007/978-981-95-1954-5_6

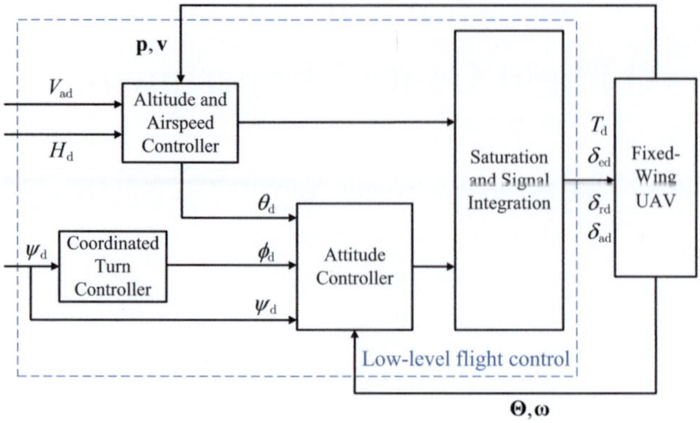

Fig. 6.1 A low-level flight control structure for fixed-wing UAVs

signals (elevator deflection δ_e, aileron deflection δ_a, rudder deflection δ_r, and throttle thrust T). Attitude control is implemented as an inner-outer loop control system, with the inner loop controlling the angular rate ω and the outer loop controlling the attitude angle $\boldsymbol{\Theta}$.

6.1.1 *Attitude Control*

The inner-outer loop control is often used in fixed-wing UAVs. The outer loop controller controls the altitude airspeed and flight path direction of the micro-small fixed-wing UAV, while the inner loop controller focuses on the attitude of the micro-small fixed-wing UAV. The outer loop controller generates commands to the inner loop controller, using the altitude channel controller's output as the reference for pitch control and the trajectory control's output as the reference for yaw. In this context, $\boldsymbol{\Theta}_d$ serves as the reference. The ultimate goal of attitude control is to achieve $\lim_{t\to\infty} ||\boldsymbol{\Theta}(t) - \boldsymbol{\Theta}_d(t)|| = 0$. The altitude and airspeed control is resolved as long as the attitude control objective is achieved effectively. This section will now outline the design of the attitude controller.

6.1.1.1 Pitch Control

1. Traditional PID Controller

The controller design for the pitch model is shown in Fig. 6.2. First, the controller is designed for the pitch and then for the pitch rate. The objective of pitch control is as follows: given a desired pitch angle θ_d, a controller is designed to ensure

$$\lim_{t\to\infty} |e_\theta(t)| = 0, e_\theta \triangleq \theta_d - \theta$$

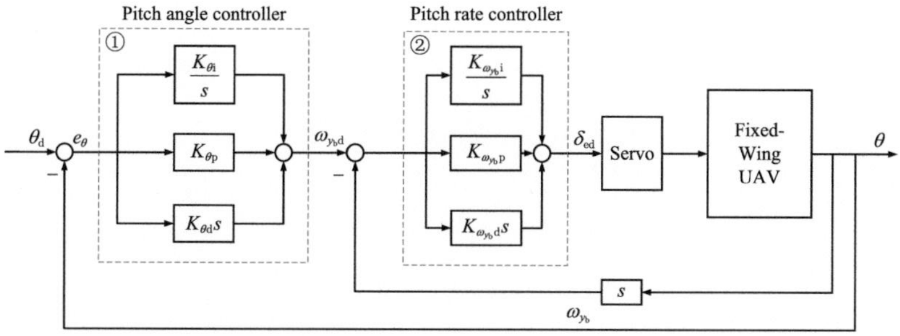

Fig. 6.2 A traditional pitch control loop

where θ_d is given by the output of the altitude control. To achieve this objective, the controller is first designed for

$$\dot{\theta} = \omega_{y_b}\cos\phi - \omega_{z_b}\sin\phi. \tag{6.1}$$

When designing the controller for the pitch channel, assuming roll angle $\phi = 0$, Eq. (6.1) is simplified to

$$\dot{\theta} = \omega_{y_b}. \tag{6.2}$$

The desired pitch angle rate $\omega_{y_b d}$ is designed as

$$\omega_{y_b d} = K_{\theta p}e_\theta + K_{\theta i}\int e_\theta + K_{\theta d}\dot{e}_\theta \tag{6.3}$$

where $K_{\theta p}$, $K_{\theta d}$, $K_{\theta i} \in \mathbb{R}$. Equations (6.2) and (6.3) form the pitch control loop, as shown in the dashed box ① in Fig. 6.2.

Assume $\dot{\theta}_d = 0$, and

$$\lim_{t\to\infty} |e_{\omega_{y_b}}(t)| = 0,\ e_{\omega_{y_b}} \triangleq \omega_{y_b d} - \omega_{y_b}.$$

Then $\lim_{t\to\infty} |e_\theta(t)| = 0$. The next objective is to design a controller for

$$J_{yy}\dot{\omega}_{y_b} = \omega_{x_b}\omega_{z_b}(J_{zz} - J_{xx}) - J_{xz}(\omega_{x_b}^2 - \omega_{z_b}^2) + m_{ay}. \tag{6.4}$$

The desired pitch moment m_{ayd} is designed as follows:

$$m_{ayd} = K'_{\omega_{y_b}p}e_{\omega_{y_b}} + K'_{\omega_{y_b}i}\int e_{\omega_{y_b}} + K'_{\omega_{y_b}d}\dot{e}_{\omega_{y_b}} \tag{6.5}$$

where $K'_{\omega_{y_b}\mathrm{p}}, K'_{\omega_{y_b}\mathrm{d}}, K'_{\omega_{y_b}\mathrm{i}} \in \mathbb{R}$. If $m_{\mathrm{a}y} = m_{\mathrm{a}y\mathrm{d}}$, then $\lim_{t\to\infty} |e_{\omega_{y_b}}(t)| = 0$. According to Eq. (4.29), the linear relationship between the pitch moment and the elevator deflection is rewritten as

$$m_{\mathrm{a}y} = K_{C_m}\delta_{\mathrm{e}} + d_{m_{\mathrm{a}y}} \tag{6.6}$$

where

$$K_{C_m} = \frac{1}{2}\rho V_{\mathrm{a}}^2 S c_{\mathrm{A}} C_{m\delta_{\mathrm{e}}}, d_{m_{\mathrm{a}y}} = \frac{1}{2}\rho V_{\mathrm{a}}^2 S c_{\mathrm{A}}(C_{m0} + C_{m\alpha}\alpha + C_{m\dot{\alpha}}\frac{c}{2V_{\mathrm{a}}}\dot{\alpha} + C_{mq}\frac{c}{2V_{\mathrm{a}}}\omega_{y_b}).$$

Therefore, the control law in (6.5) is transformed into

$$\delta_{\mathrm{ed}} = K_{\omega_{y_b}\mathrm{p}} e_{\omega_{y_b}} + K_{\omega_{y_b}\mathrm{i}} \int e_{\omega_{y_b}} + K_{\omega_{y_b}\mathrm{d}} \dot{e}_{\omega_{y_b}} \tag{6.7}$$

where $K_{\omega_{y_b}\mathrm{p}}, K_{\omega_{y_b}\mathrm{d}}, K_{\omega_{y_b}\mathrm{i}} \in \mathbb{R}$. Equations (6.4) and (6.7) form the angular rate control loop, as shown in the dashed box② in Fig. 6.2. Sometimes, the derivative terms $K_{\omega_{y_b}\mathrm{d}}\dot{e}_{\omega_{y_b}}$ and $K_{\theta\mathrm{d}}\dot{e}_{\theta}$ are removed for avoiding noise caused by differentiation. So far, the pitch controller design is completed, consisting of Eqs. (6.3) and (6.7). Its control diagram is shown in Fig. 6.2.

2. PID Controller in Autopilot

In practical situations, if $\phi \neq 0$, then $\dot{\theta} \neq \omega_{y_b}$. In this case, a more precise design is needed to make the controller more effective. Following the PX4 autopilot attitude controller design concept [1], the inner and outer loop cascade controllers are designed separately. The inner loop serves as the angular rate controller, while the outer loop serves as the angular controller. For the pitch channel, the outer loop pitch angle channel uses the pitch angle as the controlled object to obtain the desired pitch angle rate. According to the attitude motion Eq. (4.18), the desired pitch angle rate is

$$\omega_{y_b\mathrm{d}} = \cos\phi\dot{\theta}_{\mathrm{d}} + \cos\theta\sin\phi\dot{\psi}_{\mathrm{d}}. \tag{6.8}$$

In the PX4 pitch channel controller, the code for converting the desired angular rate to the body coordinate system is shown in Code 6.1. Here, "_bodyrate_setpoint" represents the desired pitch rate $\omega_{y_b\mathrm{d}}$, "ctl_data.roll" represents the actual roll angle ϕ, and "ctl_data.yaw_rate_setpoint" represents the desired yaw rate $\dot{\psi}_{\mathrm{d}}$.

Code 6.1 PX4 conversion of pitch rate to body pitch rate

```
/* Transform setpoint to body angular rates (jacobian) */
_bodyrate_setpoint = cosf(ctl_data.roll) * _rate_setpoint +
cosf(ctl_data.pitch) * sinf(ctl_data.roll) *
    ctl_data.yaw_rate_setpoint;
```

In Eq. (6.8), the desired yaw rate $\dot{\psi}_{\mathrm{d}}$ is obtained through the coordinated turn relationship. Next, the desired pitch rate $\dot{\theta}_{\mathrm{d}}$ is designed such that

$$\lim_{t\to\infty} |e_\theta(t)| = 0, e_\theta \triangleq \theta_{\mathrm{d}} - \theta.$$

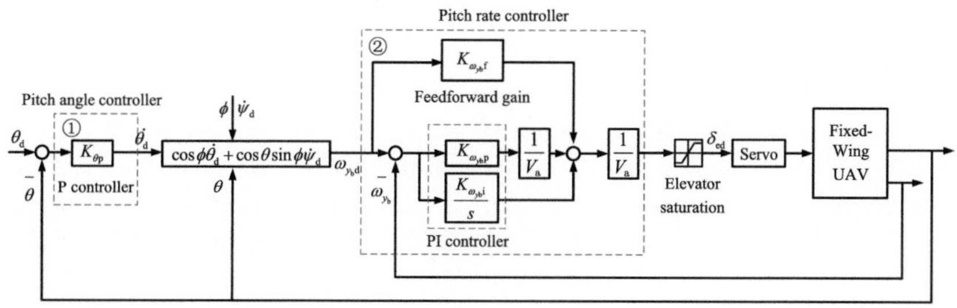

Fig. 6.3 PX4 pitch control loop in an autopilot

One straightforward design is

$$\dot{\theta}_{\mathrm{d}} = K_{\theta \mathrm{p}} e_{\theta} \tag{6.9}$$

where $K_{\theta \mathrm{p}} \in \mathbb{R}_{+}$. If $\dot{\theta} = \dot{\theta}_{\mathrm{d}}$, then

$$\dot{\theta} = K_{\theta \mathrm{p}} e_{\theta}. \tag{6.10}$$

Clearly, when θ_{d} is constant, $\lim_{t \to \infty} |e_{\theta}(t)| = 0$.

Here, the outer loop stabilizes the pitch angle, improving the damping characteristics of the UAV's phugoid mode, as depicted by the dashed box ① in Fig. 6.3. The outer loop control code for the PX4 pitch channel controller [2] is shown in Code 6.2. Here, "ctl_data.pitch_setpoint" represents the desired pitch angle θ_{d}, "ctl_data.pitch" represents the practical pitch angle θ, "pitch_error" denotes the error between the desired and actual pitch angles, and "_rate_setpoint" represents the desired pitch rate $\dot{\theta}_{\mathrm{d}}$.

Code 6.2 PX4 angle loop controller in pitch channel

```
/* Calculate the error */
float pitch_error = ctl_data.pitch_setpoint - ctl_data.pitch;
/*  Apply P controller: rate setpoint from current error and
    time constant */
_rate_setpoint = pitch_error / _tc;
```

Next, assuming that the roll rate and yaw rate are zero, i.e., $\omega_{x_{\mathrm{b}}} = \omega_{z_{\mathrm{b}}} = 0$, Eq. (4.23) is simplified to

$$\dot{\omega}_{y_{\mathrm{b}}} = \frac{\rho V_{\mathrm{a}}^{2} S c}{2 J_{yy}} \left(C_{m_0} + C_{m_{\alpha}} \alpha + C_{m_{\dot{\alpha}}} \frac{c}{2 V_{\mathrm{a}}} \dot{\alpha} + C_{m_q} \frac{c}{2 V_{\mathrm{a}}} \omega_{y_{\mathrm{b}}} + C_{m_{\delta_{\mathrm{e}}}} \delta_{\mathrm{e}} \right). \tag{6.11}$$

The inner loop targets the pitch rate as the controlled variable, aiming to design the desired elevator deflection δ_{ed} based on Equation (6.11). This is achieved using feedforward and PI control:

$$\delta_{\text{ed}} = \frac{1}{V_{\text{a}}} K_{\omega_{y_{\text{b}}}\text{f}} \omega_{y_{\text{b}}\text{d}} + \frac{1}{V_{\text{a}}^2} K_{\omega_{y_{\text{b}}}\text{p}} (\omega_{y_{\text{b}}\text{d}} - \omega_{y_{\text{b}}}) + \frac{1}{V_{\text{a}}} K_{\omega_{y_{\text{b}}}\text{i}} \int (\omega_{y_{\text{b}}\text{d}} - \omega_{y_{\text{b}}}) \tag{6.12}$$

where $K_{\omega_{y_{\text{b}}}\text{f}}, K_{\omega_{y_{\text{b}}}\text{p}}, K_{\omega_{y_{\text{b}}}\text{i}} \in \mathbb{R}$. To satisfy $\lim_{t\to\infty} |\omega_{y_{\text{b}}\text{d}}(t) - \omega_{y_{\text{b}}}(t)| = 0$, the pitch rate loop controller is depicted by the dashed box ② in Fig. 6.3. In the PX4 pitch channel controller, the inner loop control is implemented as shown in Code 6.3. Lines 2–4 capture the incremental variable $\text{d}t$ for integral action. Line 8 computes the pitch rate error $\omega_{y_{\text{b}}\text{d}} - \omega_{y_{\text{b}}}$. Lines 9-17 are used to obtain $\int(\omega_{y_{\text{b}}\text{d}} - \omega_{y_{\text{b}}})$. Lines 19 and 21 complete the pitch channel controller.

Code 6.3 Pitch channel rate loop controller

```
1  /* get the usual dt estimate */
2  uint64_t dt_micros = ecl_elapsed_time(&_last_run);
3  _last_run = ecl_absolute_time();
4  float dt = (float)dt_micros * 1e-6f;
5  /* lock integral for long intervals */
6  bool lock_integrator = ctl_data.lock_integrator;
7  if (dt_micros > 500000) {lock_integrator = true;}
8  _rate_error = _bodyrate_setpoint - ctl_data.body_y_rate;
9  if (!lock_integrator && _k_i > 0.0f) {
10 float id = _rate_error * dt * ctl_data.scaler;
11 /* anti-windup: do not allow the integrator to increase if the
       actuator is at a limit */
12 if (_last_output < -1.0f) {
13 /* only allow motion to center: increase value */
14 id = math::max(id, 0.0f);
15 } else if (_last_output > 1.0f) {
16 /* only allow motion to center: decrease value */
17 id = math::min(id, 0.0f);}
18 /* add and constrain */
19 _integrator = math::constrain(_integrator + id * _k_i,
       -_integrator_max, _integrator_max);}
20 /* Apply PI rate controller and store non-limited output */
21 _last_output = _bodyrate_setpoint * _k_ff * ctl_data.scaler
       +_rate_error * _k_p * ctl_data.scaler * ctl_data.scaler +
       _integrator; //scaler is proportional to 1/airspeed
```

In practice, to prevent excessive deflection of the elevators that could cause the UAV to pitch sharply up or down, potentially leading to damage, it is necessary to consider the saturation of the elevators. Taking saturation into account, the control law for the elevators becomes

$$\delta_{\text{ed}} = \text{sat}\left(\frac{1}{V_{\text{a}}} K_{\omega_{y_{\text{b}}}\text{f}} \omega_{y_{\text{b}}\text{d}} + \frac{1}{V_{\text{a}}^2} K_{\omega_{y_{\text{b}}}\text{p}} (\omega_{y_{\text{b}}\text{d}} - \omega_{y_{\text{b}}}) + \frac{1}{V_{\text{a}}} K_{\omega_{y_{\text{b}}}\text{i}} \int (\omega_{y_{\text{b}}\text{d}} - \omega_{y_{\text{b}}}), a_{\delta_{\text{e}}} \right). \tag{6.13}$$

The saturation function is defined as follows

$$\text{sat}(x_k, a) = \begin{cases} x_k, & |x_k| \le a \\ a \cdot \text{sign}(x_k), & |x_k| > a. \end{cases} \tag{6.14}$$

Here, the saturation parameter $a_{\delta_e} = \pi/9$. In practice, the saturation parameter $a \in \mathbb{R}_+$ can be determined according to actual requirements. Finally, the pitch control loop in the open-source autopilot is shown in Fig. 6.3.

6.1.1.2 Roll Control

Micro-small fixed-wing UAVs exhibit lateral motion through roll and yaw motion in the lateral plane. There are three primary methods for performing lateral turns: performing coordinated turns using the rudder, heading correction with ailerons while reducing dutch roll with the rudder, and performing lateral turns with a constant roll angle [3]. These three turning methods have distinct characteristics that are summarized below (Table 6.1).

The first two turning methods may not be effective in correcting large yaw deviations due to the presence of the sideslip angle, which can negatively impact flight quality and result in a distant turning path. To address this issue, it is recommended to use a coordinated turn for better control performance when turning with a constant roll angle. Therefore, to control lateral attitude motion, it is necessary to coordinate the turn by the roll and yaw motion. The control principle for lateral coordinated turning is illustrated in Fig. 6.4. As shown in the figure, during fixed-wing turning, the roll and yaw channels are coupled, satisfying the coordinated turning relationship (4.63). However, the two channels are controlled independently of each other.

1. Coordinated Turning Controller

For low-level flight control, the lateral-directional channel is given by the yaw angle ψ_d, while the roll channel control requires the desired roll angle ϕ_d. First, assuming the roll angle is relatively small, the coordinated turn formula (4.63) is simplified as

$$\dot{\psi} = \frac{g}{V_a}\phi. \tag{6.15}$$

Table 6.1 Comparison of lateral motion methods

Lateral motion	Advantages	Disadvantages
Coordinated turns using the rudder	Independent lateral control channels	1. Large sideslip angle 2. Poor coordination between airspeed vector and longitudinal axis 3. Large turning radius
Correcting heading with ailerons while reducing Dutch roll with the rudder	Stable Dutch roll mode	1. Large sideslip angle 2. Poor coordination 3. Inferior air travel experience
Lateral turns with constant roll angle	Small sideslip angle	Loss of altitude (can be improved by the pitch rate)

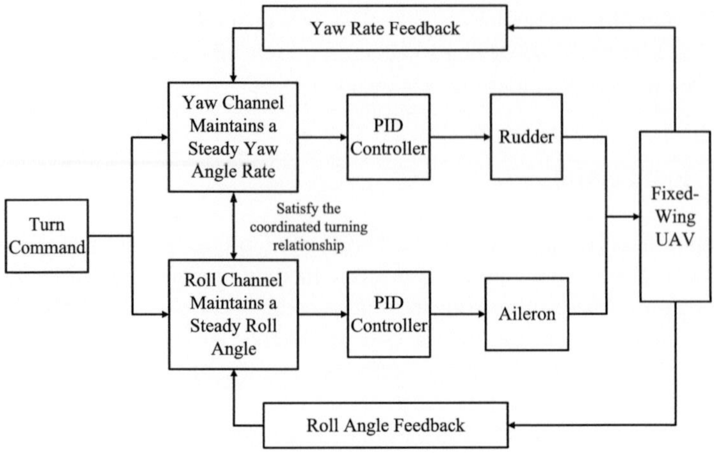

Fig. 6.4 Lateral attitude control structure in coordinated turning

Design the desired roll angle ϕ_{d} for the simplified coordinated turn formula (6.15) using proportional control:

$$\frac{g}{V_{\mathrm{a}}}\phi_{\mathrm{d}} = K_{\psi} e_{\psi} \tag{6.16}$$

where $K_{\psi} \in \mathbb{R}_{+}$, $e_{\psi} = \psi_{\mathrm{d}} - \psi$. If

$$\lim_{t \to \infty} |\phi(t) - \phi_{\mathrm{d}}(t)| = 0$$

then $\lim_{t \to \infty} |e_{\psi}(t)| = 0$. From Eq. (6.16), the desired roll angle is derived as

$$\phi_{\mathrm{d}} = \frac{V_{\mathrm{a}}}{g}(K_{\psi} e_{\psi}). \tag{6.17}$$

During actual flight, to prevent excessive maneuvers from damaging the UAV, the saturation needs to be considered to ensure the roll angle changes within a reasonable range. The control law with saturation is

$$\phi_{\mathrm{d}} = \mathrm{sat}\left(\frac{V_{\mathrm{a}}}{g}(K_{\psi} e_{\psi}), a_{\phi}\right) \tag{6.18}$$

where the saturation parameter $a_{\phi} = \pi/6$.

Note that in actual coordinated turn flight, due to the need to roll, the vertical component of the lift will decrease, resulting in a loss of altitude. To maintain a stable altitude, the angle of attack must be increased to compensate for the reduced lift increment, which requires an altitude hold controller.

2. Traditional PID Controller

The design of the roll angle channel controller is similar to that of the pitch angle channel controller. For the roll angle channel control objective, given the desired roll angle ϕ_{d}, design a controller such that

$$\lim_{t\to\infty}|e_\phi(t)|=0,\quad e_\phi\triangleq\phi_{\mathrm{d}}-\phi.$$

Here, ϕ_{d} is given by the remote controller or Eq. (6.18). From Eq. (4.17), one has

$$\omega_{x_\mathrm{b}}=\dot{\phi}-\sin\theta\dot{\psi}. \tag{6.19}$$

When designing the roll channel controller, assuming pitch angle $\theta=0$, Eq. (6.19) simplifies to

$$\dot{\phi}=\omega_{x_\mathrm{b}}. \tag{6.20}$$

Design the desired roll rate $\omega_{x_\mathrm{b}\mathrm{d}}$ as

$$\omega_{x_\mathrm{b}\mathrm{d}}=K_{\phi\mathrm{p}}e_\phi+K_{\phi\mathrm{i}}\int e_\phi+K_{\phi\mathrm{d}}\dot{e}_\phi \tag{6.21}$$

where $K_{\phi\mathrm{p}}, K_{\phi\mathrm{d}}, K_{\phi\mathrm{i}}\in\mathbb{R}$. Equations (6.20) and (6.21) form the roll angle control loop, as depicted by the dashed box ① in Fig. 6.5.

Assume $\dot{\phi}_{\mathrm{d}}=0$, and

$$\lim_{t\to\infty}|e_{\omega_{x_\mathrm{b}}}(t)|=0,\quad e_{\omega_{x_\mathrm{b}}}\triangleq\omega_{x_\mathrm{b}\mathrm{d}}-\omega_{x_\mathrm{b}},$$

then $\lim_{t\to\infty}|e_\phi(t)|=0$. The next task is to design a controller for the roll angle channel dynamic equation. From Eq. (4.50), one has

$$\begin{aligned}(J_{xx}J_{zz}-J_{xz}^2)\dot{\omega}_{x_\mathrm{b}}&=J_{zz}m_{\mathrm{a}x}+J_{xz}m_{\mathrm{a}z}+\omega_{y_\mathrm{b}}\omega_{z_\mathrm{b}}(J_{zz}J_{yy}-J_{zz}^2-J_{xz}^2)\\&\quad-\omega_{x_\mathrm{b}}\omega_{y_\mathrm{b}}(J_{yy}J_{xz}-J_{zz}J_{xz}-J_{xx}J_{xz}).\end{aligned} \tag{6.22}$$

Design the desired roll moment $m_{\mathrm{a}x\mathrm{d}}$ such that $\lim_{t\to\infty}|e_{\omega_{x_\mathrm{b}}}(t)|=0$. The desired roll moment is designed as

$$m_{\mathrm{a}x\mathrm{d}}=K'_{\omega_{x_\mathrm{b}}\mathrm{p}}e_{\omega_{x_\mathrm{b}}}+K'_{\omega_{x_\mathrm{b}}\mathrm{i}}\int e_{\omega_{x_\mathrm{b}}}+K'_{\omega_{x_\mathrm{b}}\mathrm{d}}\dot{e}_{\omega_{x_\mathrm{b}}} \tag{6.23}$$

where $K'_{\omega_{x_\mathrm{b}}\mathrm{p}}, K'_{\omega_{x_\mathrm{b}}\mathrm{d}}, K'_{\omega_{x_\mathrm{b}}\mathrm{i}}\in\mathbb{R}$. If $m_{\mathrm{a}x}=m_{\mathrm{a}x\mathrm{d}}$, then $\lim_{t\to\infty}|e_{\omega_{x_\mathrm{b}}}(t)|=0$. According to Eq. (4.31), the linear relationship between roll moment and aileron deflection can be rewritten as

$$m_{\mathrm{a}x}=K_{C_l}\delta_{\mathrm{a}}+d_{m_{\mathrm{a}x}}$$

where

$$K_{C_l} = \frac{1}{2}\rho V_a^2 b C_{l_{\delta_a}}, \quad d_{m_{ax}} = \frac{1}{2}\rho V_a^2 S b \left(C_{l_0} + C_{l_\beta}\beta + C_{l_p}\frac{b}{2V_a}\omega_{x_b} + C_{l_r}\frac{b}{2V_a}\omega_{z_b} + C_{l_{\delta_r}}\delta_r \right).$$

Therefore, the roll rate control law (6.23) is further derived as

$$\delta_{ad} = K_{\omega_{x_b}p} e_{\omega_{x_b}} + K_{\omega_{x_b}i} \int e_{\omega_{x_b}} + K_{\omega_{x_b}d} \dot{e}_{\omega_{x_b}} \tag{6.24}$$

where $K_{\omega_{x_b}p}$, $K_{\omega_{x_b}i}$, $K_{\omega_{x_b}d} \in \mathbb{R}$. Equations (6.22) and (6.24) form the angular rate control loop, as shown in the dashed box ② in Fig. 6.5. Sometimes, to avoid noise caused by the derivative term, the derivative terms $K_{\omega_{x_b}d}\dot{e}_{\omega_{x_b}}$ and $K_{\phi d}\dot{e}_\phi$ can be removed, or a filter can be added. At this point, the design of the roll angle channel controller is completed, formed by Eqs. (6.21) and (6.24). The control block diagram is shown in Fig. 6.5.

3. PID Controller in the Autopilot

In practical scenarios, if $\theta \neq 0$, then $\dot{\phi} \neq \omega_{x_b}$, requiring a more precise design to enhance the effectiveness of the controller. Below is an outline of the design approach for the roll angle channel controller in PX4. Similar to the pitch angle channel control loop, the roll angle channel control loop consists of a cascade of roll angle and roll angular rate feedback loops. The outer loop of the roll angle channel uses the roll angle as the controlled variable to obtain the desired roll angular rate. According to the attitude motion (4.18), the desired roll angular rate is

$$\omega_{x_b d} = \dot{\phi}_d - \sin\theta\dot{\psi}_d. \tag{6.25}$$

Its code is shown in line 15 of Code 6.4. In Eq. (6.25), the desired yaw rate $\dot{\psi}_d$ can be obtained through coordinated turn relationships. Next, the desired roll rate $\dot{\phi}_d$ is designed to ensure

$$\lim_{t\to\infty} |e_\phi(t)| = 0, \, e_\phi \triangleq \phi_d - \phi.$$

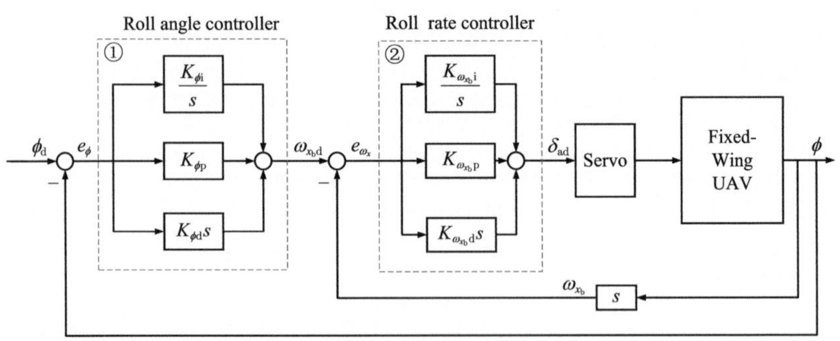

Fig. 6.5 A traditional roll control loop

Here is a simple design, using proportional control as

$$\dot{\phi}_{\mathrm{d}} = K_{\phi\mathrm{p}} e_{\phi} \tag{6.26}$$

where $K_{\phi\mathrm{p}} \in \mathbb{R}_{+}$. If $\dot{\phi} = \dot{\phi}_{\mathrm{d}}$, then

$$\dot{\phi} = K_{\phi\mathrm{p}} e_{\phi}. \tag{6.27}$$

Clearly, when ϕ_{d} is constant, $\lim_{t\to\infty} |e_{\phi}(t)| = 0$.

Here, the outer loop stabilizes the roll angle, as illustrated by the dashed box ① in Fig. 6.8. The outer loop control code for the PX4 roll angle channel is shown in lines 1-4 of Code 6.4. Next, assuming zero pitch rate and ignoring the effect of cross products, i.e., $\omega_{y_{\mathrm{b}}} = 0$, $J_{zz} = 0$, Eq. (6.22) is simplified as:

$$\dot{\omega}_{x_{\mathrm{b}}} = \frac{\rho V_{\mathrm{a}}^2 S b}{2 J_{xx}} \left(C_{l_\beta} \beta + C_{l_p} \frac{b}{2V_{\mathrm{a}}} \omega_{x_{\mathrm{b}}} + C_{l_r} \frac{b}{2V_{\mathrm{a}}} \omega_{z_{\mathrm{b}}} + C_{l\delta_{\mathrm{a}}} \delta_{\mathrm{a}} + C_{l\delta_{\mathrm{r}}} \delta_{\mathrm{r}} \right). \tag{6.28}$$

The inner loop targets the roll rate as the controlled variable. Design the desired aileron deflection δ_{ad} for Eq. (6.28) using a PI controller with a feedforward:

$$\delta_{\mathrm{ad}} = \frac{1}{V_{\mathrm{a}}} K_{\omega_{x_{\mathrm{b}}}\mathrm{f}} \omega_{x_{\mathrm{b}}\mathrm{d}} + \frac{1}{V_{\mathrm{a}}^2} K_{\omega_{x_{\mathrm{b}}}\mathrm{p}} (\omega_{x_{\mathrm{b}}\mathrm{d}} - \omega_{x_{\mathrm{b}}}) + \frac{1}{V_{\mathrm{a}}} K_{\omega_{x_{\mathrm{b}}}\mathrm{i}} \int (\omega_{x_{\mathrm{b}}\mathrm{d}} - \omega_{x_{\mathrm{b}}}) \tag{6.29}$$

to ensure $\lim_{t\to\infty} |\omega_{x_{\mathrm{b}}\mathrm{d}}(t) - \omega_{x_{\mathrm{b}}}(t)| = 0$, where $K_{\omega_{x_{\mathrm{b}}}\mathrm{f}}, K_{\omega_{x_{\mathrm{b}}}\mathrm{p}}, K_{\omega_{x_{\mathrm{b}}}\mathrm{i}} \in \mathbb{R}$. The controller for the roll rate loop is depicted by the dashed box ② in Fig. 6.8. The inner loop control code for the roll angle channel in PX4 is shown in lines 10–13 of Code 6.4.

The controller involves the roll rate $\omega_{x_{\mathrm{b}}}$ in the inner loop, thereby enhancing the damping characteristics of the short-period mode, suppressing dutch roll, and enabling a rapid response to roll rate commands for the UAV.

During actual execution, to prevent excessive aileron deflection that could lead to aggressive rolling motion and potentially damage the UAV, it is necessary to consider the saturation of the ailerons. A limitation of ±23° is set for the amplitude of the aileron at the input of the roll angle command. Considering saturation, the PID controller (6.29) becomes:

$$\delta_{\mathrm{ad}} = \mathrm{sat}\left(\frac{1}{V_{\mathrm{a}}} K_{\omega_{x_{\mathrm{b}}}\mathrm{f}} \omega_{x_{\mathrm{b}}\mathrm{d}} + \frac{1}{V_{\mathrm{a}}^2} K_{\omega_{x_{\mathrm{b}}}\mathrm{p}} (\omega_{x_{\mathrm{b}}\mathrm{d}} - \omega_{x_{\mathrm{b}}}) + \frac{1}{V_{\mathrm{a}}} K_{\omega_{x_{\mathrm{b}}}\mathrm{i}} \int (\omega_{x_{\mathrm{b}}\mathrm{d}} - \omega_{x_{\mathrm{b}}}), a_{\delta_{\mathrm{a}}} \right) \tag{6.30}$$

where the saturation parameter is set as $a_{\delta_{\mathrm{a}}} = 23\pi/180$. The code for the PX4 roll angle channel controller is shown in Code 6.4. Line 2 computes the roll angle error e_{ϕ}; Line 4 implements the outer loop P control, where 1/_tc represents the proportional control gain; Line 6 computes the roll rate error $\omega_{x_{\mathrm{b}}\mathrm{d}} - \omega_{x_{\mathrm{b}}}$; Lines 9–13 handle the feedforward and PI control for the roll rate channel; and Line 15 converts the roll rate to the body roll rate.

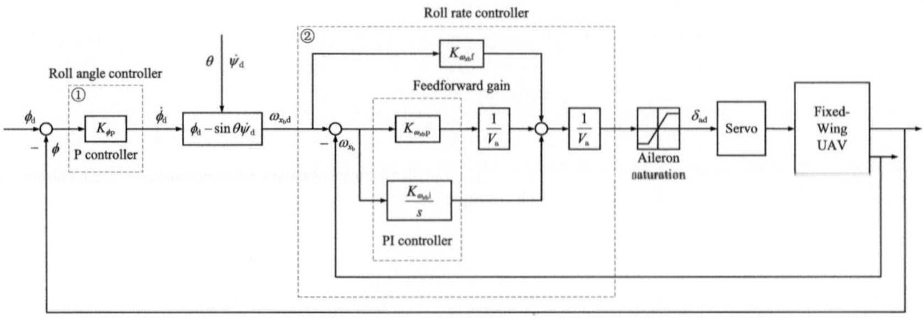

Fig. 6.6 Roll control loop in an open-source autopilot

Code 6.4 PX4 roll controller

```
/* Calculate error */
float roll_error = ctl_data.roll_setpoint - ctl_data.roll;
/* Apply P controller */
_rate_setpoint = roll_error / _tc;
/* Calculate body angular rate error */
_rate_error = _bodyrate_setpoint - ctl_data.body_x_rate;
    //body angular rate error
float id = _rate_error * dt * ctl_data.scaler;
/* add and constrain */
_integrator = math::constrain(_integrator + id * _k_i,
    -_integrator_max, _integrator_max);
/* Apply PI rate controller and store non-limited output */
_last_output = _bodyrate_setpoint * _k_ff * ctl_data.scaler +
_rate_error * _k_p * ctl_data.scaler * ctl_data.scaler
+ _integrator; //scaler is proportional to 1/airspeed
/* Transform setpoint to body angular rates (jacobian) */
_bodyrate_setpoint = ctl_data.roll_rate_setpoint -
    sinf(ctl_data.pitch) * ctl_data.yaw_rate_setpoint;
```

Finally, the control loop of the roll angle channel in the open-source autopilot is shown in Fig. 6.6.

6.1.1.3 Yaw Control

The lateral motion of micro-small fixed-wing UAVs is primarily controlled by ailerons (roll control), with a rudder (yaw control) providing stability and coordinated turns. The yaw angle change rate is set based on roll control, using a feedforward relationship (4.63).

1. Traditional PID Controller

The control of the yaw angle channel is actually the control of the yaw rate $\dot{\psi}$. The control objective of the yaw angle channel is: given the desired yaw rate $\dot{\psi}_{\mathrm{d}}$, a controller is designed as

$$\lim_{t\to\infty} |e_{\dot\psi}(t)| = 0, \quad e_{\dot\psi} \triangleq \dot\psi_{\rm d} - \dot\psi.$$

Here, $\dot\psi_{\rm d}$ is the desired yaw rate provided by the roll channel feedforward, satisfying the coordinated turn as

$$\dot\psi_{\rm d} = \frac{g}{V_{\rm a}} \tan\phi_{\rm d}.$$

If a micro-small fixed-wing UAV makes a level flight or a level turn, then $\theta \approx 0$. Then

$$\omega_{z_{\rm b}} = -\sin\phi\dot\theta + \cos\theta\cos\phi\dot\psi. \tag{6.31}$$

It can be further simplified as

$$\omega_{z_{\rm b}} = \cos\phi\dot\psi.$$

To design the desired yaw moment $m_{az{\rm d}}$ based on Eq. (4.50), such that $\lim_{t\to\infty} |e_{\dot\psi}(t)| = 0$, a PID controller is designed as follows:

$$m_{az{\rm d}} = K'_{\dot\psi{\rm p}} e_{\dot\psi} + K'_{\dot\psi{\rm i}} \int e_{\dot\psi} + K'_{\dot\psi{\rm d}} \dot e_{\dot\psi}. \tag{6.32}$$

Here, $K'_{\dot\psi{\rm p}}, K'_{\dot\psi{\rm d}}, K'_{\dot\psi{\rm i}} \in \mathbb{R}$. If $m_{az} = m_{az{\rm d}}$, then $\lim_{t\to\infty} |e_{\dot\psi}(t)| = 0$. According to Eq. (4.31), the relationship between the yaw moment and rudder deflection can be rewritten as

$$m_{az} = K_{C_n}\delta_{\rm r} + d_{m_{az}}$$

where

$$K_{C_n} = \frac{1}{2}\rho V_{\rm a}^2 b C_{n_{\delta_{\rm r}}}, \quad d_{m_{az}} = \frac{1}{2}\rho V_{\rm a}^2 Sb\left(C_{n0} + C_{n\beta}\beta + C_{n_p}\frac{b}{2V_{\rm a}}\omega_{xb} + C_{n_r}\frac{b}{2V_{\rm a}}\omega_{zb} + C_{n_{\delta_{\rm a}}}\delta_{\rm a}\right).$$

The control law for the yaw rate is expressed as:

$$\delta_{\rm rd} = K_{\dot\psi{\rm p}} e_{\dot\psi} + K_{\dot\psi{\rm i}} \int e_{\dot\psi} + K_{\dot\psi{\rm d}} \dot e_{\dot\psi} \tag{6.33}$$

where $K_{\dot\psi{\rm p}}, K_{\dot\psi{\rm d}}, K_{\dot\psi{\rm i}} \in \mathbb{R}$. The yaw rate PID controller is shown in the dashed box ② in Fig. 6.7. Equation (6.33) represents the yaw rate control loop. To avoid noise caused by differentiation, the derivative term $K_{\dot\psi{\rm d}}\, \dot e_{\dot\psi}$ can be removed. Thus, the yaw channel controller design is completed, and its control block diagram is shown in Fig. 6.7.

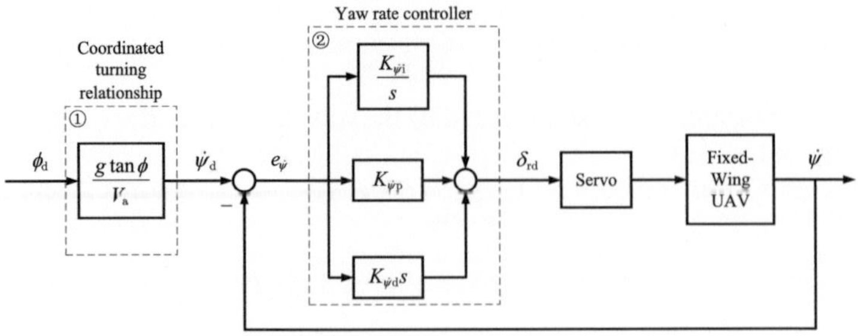

Fig. 6.7 A traditional yaw control loop

2. PID Controller in the Autopilot

The yaw angle control loop only includes a single yaw rate control loop. First, given the desired roll angle and airspeed, the desired yaw rate for coordinated turn $\dot{\psi}_{\mathrm{d}}$ can be obtained as follows[1]

$$\dot{\psi}_{\mathrm{d}} = \frac{g}{V_{\mathrm{a}}} \tan\phi \cos\theta. \tag{6.34}$$

The PX4 code for obtaining the desired yaw rate through the coordinated turn is shown in Code 6.5.

Code 6.5 Obtaining desired yaw rate via coordinated turns in PX4

```
/* Calculate desired yaw rate from coordinated turn constraint
    / (no side forces) */
_rate_setpoint = tanf(constrained_roll) * cosf(ctl_data.pitch)
    * CONSTANTS_ONE_G / (ctl_data.airspeed <
    ctl_data.airspeed_min ? ctl_data.airspeed_min :
    ctl_data.airspeed);
```

Then, the desired yaw rate is obtained through Eq. (6.31), as shown in Code 6.6, that is

$$\omega_{z_{\mathrm{b}}\mathrm{d}} = -\sin\phi\dot{\theta}_{\mathrm{d}} + \cos\theta\cos\phi\dot{\psi}_{\mathrm{d}}. \tag{6.35}$$

Code 6.6 Transforming yaw rate to body yaw rate

```
/* Transform setpoint to body angular rates (jacobian) */
_bodyrate_setpoint = -sinf(ctl_data.roll) *
    ctl_data.pitch_rate_setpoint + cosf(ctl_data.roll) *
    cosf(ctl_data.pitch) * _rate_setpoint;
```

Next, assuming the pitch rate is zero and neglecting the cross product, i.e., $\omega_{y_{\mathrm{b}}} = 0$, $J_{xz} = 0$, Eq. (4.50) is simplified as

[1] Note that the formula in the autopilot (6.34) does not match the theoretical derivation formula $\dot{\psi}_{\mathrm{d}} = \frac{g}{V_{\mathrm{a}}} \tan\phi_{\mathrm{d}}$ in Chap. 4 shown in (4.63). Here, (6.34) corresponds to the source code from the PX4 official website.

$$\dot{\omega}_{z_\mathrm{b}} = \frac{\rho V_\mathrm{a}^2 S b}{2J_{zz}}\left(C_{n_\beta}\beta + C_{n_p}\frac{b}{2V_\mathrm{a}}\omega_{x_\mathrm{b}} + C_{n_r}\frac{b}{2V_\mathrm{a}}\omega_{z_\mathrm{b}} + C_{n\delta_\mathrm{a}}\delta_\mathrm{a} + C_{n\delta_\mathrm{r}}\delta_\mathrm{r}\right). \tag{6.36}$$

For Eq. (6.36), the desired rudder deflection using feedforward and PI control is designed as:

$$\delta_\mathrm{rd} = \frac{1}{V_\mathrm{a}^2}\left(K_{\omega_{z_\mathrm{b}}\mathrm{f}}\omega_{z_\mathrm{b}\mathrm{d}} + K_{\omega_{z_\mathrm{b}}\mathrm{p}}(\omega_{z_\mathrm{b}\mathrm{d}} - \omega_{z_\mathrm{b}}) + K_{\omega_{z_\mathrm{b}}\mathrm{i}}\int(\omega_{z_\mathrm{b}\mathrm{d}} - \omega_{z_\mathrm{b}})\right). \tag{6.37}$$

The PX4 yaw angle channel controller code is shown in Code 6.7. Line 2 calculates the yaw rate error $\omega_{z_\mathrm{b}\mathrm{d}} - \omega_{z_\mathrm{b}}$, and Lines 4-11 represent the complete feedforward and PI control for the yaw rate channel.

Code 6.7 PX4 yaw channel controller

```
1  /* Calculate body angular rate error */
2  _rate_error = _bodyrate_setpoint - ctl_data.body_z_rate; //
       body angular rate error
3  $\vdots$
4  float id = _rate_error * dt;
5  $\vdots$
6  /* add and constrain */
7  _integrator = math::constrain(_integrator + id * _k_i,
       -_integrator_max, _integrator_max);
8  $\vdots$
9  /* Apply PI rate controller and store non-limited output */
10 _last_output = (_bodyrate_setpoint * _k_ff + _rate_error *
       _k_p + _integrator) * ctl_data.scaler *
11 ctl_data.scaler; //scaler is proportional to 1/airspeed
```

In practical scenarios, the saturation needs to be considered. In the yaw angle channel control, the rudder deflection δ_rd is limited to $\pm 25°$. After incorporating saturation, the control law for the rudder becomes

$$\delta_\mathrm{rd} = \mathrm{sat}\left(\frac{1}{V_\mathrm{a}^2}\left(K_{\omega_{z_\mathrm{b}}\mathrm{f}}\omega_{z_\mathrm{b}\mathrm{d}} + K_{\omega_{z_\mathrm{b}}\mathrm{p}}(\omega_{z_\mathrm{b}\mathrm{d}} - \omega_{z_\mathrm{b}}) + K_{\omega_{z_\mathrm{b}}\mathrm{i}}\int(\omega_{z_\mathrm{b}\mathrm{d}} - \omega_{z_\mathrm{b}})\right), a_{\delta_\mathrm{r}}\right) \tag{6.38}$$

where the saturation parameter is $a_{\delta_\mathrm{r}} = 5\pi/36$. Finally, the control loops for roll and yaw channels in the open-source autopilot are depicted in Fig. 6.8.

6.1.2 *Altitude and Airspeed Control*

6.1.2.1 Traditional PID Controller

1. Altitude Control

The altitude of a micro-small fixed-wing UAV varies with its flight attitude, especially in response to changes in pitch. There are two methods for controlling the altitude of

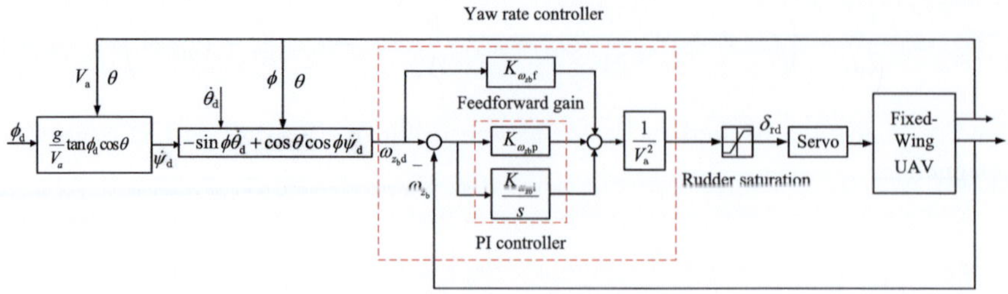

Fig. 6.8 Control loops for roll control in an open-source autopilot

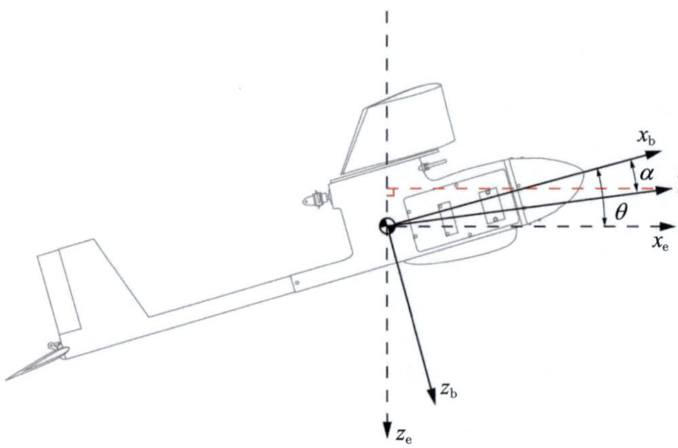

Fig. 6.9 Motion analysis in the longitudinal plane of a fixed-wing UAV

a UAV: elevator control and throttle control. Flight altitude is controlled by elevator deflection instead of changing flight speed, as changing airspeed is inefficient for throttle control.

It is crucial to analyze the reasons for changes in UAV altitude from kinematics to design a controller for UAV altitude. As shown in Fig. 6.9, the vertical component v_{z_e} of the airspeed V_a in the vertical direction ($o_e z_e$-axis) is given by

$$\dot{H} = v_{z_e} = V_a \sin(\theta - \alpha). \tag{6.39}$$

During actual flight, the altitude changes gradually. Assuming that $\theta - \alpha$ is small over a short-period, it holds that

$$\sin(\theta - \alpha) \approx \theta - \alpha. \tag{6.40}$$

This leads to the following expression:

$$\dot{H} = V_a(\theta - \alpha). \tag{6.41}$$

Airspeed V_a is controlled separately by another loop to remain constant, and the angle of attack α is also related to the pitch angle. For the altitude controller design based on Eq. (6.41), in order to achieve

$$\lim_{t\to\infty} |e_H(t)| = 0, e_H \triangleq H_d - H$$

the desired pitch angle θ_d is designed as

$$\theta_d = K_{Hp}e_H + K_{Hi}\int e_H + K_{Hd}\dot{e}_H \tag{6.42}$$

where $K_{Hp}, K_{Hd}, K_{Hi} \in \mathbb{R}$. Equations (6.3), (6.7) and (6.42) together constitute the altitude control loop, and its control structure is shown in the dashed box in Fig. 6.10.

2. Airspeed Control

There are two methods for controlling the airspeed of micro-small fixed-wing UAVs: adjusting the pitch angle via elevator deflection and controlling airspeed through throttle adjustments. When elevators are used to control airspeed, the throttle stick remains constant, and only the elevators are used to control airspeed, resulting in a limited range of airspeed adjustment. Therefore, it is recommended to use airspeed control by changing the throttle.

Before designing the airspeed control of the UAV, the forces acting on the UAV in the wind coordinate system are analyzed, as shown in Fig. 6.11. From this figure, it is observed that the dynamic equation of the UAV in the airspeed direction is represented by

$$m\dot{V}_a = T\cos\alpha - f_{ax} - mg\sin\gamma \tag{6.43}$$

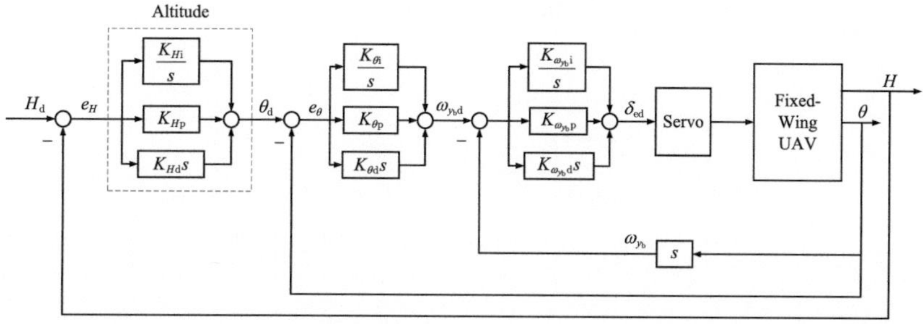

Fig. 6.10 A traditional altitude control loop

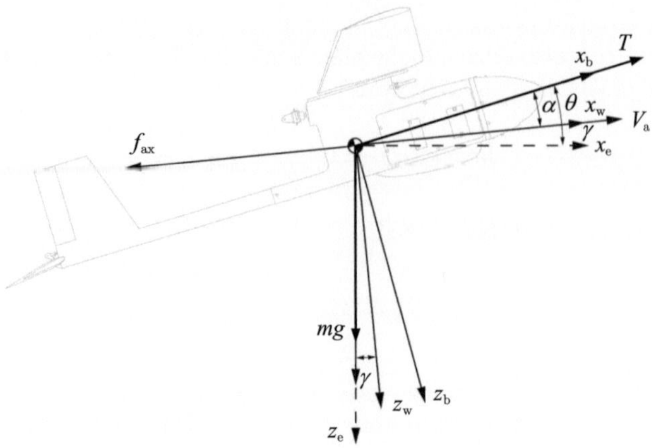

Fig. 6.11 Force analysis of a fixed-wing UAV in the longitudinal plane

where $\gamma = \theta - \alpha$. Assuming that during flight, the drag f_{ax} remains constant, and both the angle of attack α and flight path angle γ are small, an airspeed controller is designed based on Eq. (6.43). In order to achieve

$$\lim_{t \to \infty} |e_{V_a}(t)| = 0, e_{V_a} \triangleq V_{ad} - V_a$$

the desired thrust T_d is designed as

$$T_d = K_{V_a p} e_{V_a} + K_{V_a i} \int e_{V_a} + K_{V_a d} \dot{e}_{V_a} \tag{6.44}$$

where $K_{V_a p}$, $K_{V_a d}$, $K_{V_a i} \in \mathbb{R}$. Equation (6.44) represents the control law of UAV's airspeed, and the control loop is shown in Fig. 6.12.

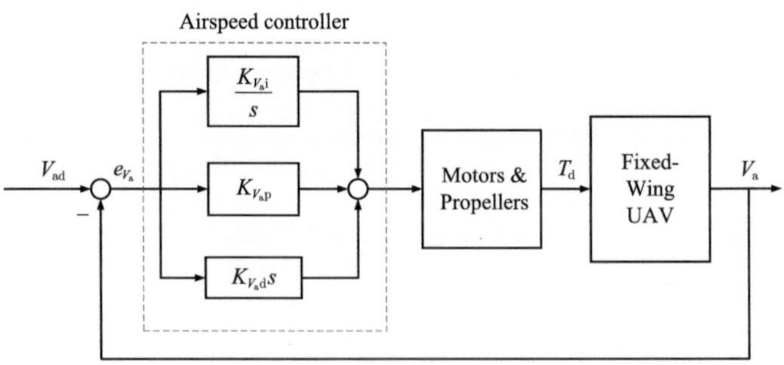

Fig. 6.12 A traditional speed control loop

The controllers for altitude and airspeed of the micro-small fixed-wing UAV have been designed. Each controller manipulates either the elevator or the throttle. It is observed from Eq. (6.41) that a coupling exists between the altitude and airspeed of the UAV. Therefore, airspeed stability is required when controlling altitude. The following section will introduce the autopilot controller design for altitude and airspeed.

6.1.2.2 Altitude-Airspeed Controller in the Autopilot

The autopilot is capable of two distinct operating states: altitude stabilization and pitch angle stabilization. During the altitude stabilization state, the airspeed vector is maintained in a horizontal position, and an increase in throttle input results in the engine thrust to accelerate the aircraft's airspeed. In a steady pitch angle, the engine thrust influences not only the aircraft's airspeed but also the angle of attack, flight path angle, and altitude. Consequently, the open-source auto-pilot controller design frequently employs a *total energy control* strategy to adjust the aircraft's altitude and airspeed concurrently.

In the case of micro-small fixed-wing UAVs, elevators are employed to regulate both short-term airspeed and flight path angle. Nevertheless, elevator control is essential for maintaining optimal airspeed in the long term. The throttle is employed to regulate the aircraft's airspeed over the course of a relatively brief period. Nevertheless, in the long term, control of the throttle is of paramount importance for the regulation of the flight path angle. It is common practice to divide the airspeed and altitude channels into two distinct single-input-single-output systems, thereby ignoring the strong coupling between airspeed and flight path. To overcome this defect, a Total Energy Control System (TECS) was proposed [4]. This method regulates total energy by adjusting the throttle (i.e., changing thrust) and controls the conversion of kinetic and potential energy by adjusting the elevator (i.e., changing the pitch angle), thereby uniquely tracking the desired kinetic and potential energies and, consequently, uniquely tracking airspeed and altitude.

1. Throttle Control

The total energy E_{T} of the UAV equals the sum of kinetic energy E_{K} and potential energy E_{P}, which is

$$E_{\mathrm{T}} = E_{\mathrm{K}} + E_{\mathrm{P}} = \frac{1}{2} m V_{\mathrm{a}}^{2} + mgH. \tag{6.45}$$

Differentiating the total energy shown in Eq. (6.45), it is obtained that

$$\dot{E}_{\mathrm{T}} = m V_{\mathrm{a}} \dot{V}_{\mathrm{a}} + mg\dot{H}. \tag{6.46}$$

The UAV's longitudinal motion equations in the ground coordinate system are

$$\begin{cases} \dot{p}_{x_e} = V_a \cos\gamma \\ \dot{H} = V_a \sin\gamma. \end{cases} \tag{6.47}$$

Substituting Eqs. (6.47) into (6.46) yields

$$\dot{E}_T = mgV_a\left(\frac{\dot{V}_a}{g} + \sin\gamma\right). \tag{6.48}$$

Thus, the derivative of UAV's total energy is defined as

$$\dot{E} = \frac{\dot{V}_a}{g} + \sin\gamma. \tag{6.49}$$

Substituting the UAV's longitudinal motion (6.43) into Eq. (6.49), the relationship for the derivative of total energy is obtained as follows:

$$\dot{E} = \frac{\dot{V}_a}{g} + \sin\gamma = \frac{T\cos\alpha - f_{ax}}{mg}. \tag{6.50}$$

From Eq. (6.50), it is observed that during flight, the change in total energy of the UAV is primarily regulated by the thrust T. Assume that the aerodynamic drag f_{ax} remains constant during flight and the angle of attack $\alpha = 0$. The effect of the thrust control increment is

$$T \approx mg\left(\frac{\dot{V}_a}{g} + \sin\gamma\right) + f_{ax}. \tag{6.51}$$

When the UAV is in steady-level flight, namely $\dot{V}_a = 0$, $\gamma = 0$, one has

$$T^* \approx f_{ax}. \tag{6.52}$$

To trim, and since $\sin\gamma \approx \gamma$, it is obtained as

$$\Delta T \approx mg\left(\frac{\dot{V}_a}{g} + \gamma\right) \tag{6.53}$$

where

$$\Delta T = T^* - T.$$

This implies that the total energy of the aircraft is directly controlled using thrust.

Given the desired airspeed and altitude V_{ad}, H_d to be tracked, the following dynamics is employed

$$\begin{aligned}\dot{V}_a &= -K_V(V_a - V_{ad}) \\ \dot{H} &= -K_H(H - H_d).\end{aligned} \tag{6.54}$$

These will make $V_a \to V_{ad}$, $H \to H_d$. In other words, it is expected that

$$\left|\frac{\dot{V}_{ad}}{g} - \frac{\dot{V}_a}{g}\right| \to 0, |\gamma_d - \gamma| \to 0$$

where

$$\dot{V}_{ad} = -K_V(V_a - V_{ad}), \gamma_d = \frac{-K_H(H - H_d)}{V_a}.$$

Let

$$\begin{aligned}e_{\dot{V}_a} &= \frac{\dot{V}_{ad}}{g} - \frac{\dot{V}_a}{g} \\ e_\gamma &= \gamma_d - \gamma.\end{aligned}$$

Then, the desired ΔT is designed as

$$\Delta T_d \approx \left(K_{Ep} + K_{Ei}\frac{1}{s}\right)(e_{\dot{V}_a} + e_\gamma) \tag{6.55}$$

where $1/s$ is the Laplace operator for integration. Because of the presence of the integrator, the steady-state error of $e_{\dot{V}_a} + e_\gamma$ is eliminated. Therefore, the controller can also be expressed as

$$T_d \approx \left(K_{Ep} + K_{Ei}\frac{1}{s}\right)(e_{\dot{V}_a} + e_\gamma). \tag{6.56}$$

In addition to the above design, there is also a modified form of total energy control. Equation (6.53) is written as

$$\dot{E}_T \approx V_a \Delta T. \tag{6.57}$$

Given the desired airspeed and altitude V_{ad}, H_d to be tracked, the desired total energy is further expressed as

$$E_{Td} = \frac{1}{2}mV_{ad}^2 + mgH_d. \tag{6.58}$$

Define $e_E = E_{Td} - E_T$. Then Eq. (6.57) becomes

$$\dot{e}_E \approx -V_a \Delta T. \tag{6.59}$$

Then the desired ΔT is

$$\Delta T_{\rm d} = \frac{1}{V_{\rm a}}\left(K_{E{\rm p}}e_E + K_{E{\rm i}}\int e_E\right). \tag{6.60}$$

Because of the presence of the integrator, the steady-state error of e_E will be eliminated [5]. Therefore, the controller can also be expressed as

$$T_{\rm d} = \frac{1}{V_{\rm a}}\left(K_{E{\rm p}}e_E + K_{E{\rm i}}\int e_E\right). \tag{6.61}$$

2. Elevator Control

Tracking total energy alone does not guarantee tracking the desired airspeed and altitude $V_{\rm ad}$, $H_{\rm d}$. It is necessary to allocate kinetic and potential energy specifically. The elevator deflection primarily affects the pitch moment of the aircraft, resulting in a change in the aircraft's flight attitude, with minimal impact on lift and drag. This is accomplished by manipulating the elevator to convert kinetic and potential energy as

$$B = E_{\rm P} - E_{\rm K}. \tag{6.62}$$

Then,

$$\dot{B} = \dot{E}_{\rm P} - \dot{E}_{\rm K} = mgV_{\rm a}\left(\sin\gamma - \frac{\dot{V}_{\rm a}}{g}\right). \tag{6.63}$$

Similar to the total energy control method, the controller for the elevator (sometimes set as the pitch angle [1]) is designed as

$$\delta_{\rm ed} \approx \left(K_{B{\rm p}} + K_{B{\rm i}}\frac{1}{s}\right)\left(e_{\dot{V}_{\rm a}} - e_\gamma\right) \tag{6.64}$$

or

$$\delta_{\rm ed} = \frac{1}{V_{\rm a}}\left(K_{B{\rm p}}e_B + K_{B{\rm i}}\int e_B\right). \tag{6.65}$$

Based on the total energy control (6.56) and total energy allocation control (6.64), the core algorithm of total energy control is obtained, with the structure shown in Fig. 6.13.

6.1.3 Principles and Tuning Steps of PID Control

6.1.3.1 Principles of PID Control

The PID control algorithm is composed of three terms: Proportional (P), Integral (I), and Derivative (D). For micro-small fixed-wing UAVs, control over five vir-

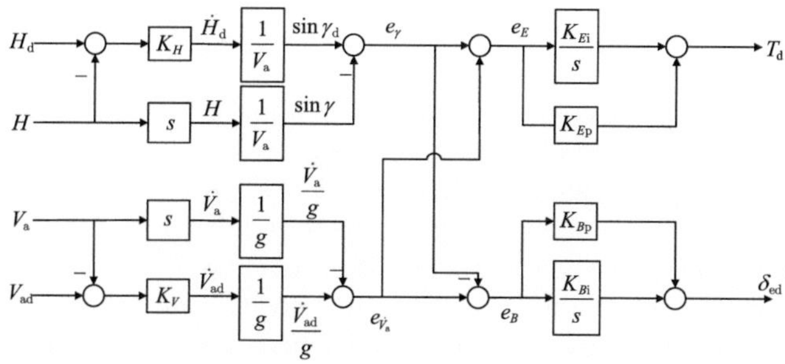

Fig. 6.13 A total energy control structure.

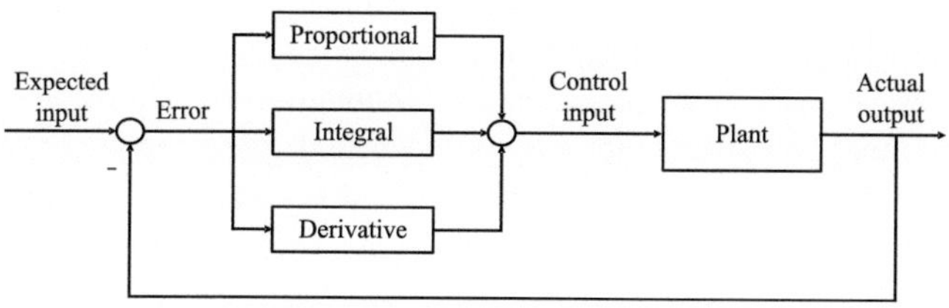

Fig. 6.14 Structure diagram of a conventional PID control system

tual variables is necessary: altitude, airspeed, pitch, roll, and yaw. The PID control algorithm calculates the control inputs by summing the proportional, integral, and derivative of the errors of the five variables. Figure 6.14 shows the structure diagram of a traditional PID control system.

The 'P' control term, as illustrated in Fig. 6.14, is employed for the purpose of correcting the error. As the value of the 'P' coefficient increases, the system demonstrates heightened sensitivity to errors, accelerated error correction speed, and enhanced system response. Nevertheless, a large 'P' coefficient may result in a considerable overshoot, thus necessitating the avoidance of excessive values. Moreover, the integral 'I' control term can potentially enhance the system's tracking precision. An increase in the 'I' coefficient results in enhanced tracking accuracy, leading to a reduction in the tracking error. However, if the value is large, it may take longer to eliminate the steady-state error. The derivative coefficient 'D' helps prevent overshoot by reflecting the error signal's rate of change and slowing the system as it approaches the desired target. It is generally recommended to avoid setting a large value for the 'D' coefficient, which may cause system oscillations.

The PID control algorithm is a widely utilized control scheme due to its straightforward structure, robust stability, high reliability, and straightforward adjustment capabilities. Its primary advantage is that it does not require the construction of a system model or identification of its parameters; instead, it can be adjusted through experimentation following the completion of the design phase. PID control algorithms are frequently employed in the control of micro-small fixed-wing UAVs due to the inherent complexity of their models.

6.1.3.2 PID Tuning

PID tuning is the process of adjusting the proportional (P), integral (I), and derivative (D) coefficients of a controller, also known as the *adjust parameters*. Proper tuning ensures that the controlled system accurately reaches the desired target. For fixed-wing UAVs, PID tuning typically follows the critical proportional method [6], combined with trial and error along with experience.

When tuning the longitudinal controller, it is divided into an altitude loop and a pitch loop. The pitch loop further includes the pitch angle loop and the pitch rate loop. The pitch rate loop is the inner loop of the pitch loop. Its tuning goal is the angular rate, which is related to the dynamic characteristics of the UAV and its ability to quickly and flexibly track the desired target. The outer loop of the pitch loop is responsible for tuning the attitude angle, which is related to the UAV's static characteristics, stability, and wind resistance. In longitudinal control, the pitch loop serves as the inner loop. The outer loop of the altitude control converts the altitude error into a pitch reference. The tuning process involves adjusting the inner loop first, followed by the outer loop.

6.1.4 Gain Scheduling Control

For control systems with large dynamic uncertainties, *gain scheduling control* is a common engineering approach. These method employs linear design techniques to address nonlinear control issues without requiring an exhaustive analytical system model. It demonstrates a high degree of responsiveness to alterations in operational parameters and has a broad range of applications, including in the domains of aircraft control and chemical process control [7].

One or more design variables are selected to represent the operating points of a system, with each point corresponding to a particular system operating state. At these operating points, the system is linearized, and corresponding linear feedback controllers are designed. The aforementioned controller is implemented as a single controller, with interpolation employed between the operating points. It is essential to monitor the schedule variables during operation to select the most appropriate controller [8]. To illustrate, consider a fixed-wing aircraft. The flight envelope, as defined in Sect. 3.1.3, is partitioned into distinct regions, each represented by a pair of altitude and airspeed operating points. Since the aerodynamic parameters of the aircraft are different at each point of operation, it is challenging to develop a single set of flight control parameters that can adequately address the flight qualities at all points within the envelope. Gain scheduling controllers are designed to vary their parameters with altitude and airspeed to achieve the desired controller performance.

Classical gain scheduling control involves the following four steps.

(1) Identify scheduling variables and linearize the system at typical operating points. The change in scheduling variables should reflect the system's operating state change.
(2) Use linearization methods to design the desired controllers for the selected series of operating points.
(3) Interpolation between two operating points is used to obtain and combine controllers to construct a gain scheduling controller.
(4) Test the performance of the closed-loop system introduced by the gain scheduling controller.

6.2 Basic Experiment

6.2.1 Experimental Objective

1. Things to prepare

(1) Software: MATLAB R2020b or above.
(2) Program: Experimental package: "e4/e4-1". The package mainly contains: micro-small fixed-wing UAV parameter initialization file "InitDatactrl.m", which can import UAV parameters into the workspace after running; micro-small fixed-wing UAV simulation model file "SmallFixedWingUAVattitude1.slx".

2. Objective
Design an attitude controller for the micro-small fixed-wing UAV and perform Software-In-the-Loop (SIL) simulation in Simulink, and plot relevant response curves.

6.2.2 Experimental Procedure

In Step 1, based on the micro-small fixed-wing UAV model designed in Chap. 4, the simulation model of a complete fixed-wing UAV system is to be built, including the low-level flight control with a traditional PID controller as designed in the experimental principle. Then, in Step 2, the control commands are changed respectively to observe the tracking performance of each fixed-wing UAV channel, and the designed controller's effectiveness is verified by experimental data.

1. Step 1: Attitude Controller Implementation
(1) Pitch channel controller implementation.
Open the Simulink file "ch4/e4/e4-1/SmallFixedWingUAVattitude.slx", as shown in Fig. 6.15. The model consists of four main parts: the dashed box ①—"RC Signal

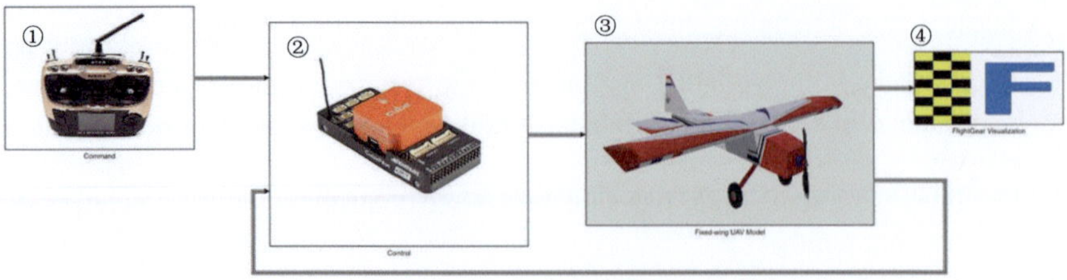

Fig. 6.15 Simulink control model for the fixed-wing UAV, from "e4/e4-1/SmallFixedWingUAVattitude.slx"

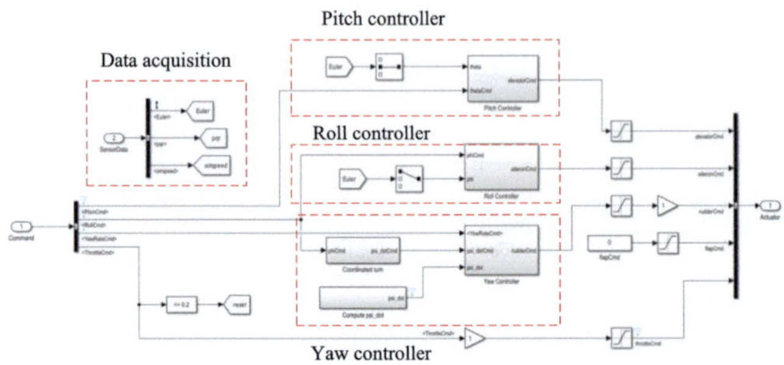

Fig. 6.16 Attitude control module, from "e4/e4-1/SmallFixedWingUAVattitude.slx"

Process", which generates command signals; the dashed box ②—"Attitude Controller", the attitude control part; the dashed box ③—"Fixed-Wing UAV System", modeling the dynamic system of the micro-small fixed-wing UAV; and the dashed box ④—"FlightGear", the SIL display module. The dynamic system model of the micro-small fixed-wing UAV, namely "Fixed Wing UAV System" module, has been previously established.

Double-click to open the "Control" module to construct the attitude controller (as shown in Fig. 6.16), and build the pitch controller in "Pitch Controller" module, which consists of two PID controllers in series, as shown in Fig. 6.17. First, obtain the current pitch angle signal of the fixed-wing UAV from the onboard sensors, i.e., get the Euler angle information from "SensorData". Then, obtain the command signal from the remote controller, i.e., getting the "PitchCmd" signal from the "Command" module.[2] Build the angle loop PID controller according to the control law in Eq. (6.3). The integral and derivative components use discrete integrators and differentiators, which facilitate HIL simulation later. The discrete integrator resets the initial state using the airspeed signal, with the trigger type "External reset" option set to "either", meaning that the integrator's initial value resets to the initial state (initial value is 0) whenever the airspeed is less than 0.2, regardless of whether it is in the rising or falling phase. This helps avoid the appearance of algebraic loops. Get the actual pitch

---

[2] In the experiment, the subscript "d" represents the desired value by autopilots, while the subscript "cmd" represents the value by the remote controller.

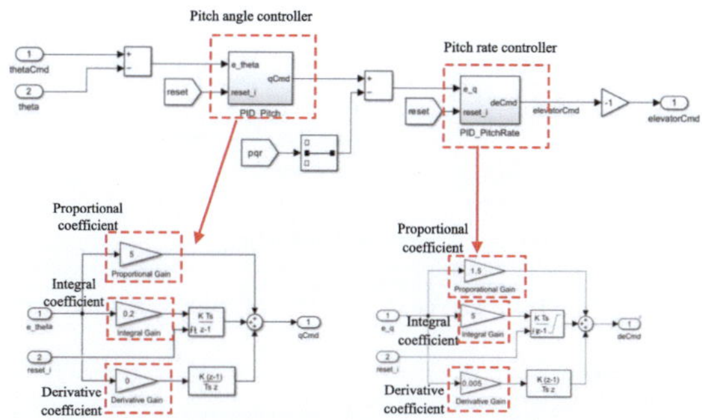

Fig. 6.17 Pitch control module, from "e4/e4-1/SmallFixedWingUAVattitude.slx"

rate from "SensorData", and the angle loop "PID_Pitch" module outputs the given pitch rate. Build the pitch rate loop PID controller according to the control law in Eq. (6.7). Finally, "PID_PitchRate" module outputs the desired elevator deflection. The fixed-wing UAV model generates a positive pitch angle when the UAV is nose-up. For the elevator, a downward deflection generates a nose-up action. Therefore, the controller output deflection value needs to include a negative gain. To prevent the fixed-wing UAV from excessive nose-up or nose-down actions that could overload the actuators and damage the UAV, the deflection output is limited. Thus, a saturation module is added after the control output to restrict the elevator movement within a reasonable range.

(2) Coordinated turning control implementation
The construction of the roll controller is similar to that of the pitch controller. In the "Roll Controller" module, the roll angle PID controller and roll rate PID controller are constructed based on the control laws in Eqs. (6.21) and (6.24), respectively. The actual roll angle signal and roll rate signal are obtained from the "SensorData" module. The "RollCH1" signal from the "Command" module is output as the desired roll angle signal, and the roll angle PID controller and roll rate PID controller are constructed in the "PID_Roll" and "PID_RollRate" modules, respectively. To prevent excessive roll movements that could damage the fixed-wing UAV and ensure that aileron deflection is within a specific range, a saturation module is added after the control output, limiting the aileron deflection to $[-23/180\pi, 23/180\pi]$ rad.

For yaw control, it is necessary to satisfy the coordinated turn equation with the roll channel. According to Eq. (4.63), the "Coordinated turn" module is constructed to obtain the desired yaw rate $\dot{\psi}_{\mathrm{d}}$ from the reference roll angle ϕ_{d}, represented by the signal "psi_dotCmd". The actual yaw rate in the fixed-wing is obtained in the "Compute psi_dot" module based on the motion equation:

$$\dot{\psi} = (\omega_{z_{\mathrm{b}}}\cos\phi + \omega_{y_{\mathrm{b}}}\sin\phi)/\cos\theta. \tag{6.66}$$

The actual yaw rate signal is denoted as "psi_dot". The PID controller for the yaw angle rate is constructed according to Eq. (6.33). In some instances, the rudder

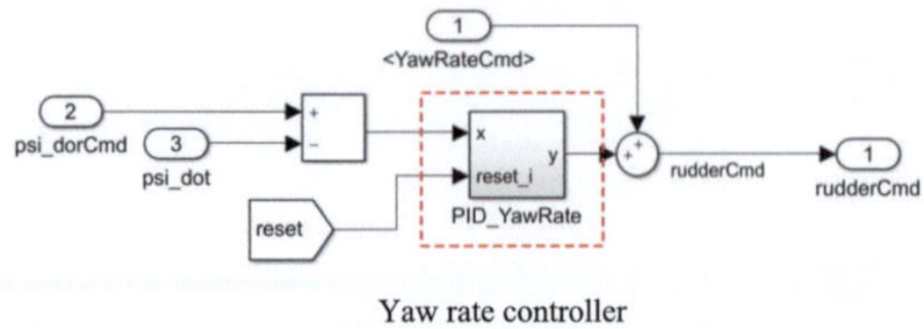

Fig. 6.18 Yaw control module, from "e4/e4-1/SmallFixedWingUAVattitude.slx"

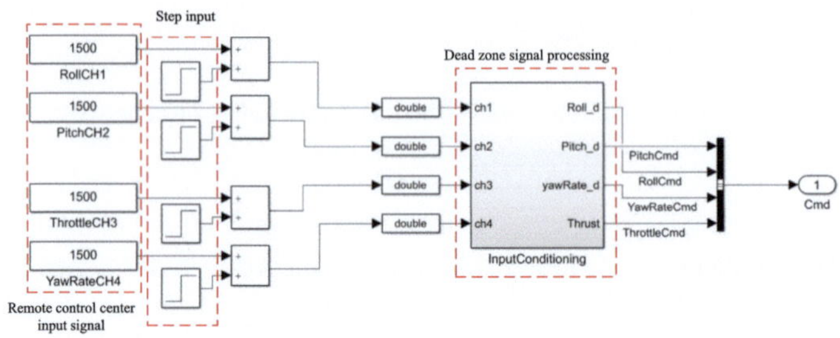

Fig. 6.19 Remote controller input module

command from the remote controller must be considered, and the rudder deflection signal should include the given yaw rate from the remote controller. Generally, the "YawRateCmd" signal from the "Command" module is zero. The yaw channel controller is shown in Fig. 6.18.

2. Step 2: Observing and Recording Data from Attitude Control Experiment
(1) Open file "e4-1/InitDatactrl.m" and run initialization file. Double-click on the "Command" module in the model, which simulates the input from the remote controller, as shown in Fig. 6.19. Each control command default value is 1500, normalized to 0 when processed, indicating the neutral position of the remote controller stick. When needing to give a control command, double-click on the step signal module for the "PitchCH2" channel. Input a step signal of 100 at the 10th second, corresponding to a step of 1600 for "ch2" channel. It is important to note that the step signal should be within the range of [1100, 1900].

(2) Run the "SmallFixedWingUAVattitude.slx" program, and click on the "Run" button on the toolbar above Simulink. After the program finishes, click on "Data Inspector". Relevant state signals are found on the left side of the "Data Inspector" interface.

(3) Observe the simulation data and plot the images. When a pitch-up command is given at the 10th second, the elevator generates a negative deflection at the 10th second. The pitch control result and elevator deflection are shown in Fig. 6.20.

(4) Repeat Steps (1) and (2). Delay the input step of "RollCH1" channel by 10 s in "Command" and set it to 1600. Observe the simulation data and plot the curves. When a positive roll motion occurs, due to coordinated turn, the yaw also moves positively. The flight status is shown in Fig. 6.21.

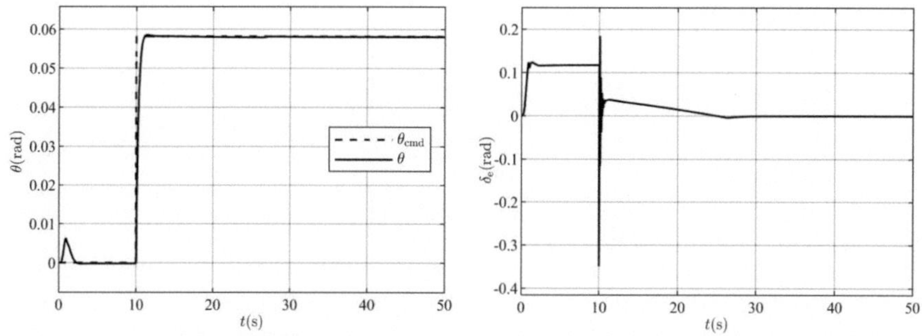

(a) Pitch control response (b) Elevator deflection changing

Fig. 6.20 Pitch control performance

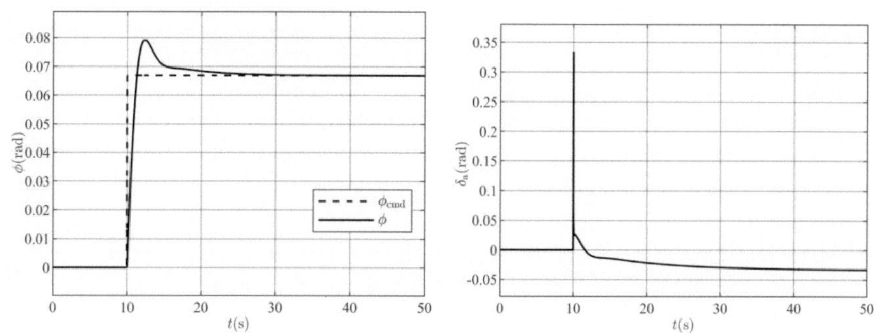

(a) Roll control response (b) Aileron deflection changing

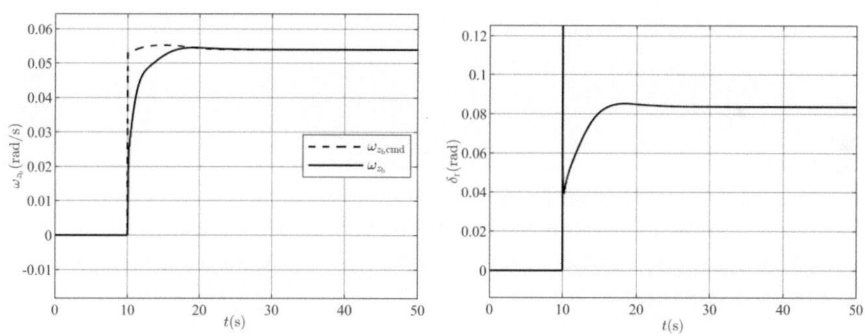

(c) Yaw control response (d) Rudder deflection changing

Fig. 6.21 Flight response under coordinated turn

6.3 Analysis Experiment

6.3.1 Experimental Objective

1. Things to prepare

(1) Software: MATLAB R2020b or above.
(2) Program: Experimental package: "e4/e4-2". The package mainly contains: micro-small fixed-wing UAV parameter initialization file "InitDatactrl.m", which can import UAV parameters into the workspace after running, micro-small fixed-wing UAV attitude simulation model file "SmallFixedWingUAVanalysis.slx" and micro-small fixed-wing UAV altitude velocity simulation model file "SmallFixedWingUAVupper.slx".

2. Objective

(1) Adjust the parameters of the pitch channel PID controller to improve control performance and analyze the effects with each PID parameter.
(2) Use the tuned parameters to sweep the system, draw Bode diagrams, observe the system's magnitude and phase frequency response curves, and analyze stability margins.
(3) Analyze the stability of coordinated turning control in lateral channel.
(4) Analyze the stability of altitude and airspeed channel control.

6.3.2 Experimental Procedure

In Step 1, taking the pitch channel as an example, the model is first initialized. In Step 2, the PID controller of the pitch channel is adjusted by the PID parameter tuning method to observe the effects with different PID parameters on the control performance. In Step 3, the designed time-domain controller is implemented, and Bode diagrams are generated. In the following steps, the parameters of the lateral attitude and altitude-airspeed channels of the fixed-wing UAV are adjusted using the same parameter tuning method to achieve good tracking performance for the entire low-level flight control.

6.3.2.1 Pitch Channel Control Analysis

As an illustrative example, consider the pitch channel of an micro-small fixed-wing UAV. The process begins with model parameter initialization, which involves setting the initial values for the model parameters. This is followed by tuning the pitch rate loop parameters and the pitch angle loop parameters. The next step is to set the input-output points for frequency sweeping. This allows the time-domain parameters of the entire pitch channel PID controller to be tuned. Finally, the frequency-domain

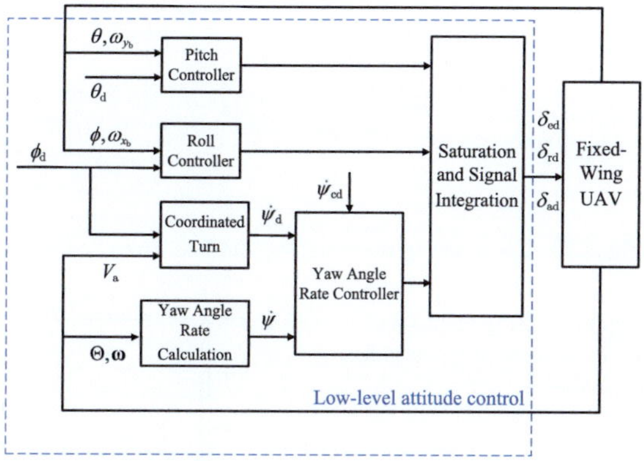

Fig. 6.22 A low-level attitude control structure for fixed-wing UAVs

characteristics of the pitch channel are observed through Bode diagrams. The simulation model currently under consideration is the attitude control model, whose control block diagram is depicted in Fig. 6.22. It should be noted that the yaw channel, in addition to the desired yaw rate (denoted as $\dot{\psi}_{\mathrm{d}}$) computed from the coordinated turn relationship, can also be assigned a desired yaw rate ($\dot{\psi}_{\mathrm{cd}}$) by the remote control.

1. Step 1: Model Initialization

To adjust the parameters of the PID controller, the environment platform in the basic experiment needs to be simplified. The modified model is found in the file "e4/e4-2/SmallFixedWingUAVanalysis.slx". The parameter adjustment method is consistent across all attitude channels of the fixed-wing UAV, so the adjustment of parameters for the pitch channel is taken as an example. The initial state for adjusting the parameters should be that the UAV is in a certain altitude flight, with the initial altitude set to 50 m, throttle value set to 0.3725, and the initial airspeed of the aircraft set to 12.12 m/s. The corresponding settings in the parameter setting file "InitDatactrl.m" are shown in Code 6.8.

Code 6.8 UAV parameters, from "e4/e4-1/InitDatactrl.m"

```
%% Initial Condition
% Start from ground:
uav.ic = struct;
uav.ic.Pos_0 = [0; 0; 50];
uav.ic.Euler_0 = [0; 0; 0];
uav.ic.PQR_0 = [0; 0; 0];
uav.ic.Vb_0 = [12.12; 0; 0];
uav.ic.gsLL = [45 120];
uav.ic.gsH = -1;
```

2. Step 2: Adjustment of Pitch Controller Parameters

(1) Adjustment of Angular Rate Loop Parameters

In the "Pitch Controller" module located under the "SmallFixedWingUAVanalysis/-Controller/Pitch Controller" in the "SmallFixedWingUAVanalysis.slx" file, replace the desired angular rate $\omega_{y_{\mathrm{b}}\mathrm{d}}$ with a step input. Also, set "q" signal line (correspond-

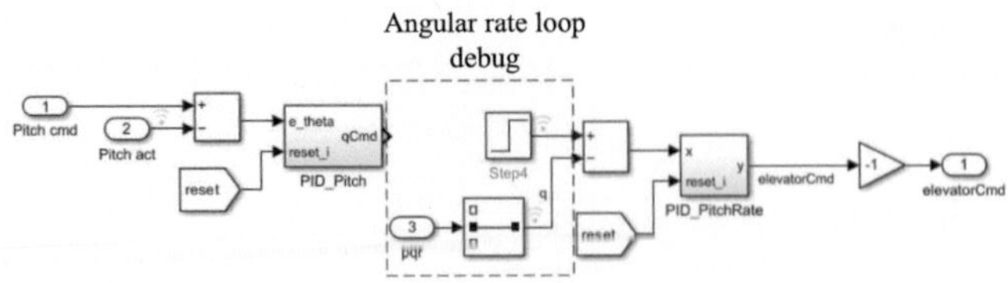

Fig. 6.23 Pitch rate loop debugging module, from "e4/e4-2/SmallFixedWingUAVanalysis.slx"

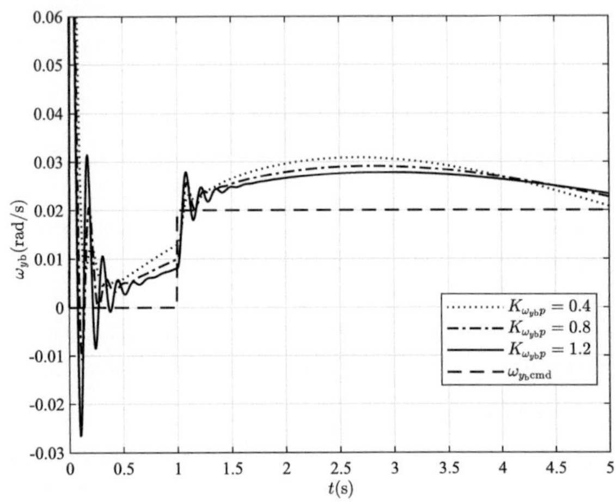

Fig. 6.24 Step responses with different proportional coefficients

ing to the actual pitch rate ω_{y_b}) and the step signal line to "Enable Data Logging" to obtain the step response curve. The pitch rate loop debugging model is shown in Fig. 6.23.

Adjust the parameters of the PID controller in the "PID_PitchRate" module. First, adjust the proportional coefficients and set the integral and derivative coefficients to 0. Then, run file "InitDatactrl.m" by clicking the "Run" button in Simulink, and observe the recorded pitch rate change curve in "Simulink Data Inspector". The proportional coefficient gradually increases during the debugging process. The step response curves for different proportional coefficients are shown in Fig. 6.24. A larger proportional coefficient makes the pitch rate closer to the desired step signal, but large proportional coefficients bring about a large overshoot and settling time. Therefore, the proportional coefficient should not be too large, even if there is a steady-state error.

Then, adjust the integral coefficient in the "Pitch Controller" module, set the proportional coefficient to 12, and keep the derivative coefficient at 0. Click the "Run" button and observe the recorded pitch rate change curve in "Simulink Data Inspector". During the debugging process, the integral coefficient is gradually increased. The step response curves for different integral coefficients are shown in Fig. 6.25. It can be observed from the figure that adjusting the integral coefficient can resolve the steady-state error. The larger the integral coefficient, the smaller the steady-state

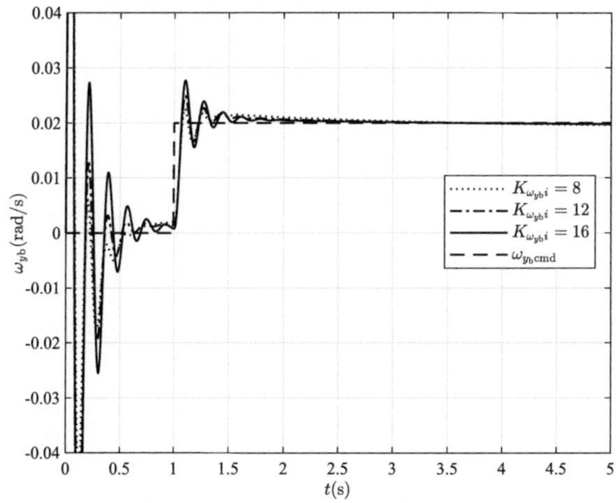

Fig. 6.25 Step responses with different integral parameters

Table 6.2 Final PID parameters for pitch attitude channel

Angle loop proportional	Angle loop integral	Angular rate loop proportional	Angular rate loop integral	Angular rate loop derivative
6	0.4	0.8	12	0.04

error of the pitch rate loop and the faster the tracking speed; however, an excessively large integral term can increase the overshoot of the tracking curve.

Then, adjust the derivative coefficient in the "Pitch Controller" module. Note that the derivative coefficient is generally small, as a large derivative term can cause oscillations in the system. Finally, fine-tune the proportional coefficient to achieve a set of satisfied coefficients as shown in Table 6.2, with their step response curves shown in Fig. 6.26.

(2) Adjustment of Angle Loop Parameters

Adjust the angle loop proportional coefficient using the angular rate loop control parameters already obtained. That is to say, the PID parameters in the "PID_PitchRate" under "SmallFixedWingUAVanalysis/Controller/Pitch Controller" are adopted from Table 6.2. In the "Pitch Controller" module, replace the desired pitch angle "Pitch cmd" with a step input and set the step input and the actual pitch angle output "Pitch act" to "Enable Data Logging" as illustrated in Fig. 6.27. Gradually increase the proportional coefficient and observe the recorded pitch angle change curve in "Simulink Data Inspector". The step response for different proportional coefficients is shown in Fig. 6.28.

By observing Fig. 6.28, it is found that after adjusting the proportional coefficients, the steady-state error between the actual pitch angle and the desired pitch angle is very small. Therefore, only a small integral coefficient is needed to achieve satisfied control results. The step response with this set of coefficients is shown in Fig. 6.29.

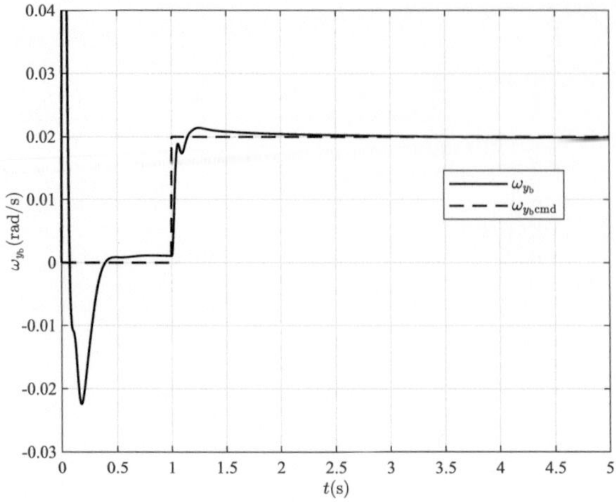

Fig. 6.26 Step response of pitch rate with a set of satisfied coefficients

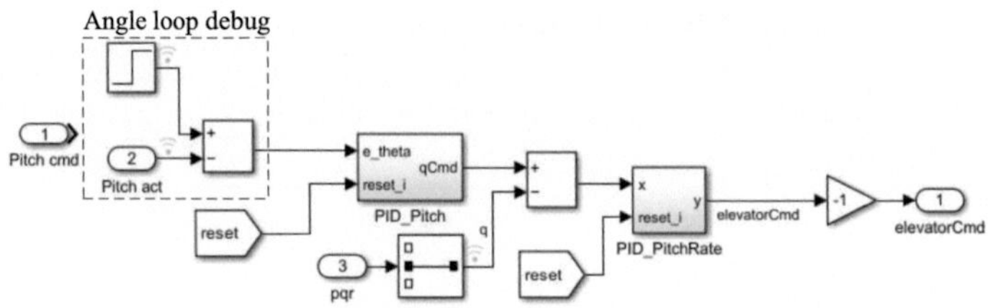

Fig. 6.27 Pitch angle loop debugging model, from "e4/e4-2/SmallFixedWingUAVanalysis.slx"

Fig. 6.28 Step responses of pitch angle with different proportional coefficients

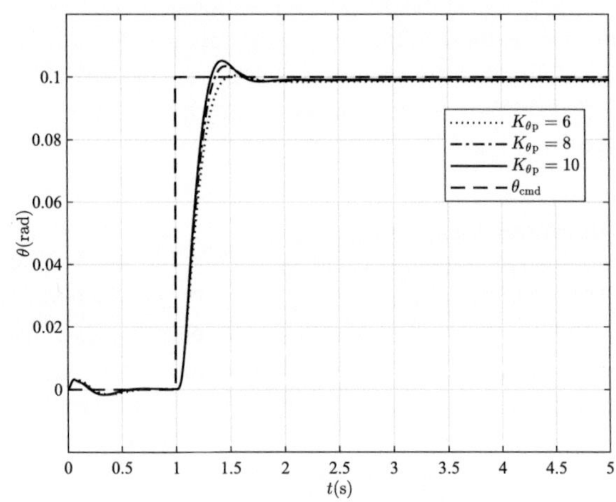

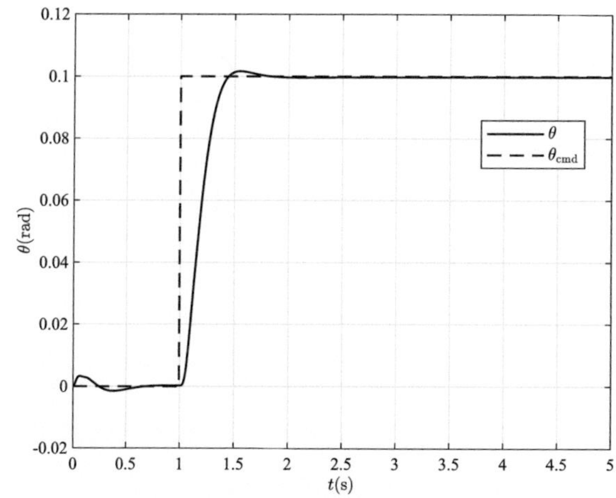

Fig. 6.29 Step response with a set of satisfied coefficients

In order to circumvent the noise impact caused by the derivative term, no derivative is applied in the angle loop. The final selected coefficients are presented in Table 6.2.

3. Step 3: Frequency Sweep to Obtain Bode Diagram

(1) Setting the input-output points for the signal.

Input point setting: Choose the pitch angle channel for frequency sweeping. Set the input point by selecting the signal line, right-clicking, and choosing "Linear Analysis Points → Open-loop Input" command from the shortcut menu.

Output point setting: Similar to the input point setting, and select "Open-loop Output" command. The configured input-output points are shown in Fig. 6.30.

(2) Obtain the Bode diagram.

First, right-click on the control module to analyze it in Simulink. Then, find "APP → Model Linearization Manager" in the Simulink menu bar. Next, select "Linear Analysis → Linearize Block..." command from the menu, as shown in Fig. 6.31. In the pop-up dialog box that appears, choose "LINEAR ANALYSIS" and then select "Model I/Os" from "Analysis I/Os" menu. Click on "Bode" to obtain the

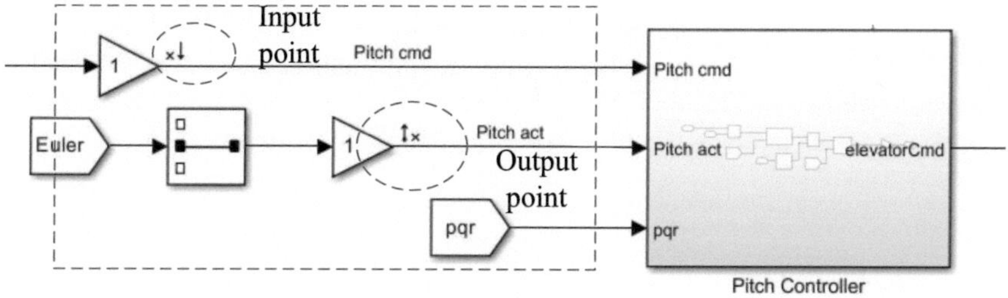

Fig. 6.30 Configured input-output points, from "e4/e4-2/SmallFixedWingUAVanalysis.slx"

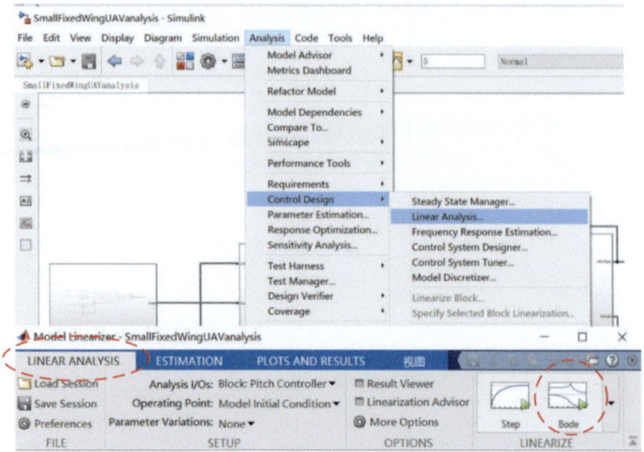

Fig. 6.31 Menu path for "LINEAR ANALYSIS"

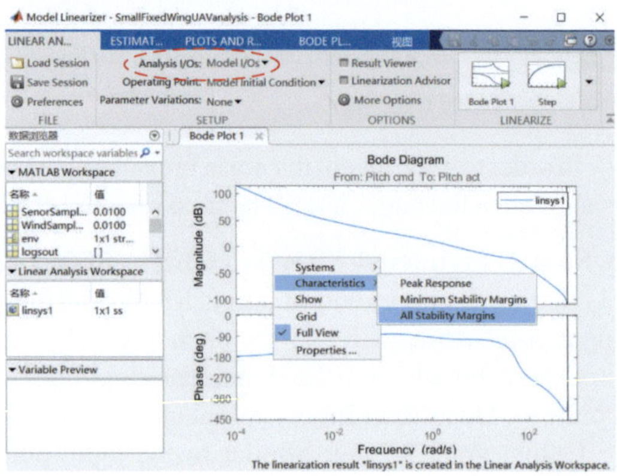

Fig. 6.32 Menu path of Bode diagram

Bode diagram, as shown in Fig. 6.32. Right-click in "Bode Diagram", then choose "Characteristic → All Stability Margins" from the pop-up shortcut menu to see the cutoff frequency, gain margin, and phase margin. Further, select "Characteristic → Minimum Stability Margins" to display the gain crossover frequency and phase crossover frequency in the diagram.

Apply the PID parameters already adjusted in the time-domain analysis in Sect. 6.3.2.1 for frequency domain analysis of the pitch channel. Follow the steps mentioned above, set "Pitch cmd" signal as the input point and "Pitch act" as the output point. The open-loop Bode diagram of the pitch angle control channel is obtained as shown in Fig. 6.33. From the Bode diagram, the amplitude margin is 14.7dB, the cutoff frequency is 64.2 rad/s, and the phase margin is 76.3°.

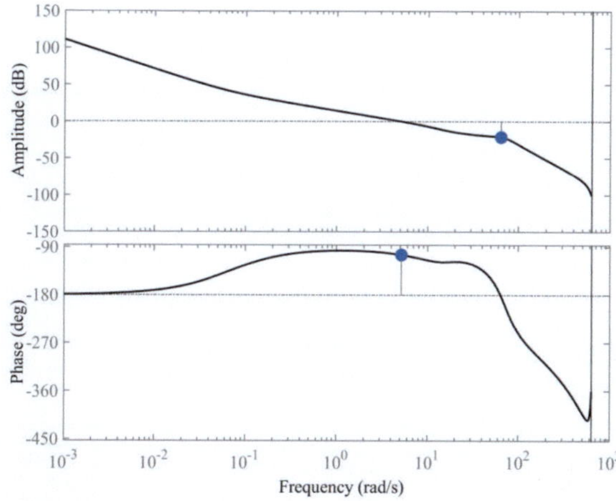

Fig. 6.33 Bode diagram of the pitch channel

6.3.2.2 Lateral Channel Control Analysis

The lateral channel includes the roll channel and the yaw angle change rate channel. The PID controllers of the lateral channel are adjusted using the parameters first for angular rate and then for attitude angle parameters. Then, reasonable control performance is ultimately achieved.

1. Step 1: Implementation of Lateral Roll Channel Controller

The PID control parameter adjustment strategy for the roll channel is similar to that of the pitch channel, employing a method of first adjusting the inner loop (roll rate loop) and then adjusting the outer loop (roll angle loop). The adjustment of control parameters follows the steps outlined in Sect. 6.3.2.1 for pitch channel parameter tuning. At the beginning, the PID control parameters in "PID_RollRate" module are adjusted, followed by adjustments to the PID control parameters in "PID_Roll" module. The final parameters for the roll channel are shown in Table 6.3. Running the Simulink file, the control results of the roll channel are observed in "Simulink Data Inspector", and its step response is shown in Fig. 6.34.

2. Step 2: Implementation of Lateral Yaw Channel Controller

In order to achieve yaw control, it is essential to satisfy the coordinated turning equation with the roll channel. Consequently, the desired roll angle provides the

Table 6.3 Final PID parameters selected for roll channel

Angle loop proportional	Angle loop integral	Roll rate loop proportional	Roll rate loop integral	Roll rate loop derivative
9	0.4	0.8	2	0.01

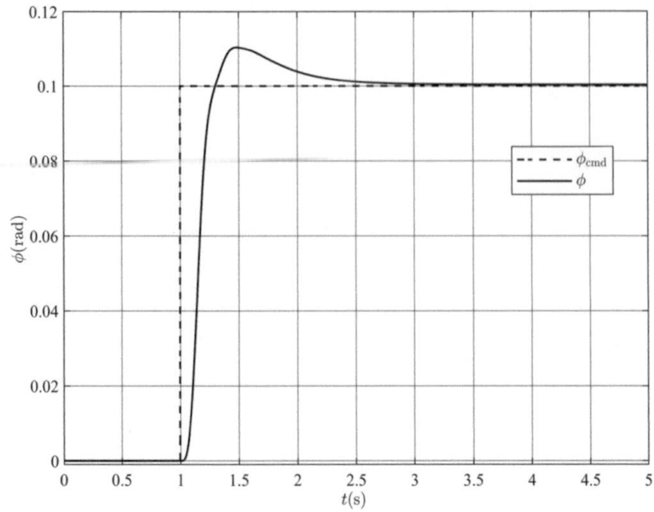

Fig. 6.34 Step response of roll control

Table 6.4 Final PID parameters selected for yaw rate channel

Yaw rate loop proportional	Yaw rate loop integral	Yaw rate loop derivative
4	3	0.3

desired yaw angle change rate. In other words, the yaw channel receives its desired input through roll angle feedforward, necessitating only the design of a yaw rate loop, i.e., only one PID controller is required. The aforementioned adjustment steps for pitch rate, as outlined in Sect. 6.3.2.1, are repeated for the parameters within the "Yaw Controller" module. The final parameter adjustment results are presented in Table 6.4. To observe the control results of the yaw angle change rate channel, the Simulink file "SmallFixedWingUAVanalysis.slx" is executed, and its response is illustrated in Fig. 6.35.

6.3.2.3 Altitude-Airspeed Channel Control Analysis

Once the attitude controller parameters have been adjusted with attitude maintained in a stable state, the parameters for the altitude-airspeed channel are adjusted using the PID parameter tuning method in order to achieve effective tracking.

1. Step 1: Implementation of Altitude Channel Controller

To ensure the attitude controller parameters are properly tuned and stability is maintained, the PID control parameters for the altitude channel are adjusted. Similar to pitch channel tuning, the proportional coefficient is adjusted first, and then the integral coefficient is adjusted. Finally, the derivative coefficient is adjusted. The

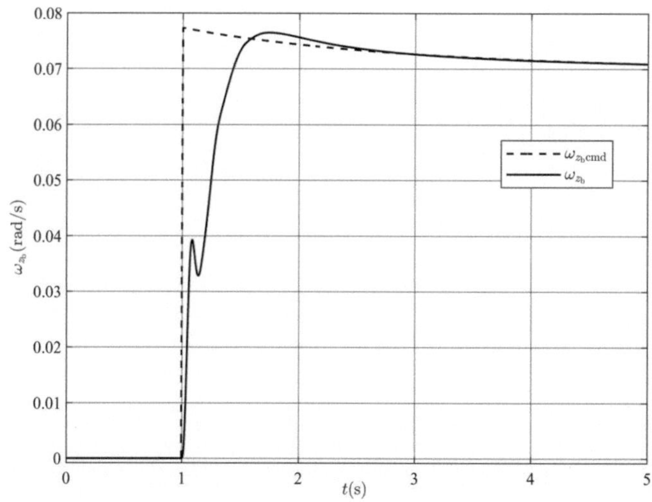

Fig. 6.35 Yaw control response

Table 6.5 Final PID parameters selected for altitude channel

Altitude proportional	Altitude integral	Altitude derivative
0.055	0	0

parameter adjustment method for the "Altitude Controller" module follows the steps outlined in Sect. 6.3.2.1. The final parameters for the altitude channel are presented in Table 6.5. First, the "InitDatactrl.m" file should be run in order to initialize the parameters. Secondly, the Simulink file "SmallFixedWingUAVupper.slx" should be run in order to observe the control results of the altitude channel in "Simulink Data Inspector". Finally, the response curve is shown in Fig. 6.36.

2. Step 2: Implementation of Airspeed Channel Controller

For the airspeed channel, only one PID controller is needed. Repeating the adjustment steps in Sect. 6.3.2.1, parameters in the "Airspeed Controller" module are adjusted, and the final parameter adjustment results are shown in Table 6.6. Run Simulink file "SmallFixedWingUAVupper.slx", observe the control results of the airspeed channel in "Simulink Data Inspector", and the response is shown in Fig. 6.37.

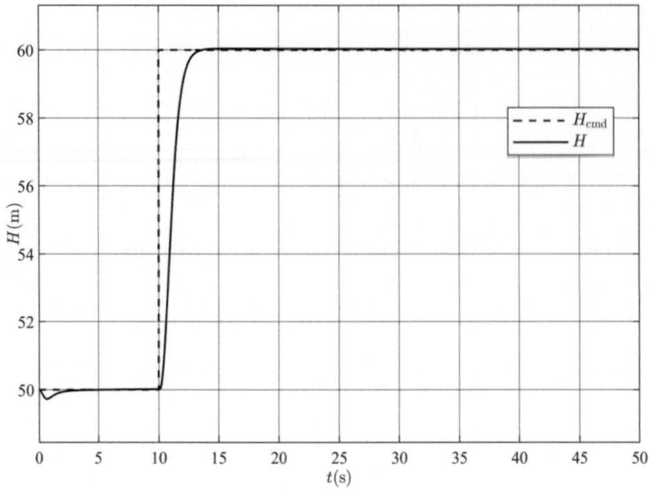

Fig. 6.36 Altitude control response

Table 6.6 Final PID parameters selected for airspeed channel

Airspeed proportional	Airspeed integral	Airspeed derivative
0.5	0.13	0

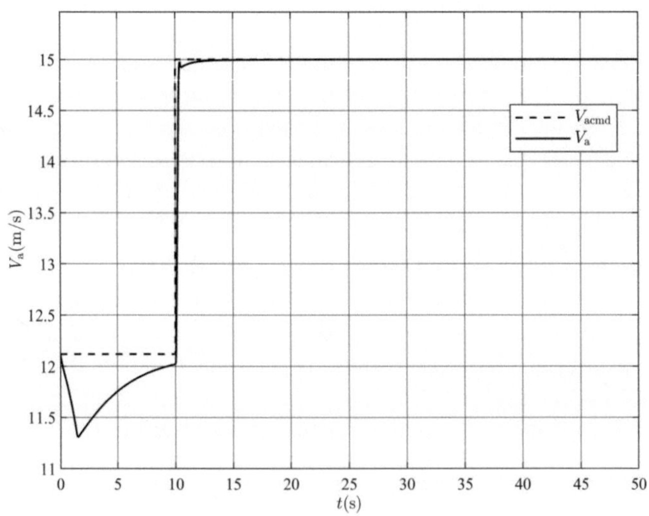

Fig. 6.37 Airspeed control response

6.4 Design Experiment

6.4.1 Experimental Objective

1. Things to prepare

(1) Software: MATLAB R2022b or above, RflySim toolchain. The RflySim toolchain provides a HIL simulation environment for micro-small fixed-wing UAVs.
(2) Program: Experimental package "e4/e4-3". The package mainly includes: initialization files for micro-small fixed-wing UAV parameters, a simulation model for micro-small fixed-wing UAV corrective control "SmallFixedWingUAVanalysis-Continuous.slx", transfer function design program "controlsystemdesigner.m".

2. Objective
After takeoff, the micro-small fixed-wing UAV reaches point A (100 m, 100 m, 50 m) as the starting point and flies in a straight line to point B (150 m, 100 m, 70 m) at an airspeed of 12.12 m/s. The goal is to design a controller that ensures stable flight with optimal flying qualities. Establish the transfer function model of the attitude control channel and design a corrective controller to ensure that the steady-state error of the attitude angular rate loop $e_{rss} \leq 1$, cut-off frequency > 10 rad/s, phase margin $> 55°$, magnitude margin > 10 dB. The cut-off frequency of the attitude angle loop should be > 5 rad/s, and a phase margin $> 50°$.

6.4.2 Experimental Design

To complete the objectives of the design experiment and achieve HIL simulation, it is necessary to follow the steps below for the experiment, as shown in Fig. 6.38.

First, determine the tracking targets for each channel based on the reference flight trajectory. The UAV flies at a constant airspeed, so $V_{\mathrm{ad}} = 12.12$ m/s. The UAV performs a straight-line climbing flight, so the yaw angle is constant, that is, $\dot{\psi} = 0$. The problem of correcting the yaw angle after a deviation from the trajectory is not considered, so $\psi_{\mathrm{d}} = 0$ is set. The trajectory tracking problem will be introduced in detail in the next chapter. According to the coordinated turn equation, one has

$$\phi_{\mathrm{d}} = \arctan \frac{V_{\mathrm{a}}}{g} \dot{\psi}_{\mathrm{d}} = 0.$$

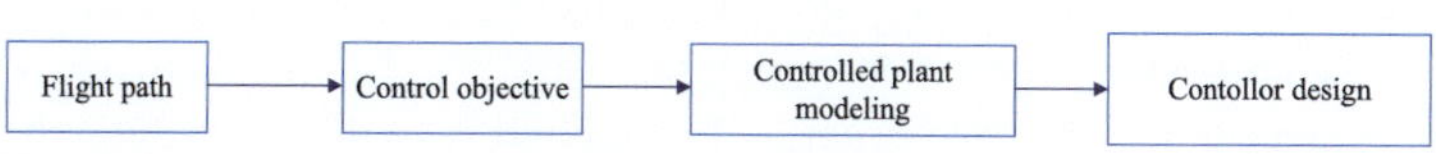

Fig. 6.38 Design experiment flowchart

The tracking target for the altitude channel is given by

$$\begin{cases} H_{\mathrm{d}} = 50, & 0 \leqslant t < 10 \\ H_{\mathrm{d}} = t + 50, & 10 \leqslant t \leqslant 20 \\ H_{\mathrm{d}} = 70, & \text{others.} \end{cases}$$

Then, the controllers for each channel will be designed to achieve the desired flying qualities and complete the UAV's flight mission.

6.4.3 *Experimental Procedure*

The experimental procedure for Design Experiment is given in the following. In Step 1, the pitch channel of the micro-small fixed-wing UAV is taken as an example, and the simulation model of the fixed-wing UAV is simplified for subsequent frequency domain analysis. In Step 2, the pitch rate loop is analyzed to obtain the transfer function in the form of zeros and poles. In Step 3, the open-loop transfer function gain is adjusted to meet the design requirements. In Step 4, the pitch channel is scaled to obtain a normalized transfer function. In Step 5, the designed angular rate controller is incorporated. In Step 6, open-loop zeros and poles are introduced to ensure flying qualities for the pitch angle loop. Finally, the designed attitude controller is implemented into the pitch channel.

6.4.3.1 Design Experiment

In designing the corrective controller for the attitude channel, the pitch channel is taken as a case study. The corrective controller for pitch rate is first designed, followed by the pitch angle. For the time being, the tracking reference for the pitch channel is set as a pulse input:

$$\begin{cases} \theta_{\mathrm{d}} = 0.156 & 5 \leqslant t \leqslant 20, \\ \theta_{\mathrm{d}} = 0 & \text{else.} \end{cases}$$

In order to achieve the desired frequency-domain design objective, it is first necessary to obtain the open-loop transfer functions of the pitch rate loop and the pitch angle loop. The second step is to adjust the correction loop's open-loop gain to meet the steady-state error requirement and to improve the cut-off frequency and phase margin by introducing the corrector (lead-lag corrector) to achieve the desired criterion. The improvement of the frequency-domain performance is observed through Bode diagrams. Two methods exist for designing the corrector. The first is to calculate the corrector based on performance criteria. The second is to directly adjust the open-loop gain in the Bode diagram and bring open-loop zeros and poles to achieve the desired criterion.

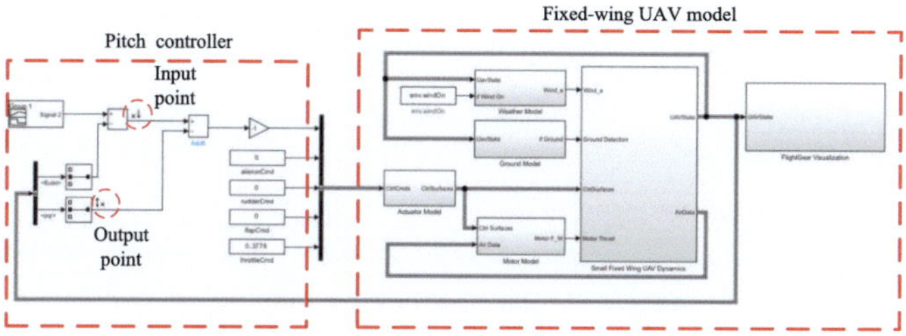

Fig. 6.39 A simplified model of the pitch control system, from "e4/e4-3/SmallFixedWingUAVanalysisContinuous.slx"

1. **Step 1: Simplification of the Overall Structure**
Open file "e4/e4-3/SmallFixedWingUAVanalysisContinuous.slx". Only the longitudinal pitch channel is analyzed. The simplified model of the pitch control system is shown in Fig. 6.39. This simulation model is similar to Fig. 6.15, except that all commands except the elevator control command are set to 0, and the throttle command is set to 0.3724. Only the pitch channel is considered for designing and analyzing the pitch controller.

2. **Step 2: Analysis of Angular Rate Loop**
Set the desired pitch rate as the input and the actual pitch rate as the output. Subsequently, the input-output points are set, and a Bode diagram is generated. The variable "linsys1" will appear in "Linear Analysis Workspace" on the left side. First, select "PLOT AND RESULTS" from the main menu to obtain the transfer function model. Then, navigate to the submenu of "RESULTS" and select "Result Viewer". Finally, choose "Zero-Pole-Gain" from the pop-up window to complete the process. Figure 6.40 illustrates the aforementioned operational step and the resulting transfer function, which is as follows:

Fig. 6.40 Linear analysis tool

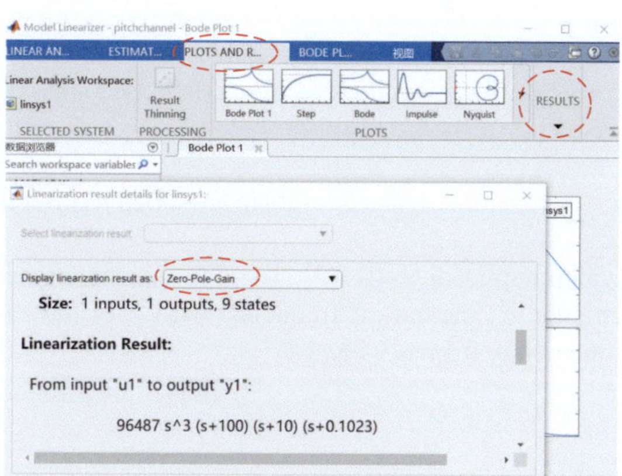

$$G(s) = \frac{82640\,s(s+100)(s+10)(s+2.308*10^{18})(s^2-1.192*10^{-7}s+6.904*10^{-14})}{(s+21.94)(s+10.28)(s+54.13)(s+55.84)(s+100)(s+0.1236)(s+6.398*10^{-7})(s-6.398*10^{-7})(s+2.308*10^{18})}. \tag{6.67}$$

It is simplified as

$$G(s) = \frac{1.3168*10^{11}s(0.1s+1)(4.3328*10^{-19}s+1)(1.4484*10^{13}s^2-1.7265*10^{6}s+1)}{7.9608*10^{10}(0.0456s+1)(0.0973s+1)+(0.0185s+1)(8.0906s+1)(1.5630*10^{6}s+1)(1.5630*10^{6}s-1)(4.3328*10^{-19}s+1)}. \tag{6.68}$$

3. Step 3: Open-Loop Gain Adjustment
The steady-state error of the system under the input $r(t) = t$ is obtained using the final value theorem:

$$e_{rss} = \frac{1}{K}.$$

Adjusting the open-loop gain based on the steady-state error, the open-loop gain of the system without correction is:

$$K = \lim_{s\to 0}\frac{1}{s}\frac{1.3168*10^{11}s(0.1s+1)(4.3328*10^{-19}s+1)(1.4484*10^{13}s^2-1.7265*10^{6}s+1)}{7.9608*10^{10}(0.0456s+1)(0.0973s+1)+(0.0185s+1)(8.0906s+1)(1.5630*10^{6}s+1)(1.5630*10^{6}s-1)(4.3328*10^{-19}s+1)}$$
$$= 1.6541.$$

To satisfy $e_{rss} \leq 1$, it is required that $K \geq 1$. Currently, $K = 1.6541$, which meets the design requirement. For better quality in the frequency domain, the open-loop gain can also be increased if the open-loop gain is increased by four times. The new open-loop gain becomes $K' = 6.6164$.

4. Step 4: Angular Rate Loop Corrector Design
Plot the open-loop Bode diagram of the rate loop, as shown in Fig. 6.41. The phase margin is 25.6°, and the cut-off frequency is $\omega_c = 52.6\text{rad/s}$, which does not meet the desired criterion. By using phase lag correction, the phase margin is increased, and the steady-state performance of the system is improved.

Setting the cut-off frequency to $\omega_c^{'} = 20$ rad/s, it is observed from the Bode diagram that the phase margin at this frequency is 98.7°, meeting the design requirements. From the magnitude-frequency characteristic curve of the Bode diagram, the magnitude response at this frequency is 10.2 dB. At $\omega_c^{'} = 20$ rad/s, after adding the corrector, the magnitude characteristic becomes 0dB. Based on the lagging phase frequency response, it follows that:

$$20\lg h + 20\lg|G(j\omega_c^{'})| = 0$$

which implies $h = 10^{-0.51}$. To minimize the impact of the lagging phase characteristic on $\omega_c^{'}$, the cut-off frequency $(h\tau)^{-1}$ of the corrector should be positioned ten times away from $\omega_c^{'}$, i.e.,

$$(h\tau)^{-1} = 0.1\omega_c^{'}.$$

This obtains $\tau = 1.618$. At this point, the transfer function of the corrector is:

$$G_c(s) = \frac{1 + h\tau s}{1 + \tau s} = \frac{1 + 0.5s}{1 + 1.618s}.$$

5. Step 5: Integrating the Corrector into the Angular Rate Loop

An additional gain amplifier is incorporated into the model following the angular rate error, with the objective of enhancing the open-loop gain. So far, the open-loop gain has been augmented by a factor of four. Subsequently, the corrector is incorporated following the augmentation of the open-loop gain. The angular rate control loop after corrector is illustrated in Fig. 6.42. The open-loop Bode diagram is obtained by sweeping the corrected pitch rate loop, as shown in Fig. 6.43. Compared to the uncorrected case in Fig. 6.41, the frequency-domain characteristics have been enhanced, exhibiting a magnitude margin of 16 dB and a phase margin of 102°.

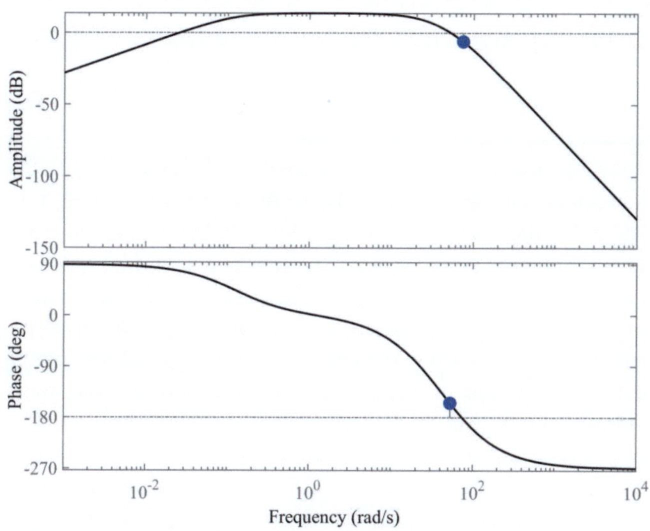

Fig. 6.41 Open-loop Bode diagram of the uncorrected angular rate loop

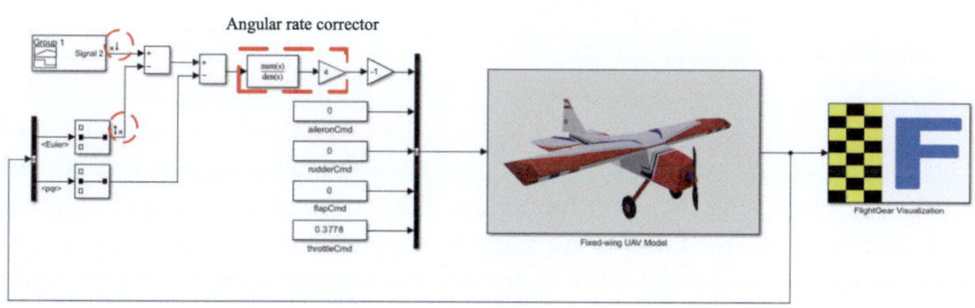

Fig. 6.42 Pitch channel control with a corrector added

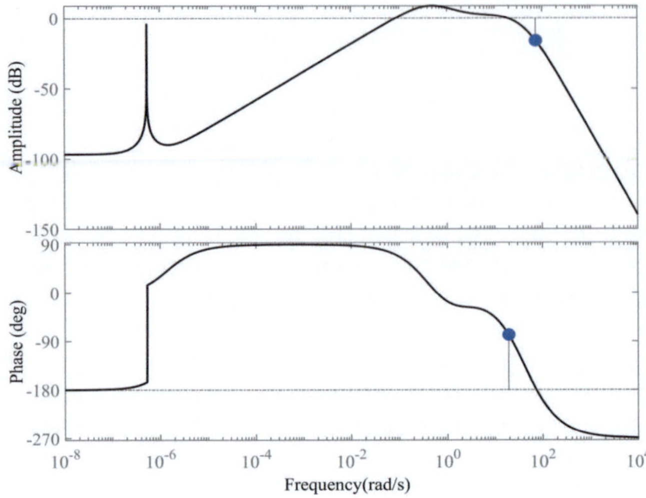

Fig. 6.43 Open-loop Bode diagram of the corrected pitch rate loop

6. Step 6: Attitude Loop Corrector Design

The input of the attitude loop is the desired pitch angle, and the output is the actual pitch angle, as shown in Fig. 6.42. Similar to the angular rate loop, the open-loop Bode diagram of the uncorrected pitch angle loop is obtained, as shown in Fig. 6.44.

From the Bode diagram, it can be observed that the cut-off frequency is 41.9 rad/s, the phase margin is 91.4°, and the magnitude margin is 27 dB, which meets the design requirements. Repeating the aforementioned steps, the open-loop transfer function of the pitch angle loop is obtained. Utilizing the transfer function, a ".m" file is created, as illustrated in Code 6.9. Upon running this file, the dialog box "ControlSystemDesigner" with plots is displayed, as shown in Fig. 6.45.

Code 6.9 Transfer function design, from "e4/e4-3/controlsystemdesigner.m"

```
num=[2.369e23 2.654e25 2.891e26 4.739e26 7.972e20
    -1.755e17];
den=[1 2.308e18 5.606e20 4.417e22 1.536e24 4.581e25
    4.18e26 5.5e26 8.607e25 3.521e12 5.681e09];
G=tf(num,den);
controlSystemDesigner(G);
```

According to the step response in Fig. 6.46, it is observed that the system has a long response time and significant overshoot. To address this issue, the Bode diagram is adjusted by increasing the open-loop gain, eliminating overshoot and shortening the response time. The resulting Bode diagram and step response after increasing the open-loop gain are presented in Fig. 6.46. The final phase margin is 81.2°, and the cut-off frequency is 32.7 rad/s.

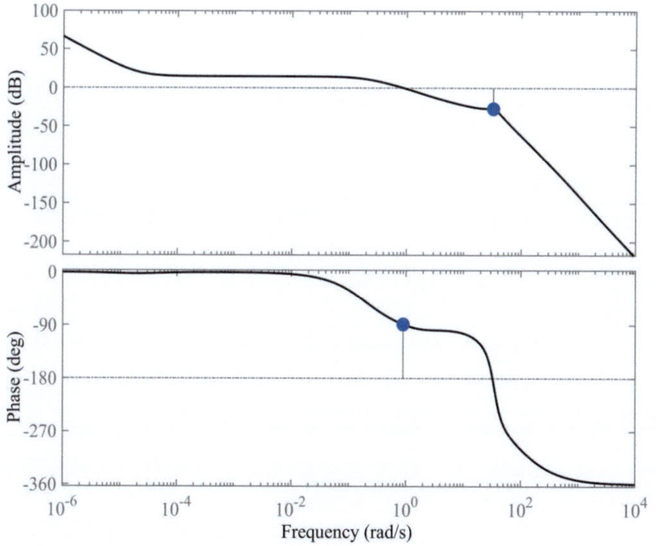

Fig. 6.44 Open-loop Bode diagram of the uncorrected pitch angle loop

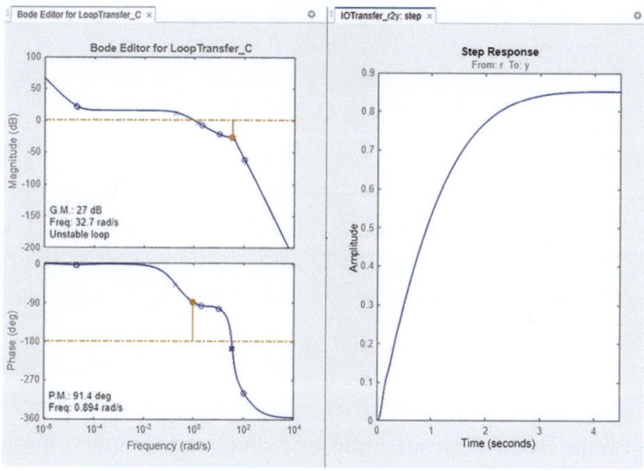

Fig. 6.45 Bode diagram and step response of pitch angle loop

Although increasing the open-loop gain can improve the response speed, it may decrease the phase margin. A lead corrector has been added to maintain the phase margin while improving the response speed. Right-click on the Bode diagram, choose "Add Pole or Zero → Lead" command from the shortcut menu, click the left mouse button on the Bode curve to add a pole or zero, and then drag the pole or zero to observe the response curve and obtain a suitable corrector. The corrected result is shown in Fig. 6.47.

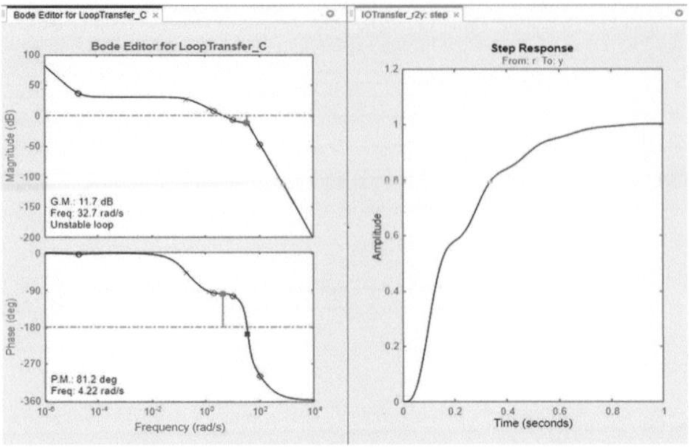

Fig. 6.46 Bode diagram and step response after increasing the open-loop gain

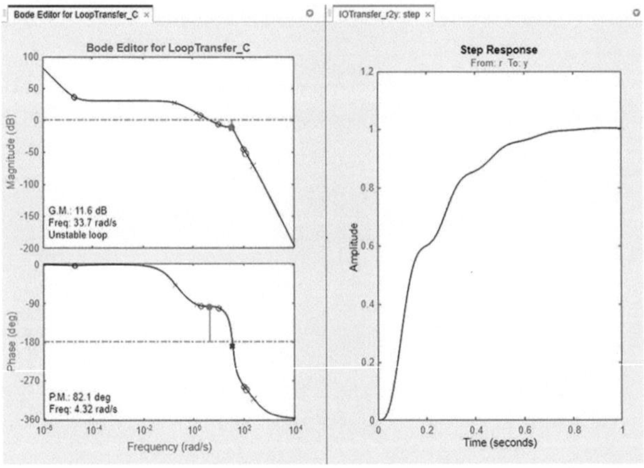

Fig. 6.47 Bode diagram and step response with a lead corrector

Right-click on the Bode diagram, choose "Edit Compensator" command from the shortcut menu, and examine the corrector obtained in the dialog box:

$$G_c(s) = 5.9288\frac{1 + 0.008s}{1 + 0.0042s}.$$

6.5 HIL Experiment

6.5.1 Experimental Objective

1. Things to prepare

(1) Software: MATLAB R2022b or above, RflySim toolchain. The RflySim toolchain provides a HIL simulation environment for micro-small fixed-wing UAVs.
(2) Program: Experimental package "e4/e4-4". The package mainly includes: micro-small fixed-wing UAV parameter initialization file "InitDatactrl.m", Z-transformation program "c2d.m", micro-small fixed-wing UAV simulation model "SmallFixedWingUAVupperHIL.slx" for the HIL experiment; HIL simulation file "SmallFixedWingattitudeUAVHIL_pitch.slx" for introducing compensator.
(3) Hardware: a CubePilot/Pixhawk autopilot, a remote controller, and a receiver.

2. Objective

(1) Implement HIL simulation of PID control for altitude and airspeed channels of the micro-small fixed-wing UAV, observe the tracking performance of the given step response, and display it in RflySim.
(2) Implement HIL simulation of corrective control as designed in the designed experiment, and observe the control performance.

6.5.2 Experimental Procedure

6.5.2.1 Experimental Preparation

It is paramount to ensure that the remote controller, receiver, CubePilot/Pixhawk autopilot, and computer are communicating usually when conducting HIL simulation for fixed-wing UAVs. Once this has been established, the simulation model code can be compiled and burned into the autopilot. The UAV must be disarmed using the remote controller and operated using the joystick. Then, the flight status of the fixed-wing UAV can be observed in RflySim3D software.

In the context of HIL experiments, two distinct control modes are established through the command input module, as illustrated in Fig. 6.48. These are the manual control mode and the altitude hold flight mode. Activation of "ch5" switch in the low position (i.e., "ch5" $<$ 1500) enables the manual control mode. In the absence of disarming, the fixed-wing UAV maintains a constant altitude flight at 30 m. The left joystick, operated via "ch3" channel, controls the throttle. The center position corresponds to a airspeed of 20 m/s, and upward movement accelerates the aircraft. The left joystick, operated via "ch4" channel, controls yaw. The right joystick, operated via "ch2" channel, controls pitch, thereby adjusting the altitude. The up and down movement control of joystick is used to roll("ch1" channel).

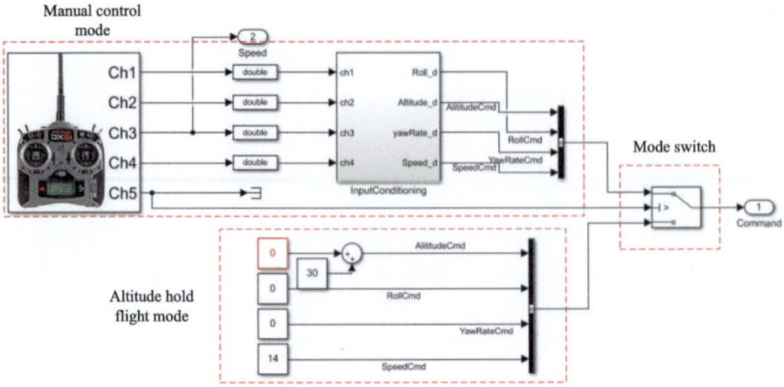

Fig. 6.48 Mode switch module, from "e4/e4-4/SmallFixedWingUAVupperHIL.slx"

6.5.2.2 HIL Simulation Experiment in Altitude-Airspeed Channel

The experimental procedure for an altitude-airspeed channel PID control in HIL simulation is as follows. In Step 1, HIL programs are prepared. In Step 2, a hardware connection is established to facilitate communication between the computer and the autopilot shown in Fig. 6.50. In Step 3, the code is compiled and downloaded to the autopilot. In Step 4, the remote controller operates the fixed-wing UAV to complete the specified flight task.

1. Step 1: Open Simulink Simulation Model
Open file "e4/e4-4/InitDatactrl.m", run parameter initialization, and open file "SmallFixedWingUAVupperHIL.slx" as shown in Fig. 6.49. In this figure, the dashed box ① represents the command input module, including remote controller command input mode (manual control mode) and altitude-airspeed given by software (altitude hold flight mode); the dashed box ② represents the controller, analyzing the pre-designed attitude controller and altitude-airspeed controller in the experiment; the dashed box ③ represents the PWM signal conversion module, converting the control output commands into PWM signal applied in the autopilot; the dashed box ④ represents the message pool, writing time information and actuator action information into the message pool for the convenience of the fixed-wing controlled dynamic system to subscribe.

2. Step 2: Hardware Connection
Connect a CubePilot/Pixhawk autopilot to the receiver, and connect the CubePilot/Pixhawk autopilot to the computer using a USB data cable, shown in Fig. 6.50.

3. Step 3: Code Compilation and Upload
In the toolbar of the "SmallFixedWingUAVupperHIL.slx" file, select the "Compile" button, as illustrated in Fig. 6.51. Once the compilation process is complete, select "Code" from the toolbar and choose "PX4 PSP Upload code to Px4FMU" to initiate the download and transfer the code to the CubePilot/Pixhawk autopilot. The completion status of the download is indicated on the left side of Fig. 6.51.

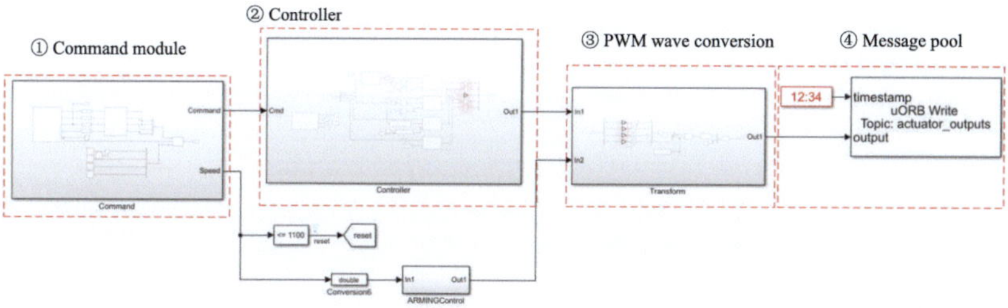

Fig. 6.49 HIL simulation model for validating the low-level flight control, from "e4/e4-4/SmallFixedWingUAVupperHIL.slx"

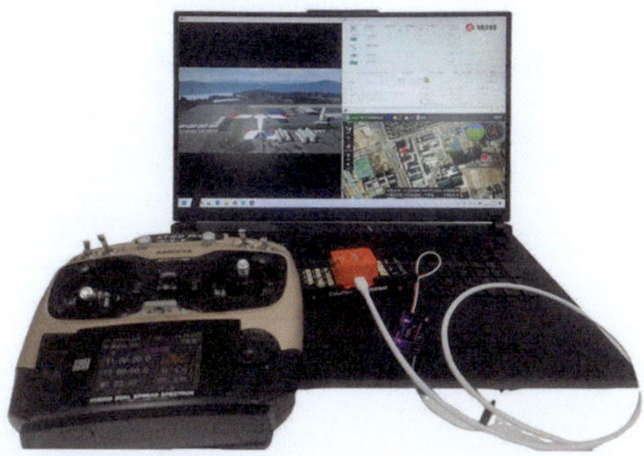

Fig. 6.50 Hardware layout of HIL simulation

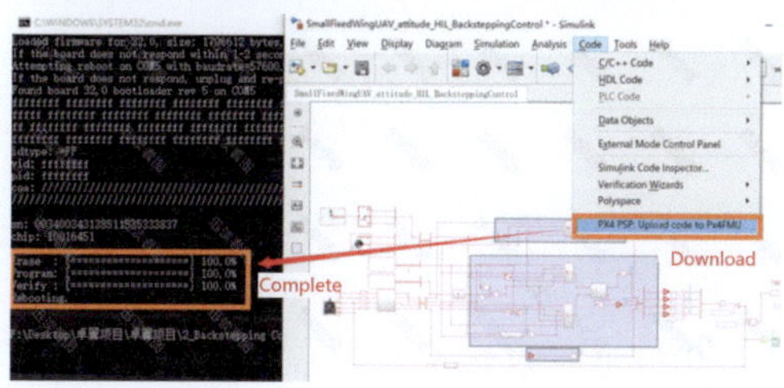

Fig. 6.51 Code download process.

4. Step 4: Observing Simulation Results

Open the "CopterSim" software interface, select the DLL model file, choose the simulation scene in "3D Display Scene", click "Start Simulation", and wait for initialization. Disarm the fixed-wing UAV with the remote controller, keep the "ch5" joystick high, and achieve manual control. Push the throttle joystick up to achieve a slow takeoff of the fixed-wing UAV. After reaching a certain height, release the throttle joystick to keep the UAV flying horizontally. Observe the changes in the aircraft flight status on the "RflySim3D" software interface and real-time flight data on the "CopterSim" software interface. The flight results of the altitude and airspeed channels are shown in Fig. 6.52.

Fig. 6.52 Comparison between SIL and HIL simulation

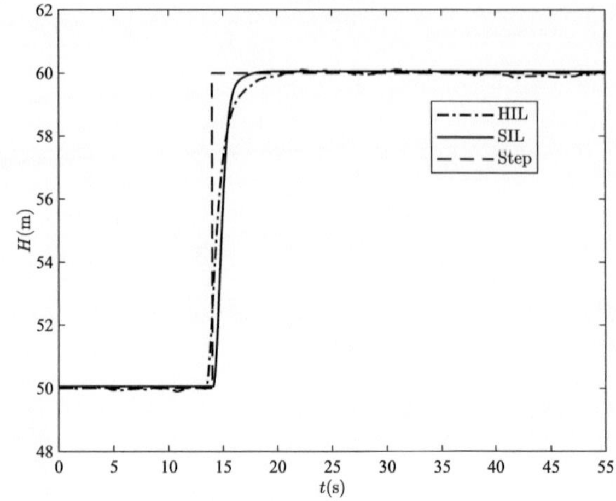

(a) Altitude response

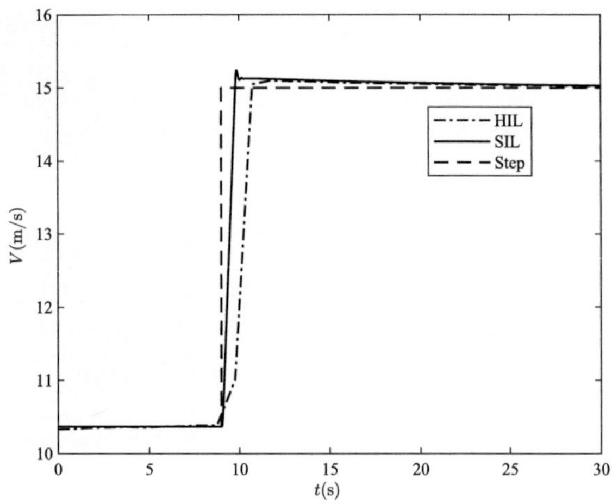

(b) Airspeed response

From Fig. 6.52, it is observed that the HIL simulation results are approximately consistent with the SIL results, and there is no apparent oscillation, demonstrating the effectiveness of the proposed controller.

6.5.2.3 HIL Simulation Experiment of Corrective Control

The preparation for the corrective control of the HIL experiment includes the following: in Step 1, the designed continuous corrective controller is discretized, and the sampled-data controller transfer function is obtained. In Step 2, the discrete controller replaces the original pitch channel PID controller. In Step 3, connect the autopilot, complete the code compilation and download, and then make the fixed-wing UAV complete the designated flight task through the remote controller.

1. Step 1: Discretization of the Continuous Corrective Controller
In this step, the designed corrective controller from Experiment 1 (Sect. 6.4.3.1) is sampled to obtain a discrete transfer function. The "c2d" function is used to convert the transfer function from the Laplace domain (s-domain) to the discrete-time domain (z-domain), as shown in Code 6.10. "Ts" represents the simulation time step.

Code 6.10 Discretization of transfer function from s-Domain to z-Domain, from "e4/e4-4/c2d.m"

```
%%pitchrate G(s)-G(z)
num=[0.5 1];
den=[1.618 1];
G=tf([num],[den]);
Ts=0.005;
Gd=c2d(G,Ts,'zoh');
```

The transfer function used to replace the continuous corrective controller of the pitch rate loop is as follows:

$$G_c(s) = \frac{1+0.5s}{1+1.618s} \rightarrow G(z) = \frac{0.309z-0.3059}{z-0.9969}.$$

The transfer function used to replace the continuous corrective controller of the pitch angle loop is as follows:

$$G_c(s) = 5.9288\frac{1+0.008s}{1+0.0042s} \rightarrow G(z) = 5.9288\frac{1.905z-1.209}{z-0.3041}.$$

2. Step 2: Replacing the Control Model
Replace the PID controllers in the simulation model of "SmallFixedWingUAVattitudeHIL_pitch.slx" with discrete modules, as shown in Fig. 6.53.

3. Step 3: HIL Simulation
Proceed with the HIL simulation following the operational preparations outlined in Sect. 6.5.2.2. When the fixed-wing UAV reaches a certain flight altitude, release the joystick, and the fixed-wing UAV maintains level flight. The joystick channel "ch5" switches to the up position, and the fixed-wing UAV enters the specified flight mode. At this time, joystick movements cannot change the flight states of the fixed-wing UAV. The HIL simulation results with the corrective controller added are shown in Fig. 6.54, demonstrating compliance with frequency-domain design requirements and ensuring stable flight.

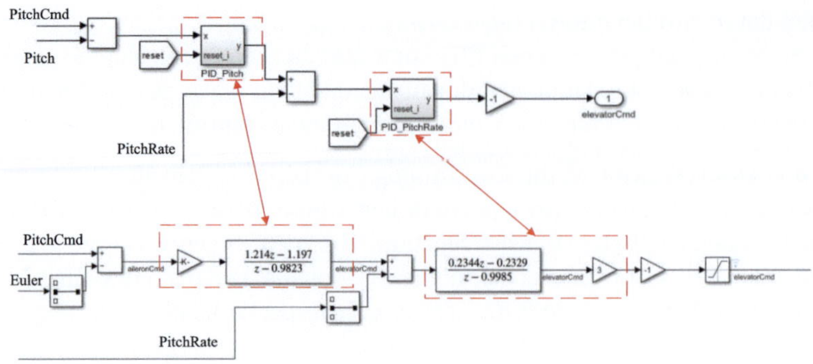

Fig. 6.53 Replacement of the PID controller with the corrective controller, from "e4/e4-4/SmallFixedWingUAVattitudeHIL_pitch.slx"

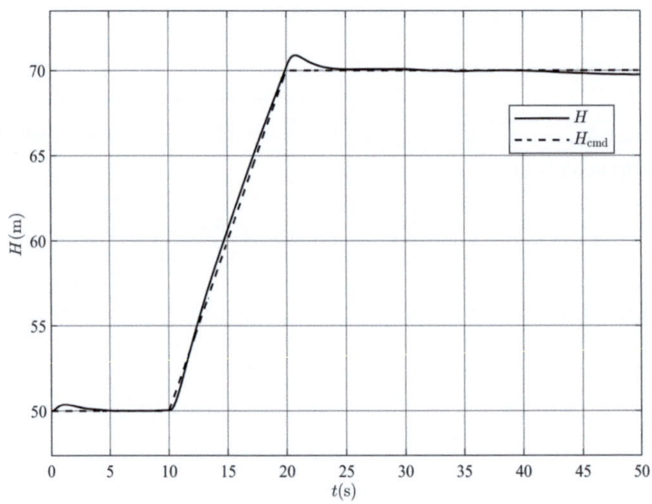

Fig. 6.54 Performance of HIL simulation with the corrective controller

6.6 Exercise

1. Previously, it is considered that

$$\phi_{\mathrm{d}} = \arctan \frac{V_{\mathrm{a}}}{g} \dot{\psi}_{\mathrm{d}} \tag{6.69}$$

to compute the roll angle corresponding to the desired yaw. However, this feed-forward approach cannot effectively compensate for uncertainties, thus cannot accurately track the desired yaw. Based on (4.63), can you consider the roll angle as the control input, design a controller to precisely track the yaw angle, and

ultimately achieve a fixed-wing UAV flying at 60 m altitude with the airspeed of 18 m/s along a circular trajectory with a radius of 60 m?
2. The elevator controls the aircraft's altitude, and the throttle controls the aircraft's speed. However, changing altitude solely through the elevator will result in a decrease or increase in speed, and changing airspeed only through the throttle will cause a change in altitude. Therefore, by coordinating the throttle and elevator, the aircraft's altitude and airspeed are controlled to achieve preset values. Total energy control involves decoupling the aircraft's flight altitude and airspeed using energy control and distribution methods. In the existing fixed-wing simulation model framework, based on the simulation model "SmallFixedWingUAVanalysis.slx" in the program file "e4-2", replace the PID control in "Control" module with a total energy controller. Realize total energy control through tuning, and analyze the performance of the total energy controller (e.g., settling time, overshoot, and steady-state error).
3. When the flight altitude is certain, linearize the fixed-wing at different flight airspeeds to obtain different models and adjust the parameters of the PID controller accordingly. Subsequently, using gain scheduling principles, design a comprehensive controller to ensure that the fixed-wing UAV exhibits a similar dynamic response at different airspeeds.
4. Replace the PID controller used in this chapter with some advanced controllers and analyze their performance.
5. In actual flight, there are many uncertainties, such as wind disturbances, disturbances in weight, and moments of inertia. Improve existing controllers for coordinated turning flight, minimizing the impact of sideslip angles as much as possible.

References

1. *PX4 autopilot user guide*, Accessed 28 Sept 2023 [Online]. Available: https://docs.px4.io/main/zh/flight_stack/controller_diagrams.html
2. *PX4 autopilot softare*, Accessed 20 Jan 2024 [Online]. Available: https://github.com/PX4/PX4-ECL/blob/1a1106878d6ab1cc48565bef9b25b07d626c5d0b/attitude_fw/ecl_pitch_controller.cpp
3. R.F. Stengel, *Flight Dynamics* (Princeton University Press, NJ, 2004)
4. A.A. Lambregts, *Vertical Flight Path and Speed Control Autopilot Design Using Total Energy Principles* (Gatlinburg, TN, USA), pp. 559–572 (Vintage, 1983)
5. Q. Quan, K.-Y. Cai, Repetitive control for nonlinear systems: an actuator-focussed design method. Int. J. Control **94**(5), 1401–1425 (2021)
6. K.J. Åström, T. Hägglund, *Advanced PID Control* (Instrument Society of America, Research Triangle Park, 1995)
7. W.J. Rugh, J.S. Shamma, Research on gain scheduling. Automatica **36**, 1401–1425 (2000)
8. H.K. Khalil, *Nonlinear Systems*, 3rd edn. (Prentice Hall, London, 2002)

Chapter 7
Path Following and Planning Experiment

Wenhan Gao, Xinquan Chen, Xunhua Dai, and Quan Quan

Path planning and path following are critical for fixed-wing Unmanned Aerial Vehicles (UAVs) to execute flight missions successfully. This chapter presents Dubins-curve-based path-planning algorithms and commonly employed path-following control algorithms. Three types of experiments are included: basic, analysis, and design experiments. These experiments will provide readers with a deeper understanding of path following and path planning for micro-small fixed-wing UAVs. In the basic experiment, readers will design path-following control algorithms for micro-small fixed-wing UAVs using MATLAB and Simulink. In the analysis experiment, readers will apply the principles of Dubins curves to path planning. In the design experiment, readers will integrate path-following and path-planning algorithms to accomplish flight missions of micro-small fixed-wing UAVs. The effectiveness of the algorithms will be further verified through Hardware-In-the-Loop (HIL) simulation using the RflySim toolchain. For more information on this chapter, please refer to the link

https://rflysim.com/doc/en/C/5.FW_Book.html.

7.1 Preliminary

Autonomous path flight of micro-small fixed-wing UAVs involves two main challenges: path following control and path planning. For micro-small fixed-wing UAVs, paths can be in the form of straight-line segments or circular arc segments. The L1 guidance law is commonly used for following straight-line segments, while algorithms based on lateral error are widely used for following circular arc paths. For the path planning problem, a planning algorithm based on Dubins curves can determine the shortest path, which consists of straight lines and circular arcs, between the starting point and the endpoint.

Q. Quan et al., *Micro-Small Fixed-Wing UAV Control*,
https://doi.org/10.1007/978-981-95-1954-5_7

7.1.1 L1 Straight-Line Following

The principle of the L1 straight-line path-following algorithm is illustrated in Fig. 7.1 [1]. Here, d represents the perpendicular distance from the micro-small fixed wing UAV to the straight line, a_{cmd} denotes the commanded centripetal acceleration, L_1 indicates the distance between the current position and the L1 reference point, and η is the angle between the UAV's velocity vector and the line connecting its current position to the L1 reference point. Based on the centripetal acceleration calculation, the following equation holds:

$$a_{\mathrm{cmd}} = 2\frac{V^2}{L_1}\sin\eta. \tag{7.1}$$

This is the general equation for the L1 guidance law. The expected motion of the micro-small fixed-wing UAVs at the current moment is a circular motion with a radius of R and a velocity of V.

For the course angle η in Eq. (7.1), it is considered as composed of two parts, η_1 and η_2. Under the small angle assumption, it holds that

$$\eta_1 \approx \sin(\eta_1) = \frac{d}{L_1} \tag{7.2}$$

$$\eta_2 \approx \sin(\eta_2) = \frac{\dot{d}}{V}. \tag{7.3}$$

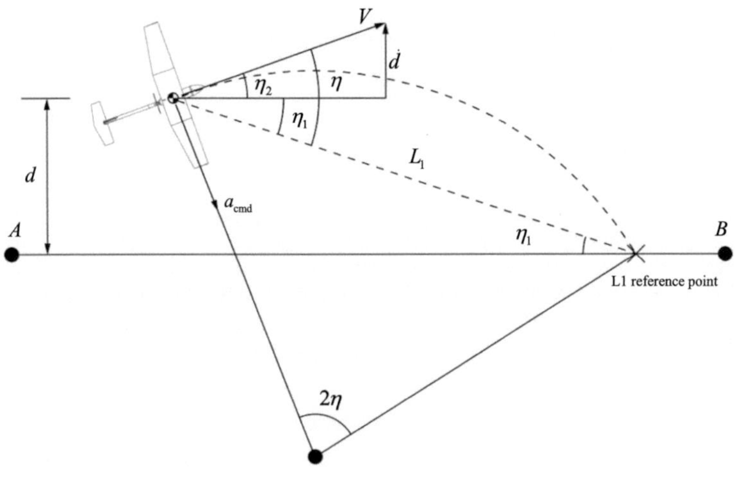

Fig. 7.1 Illustration of L1 path following algorithm

Furthermore, the acceleration control command is expressed as

$$a_{\text{cmd}} = 2\frac{V^2}{L_1}\sin(\eta_1 + \eta_2) \approx 2\frac{V}{L_1}(\dot{d} + \frac{V}{L_1}d). \tag{7.4}$$

It can be taken as a PD controller, and its parameters are determined by V and L_1. Under the assumption of no inner-loop dynamic and small η_2, $a_{\text{cmd}} \approx -\ddot{d}$ holds. So, Eq. (7.4) is equivalent to the following second-order system

$$\ddot{d} + 2\xi\omega_n\dot{d} + \omega_n^2 d = 0. \tag{7.5}$$

Here, the damping ratio of the second-order system is $\xi = 1/\sqrt{2}$, and the natural frequency is $\omega_n = \sqrt{2}V/L_1$, which depends on the ratio of the aircraft's velocity to the distance to the next reference point. In other words, if the parameter L_1 is fixed, the natural frequency of this second-order system will vary with flight speed, which cannot ensure optimal tracking performance across different speeds. Therefore, in the autopilot system, by specifying the period $L_{1,\text{P}}$ and the damping ratio $L_{1,\text{D}}$ of the path-following algorithm, L_1 can be dynamically adjusted to configure the frequency and damping of the second-order system as follows:

$$L_1 = L_{1,\text{P}}\frac{L_{1,\text{D}}V}{\pi}. \tag{7.6}$$

Substituting it into the natural frequency of the straight-line path, namely $\omega_n = \sqrt{2}V/L_1$, yields

$$\omega_n = \frac{\sqrt{2}V}{(L_{1,\text{P}}L_{1,\text{D}}\frac{V}{\pi})} = \frac{\sqrt{2}\pi}{L_{1,\text{P}}L_{1,\text{D}}}. \tag{7.7}$$

7.1.2 Circular Path Following

The idea behind the circular path following algorithm is relatively simple, as shown in Fig. 7.2. In the figure, d represents the offset distance from the UAV to the center of the circle, with the expectation that d is equal to the radius R. Therefore, the difference between the distance d and the radius R, denoted as $e_\text{d} = d - R$ (lateral error), is introduced as an error term for the proportional control. With the effect of the circular motion, the centripetal force, the acceleration command is designed as

$$a_{\text{cmd}} = K_\text{p}e_\text{d} + \frac{V^2}{R}. \tag{7.8}$$

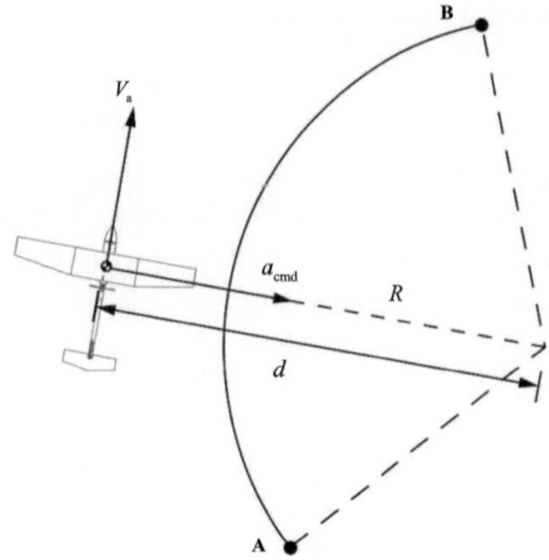

Fig. 7.2 Illustration of circular path following based on offset distance

In practice, introducing only the proportional term causes oscillations in e_d. To eliminate this oscillation, acceleration command with differential term of the error is introduced as:

$$a_{\mathrm{cmd}} = K_{\mathrm{p}} e_{\mathrm{d}} + K_{\mathrm{d}} \dot{e}_{\mathrm{d}} + \frac{V^2}{R}. \tag{7.9}$$

7.1.3 Desired Acceleration Performed by the Low-Level Flight Controller

The desired acceleration commands can be designed based on the L1 straight-line and circular path following algorithms. To achieve the desired acceleration by the low-level flight controller in Chap. 6, it must be converted into the desired roll angle commands. When micro-small fixed-wing UAVs follow a path under coordinated turning conditions, as described by Eq. (4.61), the force analysis is shown in Fig. 7.3.

Thus, the relationship between the desired roll angle ϕ_d and the desired acceleration a_{cmd} is given by

$$\phi_{\mathrm{d}} = \arctan(\frac{a_{\mathrm{cmd}}}{g}). \tag{7.10}$$

Furthermore, according to Eq. (4.63), the yaw rate $\dot{\psi}_d$ and the roll angle ϕ_d are related as follows:

$$\dot{\psi}_{\mathrm{d}} = \frac{g}{V_{\mathrm{a}}} \tan \phi_{\mathrm{d}}. \tag{7.11}$$

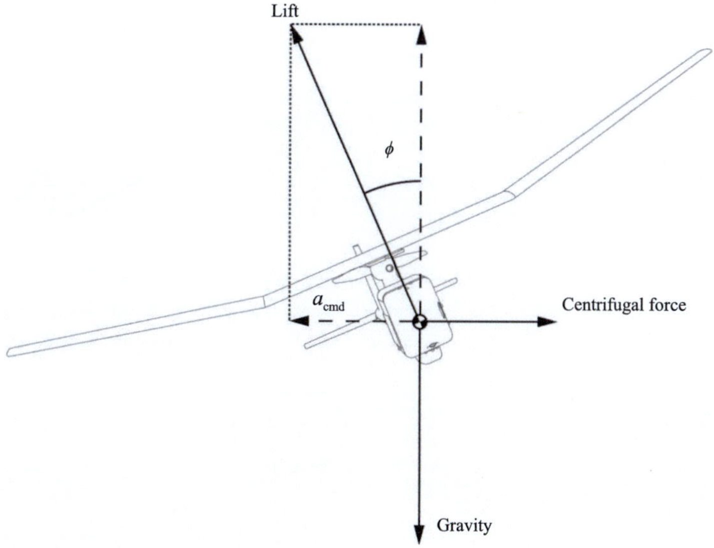

Fig. 7.3 Force analysis in coordinated turn

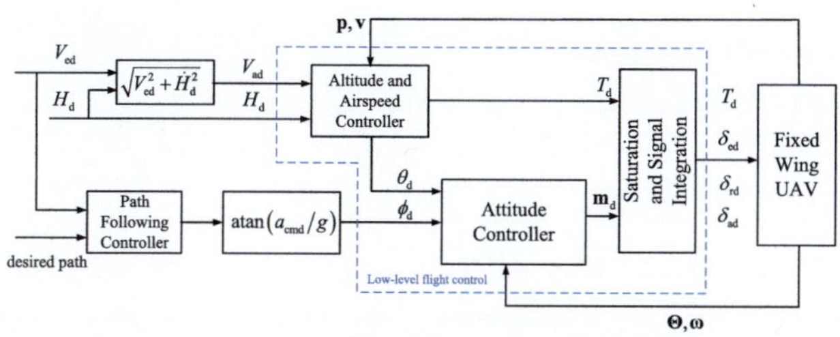

Fig. 7.4 Path following structure based on a low-level flight controller

The desired roll angle ϕ_{d} obtained from Eq. (7.10) and the desired yaw rate $\dot{\psi}_{\text{d}}$ obtained from Eq. (7.11) are then fed into the attitude controller to achieve the path following performance for micro-small fixed-wing UAVs. The relationship between the path-following controller and the low-level flight controller is shown in Fig. 7.4.

7.1.4 Dubins Curves

Dubins curves represent the shortest path between two points in a plane with specified initial and final tangents, subject to curvature constraints and assuming forward-only motion. In 1957, Lester Eli Dubins demonstrated that any such path consists of arcs and straight lines, ensuring a path always exists between two points [2]. The following content on Dubins path planning is referenced and adapted from [3] for completeness.

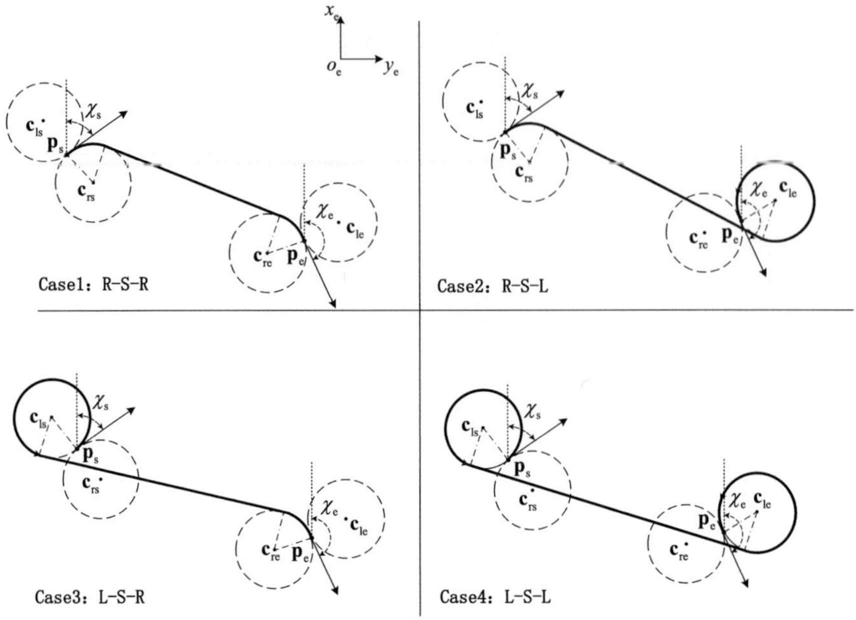

Fig. 7.5 Four cases of Dubins paths

The radius of the arcs of Dubins paths is denoted by the symbol R, where it is assumed that R is at least as large as the minimum turning radius of the UAV. In this section, the state of a micro-small fixed-wing UAV is defined by $(\mathbf{p}, \chi)$, where $\mathbf{p}$ represents the position of the micro-small fixed-wing UAV in the inertial coordinate system, and χ represents the course angle.

Given the starting state represented as $(\mathbf{p}_s, \chi_s)$ and the ending state represented as $(\mathbf{p}_e, \chi_e)$, a Dubins path consists of an initial arc followed by a straight-line segment, and ends with another arc to reach the final state. As shown in Fig. 7.5, for any given starting and ending states of a micro-small fixed-wing UAV, four possible paths are composed of an arc followed by a straight-line segment and then another arc [3].

Case 1 (R-S-R): The right arc is followed by a straight line segment and another right arc. Case 2 (R-S-L): The right arc followed by a straight line segment followed by a left arc. Case 3 (L-S-R): The left arc followed by a straight line segment followed by a right arc. Case 4 (L-S-L): The left arc followed by a straight line segment followed by another left arc. A Dubins path is defined as the shortest path among these four cases.

7.1.5 Path Length Computation

To determine a Dubins path, it is necessary to compute the path lengths for the four cases shown in Fig. 7.5. This subsection will derive explicit equations for the path

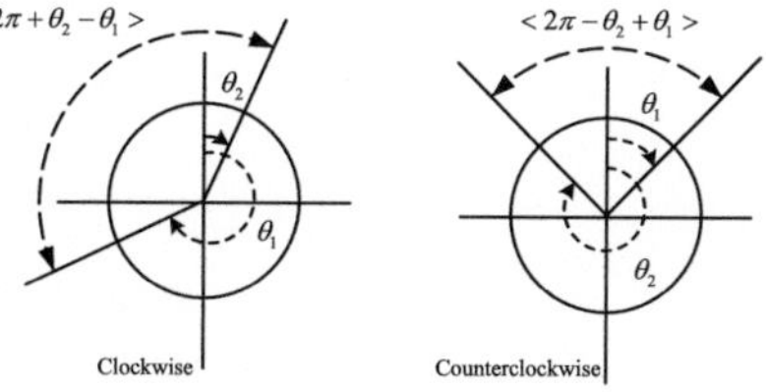

Fig. 7.6 Angular distance between θ_1 and θ_2 clockwise (CW) and counterclockwise (CCW)

lengths in each case. Given the position $\mathbf{p}$, course angle χ, and radius R, the centers of the right and left turning arcs are given by:

$$\mathbf{c}_\mathrm{r} = \mathbf{p} + R\left[\cos(\chi + \tfrac{\pi}{2})\ \sin(\chi + \tfrac{\pi}{2})\ 0\right]^\mathrm{T} \tag{7.12}$$

$$\mathbf{c}_\mathrm{l} = \mathbf{p} + R\left[\cos(\chi - \tfrac{\pi}{2})\ \sin(\chi - \tfrac{\pi}{2})\ 0\right]^\mathrm{T}. \tag{7.13}$$

To compute the path lengths for different trajectories, a general equation for angular distance on a circle is needed. Figure 7.6 illustrates the geometries of clockwise (CW) and counterclockwise (CCW) circles.

Assuming that θ_1 and θ_2 are between 0 and 2π, for the clockwise circle, the angular distance between θ_1 and θ_2 is given by

$$|\theta_2 - \theta_1|_\mathrm{CW} \triangleq < 2\pi + \theta_2 - \theta_1 > \tag{7.14}$$

where $<\phi> = \phi \bmod 2\pi$. Similarly, for the counterclockwise circle, it holds that

$$|\theta_2 - \theta_1|_\mathrm{CCW} \triangleq < 2\pi - \theta_2 + \theta_1 > . \tag{7.15}$$

7.1.5.1 Case 1: R-S-R

The Dubins curve for each case is represented as the sum of the distance between the centers and the arc length. In Case 1, the geometry is shown in Fig. 7.7, where θ is the angle formed between the line $\overline{\mathbf{c}_\mathrm{rs}\,\mathbf{c}_\mathrm{re}}$ and $o_\mathrm{e}x_\mathrm{e}$. Using Eq. (7.14), the angular distance around $\mathbf{c}_\mathrm{rs}$ is

$$R < 2\pi + < \theta - \frac{\pi}{2} > - < \chi_s - \frac{\pi}{2} >> . \tag{7.16}$$

Similarly, using Eq. (7.14), the angular distance around $\mathbf{c}_\mathrm{re}$ is

$$R < 2\pi + < \chi_\mathrm{e} - \frac{\pi}{2} > - < \theta - \frac{\pi}{2} >> . \tag{7.17}$$

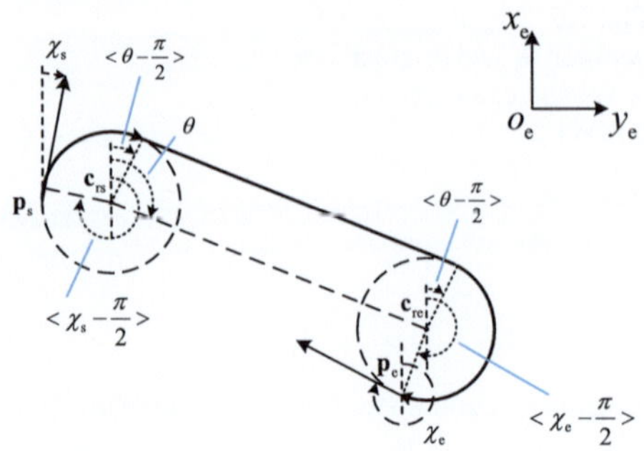

Fig. 7.7 Dubins path in Case 1

Thus, the total path length for Case 1 is given by

$$L_1 = ||\mathbf{c}_{rs} - \mathbf{c}_{re}|| + R < 2\pi + < \theta - \frac{\pi}{2} > - < \chi_s - \frac{\pi}{2} >> \\ + R < 2\pi + < \chi_e - \frac{\pi}{2} > - < \theta - \frac{\pi}{2} >> . \tag{7.18}$$

7.1.5.2 Case 2: R-S-L

In Case 2, the geometry is shown in Fig. 7.8, where θ is the angle formed between the line $\overline{\mathbf{c}_{rs}\,\mathbf{c}_{le}}$ and $o_e x_e$. The distance is $l = ||\mathbf{c}_{le} - \mathbf{c}_{rs}||$, and the triangle $\Delta\mathbf{c}_1\mathbf{c}_{rs}\mathbf{c}_2$ in the figure is a right triangle so that the following angle relationship can be obtained

$$\sin\left(\frac{\pi}{2} - \theta + \theta_2\right) = \frac{R}{l/2}. \tag{7.19}$$

Arranging the above equation, θ_2 can be calculated as follows:

$$\theta_2 = \theta - \frac{\pi}{2} + \arcsin\left(\frac{2R}{l}\right). \tag{7.20}$$

Using Eq. (7.14), the angular distance around $\mathbf{c}_{rs}$ is

$$R < 2\pi + < \theta_2 > - < \chi_s - \frac{\pi}{2} >> . \tag{7.21}$$

Similarly, using Eq. (7.15), the angular distance around $\mathbf{c}_{re}$ is

$$R < 2\pi + < \theta_2 + \pi > - < \chi_e + \frac{\pi}{2} >> . \tag{7.22}$$

Fig. 7.8 Dubins path in Case 2

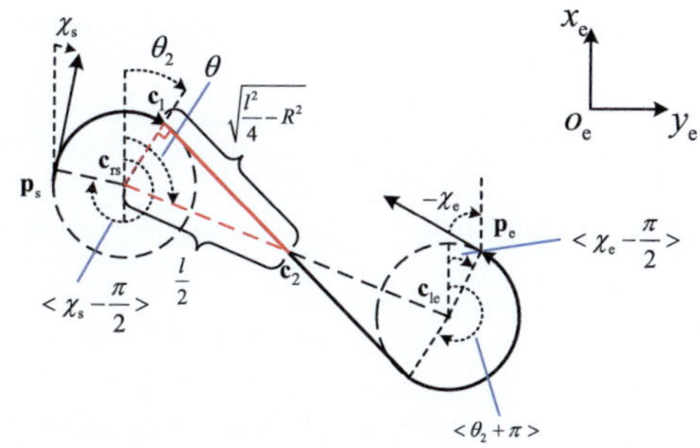

Thus, the total path length for Case 2 is given by

$$L_2 = \sqrt{l^2 - 4R^2} + R < 2\pi + < \theta_2 > - < \chi_s - \frac{\pi}{2} >> \\ + R < 2\pi + < \theta_2 + \pi > - < \chi_e + \frac{\pi}{2} >> . \tag{7.23}$$

7.1.5.3 Case 3: L-S-R

In Case 3, the geometry is shown in Fig. 7.9, where θ is the angle formed between the line $\overline{\mathbf{c}_{ls}\mathbf{c}_{re}}$ and $o_e x_e$. The distance is $l = ||\mathbf{c}_{re} - \mathbf{c}_{ls}||$, and

$$\theta_2 = \arccos\left(\frac{2R}{l}\right). \tag{7.24}$$

Using Eq. (7.15), the angular distance around $\mathbf{c}_{ls}$ is

$$R < 2\pi + < \chi_s + \frac{\pi}{2} > - < \theta + \theta_2 >> . \tag{7.25}$$

Fig. 7.9 Dubins path in Case 3

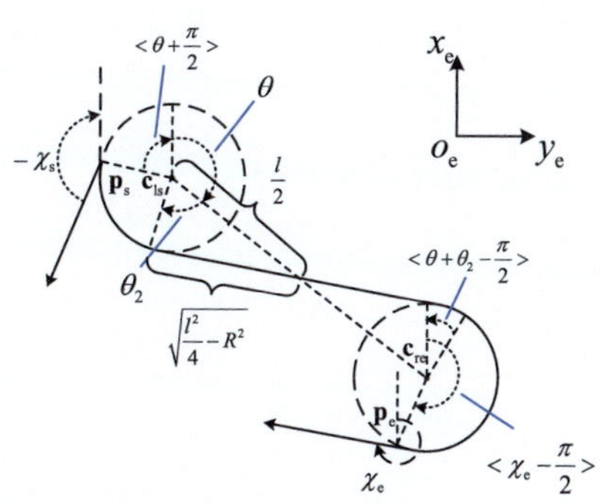

Similarly, using Eq. (7.14), the angular distance around $\mathbf{c}_{\text{re}}$ is

$$R < 2\pi + < \chi_{\text{e}} - \frac{\pi}{2} > - < \theta + \theta_2 - \pi >> . \tag{7.26}$$

Thus, the total path length for Case 3 is given by

$$L_3 = \sqrt{l^2 - 4R^2} + R < 2\pi + < \chi_{\text{s}} + \frac{\pi}{2} > - < \theta + \theta_2 >> \\ + R < 2\pi + < \chi_{\text{e}} - \frac{\pi}{2} > - < \theta + \theta_2 - \pi >> . \tag{7.27}$$

7.1.5.4 Case 4: L-S-L

In Case 4, the geometry is shown in Fig. 7.10, where θ is the angle formed between the line $\overline{\mathbf{c}_{\text{ls}}\mathbf{c}_{\text{le}}}$ and $o_{\text{e}}x_{\text{e}}$. Using Eq. (7.15), the angular distance around $\mathbf{c}_{\text{ls}}$ is

$$R < 2\pi + < \chi_{\text{s}} + \frac{\pi}{2} > - < \theta + \frac{\pi}{2} >> . \tag{7.28}$$

Similarly, using Eq. (7.15), the angular distance around $\mathbf{c}_{\text{le}}$ is

$$R < 2\pi + < \theta + \frac{\pi}{2} > - < \chi_{\text{e}} + \frac{\pi}{2} >> . \tag{7.29}$$

Thus, the total path length for Case 4 is given by

$$L_4 = ||\mathbf{c}_{\text{rs}} - \mathbf{c}_{\text{re}}|| + R < 2\pi + < \chi_{\text{s}} + \frac{\pi}{2} > - < \theta + \frac{\pi}{2} >> \\ + R < 2\pi + < \theta + \frac{\pi}{2} > - < \chi_{\text{e}} + \frac{\pi}{2} >> . \tag{7.30}$$

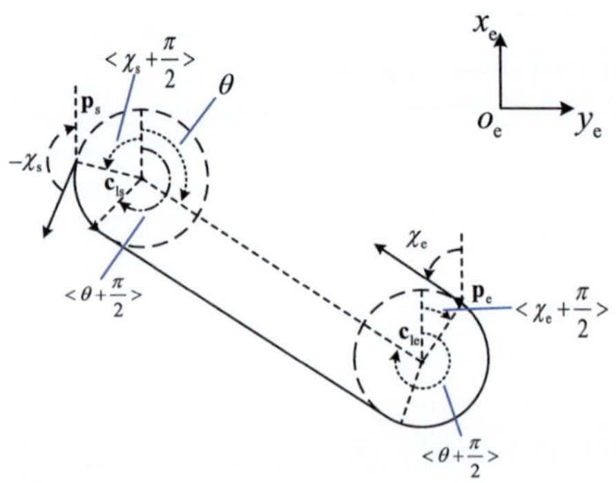

Fig. 7.10 Dubins path in Case 4

7.1.6 Related Development

Path planning is determining the optimal or near-optimal route from a given starting point to a specified destination within a given environment. In the context of UAVs, researchers have introduced various methods and strategies for path planning. The following section outlines some of the most significant recent developments and pioneering methodologies.

The first approach to path planning is based on deep reinforcement learning. Deep reinforcement learning is a machine learning method that combines deep neural networks with reinforcement learning, thereby enabling end-to-end path planning in complex dynamic environments without requiring manual feature design or prior knowledge. Among these, value function-based methods [4] select the optimal actions using greedy policies. Policy-based methods [5] employ multiple parallel agents to explore and learn simultaneously, thereby enhancing learning efficiency and stability. Model-based methods [6] utilize models of environmental dynamics to predict future states and rewards, selecting optimal actions by optimizing objective functions.

Classical graph search algorithms such as Dijkstra and A* have been widely applied to path-planning problems. However, recent studies have shown that they have lower computational efficiency when dealing with large-scale graphs. Therefore, improvements have been made in recent years, including graph neural network-based path planning methods that can handle graph-structured data [7]. Graph neural networks are a deep learning method that can process graph-structured data, extracting topological information and node features for decision-making and prediction in path planning. Adversarial learning is used for graph search tasks [8] by utilizing generators to create data from random noise and discriminators to classify both real and generated data. This approach achieves path planning generation and optimization by minimizing the generator's log loss and maximizing the discriminator's log loss. Additionally, research is being conducted on multimodal graph searches for various data types, including text, images, and videos.

Sampling-based search algorithms, such as PRM, RRT, and RRT*, are highly effective for path planning in high-dimensional spaces. The basic principle involves obtaining random sample points in the free configuration space and connecting them according to certain rules to generate paths. The advantage of these methods is their ability to scale to high-dimensional spaces. However, their disadvantage is that they cannot guarantee finding an optimal path due to their probabilistic completeness. Therefore, recent research mainly focuses on improving their success rate and efficiency [9]. Metaheuristic algorithms, such as the Particle Swarm Optimization algorithm, Genetic Algorithm, and Ant Colony Optimization algorithm, mimic natural behaviors to find approximately optimal solutions through iterative search. For instance, the Genetic Algorithm performs well in multi-objective path planning and can find efficient paths. However, this method has a relatively high number of iterations and lower computational efficiency compared to other methods. Therefore, it may not be suitable for real-time path planning of UAVs with limited online computing resources. Recent research has focused on improving the convergence speed and stability of algorithms. For example, the Hybrid Particle Swarm Optimization algorithm (SDPSO) provides each particle with information about the global optimal solution, improving the algorithm's convergence rate [10]. Extended versions of the Ant Colony Algorithm have achieved similar efficiency to linear programming in multi-objective path planning problems [11].

In summary, the field has undergone dynamic development over the past decade, mainly reflected in several aspects: consideration of complexity and uncertainty, multi-UAV collaboration, real-time adjustment, and optimization. To address these challenges, numerous studies have combined the latest tools, including reinforcement learning, to achieve efficient and stable path planning.

7.2 Basic Experiment

7.2.1 Experimental Objective

1. Things to prepare

(1) Software: MATLAB R2022b or above.
(2) Program: Experimental package "e5/e5-1". In the package "e5/e5-1", there is: the initial state file "InitDatactrl.m" for a fixed-wing UAV, the Simulink simulation model "SmallFixedWingUAVfollow.slx" for fixed-wing UAV path following, the circle center calculation function "circlecenter.m", and the files for drawing flight paths and desired paths "Drawxy.m".

2. Objective

In the existing experimental package "e4/e4-1" for basic flight control of micro-small fixed-wing UAVs, a complete "Path Follow" module is implemented by incorporating the L1 straight-line following algorithm and the circular path-following algorithm, followed by the Software-In-the-Loop (SIL) simulation.

7.2.2 Experimental Procedure

The basic experiment focuses on developing the "Path Follow" module within the "SmallFixedWingUAVfollow.slx" file, incorporating the L1 straight-line following module ("L1 Navigation") and the circular path following module ("PD Navigation"). The steps are outlined as follows: Step 1 creates the "Path Follow" module; Step 2 integrates the L1 Navigation module; Step 3 integrates the PD Navigation module; and Step 4 provides the specified trajectory for the SIL simulation to verify accuracy.

1. Step 1: Building the framework of the "Path Follow" module

In file "e5/e5-1/SmallFixedWingUAVfollow.slx", build the "Path Follow" module. The overall framework is shown in Fig. 7.12. Open "Simulink Library Browser" as shown in Fig. 7.11, and select the "MATLAB Function" module to establish: ①—L1 path following algorithm module "L1 Navigation" on the left side of Fig. 7.12 and ②—the PD path following algorithm module "PD Navigation" on the right side. "MATLAB Function" is a user-defined function module where users can configure input-output information to encapsulate custom algorithms in Simulink. Then, configure the inputs for these two modules. According to the principles of the experiment in Sect. 7.1, it is known that the feedback values required for the L1 straight-line following algorithm are the current position, yaw angle, and current airspeed, while the PD arc path following algorithm requires the current position and current airspeed as feedback values. Obtain the relevant data from the "SensorData" bus signal output by the model to establish ③—the state feedback module. Finally, both modules

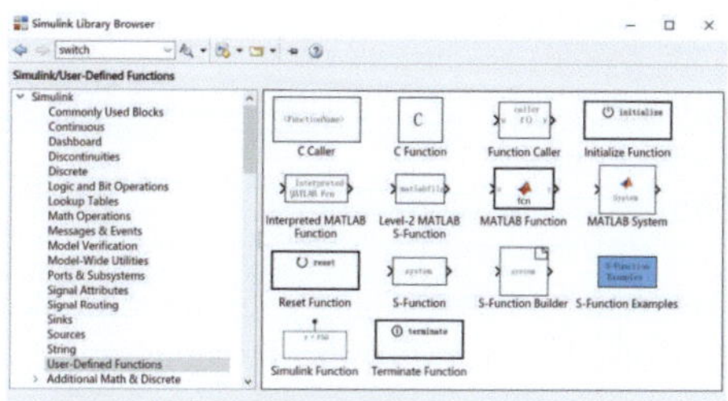

Fig. 7.11 Selecting "MATLAB Function" module

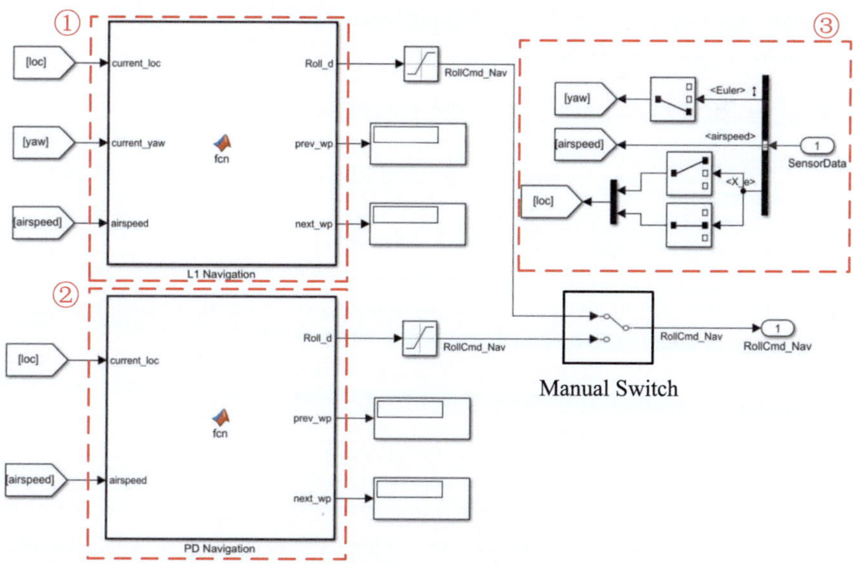

Fig. 7.12 Path follow module, from"e5/e5-1/SmallFixedWingUAVfollow.slx"

output the navigation roll command "RollCmd_Nav", which is then limited by the maximum roll angle. The "Manual Switch" block can switch between the output of straight-line flight experiments and arc flight experiments during simulation. The navigation roll command "RollCmd_Nav" is fed to the fixed-wing controller module as the desired roll angle. The controller module calculates the deflection angle of the fixed-wing aircraft's control surface based on the desired command and the current state, thereby controlling the fixed-wing aircraft to fly along the given trajectory. Concurrently, the preceding and subsequent waypoint data of the two modules are transmitted to ascertain the present path segment information within this module.

2. Step 2: Writing the L1 straight-Line following algorithm in "L1 Navigation" module

(1) First, define a straight-line path, for instance, from coordinate point $x = 100$ m, $y = 100$ m to coordinate point $x = 500$ m, $y = 500$ m, as shown in Code 7.1. In Code 7.1, "prev_wp" represents the previous coordinate point, and here it is also the starting coordinate point; "next_wp" represents the next coordinate point, and here it is the termination coordinate point.

Code 7.1 Waypoint coordinates, from"e5/e5-1/SmallFixedWingUAVfollow.slx"

```
prev_wp = [100; 100];
next_wp = [500; 500];
```

(2) Then, set up the relevant parameters and constant variables required in the L1 algorithm, as shown in Code 7.2, including the damping ratio $L_{1,\mathrm{D}}$ ("L1_damping"), period $L_{1,\mathrm{P}}$ ("L1_period"), etc. The parameter "dist_min" is used to limit the final calculated L1 length, i.e., the L1 length cannot be less than this value. Since the L1 length is directly related to the lateral acceleration, and the lateral acceleration that the fixed-wing UAVs can produce is limited, "dist_min" is introduced here to ensure the stability of the algorithm.

Code 7.2 Settings for L1 algorithm, from"e5/e5-1/SmallFixedWingUAVfollow.slx"

```
% Parameters initialization
dist_min = 30; % Minimum L1 length
L1_damping = 3; % L1 damping ratio param
L1_period = 8; % L1 period param
ISA_g = 9.80665;
```

(3) Calculate the current position's distance to the L1 point on the straight line based on the airspeed data in the feedback, as shown in Code 7.3.

Code 7.3 Calculation of L1 length, from"e5/e5-1/SmallFixedWingUAVfollow.slx"

```
% Calculate L1 distance
L1_dist = max(L1_damping * L1_period * airspeed / pi,
    dist_min);
```

(4) The key aspect of the L1 algorithm is to calculate the angle η_1 between the L1 vector and the path vector, represented as "Nu1" in the code, and the angle η_2 between the

velocity vector and the path vector, represented as "Nu2", both determined through geometric relationships. The calculation of angles η_1 and η_2 is implemented according to Eqs. (7.2) and (7.3), as shown in Code 7.4. In Code 7.4, "AB" represents the path vector, "Vertical_d" corresponds to d in Eq. (7.2), indicating the perpendicular distance from the current UAV position to the path, and "V_vetor" represents the projection vector of the current UAV's velocity in the horizontal direction.

Code 7.4 Calculation of angles η_1 and η_2, from"e5/e5-1/SmallFixedWingUAVfollow.slx"

```
% Calculate angle Nu1 between L1 and AB
% Calculate the ratio of vertical distance Vertical_d to L1_dist, which is sin_Nu1
sin_Nu1 = Vertical_d / L1_dist;
if sin_Nu1 >= 0.9999
sin_Nu1 = 1;
elseif sin_Nu1 <= -0.9999
sin_Nu1 = -1;
else
sin_Nu1 = sin_Nu1;
end
Nu1 = asin(sin_Nu1); % get the radian value of Nu1
if Nu1 > 0.7854
Nu1 = 0.7854;
elseif Nu1 < -0.7854
Nu1 = -0.7854;
end
% Calculate angle Nu2 between V and AB
% If the norm of V_vetor is less than or equal to 1e-3
if norm(V_vetor) <= 1e-3
Nu2 = 0;
else
cos_Nu2 = dot(AB,V_vetor) / (norm(AB) * norm(V_vetor));
% Calculates the dot product of AB and V_vetor divided by the product of their module lengths, which is cos_Nu2
if cos_Nu2 >= 0.9999
cos_Nu2 = 1;
elseif cos_Nu2 <= -0.9999
cos_Nu2 = -1;
else
cos_Nu2 = cos_Nu2;
end
Nu2 = acos(cos_Nu2); % get the radian value of Nu2
end
```

(5) Finally, the lateral acceleration is obtained through the above angles, thereby solving the expected roll command. The example code in Code 7.5 calculates angle η in lines 1–6 and restricts the magnitude of angle η to be within [$-\pi/2,\pi/2$]. In line 7, the lateral acceleration is calculated according to Eq. (7.1). Here, the calculation is slightly different from Eq. (7.1) but essentially the same, except for adding the damping ratio as an adjustable parameter. The desired roll command is obtained in line 8 through Eq. (7.10).

Code 7.5 Obtaining desired roll command, from"e5/e5-1/SmallFixedWingUAVfollow.slx"

```
Nu = Nu1 + Nu2;
if Nu > 1.5708
Nu = 1.5708;
elseif Nu < -1.5708
Nu = -1.5708;
end
latAcc = L1_damping * L1_damping * airspeed * airspeed *
    sin(Nu) / L1_dist;
Roll_d = atan(latAcc/ISA_g);
```

Step 3: Writing the circular path following algorithm in "PD Navigation" module

(1) First, set the current arc path. The arc path must be pre-planned to ensure its correctness, for example, from the starting point with coordinates "$x = 0$ m, $y = 0$ m" to the endpoint with coordinates "$x = 0$ m, y = 1000 m", with a turning radius of 500 m, and clockwise rotation (denoted by 1). Initialize the static variable "rdist_prev" to save the previous offset distance for calculating the differential term of the offset, as shown in Code 7.6.

Code 7.6 Circular path following algorithm parameter setting, from"e5/e5-1/SmallFixedWingUAVfollow.slx"

```
prev_wp = [0; 0];
next_wp = [0; 1000];
r_wp = 500;
turn_wp = 1;
ISA_g = 9.80665;
persistent rdist_prev;
```

(2) Using the current arc path information, first determine the center of the arc. The center is uniquely defined by the given starting point, endpoint, radius, and direction of rotation, as illustrated in Fig. 7.13. This problem can be described as

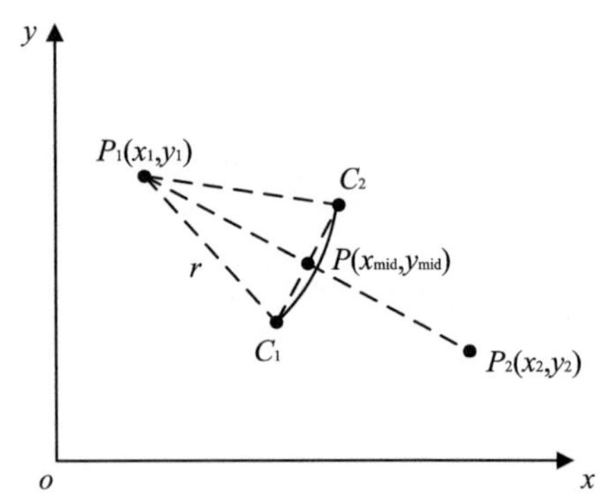

Fig. 7.13 Example of finding the center of the circle

follows: given the coordinates of points P_1 and P_2 as (x_1, y_1) and (x_2, y_2) in the plane coordinate system, find the coordinates of the circle center C_1 and C_2 with radius r passing through points P_1 and P_2. It can be easily observed that the coordinates of point P are $x_{\text{mid}} = (x_1 + x_2)/2$, $y_{\text{mid}} = (y_1 + y_2)/2$, and the slope of line P_1P_2 is $k_1 = (y_2 - y_1)/(x_2 - x_1)$, so the analysis expression of line C_1C_2 is:

$$y = k_2(x - x_{\text{mid}}) + y_{\text{mid}} \tag{7.31}$$

where $k_2 = -1/k_1$. So, the centers C_1 and C_2 are the intersection points of the line and the circle with radius r centered at point P_1. The analytical expression of the circle is

$$(x - x_1)^2 + (y - y_1)^2 = r^2. \tag{7.32}$$

Solving Eqs. (7.31) and (7.32) simultaneously gives the coordinates of the center:

$$x_c = \frac{-b \pm \sqrt{b^2 - 4ac}}{2a}$$
$$y_c = k_2(x_c - x_{\text{mid}}) + y_{\text{mid}}$$

where

$$a = 1 + k_2^2$$
$$b = -2x_{\text{mid}} - k_2^2(x_1 + x_2)$$
$$c = x_{\text{mid}}^2 + k_2^2\frac{(x_1 + x_2)^2}{4} - r^2 - (x_{\text{mid}} - x_1)^2 + (y_{\text{mid}} - y_1)^2.$$

The specific implementation is shown in Code 7.7. The vectors "center1" and "center2" represent the two solved centers. Lines 4–16 of the example code implement the method of calculating the center when the line connecting the starting point and the endpoint is perpendicular, as shown in the case represented by line 4 of the code. When the x-coordinate difference of the line is less than "1e-3", it is considered perpendicular. Lines 18–23 of the example code implement the method of calculating the center when the line connecting the starting point and the endpoint is horizontal, corresponding to the case represented by the 19th line of the code, where the slope of the line is zero. Lines 25–36 of the example code implement the method of calculating the center when the line connecting the starting point and the endpoint is in a general position, corresponding to Eqs. (7.31) and (7.32).

Code 7.7 Circle center calculation method, from"e5/e5-1/SmallFixedWingUAVfollow.slx"

```
% Calculate circle center
center1 = [0;0];
center2 = [0;0];
if abs(p2(1) - p1(1)) < 1e-3 % In the special case, k is
    infinite
k_verticle = 0;
mid_x = (p1(1) + p2(1)) / 2.0;
mid_y = (p1(2) + p2(2)) / 2.0;
a = 1.0 + k_verticle * k_verticle;
b = -2 * mid_x - k_verticle * k_verticle * (p1(1) + p2(1));
c = mid_x * mid_x + k_verticle * k_verticle * (p1(1) +
    p2(1)) * (p1(1) + p2(1)) / 4.0 - ...
(dRadius * dRadius - ((mid_x - p1(1)) * (mid_x - p1(1)) +
    (mid_y - p1(2)) * (mid_y - p1(2))));

center1(1) = (-1.0 * b + sqrt(b * b -4 * a * c)) / (2 * a);
center2(1) = (-1.0 * b - sqrt(b * b -4 * a * c)) / (2 * a);
center1(2) = k_verticle * center1(1) - k_verticle * mid_x
    + mid_y;
center2(2) = k_verticle * center2(1) - k_verticle * mid_x
    + mid_y;
else
k = (p2(2) - p1(2)) / (p2(1) - p1(1));
if k == 0
center1(1) = (p1(1) + p2(1)) / 2.0;
center2(1) = (p1(1) + p2(1)) / 2.0;
center1(2) = p1(2) + sqrt(dRadius * dRadius -(p1(1) -
    p2(1)) * (p1(1) - p2(1)) / 4.0);
center2(2) = p2(2) - sqrt(dRadius * dRadius -(p1(1) -
    p2(1)) * (p1(1) - p2(1)) / 4.0);
else
k_verticle = -1.0 / k;
mid_x = (p1(1) + p2(1)) / 2.0;
mid_y = (p1(2) + p2(2)) / 2.0;
a = 1.0 + k_verticle * k_verticle;
b = -2 * mid_x - k_verticle * k_verticle * (p1(1) + p2(1));
c = mid_x * mid_x + k_verticle * k_verticle * (p1(1) +
    p2(1)) * (p1(1) + p2(1)) / 4.0 - ...
(dRadius * dRadius - ((mid_x - p1(1)) * (mid_x - p1(1)) +
    (mid_y - p1(2)) * (mid_y - p1(2))));

center1(1) = (-1.0 * b + sqrt(b * b -4 * a * c)) / (2 * a);
center2(1) = (-1.0 * b - sqrt(b * b -4 * a * c)) / (2 * a);
center1(2) = k_verticle * center1(1) - k_verticle * mid_x
    + mid_y;
center2(2) = k_verticle * center2(1) - k_verticle * mid_x
    + mid_y;
end
end
```

So far, the coordinates of the two centers, "center1" and "center2" are computed. Code 7.8 selects the appropriate center for rotation based on the specified direction, either clockwise or counterclockwise. Lines 2–6 of the example code calculate the rotation direction of the micro-small fixed-wing UAV around "center1" by using the cross product of vectors. Lines 8–12 of the example code calculate the direction of rotation of the micro-small fixed-wing UAV around the center "center2" using the cross product of vectors. Lines 14–19 of the example code determine the consistency of the rotation direction with the given rotation direction "turn_wp", thereby selecting the rotation center of the micro-small fixed-wing UAV "center".

Code 7.8 Determining the rotation center, from"e5/e5-1/SmallFixedWingUAVfollow.slx"

```
1  % Turning direction center1
2  P1C1 = p1 - center1;
3  P2C1 = p2 - center1;
4  P1C10 = [P1C1(1),P1C1(2),0];
5  P2C10 = [P2C1(1),P2C1(2),0];
6  CrossC1 = cross(P1C10,P2C10);
7  % Turning direction center2
8  P1C2 = p1 - center2;
9  P2C2 = p2 - center2;
10 P1C20 = [P1C2(1),P1C2(2),0];
11 P2C20 = [P2C2(1),P2C2(2),0];
12 CrossC2 = cross(P1C20,P2C20);
13 % Value turn anticlockwise = -1, clockwise = 1
14 if CrossC1(3) ==0
15 center = center1;
16 elseif sign(CrossC1(3)) == sign(turn)
17 center = center1;
18 else
19 center = center2;
20 end
```

(3) Finally, set the parameters K_p as "K_rdist_P" and K_d as "K_rdist_D", and calculate the expected acceleration based on the offset error, and further calculate the expected roll command. Lines 2–3 of the example code set "P" and "D" values of the PD controller. Lines 4–13 of the example code design the PD controller of the circular path following algorithm based on Eq. (7.9). Line 15 of the example code calculates the expected roll command based on geometric relationships.

Code 7.9 Obtaining the expected roll command, from"e5/e5-1/SmallFixedWingUAVfollow.slx"

```
% PD tracking
K_rdist_P = 0.015;
K_rdist_D = 0.05;
if isempty(rdist_prev)
rdist_prev = 0;
end
rdist = norm(current_loc-center) - dRadius;
rdist_error = rdist - rdist_prev;
if turn > 0
latAcc = airspeed * airspeed / dRadius + K_rdist_P * rdist
    + K_rdist_D * rdist_error * 250;
else
latAcc = airspeed * airspeed / -dRadius + -K_rdist_P *
    rdist + -K_rdist_D * rdist_error * 250;
end
rdist_prev = rdist;
Roll_d = atan(latAcc/ISA_g);
```

4. Step 4: Setting initial conditions for SIL simulation

(1) Modify the initialization script "InitDataCtrl" in "Initial Condition" section, as shown in Code 7.10, to set the initial state to fly at a speed of 20 m/s in the north direction (i.e., along the $o_e x_e$ axis).

Code 7.10 Parameter setting , from"e5/e5-1/SmallFixedWingUAVfollow.slx"

```
%% Initial Condition
% test
uav.ic = struct;
uav.ic.Pos_0 = [0; 0; 50];
uav.ic.Euler_0 = [0; 0; 0];
uav.ic.Omega_0 = [0; 0; 0];
uav.ic.PQR_0 = [0; 0; 0];
uav.ic.Vb_0 = [20; 0; 0];
uav.ic.gsLL = [45 120];
uav.ic.gsH = 0;
```

(2) Set up a module for switching between the manual remote control command and the path following command, as shown in Fig. 7.14. When the value is greater than or equal to 1500, select the upper signal command through ③—"Switch Module", i.e., output command through ①—"Manual Remote Control Module"; when the value is less than 1500, select the lower signal command through ③—"Switch Module", indicating that it is taken over by ②—"Path Following Module" command. In the step signal module, design a step signal that occurs after 1 s, allowing for a straight-line flight at a constant speed for 1 s, and then the path will follow the next command.

(3) Set "stop time" to around 80 s. By double-clicking on "Manual Switch", the straight trajectory output and circular trajectory output can be switched. Run this program separately and record the desired path and flight data as shown in Fig. 7.15.

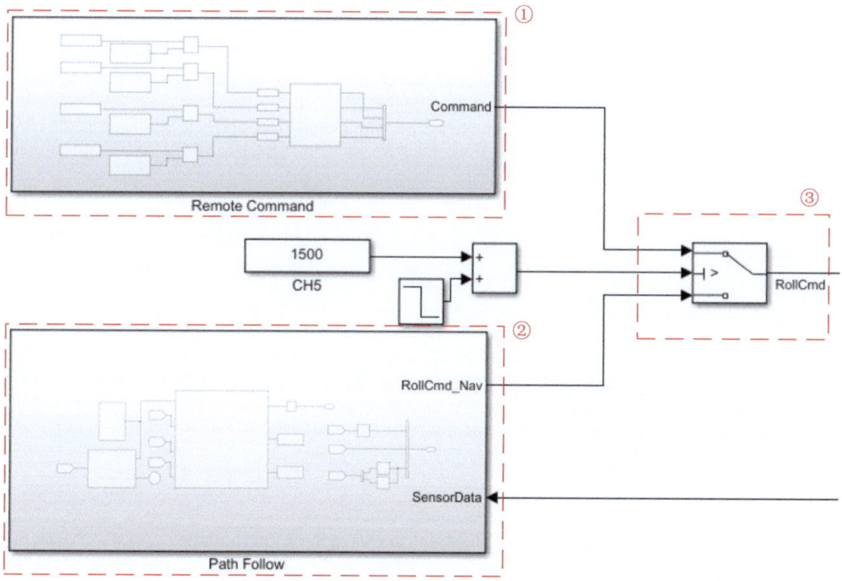

Fig. 7.14 Automatic switching module, from"e5/e5-1/SmallFixedWingUAVfollow.slx

7.3 Analysis Experiment

7.3.1 Experimental Objective

1. Things to prepare

(1) Software: MATLAB R2022b or above.
(2) Program: Experimental package "e5/e5-2". The package "e5/e5-2" contains the main path planning program "path_plan.m", coordinate point generation program "myline.m" and Dubins curve generation program "dubins_path.m."

2. Objective

(1) Analyze the Dubins path planning algorithm in the following two scenarios:

- Need to pass through multiple intermediate circles between the starting point and the ending point;
- Need to pass through a straight-line tunnel between the starting point and the ending point.

(2) Write MATLAB programs to provide a reasonable Dubins path given parameters such as starting point, ending point, intermediate obstacles, and straight-line tunnels.

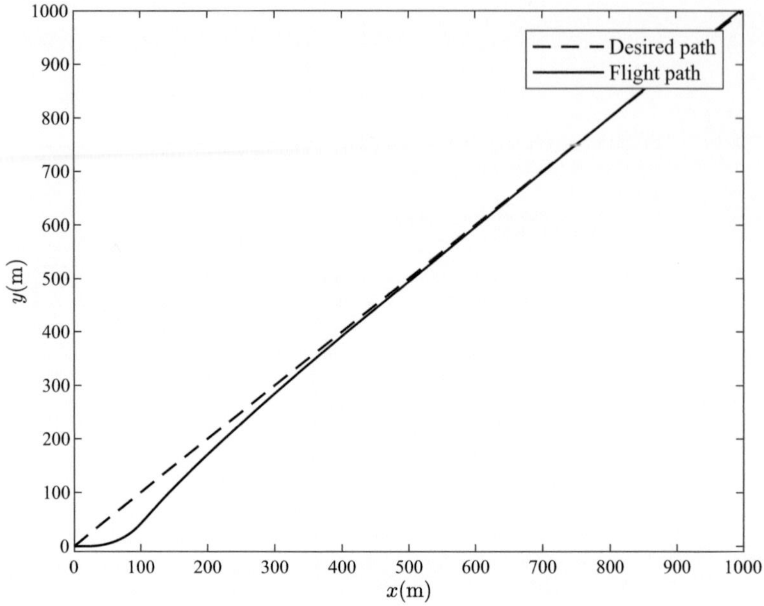

(a) Straight path following performance

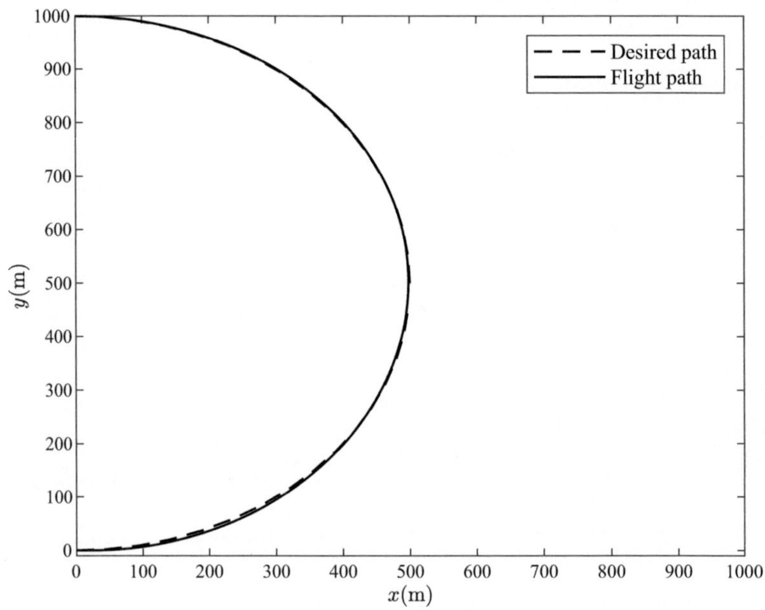

(b) Circular path following performance

Fig. 7.15 Path following

7.3.2 Experimental Procedure

The analysis experiment is mainly based on the introduction in Sect. 7.1.4. First, in Step 1, an algorithm is written to generate Dubins curves that need to pass through multiple intermediate circles between the starting point and the ending point. Then, in Step 2, an algorithm is written to generate Dubins curves that need to pass through straight-line tunnels between the starting point and the ending point. Next, in Step 3, path segment information vectors are designed. Finally, in Step 4, algorithms are written in MATLAB and integrated to generate Dubins curves under various scenarios.

1. Step 1: Analysis of the Dubins path planning algorithm requiring passing through multiple intermediate circles.

Here, the regions between the starting and ending points are modeled as several obstacle circular regions $\{O_1, O_2, ..., O_k\}$. First, intermediate circles are defined. Take several circles $\{C_1, C_2, ..., C_n\}$ from the obstacle circles as intermediate circles. The definition of intermediate circles is as follows:

- The aircraft must pass around the intermediate circles $\{C_1, C_2, ..., C_n\}$ sequentially after departing from the starting point.
- Enter the edge of the intermediate circle from the tangent direction, fly along the edge for a certain distance, and finally leave the intermediate circle in another tangent direction.

As shown in Fig. 7.16, with the positions of the starting point and the ending point given, the Dubins path planning algorithm passing through intermediate circles is pseudo-coded as Algorithm 7.1.

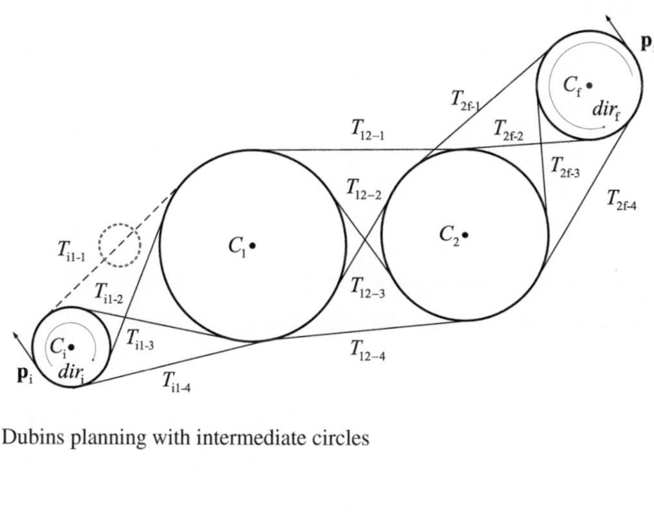

Fig. 7.16 Dubins planning with intermediate circles

Algorithm 7.1 Path following avoidance Dubins path planning algorithm with intermediate circles

Input: Initial point $\mathbf{p}_\mathrm{i}$, course angle α, turning radius r_i, turning direction dir_i, final position $\mathbf{p}_\mathrm{f}$, course angle β, turning radius r_f, turning direction dir_f, set of intermediate circles $\{C_1, C_2, ..., C_n\}$

Output: Set of all possible paths $PATH\_P$

Determine starting circle C_i and ending circle C_f

Determine tangents between C_i and C_1, C_1 and C_2, ..., C_n and C_f, excluding obstacles

From obstacle-free tangents, sequentially select a set of tangents to form combinations of tangents that a possible path from the starting point to the ending point must pass through.

Record all such tangent combinations and denote as $path$, let m be the number of tangent combinations in set $path$

for $i = 1 \to m$ **do**

  if tangents in $path$ meet the requirements of matching turning direction between consecutive arcs **then**

    if any adjacent two tangents in $path$ meet the requirement of consecutive connection **then**

      The tangent combination $path\_P$ is a feasible path

    end if

  end if

end for

Record all feasible paths $path\_P$ as the set of all possible paths $PATH\_P$

2. Step 2: Analyzing the Dubins path planning algorithm between the starting point and the ending point that may pass through a straight tunnel.

As shown in Fig. 7.17, assume the starting point of the tunnel is $\mathbf{p}_\mathrm{ti}$, the ending point is $\mathbf{p}_\mathrm{tf}$, and the direction angle from the starting point to the ending point is θ. The path is divided into three segments for planning:

- The first segment starts at $\mathbf{p}_\mathrm{i}$ with a course angle α, turning radius r_i, turning direction dir_i, and ends at $\mathbf{p}_\mathrm{ti}$ with a velocity direction angle θ, minimum turning radius of the aircraft r_min, and turning direction dir_ti.
- A straight path from $\mathbf{p}_\mathrm{ti}$ to $\mathbf{p}_\mathrm{tf}$.
- The second segment starts at $\mathbf{p}_\mathrm{tf}$ with a velocity direction angle θ, minimum turning radius of the aircraft r_min, turning direction dir_tf, and ends at $\mathbf{p}_\mathrm{tf}$ with a course angle β, turning radius r_f, and turning direction dir_f.

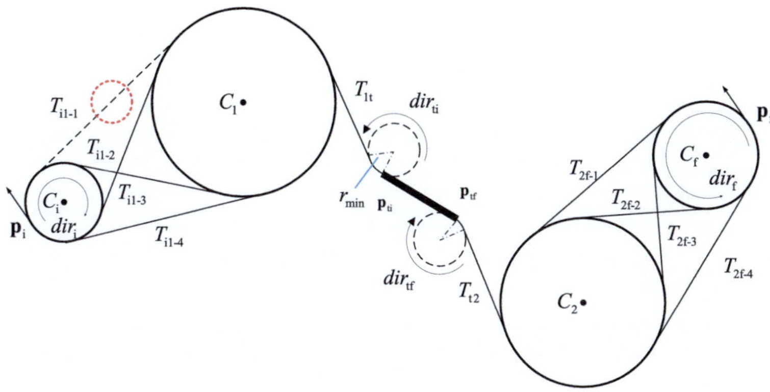

Fig. 7.17 Dubins path planning through intermediate circles and tunnels

Table 7.1 Path segment information vector

L	L(1)	L(2)	L(3)	L(4)	L(5)	L(6)	L(7)
Straight path information vector							
Example	−1	1.5	2	5	6	0	0
Meaning	Any negative number	Starting point x-coordinate	Starting point y-coordinate	Ending point x-coordinate	Ending point y-coordinate	Any number	Any number
Circular arc path information vector							
Example	pi/3	1.5	2	5	6	2	1
Meaning	Direction angle of the circular arc	Starting point x-coordinate	Starting point y-coordinate	Ending point x-coordinate	Ending point y-coordinate	Radius of the circular arc	Rotation direction

3. Step 3: Designing the path segment information vector.
Design a unified path segment information vector to represent both straight and circular arc path information, as shown in Table 7.1, where the direction angle for entering the circular arc ranges from $[0, 2\pi)$, with a rotation direction greater than zero indicating counterclockwise rotation. Otherwise, it is clockwise.

4. Step 4: Implementing the Dubins path planning algorithm in MATLAB.
(1) In the main program "path_plan.m" under "e5/e5-2", set the output data of the planning algorithm to include the starting point, ending point, obstacles, and tunnels. See Code 7.11 for the obstacle information and tunnel information setup.

Code 7.11 Obstacle setting, from"e5/e5-2/path_plan.m"

```
rmin=30; % Minimum turning radius
% Starting point
xi=0;
yi=0;
alpha=pi/6;
ri=rmin;
diri=1; % greater than 0 counterclockwise rotation, otherwise clockwise rotation
% ending point
xf=600;
yf=600;
beta=0;
rf=rmin;
dirf=-1;
% Obstacle info
obs_in=[
100 100 70;
300 300 60;
490 440 60
];
% Tunnel infor
tunnel=[200 200 0 40];
diri_tunnel=-1;
dirf_tunnel=1;
```

(2) Combining the algorithm analysis from Steps 1 and 2, the algorithm framework is built as shown in Code 7.12. If a straight tunnel exists, the path is divided into two parts for planning. The Dubins curves are obtained for the segments before and after the tunnel using the "dubins_path" function, then the three trajectory segments are connected to obtain the final trajectory. If no straight tunnel exists, the entire path is solved directly.

Code 7.12 Solving waypoints, from"e5/e5-2/path_plan.m"

```
if(~isempty(tunnel))
[XTU,YTU]=myline(tunnel(1),tunnel(2),tunnel(3),tunnel(4));
[L1,D1,X1,Y1]=dubins_path(xi, yi, alpha, ri,XTU(1), YTU(1), tunnel(3), rmin, diri, diri_tunnel, obs_in,tunnel);
LTU=[-1 XTU(1) YTU(1) XTU(length(XTU)) YTU(length(YTU)) 0 0];
[L2,D2,X2,Y2]=dubins_path(XTU(length(XTU)), YTU(length(YTU)), tunnel(3),rmin, xf, yf, beta, rf, dirf_tunnel, dirf, obs_in, tunnel);
L=[L1;LTU;L2];
Plan_X=[X1 XTU X2];
Plan_Y=[Y1 YTU Y2];
lenP=D1+D2; % Path length
else
[L,lenP,Plan_X,Plan_Y]=dubins_path(xi,yi,alpha,ri,xf,yf,beta,rf,diri,dirf,obs_in,[]);
end
```

(3) Run the main program "path_plan.m" to obtain the path plot and path segment information vector as shown in Fig. 7.18.

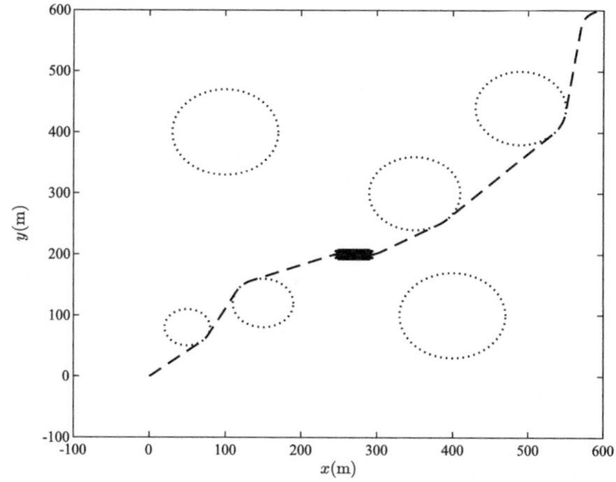

Fig. 7.18 Generated path

7.4 Design Experiment

7.4.1 Experimental Objective

1. Things to prepare

(1) Software: MATLAB R2022b or above, RflySim toolchain. The RflySim toolchain provides a HIL simulation environment for small fixed-wing UAVs.
(2) Program: Experimental packages "e5/e5-3" and "e5/e5-4". The package "e5/e5-3" contains the path planning file "Path_plan", the initialization file for a fixed-wing UAV "InitDatactrl.m", and the Simulink simulation model file for the fixed-wing UAV "SmallFixedWingUAVplan.slx", used for SIL simulation experiments. The package "e5/e5-4" contains the initialization file for the fixed-wing UAV "InitDatactrl.m" and the Simulink simulation model file for the fixed-wing UAV "SmallFixedWingUAVplanHIL.slx", used for HIL simulation experiment.
(3) Hardware: a CubePilot/Pixhawk autopilot, a remote controller, and a receiver.

2. Objective

The micro-small fixed-wing UAV is tasked with completing a low-altitude reconnaissance mission, as illustrated in Fig. 7.19, with map data provided in Table 7.2. During the mission, the following objectives must be met:

- Depart from the takeoff point to start the mission and land at the recovery point to end the mission;
- Avoid restricted zones and obstacles during the flight;
- Reconnaissance of valley areas is required;
- Due to the limited battery power of the UAV, choose the shortest path.

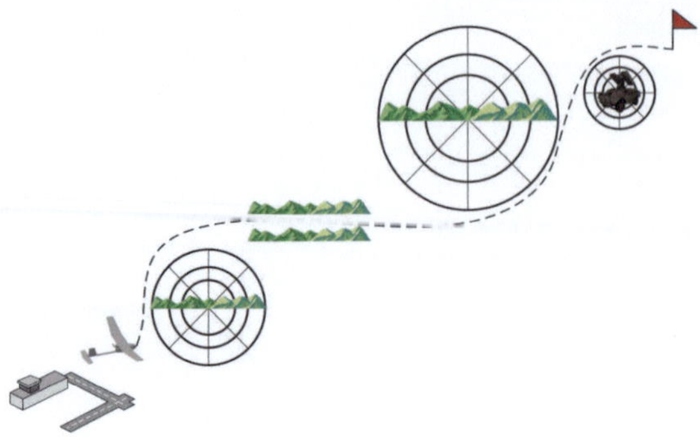

Fig. 7.19 Mission map

Table 7.2 Map data

Type	Data
Starting Point	$x = 300$ m, $y = 300$ m, α = pi/4
Ending point	$x = 1800$ m, $y = 1800$ m, $\alpha = 0$
Obstacles	$x_1 = 500$ m, $y_1 = 600$ m, $r_1 = 150$ m
	$x_2 = 1200$ m, $y_2 = 1300$ m, $r_2 = 200$ m
	$x_3 = 1600$ m, $y_3 = 1500$ m, $r_3 = 80$ m
Tunnel (Valley)	$x = 700$ m, $y = 1000$ m, $\theta = 0$, $L = 200$ m

Design the algorithm for this mission and perform SIL simulation in MATLAB/Simulink. Finally, the simulation program will be downloaded into the CubePilot/Pixhawk autopilot through code generation, and the HIL simulation will be conducted on the RflySim toolchain.

7.4.2 Experimental Design

This experiment requires corresponding experimental designs. In Step 1, the state transition relationship of the micro-small fixed-wing UAV is designed. In Step 2, the trajectory switching algorithm for the UAV is designed for waypoint detection and switching, which is based on the method presented in [3].

1. Step 1: Design of UAV state transition relationship
To accomplish the experimental objectives and achieve HIL simulation, a comprehensive state transition logic for the micro-small fixed-wing UAV must be designed, enabling the UAV to follow the planned trajectory. The state transition relationship of this experiment is shown in Fig. 7.20, where each state has the following specific meanings.

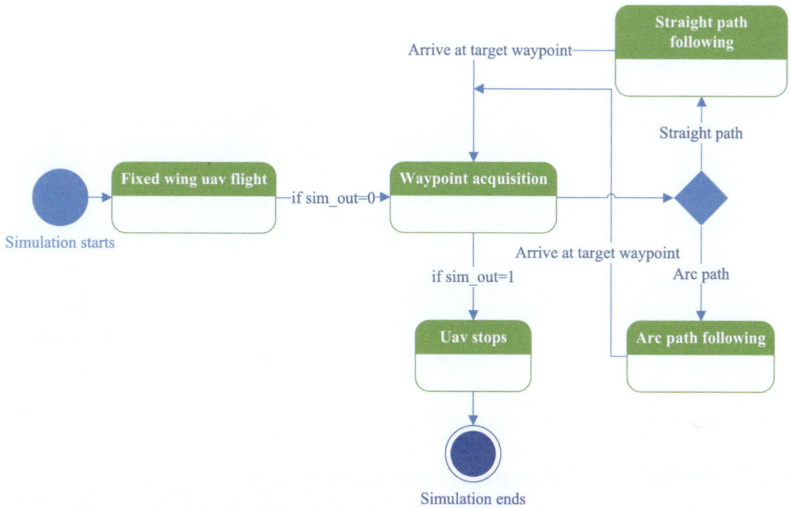

Fig. 7.20 Fixed-wing state transition

- Micro-small fixed-wing UAV Flight: The micro-small fixed-wing UAV takes off from the current position.
- Waypoint Information Acquisition: The micro-small fixed-wing UAV obtains the target waypoint information from the waypoint information table based on the current position.
- Straight Path Following: The micro-small fixed-wing UAV executes the L1 straight path following algorithm in the "L1 Navigation" module.
- Arc Path Following: The micro-small fixed-wing UAV executes the Dubins Curve path following algorithm in the "PD Navigation" module.
- UAV Stops Flying: The micro-small fixed-wing UAV stops flying at the current position.

During simulation, the micro-small fixed-wing UAV first enters the "Micro-small fixed-wing UAV Flight" state. When "sim_out" value is set to zero, the micro-small fixed-wing UAV enters the "Waypoint Information Acquisition" state. In this state, the UAV retrieves the information of the next waypoint corresponding to the current position. After obtaining the waypoint information, the UAV selects to enter the "Straight Path Following" state or the "Arc Path Following" state based on the trajectory. When the UAV reaches the target waypoint, it will enter the "Waypoint Information Acquisition" state again until all waypoints are retrieved. When "sim_out" value is 1, the UAV enters the "UAV Stops Flying" state, and the entire simulation is completed.

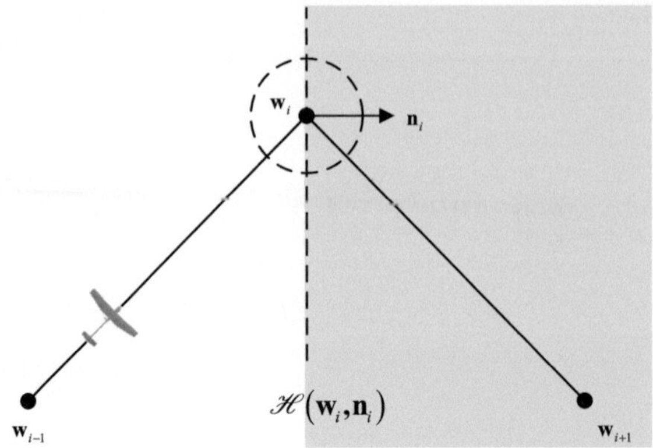

Fig. 7.21 Illustration of fixed-wing waypoint switching

2. Step 2: Designing a UAV path switching algorithm

In actual flight, the UAV may not accurately reach the designated waypoints. Thus, waypoint detection and switching are important. The UAV's path as an ordered sequence of waypoints is defined as

$$\mathscr{W} = \{\mathbf{w}_1, \mathbf{w}_2, \ldots, \mathbf{w}_N\} \tag{7.33}$$

where $\mathbf{w}_i \in \mathbb{R}^3, i = 1, \ldots, N$. Considering the scenario shown in Fig. 7.21, when the UAV needs to reach $\mathbf{w}_i$, the guidance algorithm is switched so that the UAV can track the path from $\mathbf{w}_i$ to $\mathbf{w}_{i+1}$. As shown in Eq. (7.34), a spherical region with radius b centered at waypoint $\mathbf{w}_i$ is defined, and the UAV switches its following path when it enters this region, shown as

$$||\mathbf{p}(t) - \mathbf{w}_i|| \leq b \tag{7.34}$$

where $\mathbf{p}(t) \in \mathbb{R}^3$ represents the UAV's position. However, this method has obvious drawbacks in practical applications. If wind interference or b is chosen too small, the UAV may never enter the spherical region. Therefore, a switching method insensitive to tracking errors is adopted, namely, the half-plane switching criterion. For a given position $\mathbf{r} \in \mathbb{R}^3$ and normal vector $\mathbf{n} \in \mathbb{R}^3$, a half-plane is defined as follows:

$$\mathscr{H}(\mathbf{r}, \mathbf{n}) = \left\{\mathbf{p} \in \mathbb{R}^3 : (\mathbf{p} - \mathbf{r})^{\mathrm{T}} \mathbf{n} \geq 0\right\}. \tag{7.35}$$

Here,

$$\mathbf{n}_i = \frac{\mathbf{q}_{i-1} + \mathbf{q}_i}{||\mathbf{q}_{i-1} + \mathbf{q}_i||} \tag{7.36}$$

where $\mathbf{q}_i$ is the unit vector between waypoints $\mathbf{w}_i$ and $\mathbf{w}_{i+1}$:

$$\mathbf{q}_i = \frac{\mathbf{w}_{i+1} - \mathbf{w}_i}{||\mathbf{w}_{i+1} - \mathbf{w}_i||}. \tag{7.37}$$

The UAV switches its path when it enters the half-plane $\mathscr{H}(\mathbf{r}, \mathbf{n})$. Pseudocode for UAV waypoint trajectory switching is provided in Algorithm 7.2.

Algorithm 7.2 UAV waypoint trajectory switching algorithm

Input: Waypoints $\mathscr{W} = \{\mathbf{w}_1, \mathbf{w}_2, \ldots, \mathbf{w}_N\}$, UAV position $\mathbf{p}$
Output: Path followed by the UAV

if Receive waypoints $\mathscr{W} = \{\mathbf{w}_1, \mathbf{w}_2, \ldots, \mathbf{w}_N\}$ **then**
  Initialize $i = 2$
end if
$\mathbf{q}_{i-1} = \dfrac{\mathbf{w}_i - \mathbf{w}_{i-1}}{||\mathbf{w}_i - \mathbf{w}_{i-1}||}$
$\mathbf{q}_i = \dfrac{\mathbf{w}_{i+1} - \mathbf{w}_i}{||\mathbf{w}_{i+1} - \mathbf{w}_i||}$
$\mathbf{n}_i = \dfrac{\mathbf{q}_{i-1} + \mathbf{q}_i}{||\mathbf{q}_{i-1} + \mathbf{q}_i||}$
if $\mathbf{p} \in \mathscr{H}(\mathbf{w}_i, \mathbf{n}_i)$ **then**
  $i = i + 1$ until $i = N - 1$
end if
Return Path followed by the UAV from waypoint $\mathbf{w}_{i-1}$ to $\mathbf{w}_i$

The pre-planned path in this chapter does not include transitions between two straight-line paths. The actual flight is as shown in Fig. 7.22, thus half-planes $\mathscr{H}_i$ and $\mathscr{H}_{i+1}$ are represented as follows,

$$\begin{aligned} \mathscr{H}_i &= \mathscr{H}\left(\mathbf{w}_i, \mathbf{q}_{i-1}\right) = \left\{\mathbf{p} \in \mathbb{R}^3 : \left(\mathbf{p} - \mathbf{w}_i\right)^{\mathrm{T}} \mathbf{q}_{i-1} \geq 0\right\} \\ \mathscr{H}_{i+1} &= \mathscr{H}\left(\mathbf{w}_{i+1}, \mathbf{q}_{i+1}\right) = \left\{\mathbf{p} \in \mathbb{R}^3 : \left(\mathbf{p} - \mathbf{w}_{i+1}\right)^{\mathrm{T}} \mathbf{q}_{i+1} \geq 0\right\}. \end{aligned} \tag{7.38}$$

Similarly, the UAV switches its path when it enters a new half-plane.

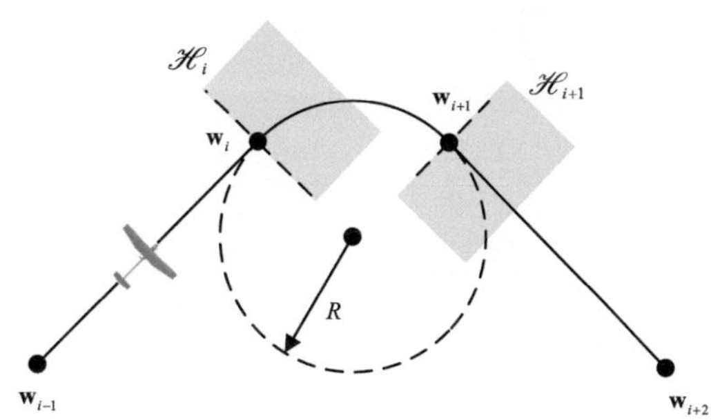

Fig. 7.22 Illustration of fixed-wing waypoint switching

7.4.3 *Experimental Procedure*

The primary objective of the experiment is to fulfill the specified task requirements. In Step 1, a state machine is designed for waypoint switching, and a specialized algorithm generates Dubins curves to define the experimental path. In Step 2, the path-following algorithm from the basic experiment is applied to achieve path-following, which is verified through software simulation. In Step 3, the RflySim toolchain, along with hardware such as the CubePilot/Pixhawk autopilot, remote controller, and receiver, is used for HIL simulation. This step validates the feasibility and accuracy of the designed algorithm by integrating real hardware with the simulation environment. The entire experimental process involves integrating SIL and HIL simulations to validate the effectiveness of the algorithm design comprehensively. This design approach will facilitate the evaluation of the UAV system's performance and provide reliable technical support and guidance for practical applications.

1. Step 1: Designing a state machine and a path following algorithm
(1) In SIL experimental package "e5/e5-3", "Simple State Machine" is designed using the "MATLAB Function" module within the "SmallFixedWingUAVplan.slx" model, as shown in Code 7.13. In the example code, "sim_stop" is set as the termination flag. It is assigned the value 1 to terminate the simulation when the entire flight path segment is completed. Lines 6–8 initialize the state variable "state" to 1 when the state is empty. A new vector table is created for the path segment information. Lines 10-36 update the waypoints followed by the UAV, continuously checking whether the current position of the micro-small fixed-wing UAV has entered the current switching region. If not, the output remains unchanged; if the UAV enters the switching region, it switches to the next "Loop" vector and updates the state value.
(2) In the "Navigation" module, combining linear and circular algorithms, the current type of path (linear or circular) is determined based on the first position data "Loop(1)" of the vector information. The algorithm framework is shown in Code 7.13.

Code 7.13 State machine design, from"e5/e5-3/SmallFixedWingUAVplan.slx"

```
1  % Stop simulation
2  sim_stop=0;
3  
4  % State machine
5  persistent state;
6  if isempty(state)
7  state = 1;
8  Loop = zeros(1,7);
9  else
10 index = state;
11 if(index < length(L(:,1)))
12 Loop = L(index,:);
13 if Loop(1)<0
14 r=[Loop(4) Loop(5)];
15 n=([Loop(4) Loop(5)]-[Loop(2) Loop(3)])/norm([Loop(4)
       Loop(5)]-[Loop(2) Loop(3)]);
16 p=[current_loc(1) current_loc(2)];
```

```
else
r=[Loop(4) Loop(5)];
n=([L(index+1,4) L(index+1,5)]-[Loop(4) Loop(5)])/norm([L(index+1,4)
    L(index+1,5)]-[Loop(4) Loop(5)]);
p=[current_loc(1) current_loc(2)];
end
if dot((p-r)/norm(p-r),n)>=0
state = state + 1;
else
state = state;
end
elseif(index == length(L(:,1)))
Loop = L(index,:);
if sqrt((current_loc(1)-L(index,4))^2+(current_loc(2)-L(index,5))^2)
    < 20
state = state + 1;
else
state = state;
end
else
Loop = zeros(1,7);
sim_stop = 1;
end
end
```

(2) In the “Navigation” module, the type of path (straight line or circular arc) is determined based on the first position data “Loop(1)” of the vector information. The algorithm framework is shown in Code 7.14.

Code 7.14 Path following algorithm selection, from“e5/e5-3/SmallFixedWingUAVplan.slx"

```
function [Roll_d,prev_wp,next_wp] = fcn(Loop, current_loc,
    current_yaw, airspeed)

flag = Loop(1);
prev_wp = [Loop(2);Loop(3)];
next_wp = [Loop(4);Loop(5)];
ISA_g = 9.80665;
persistent rdist_prev;
```

2. Step 2: SIL simulation in MATLAB/Simulink

(1) Set the starting point, ending point, obstacle information, etc. in “e5/e5-3/Path_plan/path_plan.m”. Running this yields a list of waypoints, and then put the path segment information into “Simple State Machine” state machine module, as shown in Code 7.15.

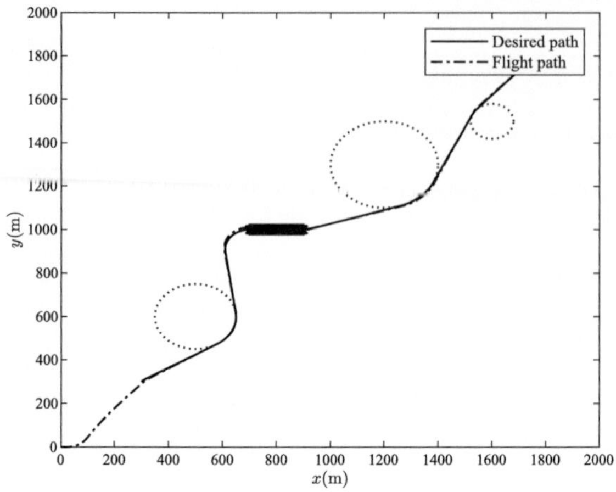

Fig. 7.23 Comparison of flight trajectory and planned trajectory

Code 7.15 Waypoint information, from"e5/e5-3/SmallFixedWingUAVplan.slx"

```
L =  1.0e+03 *[
-1         0         0         0.3       0.3       0         0;
0.0008    0.3000    0.3000    0.3260    0.3551    0.0900    0.0010;
-0.0010    0.3260    0.3551    0.3507    0.6142         0         0;
0.0015    0.3507    0.6142    0.3889    0.7007    0.1500   -0.0010;
-0.0010    0.3889    0.7007    0.6333    0.9704         0         0;
0.0008    0.6333    0.9704    0.7000    1.0000    0.0900   -0.0010;
-0.0010    0.7000    1.0000    0.9000    1.0000         0         0;
0    0.9000    1.0000    0.9271    1.0042    0.0900    0.0010;
-0.0010    0.9271    1.0042    1.2602    1.1093         0         0;
0.0003    1.2602    1.1093    1.3817    1.2165    0.2000    0.0010;
-0.0010    1.3817    1.2165    1.5273    1.5334         0         0;
0.0011    1.5273    1.5334    1.5402    1.5531    0.0800   -0.0010;
-0.0010    1.5402    1.5531    1.7327    1.7698         0         0;
0.0008    1.7327    1.7698    1.8000    1.8000    0.0900   -0.0010];
```

(2) Run "SmallFixedWingUAVplan.slx" and compare the flight trajectory with the planned trajectory as shown in Fig. 7.23. The figure shows that the actual flight path closely follows the planned path, especially when following a straight path, demonstrating the algorithm's effectiveness.

3. Step 3: HIL simulation on RflySim

The HIL simulation layout is shown in Fig. 7.24.

(1) First, run the initialization file "InitDatactrl.m", open "SmallFixedWingUAVplan-HIL.slx," and click on "Build" in the toolbar to compile the current program, as shown in Fig. 7.25.

(2) After successful compilation, connect the flight controller to the computer, click on "code" in the toolbar shown in Fig. 7.25, select "PX4 PSP Upload code to PX4FMU", and burn the code to the CubePilot/Pixhawk flight controller.

Fig. 7.24 Hardware layout of HIL simulation

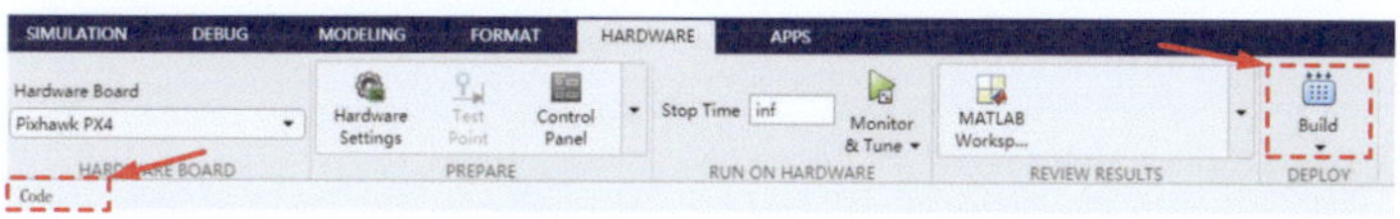

Fig. 7.25 Toolbar

(3) In the directory "/RflySimAPIs/OtherVehicleTypes/AircraftMathworks" under the RflySim toolchain installation directory, double-click on "AircraftMathworksHITLRun" script to start the fixed-wing HIL simulation environment, and select the port connected to the current CubePilot/Pixhawk flight controller.

(4) In the "SmallFixedWingUAVplanHIL.slx" simulation model, channels "ch5" and "ch6" are assigned, respectively, to the "unlock" and "automatic waypoint flight" functions. When in the up position, "ch5" issues the unlock signal and "ch6" initiates the automatic waypoint mission. Prior to commencing the simulation, both channels must be in the down position. Once the HIL environment is active, "ch5" should first be moved to the up position to unlock the flight controller, and then "ch6" should be moved to the up position to commence the autonomous flight task. To render the flight trajectory in QGroundControl (QGC), the autopilot's unlock signal must be asserted; under this condition, the trajectory displayed in QGC (see Fig. 7.26) corresponds exactly to that produced by the software simulation.

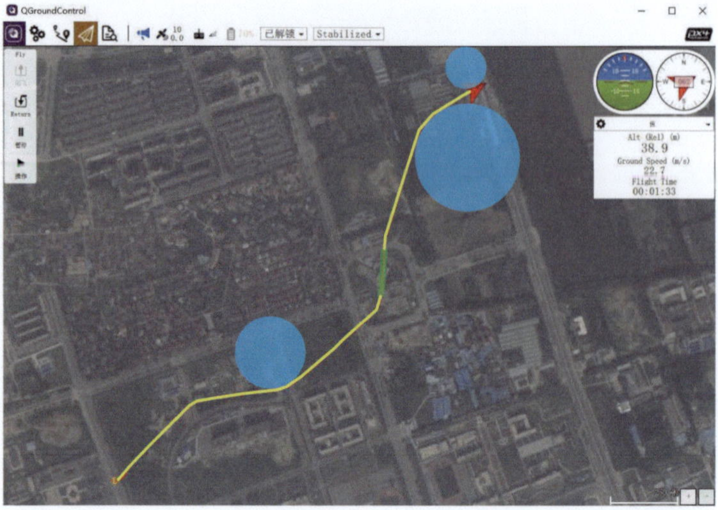

Fig. 7.26 Actual flight path

7.5 Exercise

1. When executing L1 straight-line path following, this section provides the period $L_{1,\mathrm{P}}$ and damping ratio $L_{1,\mathrm{D}}$ of the path-following algorithm. Readers are encouraged to modify these parameters and analyze their impact on path-following performance. The analysis can be approached from various perspectives, such as stability, responsiveness, and disturbance rejection. This exercise aims to enhance the understanding of how L1 guidance parameters influence path-following performance.
2. In the experiments of path following and planning design, only one tunnel is included in the given obstacles that require micro-small fixed-wing UAVs to cross. Now, there is a requirement to add a new tunnel with parameters $x = 100$ m, $y = 200$ m, $\theta = \pi/4$, $L = 100$ m. The position and orientation of this new tunnel have been defined, and it is necessary to modify the original planning so that micro-small fixed-wing UAVs can adapt to the new obstacle and pass through these two tunnels. To complete this experiment, it is necessary to replan Dubins curves to ensure that UAVs can fly smoothly, taking into account the two tunnels. After modifying the plan, the experiment needs to be conducted in HIL testing, which involves verifying the new path planning scheme using actual hardware devices to ensure its feasibility and effectiveness in real-world environments. This experimental design presents new challenges to the robustness and adaptability of planning algorithms, while also facilitating the evaluation of the practical application capabilities of UAV systems in complex environments.
3. As stated in Eq. (4.63) in Chap. 4 on low-level flight control, coordinated turns can regulate yaw by controlling roll and simultaneously managing heading

approach through roll control. Design roll angle commands for micro-small fixed-wing UAVs to ensure the convergence of the yaw angle and analyze the control parameters and tracking performance.

4. In aircraft path planning and speed control, the safety flying radius is critical. As flight speed increases, the safety radius must expand to prevent collisions. Obstacles, represented by obstacle circles, must have a radius equal to or greater than the safety flying radius and obstacle size. Thus, a strategy for controlling speed and planning trajectories along Dubins curves is essential to minimize travel time. Such a design necessitates the integration of optimization strategies for path planning and speed control, enabling the aircraft to reach the destination in the shortest possible time while accounting for obstacles. This may entail the use of heuristic algorithms, optimization methods, or model-based control strategies to achieve the shortest flight time while ensuring safety. This design holds great significance for enhancing the efficiency and applicability of aircraft.
5. In the field of path following, several algorithms are employed. One common method is constructing a vector field, which can be used for UAV navigation and guidance. For straight-line following, path planning is achieved by defining the desired track angle. The vector field generated by this method provides navigation guidance for the aircraft, as illustrated in Fig. 7.27. The specific expression is:

$$\chi_{\mathrm{d}}(y) = -\chi^{\infty}\frac{2}{\pi}\arctan(ky)$$

where $k > 0$ represents a positive constant, and y represents the lateral distance between the UAV and the path. Similarly, arc following can be achieved using vector fields. For further details, readers may refer to literature [12], which discusses vector fields in aircraft path planning, particularly their construction

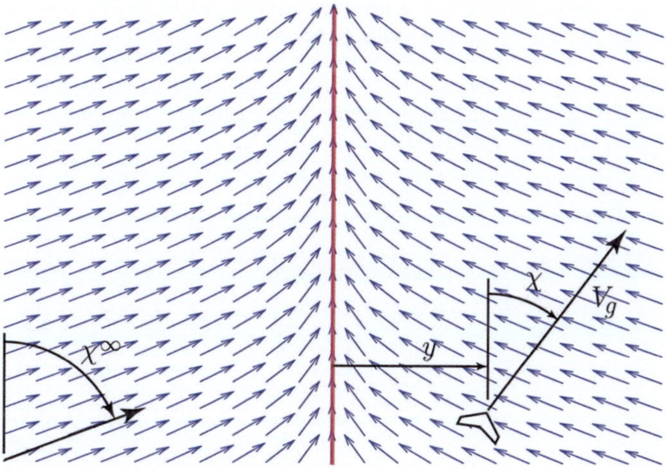

Fig. 7.27 Straight-line following vector field [12]

for the entire flight domain and use in navigating micro-small fixed-wing UAVs. These methods introduce innovative approaches that improve path-following accuracy and overall performance.

References

1. S. Park, J. Deyst, J.P. How, A new nonlinear guidance logic for trajectory tracking, in *Proceedings of the AIAA Guidance, Navigation and Control Conference* (2004), p. 4900
2. L. Dubins, On curves of minimal length with a constraint on average curvature, and with prescribed initial and terminal positions and tangents. Am. J. Math. **79**, 497–516 (1957)
3. R.W. Beard, T.W. McLain, *Small Unmanned Aircraft: Theory and Practice* (Princeton University Press, 2012)
4. V. Mnih, K. Kavukcuoglu, D. Silver, A.A. Rusu, J. Veness, M.G. Bellemare, A. Graves, M. Riedmiller, A.K. Fidjeland, G. Ostrovski et al., Human-level control through deep reinforcement learning. Nature **518**(7540), 529–533 (2015)
5. V. Mnih, A.P. Badia, M. Mirza, A. Graves, T. Lillicrap, T. Harley, D. Silver, K. Kavukcuoglu, Asynchronous methods for deep reinforcement learning, in *International Conference on Machine Learning* (PMLR, 2016), pp. 1928–1937
6. K. Chua, R. Calandra, R. McAllister, S. Levine, Deep reinforcement learning in a handful of trials using probabilistic dynamics models, in *Advances in Neural Information Processing Systems*, vol. 31 (2018)
7. D. Baranchuk, D. Persiyanov, A. Sinitsin, A. Osokin, Learning to route in similarity graphs, in *International Conference on Machine Learning* (PMLR, 2019), pp. 475–484
8. S. Peng, Multi-robot path planning combining heuristics and multi-agent reinforcement learning (2023). arXiv:2306.01270
9. W. Zhang, L. Shan, L. Chang, Z. Zhang, S. Zhang, Z. Wang, SVF-RRT*: A stream-based VF-RRT* for USVs path planning considering ocean currents. IEEE Robot. Autom. Lett. **8**(4), 2413–2420 (2023)
10. Z. Yu, Z. Si, X. Li, Z. Wang, Z. Zhang, S. Zhang, A novel hybrid particle swarm optimization algorithm for path planning of UAVs. IEEE Internet Things J. **9**(22), 22 547–22 558 (2022)
11. K. Narasimha, E. Kivelevitch, B. Sharma, K. Srinivas, An ant colony optimization technique for solving min-max multi-depot vehicle routing problem. Swarm Evol. Comput. **13**, 63–73 (2013)
12. D. Nelson, D. Barber, T. McLain, R. Beard, Vector field path following for miniature air vehicles. IEEE Trans. Rob. **23**(3), 519–529 (2007)

Chapter 8
Trajectory Planning and Differential Flatness Control Experiment

Shuli Lv, Xunhua Dai, and Quan Quan

In the field of aeronautics, trajectory planning and control have always been important research areas. Trajectory planning, unlike path planning, involves assigning time information to the path. Fixed-wing Unmanned Aerial Vehicles (UAVs) must follow specific trajectories during flight to accomplish tasks such as cruising, landing, and obstacle avoidance. Differential flatness control is considered a suitable method for low-level control of fixed-wing UAVs, as it enables efficient control of predetermined trajectories. This chapter examines the topic of trajectory planning and differential flatness control for fixed-wing UAVs. First, the coordinated flight model for fixed-wing aircraft is introduced. Next, the principles of trajectory tracking using the coordinated flight model and differential flatness control are explored. Finally, a trajectory generation method utilizing B-splines is presented. The experimental section consists of a basic experiment, an analysis experiment, and a design experiment, including the construction of the differential flatness control module, properties of B-spline trajectory generation, and B-spline trajectory differential flatness control tracking. For more information on this chapter, please refer to the link

https://rflysim.com/doc/en/C/5.FW_Book.html.

8.1 Preliminary

When considering the trajectory of a fixed-wing aircraft, especially during fast and agile flight, it is crucial to account for the aircraft's dynamics and control input constraints to ensure that the generated trajectories are dynamically feasible. In the context of trajectory generation, it is worth noting that differential flatness transforms trajectories from a flat output space to the state and control input space. By defining the trajectory in a flat output space, the feedforward control input of the fixed-wing UAV is obtained, thereby achieving efficient trajectory tracking.

Q. Quan et al., *Micro-Small Fixed-Wing UAV Control*,
https://doi.org/10.1007/978-981-95-1954-5_8

8.1.1 Coordinated Flight Model

Coordinated flight (see Chap. 4), as illustrated in Fig. 8.1, is defined as the flight of an aircraft where its body velocity is restricted to the longitudinal plane, meaning the sideslip angle $\beta = 0$. This model describes the translational dynamics and attitude kinematics of the aircraft. The normal acceleration, axial acceleration, and roll rate determine the aircraft's motion. As described in Eq. (4.20) of Chap. 4, the aircraft typically generates accelerations within the longitudinal plane due to the combined contributions of lift, drag, and thrust. The restriction of non-gravitational accelerations to the longitudinal plane, in conjunction with the enforcement of coordinated flight, has been demonstrated to yield the following results.

Use $\mathbf{x}(t) \in \mathbb{R}^3$ to represent the position of the fixed-wing aircraft's center of mass, and $\mathbf{R}_{\mathrm{s}}^{\mathrm{e}}$ as the rotation matrix describing the stable coordinate system of the fixed-wing aircraft (with sideslip angle $\beta = 0$), where $\dot{\mathbf{R}}_{\mathrm{s}}^{\mathrm{e}} = \mathbf{R}_{\mathrm{s}}^{\mathrm{e}}\hat{\omega}_{\times}$, and

$$\hat{\omega}_{\times} = \begin{bmatrix} 0 & -\omega_z & \omega_y \\ \omega_z & 0 & -\omega_x \\ -\omega_y & \omega_x & 0 \end{bmatrix}. \tag{8.1}$$

Here, $\omega = \left[\omega_x \ \omega_y \ \omega_z\right]^{\mathrm{T}}$ represents the instantaneous angular velocity in the stable coordinate system, corresponding to $o_{\mathrm{b}}x_{\mathrm{s}}$, $o_{\mathrm{b}}y_{\mathrm{s}}$, and $o_{\mathrm{b}}z_{\mathrm{s}}$ axes, respectively. Therefore, the model [1] is expressed as

$$\ddot{\mathbf{x}} = \mathbf{g} + \mathbf{R}_{\mathrm{s}}^{\mathrm{e}}\mathbf{a}_v \tag{8.2}$$

where $\mathbf{g} = [0\ 0\ g]^{\mathrm{T}}$, and $\mathbf{a}_v = [a_{v1}\ a_{v2}\ a_{v3}]^{\mathrm{T}}$ contains axial and normal accelerations in the stable coordinate system. Since only the accelerations in the longitudinal plane are considered, $a_{v2} = 0$, as shown in Fig. 8.1.

Since $\mathbf{R}_{\mathrm{s}}^{\mathrm{e}}$ transforms vectors from the stable coordinate system to the inertial coordinate system, the coordinated flight condition is expressed as

$$\dot{\mathbf{x}} = V_{\mathrm{a}}\mathbf{R}_{\mathrm{s}}^{\mathrm{e}}\mathbf{e}_1 \tag{8.3}$$

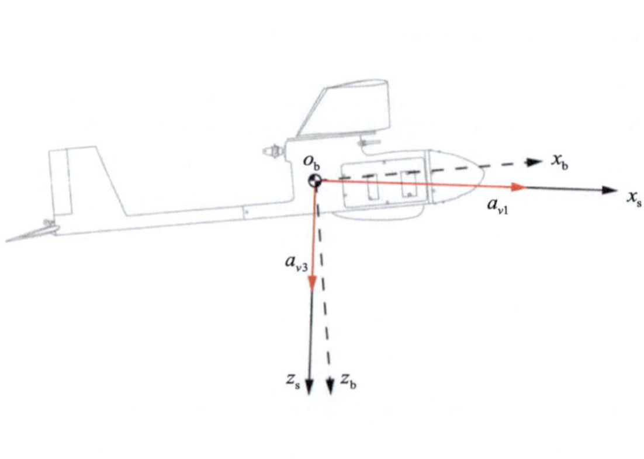

Fig. 8.1 Longitudinal plane control under coordinated flight conditions

where $\mathbf{e}_1 = [1\ 0\ 0]^{\mathrm{T}}$ and $V_{\mathrm{a}} = \|\dot{\mathbf{x}}\|$. Differentiating Eq. (8.3) and substituting into Eq. (8.2),

$$\begin{aligned} \mathbf{g} + \mathbf{R}_{\mathrm{s}}^{\mathrm{e}} \left(a_{v1}\mathbf{e}_1 + a_{v3}\mathbf{e}_3\right) &= \dot{V}_{\mathrm{a}}\mathbf{R}_{\mathrm{s}}^{\mathrm{e}}\mathbf{e}_1 + V_{\mathrm{a}}\mathbf{R}_{\mathrm{s}}^{\mathrm{e}} \left(\omega \times \mathbf{e}_1\right) \\ &= \dot{V}_{\mathrm{a}}\mathbf{R}_{\mathrm{s}}^{\mathrm{e}}\mathbf{e}_1 + V_{\mathrm{a}}\mathbf{R}_{\mathrm{s}}^{\mathrm{e}} \left(\omega_z\mathbf{e}_2 - \omega_y\mathbf{e}_3\right) \end{aligned} \tag{8.4}$$

is obtained, where $\mathbf{e}_2 = [0\ 1\ 0]^{\mathrm{T}}$ and $\mathbf{e}_3 = [0\ 0\ 1]^{\mathrm{T}}$. Therefore, to maintain coordinated flight, the angular velocities (pitch and yaw) need to satisfy the constraints

$$\begin{aligned} \omega_y &= -\frac{a_{v3} + g_{v3}}{V_{\mathrm{a}}} \\ \omega_z &= \frac{g_{v2}}{V_{\mathrm{a}}} \end{aligned} \tag{8.5}$$

where $\mathbf{g}_v = \mathbf{R}_{\mathrm{e}}^{\mathrm{s}}\mathbf{g} = [g_{v1}\ g_{v2}\ g_{v3}]^{\mathrm{T}}$.

8.1.2 *Differential Flatness Control and Trajectory Tracking*

In modern control theory, a controlled system is commonly represented as

$$\begin{aligned} \dot{\mathbf{x}} &= \mathbf{f}\left(\mathbf{x}, \mathbf{u}\right) \\ \mathbf{y} &= \mathbf{g}\left(\mathbf{x}\right) \end{aligned} \tag{8.6}$$

where $\mathbf{x}$ represents the system state, $\mathbf{u}$ represents the system input, and $\mathbf{y}$ represents the system output. The core of differential flatness is defined as the ability to take the flat output $\mathbf{y}$ and its derivatives $\dot{\mathbf{y}}, \ddot{\mathbf{y}}, \ldots, \mathbf{y}^{(k)}$ as differential parameters to represent the original state variables as basis vectors algebraically. That is, when a system is differentially flat, there exists a function relation $\mathbf{h}$ such that

$$(\mathbf{x}, \mathbf{u}) = \mathbf{h}\Big(\underbrace{\mathbf{y}, \dot{\mathbf{y}}, \ddot{\mathbf{y}}, \ldots, \mathbf{y}^{(k)}}_{\bar{\mathbf{y}}}\Big). \tag{8.7}$$

The trajectories of differentially flat systems are easy to describe: choose any output trajectory $\mathbf{y}\left(t\right), t > 0$ and the state control trajectory is then written as $\left(\mathbf{x}\left(t\right), \mathbf{u}\left(t\right)\right) = \mathbf{h}\left(\bar{\mathbf{y}}\left(t\right)\right)$ [2]. This characteristic can guide the selection of output variables for motion planning in robotics by mapping all trajectory constraints to a flat output space and planning the optimal trajectory within this space. Subsequently, one returns to the initial state and input space to obtain control inputs.

The coordinated flight model of fixed-wing aircraft is differentially flat within a reasonable operating range [3]. Differentiating Eq. (8.2) yields

$$
\begin{aligned}
\mathbf{x}^{(3)} &= \mathbf{R}_{\mathrm{s}}^{\mathrm{e}} \left(\omega \times \mathbf{a}_v + \dot{\mathbf{a}}_v\right) \\
&= \mathbf{R}_{\mathrm{s}}^{\mathrm{e}} \begin{bmatrix} \omega_y a_{v3} \\ \omega_z a_{v1} \\ -\omega_y a_{v1} \end{bmatrix} + \mathbf{R}_{\mathrm{s}}^{\mathrm{e}} \begin{bmatrix} \dot{a}_{v1} \\ -a_{v3}\omega_x \\ \dot{a}_{v3} \end{bmatrix}.
\end{aligned} \tag{8.8}
$$

The coordinated flight imposes no constraints on the roll rate ω_x, as shown in Eqs. (8.5) and (8.8). Therefore, defined on $\mathbb{R}^6 \times \mathrm{SO}(3)$, coordinated flight consists of position, velocity, and orientation variables, determined by three dimensions of input space (axial acceleration a_{v1}, normal acceleration a_{v3}, and roll rate ω_x). Here, SO (3) represents a special orthogonal group containing rotation matrices, known as the *three-dimensional rotation group*, which is used to describe the attitude rotation motion of the aircraft.

Since the matrix $\mathbf{R}_{\mathrm{s}}^{\mathrm{e}} \cdot \mathrm{diag}(1, -a_{v3}, 1)$ is nonsingular when $a_{v3} \neq 0$, letting the flat output of the fixed-wing $\mathbf{y} = \mathbf{x}$, Eq. (8.8) is transformed into

$$
\begin{bmatrix} \dot{a}_{v1} \\ \omega_x \\ \dot{a}_{v3} \end{bmatrix} = \begin{bmatrix} -\omega_y a_{v3} \\ \frac{\omega_z a_{v1}}{a_{v3}} \\ \omega_y a_{v1} \end{bmatrix} + \begin{bmatrix} 1 & 0 & 0 \\ 0 & -\frac{1}{a_{v3}} & 0 \\ 0 & 0 & 1 \end{bmatrix} \mathbf{R}_{\mathrm{e}}^{\mathrm{s}} \mathbf{y}^{(3)}. \tag{8.9}
$$

From Eq. (8.9), it is observed that the states $\dot{a}_{v1}, \dot{a}_{v3}, \omega_x$ in the input space are obtained from the output trajectory $\mathbf{y}$ and its derivatives, corresponding to Eq. (8.7). The various states on the right-hand side of Eq. (8.9) are typically obtained from onboard sensors; thus, given the desired output space and the state variables of the aircraft, the control input space is determined. For the desired state trajectory $\mathbf{x}_{\mathrm{d}}(t), t > 0$, to precisely track $\mathbf{x}_{\mathrm{d}}(t)$, the desired flat output $\mathbf{y}_{\mathrm{d}}$ is designed as

$$
\mathbf{y}_{\mathrm{d}}^{(3)} = \mathbf{x}_{\mathrm{d}}^{(3)} + k_1 \left(\ddot{\mathbf{x}}_{\mathrm{d}} - \ddot{\mathbf{x}}\right) + k_2 \left(\dot{\mathbf{x}}_{\mathrm{d}} - \dot{\mathbf{x}}\right) + k_3 \left(\mathbf{x}_{\mathrm{d}} - \mathbf{x}\right). \tag{8.10}
$$

Thus, the closed-loop dynamic system is represented as (8.10), where the selection of k_1, k_2, k_3 ensures that the polynomial $s^3 + k_1 s^2 + k_3 s + k_3$ is Hurwitz. Substituting Eqs. (8.10) into (8.9) yields expressions for $\dot{a}_{v1}, \dot{a}_{v3}, \omega_x$ as

$$
\begin{aligned}
\begin{bmatrix} \dot{a}_{v1} \\ \omega_x \\ \dot{a}_{v3} \end{bmatrix} &= \begin{bmatrix} -\omega_y a_{v3} \\ \omega_z a_{v1}/a_{v3} \\ \omega_y a_{v1} \end{bmatrix} + \begin{bmatrix} 1 & 0 & 0 \\ 0 & -1/a_{v3} & 0 \\ 0 & 0 & 1 \end{bmatrix} \mathbf{R}_{\mathrm{e}}^{\mathrm{s}} \mathbf{y}_{\mathrm{d}}^{(3)} \\
&= \begin{bmatrix} -\omega_y a_{v3} \\ \omega_z a_{v1}/a_{v3} \\ \omega_y a_{v1} \end{bmatrix} + \begin{bmatrix} 1 & 0 & 0 \\ 0 & -1/a_{v3} & 0 \\ 0 & 0 & 1 \end{bmatrix} \mathbf{R}_{\mathrm{e}}^{\mathrm{s}} \left(\mathbf{x}_{\mathrm{d}}^{(3)} + k_1 \left(\ddot{\mathbf{x}}_{\mathrm{d}} - \ddot{\mathbf{x}}\right) + k_2 \left(\dot{\mathbf{x}}_{\mathrm{d}} - \dot{\mathbf{x}}\right) + k_3 \left(\mathbf{x}_{\mathrm{d}} - \mathbf{x}\right)\right).
\end{aligned} \tag{8.11}
$$

Integrating $\dot{a}_{v1}, \dot{a}_{v3}$ provides the inputs a_{v1}, a_{v3}. This trajectory tracking approach combines feedforward and feedback to enhance tracking performance.

8.1.3 Trajectory Planning

Robot trajectory planning is a critical task in the field of robotics. It involves the perception, planning, and control of robots in complex environments to ensure that robots can complete tasks safely and efficiently [4]. Trajectory planning includes generating paths and executing them using control techniques such as differential flatness control.

As illustrated in Fig. 8.2, the initial stages of robot trajectory planning entail perception and mapping, which involve the robot's awareness of its surroundings and the subsequent modeling of this information [5]. This process typically includes three key components:

(1) Sensor data acquisition. Robots gather information about the surrounding environment through a variety of sensors, including LIDAR, cameras, ultrasonic sensors, etc. The selection of sensors depends on the task's requirements and the characteristics of the environment.
(2) Data fusion. Data obtained from different sensors may vary in type and quality. Therefore, fusion techniques are employed to integrate this data, enhancing the overall comprehension of the environment. Methods like the Kalman filter facilitate the integration of information from multiple sensors into a coherent state estimation.
(3) Map construction. Utilizing sensor data, robots must construct a map of the environment. Maps include both two-dimensional static maps and dynamic maps that incorporate temporal dimensions. While static maps typically represent obstacles that remain unchanged, dynamic maps account for objects that evolve, such as moving individuals or other robots.

As discussed in Chap. 7, path planning is integral to trajectory planning as it involves identifying suitable paths for the robot within the environment. This abstraction entails transforming the environment into a state space, where each state signifies the potential positions and orientations of the robot. The choice between continuous and discrete state spaces depends on the specific task and problem at hand. The primary objective of path planning is to ascertain a route from the current position to the

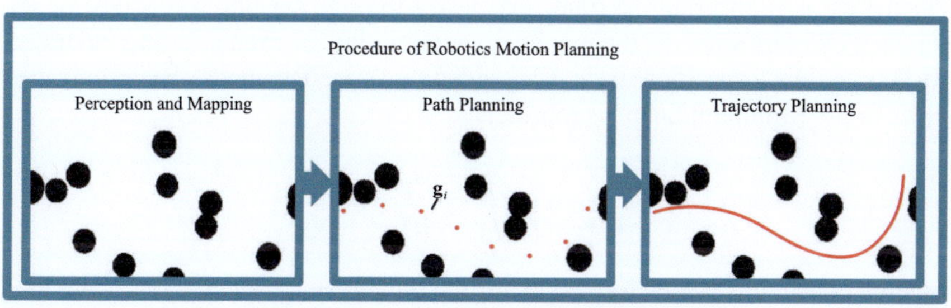

Fig. 8.2 A procedure of robotic motion planning in a clustered environment

designated target position, which may be a specific location, region, or defined condition. Path planning methodologies encompass various techniques, including graph search methods like Dijkstra's algorithm and A*, sampling-based methods such as RRT, RRT*, and PRM algorithms, energy-based approaches like the artificial potential field method, heuristic methods exemplified by the particle swarm algorithm, the genetic algorithm, ant colony algorithm, and algorithms rooted in deep reinforcement learning [6].

Trajectory planning involves determining how the robot moves along a given path. To achieve trajectory planning, a common approach is to parameterize trajectories using polynomials, such as nth-order polynomials:

$$\begin{aligned} \mathbf{p}(t) &= \mathbf{p}_0 t^0 + \mathbf{p}_1 t^1 + \mathbf{p}_2 t^2 + \mathbf{p}_3 t^3 + \cdots + \mathbf{p}_n t^n \\ &= \mathbf{P}\left[1\ t\ t^2\ \ldots t^n\right]^{\mathrm{T}} \end{aligned} \tag{8.12}$$

where t is the polynomial variable, $\mathbf{p}_0, \mathbf{p}_1, \ldots, \mathbf{p}_n$ represent the trajectory parameters of the polynomial, $\mathbf{P} = \left[\mathbf{p}_0\ \mathbf{p}_1\ \ldots\ \mathbf{p}_n\right]$ represents the coefficient matrix. Currently, the classical and effective polynomial planning method is "Minimum Snap" [7]. Based on this, numerous algorithms with diverse optimization objectives have been developed. Based on parameterized polynomials, trajectory planning aims to find the polynomial parameters $\mathbf{p}_0, \mathbf{p}_1, \ldots, \mathbf{p}_n$. A single polynomial curve is too simple, and it is challenging to represent a complex trajectory with a single polynomial. Therefore, the trajectory is divided into multiple segments over time, each represented by a polynomial curve:

$$\mathbf{p}(t) = \begin{cases} \mathbf{P}_1\left[1\ t\ t^2\ \cdots t^n\right]^{\mathrm{T}},\ t_0 \le t < t_1 \\ \mathbf{P}_2\left[1\ t\ t^2\ \cdots t^n\right]^{\mathrm{T}},\ t_1 \le t < t_2 \\ \cdots \\ \mathbf{P}_n\left[1\ t\ t^2\ \cdots t^n\right]^{\mathrm{T}},\ t_{n-1} \le t < t_n. \end{cases} \tag{8.13}$$

When performing trajectory planning, certain constraints must be satisfied, such as setting the position, velocity, or acceleration of the start and end points, ensuring smoothness at adjacent trajectory connections (continuous position, continuous velocity, etc.), or ensuring the trajectory passes through specific waypoints, or setting the maximum velocity, the maximum acceleration, or even ensuring the trajectory stays within a specified space. For polynomial-based methods, the optimization objective of "Minimum Snap" is represented as

$$\min_{\mathbf{P}_0, \mathbf{P}_1, \cdots, \mathbf{P}_n} \int_0^T \| \frac{\mathrm{d}^4\mathbf{p}}{\mathrm{d}t^4} \|^2 \mathrm{d}t. \tag{8.14}$$

Equation constraints can be formed by setting the position, velocity, or acceleration of a point to a specific value or by ensuring higher-order continuity of polynomials.

Constraints on spatial location, such as flight corridors [8], can form inequality constraints. Thus, an optimization problem is formulated as

$$\begin{aligned} &\min_{\hat{\mathbf{P}}} \ f(\hat{\mathbf{P}}) \\ &\text{s.t. } \mathbf{A}_{\text{eq}}\hat{\mathbf{P}} = \mathbf{b}_{\text{eq}} \\ &\qquad \mathbf{A}_{\text{ieq}}\hat{\mathbf{P}} \leq \mathbf{b}_{\text{ieq}} \end{aligned} \tag{8.15}$$

where $\hat{\mathbf{P}} = [\mathbf{P}_1 \ \mathbf{P}_2 \ \ldots \ \mathbf{P}_n]^{\text{T}}$, $\mathbf{A}_{\text{eq}}, \mathbf{b}_{\text{eq}}$ are the coefficients of linear equality constraints, and $\mathbf{A}_{\text{ieq}}, \mathbf{b}_{\text{ieq}}$ are the coefficients of linear inequality constraints. In addition to polynomial-based trajectory planning, trajectory planning using B-splines is also an important method. In the following sections, B-spline generation and related properties will be discussed in detail.

8.1.4 B-Spline Trajectory Generation

In Sect. 8.1.2, the trajectory tracking control design process was provided for a known desired trajectory $\mathbf{x}_{\text{d}}$. Various mature methods exist for generating desired trajectories, and this section presents the Bezier curve and B-spline trajectory generation methods, along with their respective properties, for reference.

8.1.4.1 Bezier Curve

The Bezier method is a classic concept in computer graphics used for designing smooth curves and surfaces. The positions of control points determine the shape of Bezier curves, and the number of control points determines their order of complexity. An nth-order Bezier curve corresponds to $n+1$ control points $\mathbf{p}_0, \mathbf{p}_1, \ldots, \mathbf{p}_n$. For $t \in [0, 1]$, which is understood as the time the point on the Bezier curve slides, the parameter equation of the Bezier curve is defined as follows:

$$\mathbf{b}(t) = \sum_{i=0}^{n} \mathbf{p}_i b_{i,n}(t), t \in [0, 1] \tag{8.16}$$

where $b_{i,n}(t) = \binom{n}{i}(1-t)^{n-i} t^i$ is the Bernstein basis function, and the moving point on the Bezier curve at each instant is a linear combination of these basis functions. The sequence of vectors connecting the control points in order is called the *control polygon*. Bezier curves are constructed by starting with linear curves and progressively building higher-order curves. Since the basis functions have recursion, the De Casteljau algorithm [9] is commonly used to compute Bezier curves. Due to the enormous computational complexity of the defined method and the significant reduction in computational effort offered by the De Casteljau recursive algorithm, the latter has become the standard algorithm for computing curves. The recursive algorithm proposed by De Casteljau is stated as follows:

$$\mathbf{b}_{i,n}(t) = (1-t)\,\mathbf{b}_{i,n-1}(t) + t\mathbf{b}_{i-1,n-1}(t),\ i = 1, 2, \ldots, n. \tag{8.17}$$

The De Casteljau recursive algorithm is stable, reliable, intuitive, and convenient in programming, making it advantageous for computing curves. In the following, the recursive process of the De Casteljau algorithm will be demonstrated through examples.

(1) **First-order Bezier Curve (Linear Formula)**: Given control points $\mathbf{p}_0, \mathbf{p}_1$, the linear Bezier curve is a straight line between two points. The bold line in Fig. 8.3 represents the Bezier curve, and point $\mathbf{b}_{1,0}$ represents the point on the line segment $[\mathbf{p}_0, \mathbf{p}_1]$ corresponding to the parameter t, where $t \in [0, 1]$ corresponds to the point along the line segment from $\mathbf{p}_0$ to $\mathbf{p}_1$. This curve can be represented as

$$\mathbf{b}(t) = \mathbf{p}_0 + (\mathbf{p}_1 - \mathbf{p}_0)\,t = (1-t)\,\mathbf{p}_0 + t\mathbf{p}_1, \quad t \in [0, 1]. \tag{8.18}$$

(2) **Second-order Bezier Curve**: The path of the second-order Bezier curve is generated by given points $\mathbf{p}_0, \mathbf{p}_1, \mathbf{p}_2$, as shown in the bold line in Fig. 8.4. Point $\mathbf{b}_{1,0}$ and point $\mathbf{b}_{1,1}$ are respectively the points corresponding to parameter t on the first-order Bezier curves $\overline{\mathbf{p}_0\mathbf{p}_1}$ and $\overline{\mathbf{p}_1\mathbf{p}_2}$. By using $\mathbf{b}_{1,0}$ and point $\mathbf{b}_{1,1}$ as new control points, the second-order Bezier curve is defined as:

$$\mathbf{b}_{2,0}(t) = (1-t)\,\mathbf{b}_{1,0} + t\mathbf{b}_{1,1}, \quad t \in [0, 1], \tag{8.19}$$

Fig. 8.3 A first-order Bezier curve

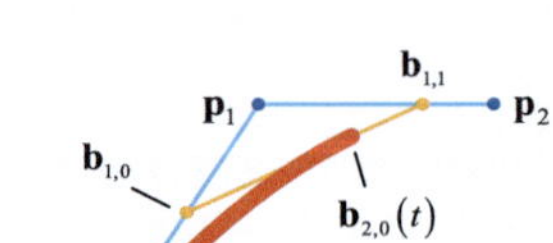

Fig. 8.4 A second-order Bezier curve

where $\mathbf{b}_{1,0} = (1-t)\,\mathbf{p}_0 + t\mathbf{p}_1$ and $\mathbf{b}_{1,1} = (1-t)\,\mathbf{p}_1 + t\mathbf{p}_2$. Thus, it is observed that the second-order Bezier curve undergoes two levels of recursion, and the curve expression $\mathbf{b}(t)$ is further written as:

$$\mathbf{b}(t) = (1-t)^2\,\mathbf{p}_0 + 2t\,(1-t)\,\mathbf{p}_1 + t^2\mathbf{p}_2, \quad t \in [0,1]. \tag{8.20}$$

(3) **Third-order Bezier Curve**: The path of the third-order Bezier curve is generated by given points $\mathbf{p}_0, \mathbf{p}_1, \mathbf{p}_2, \mathbf{p}_3$. Points $\mathbf{b}_{1,0}$, $\mathbf{b}_{1,1}$, and $\mathbf{b}_{1,2}$ are, respectively, the points corresponding to parameter t on the first-order Bezier curves $\overline{\mathbf{p}_0\mathbf{p}_1}$, $\overline{\mathbf{p}_1\mathbf{p}_2}$, and $\overline{\mathbf{p}_2\mathbf{p}_3}$. Points $\mathbf{b}_{2,0}$ and $\mathbf{b}_{2,1}$ are respectively the points corresponding to parameter t on the second-order Bezier curves $\overline{\mathbf{b}_{1,0}\mathbf{b}_{1,1}}$ and $\overline{\mathbf{b}_{1,1}\mathbf{b}_{1,2}}$. By giving $\mathbf{b}_{2,0}$ and $\mathbf{b}_{2,1}$ as new control points, the third-order Bezier curve is obtained:

$$\mathbf{b}_{3,0}(t) = (1-t)\,\mathbf{b}_{2,0} + t\mathbf{b}_{2,1}, \quad t \in [0,1]. \tag{8.21}$$

The bold line in Fig. 8.5 represents the Bezier curve, and it is observed that the third-order Bezier curve undergoes three levels of recursion. The curve expression $\mathbf{b}(t)$ can similarly be written as:

$$\mathbf{b}(t) = (1-t)^3\,\mathbf{p}_0 + 3t\,(1-t)^2\,\mathbf{p}_1 + 3t^2\,(1-t)\,\mathbf{p}_2 + t^3\mathbf{p}_3, \quad t \in [0,1]. \tag{8.22}$$

(4) ***n*th-order Bezier Curve**: The path of the nth-order Bezier curve is generated by given points $\mathbf{p}_0, \mathbf{p}_1, \mathbf{p}_2, \ldots, \mathbf{p}_n$. Thus, the definition of an nth-order Bezier curve is:

$$\mathbf{b}(t) = \sum_{i=0}^{n} \mathbf{p}_i \binom{n}{i} (1-t)^{n-i}\,t^i = \sum_{i=0}^{n} \mathbf{p}_i b_{i,n}(t). \tag{8.23}$$

Bezier curves have the following characteristics.

(1) **Globality**: Bezier methods cannot control the shape of the curve locally; adjusting any control point affects the entire curve.
(2) **Endpoint Interpolation**: The curve generally only passes through the two endpoint control points (start and end) and approximates all other points without passing through them.

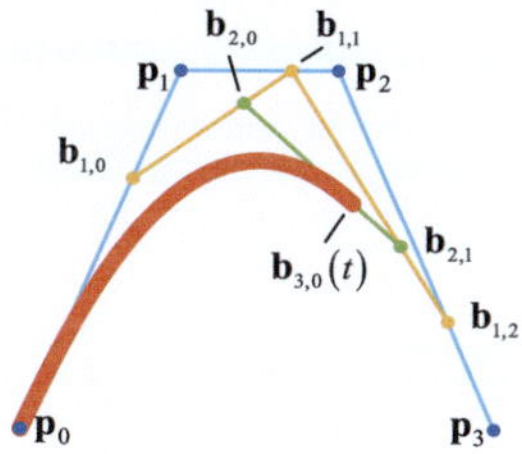

Fig. 8.5 A third-order Bezier curve

(3) **Geometric Invariance**: The geometric properties of the curve remain unchanged under coordinate transformations. The shape is dependent only on the relative positions of the control polygon vertices and is independent of the choice of coordinate system.
(4) **Derivative Properties**: The derivative of a Bezier curve remains a Bezier curve.
(5) **Convex Hull**: The convex hull is the smallest convex polygon containing all control points. Extending any edge of a convex polygon ensures that all other edges on its side are also extended. Bezier curves always lie within the minimum convex polygon, which restricts the range of the planned curve to the convex hull of the control points.

8.1.4.2 B-Spline Curve

While Bezier's methods perform well, they also have some limitations.

(1) The number of control points determines the order (degree) of Bezier curves. Increasing the number of control points can increase computational complexity, and in some cases, it may not meet the requirements for curve accuracy and smoothness.
(2) When multiple Bezier curves are spliced, the first-order derivative (tangent) of the Bezier curve may exhibit a discontinuity at the splice points. Consequently, the resulting curve may lack the requisite smoothness at these points. This may manifest as conspicuous discontinuities or corner points when connecting two curve segments.
(3) The shape of Bezier curves is constrained by the positions of the control points, such that adjustments to the curve may impact the overall shape. This implies that modifying a specific portion of the curve may inadvertently affect other parts, leading to suboptimal local controllability.

B-spline curves are a generalization of Bezier curves and are a linear combination of B-spline basis functions. In simple terms, a spline is a piecewise continuous polynomial, and it is continuously differentiable/continuous over the entire curve. The B-spline, as a whole curve, has a complete expression and is smooth within a finite order, fully satisfying human intuitive aesthetics; however, it is internally segmented. Consequently, it is capable of overcoming oscillations, possesses a low curve order, and exhibits uniform expressions and algorithms.

For $n + 1$ control points $\mathbf{p}_i, i = 0, 1, \ldots, n$ and a knot vector $\mathbf{u} = [u_0\ u_1\ \ldots\ u_m]^{\mathrm{T}}$, satisfying

$$m = n + p + 1 \tag{8.24}$$

where p is the degree of the B-spline curve.

The definition of a B-spline is:

$$\mathbf{s}(u) = \sum_{i=0}^{n} \mathbf{p}_i s_{i,p}(u) \tag{8.25}$$

where $u \in [u_0, u_m)$, and $s_{i,p}(u)$ is the pth-degree (or $p+1$ order) B-spline basis function [10]. Compared with Eq. (8.16), it is observed that the subscript n of the Bernstein basis function becomes p, which indicates that the degree of the polynomial of the B-spline and the number of control points are not related but are custom-defined. This reflects the local nature of B-spline trajectories, i.e., adjusting $p+1$ control points can adjust the trajectory contained within these $p+1$ control points.

B-spline basis functions are defined in various ways, but the most widely used and easiest to understand is the De Boor-Cox recursive definition [10]. Its principle is to construct a recursive formula for pth-degree B-spline basis functions: creating the first degree from zero degrees, the second degree from the first degree, the third degree from the second degree, and so on, similar to a pyramid, as shown in Fig. 8.6. Mathematically, it is expressed as:

$$\begin{aligned} s_{i,0}(u) &= \begin{cases} 1, u_i < u < u_{i+1} \\ 0, \text{otherwise} \end{cases} \\ s_{i,p}(u) &= \frac{u - u_i}{u_{i+p} - u_i} s_{i,p-1}(u) + \frac{u_{i+p+1} - u}{u_{i+p+1} - u_{i+1}} s_{i+1,p-1}(u). \end{aligned} \tag{8.26}$$

According to Eq. (8.26), if the ith pth-degree B-spline basis function $s_{i,p}(u)$ is to be determined, $p+1$ knots starting from u_i are required. This interval, as shown in Fig. 8.6, is called the *support interval* of this basis function. From the definition in Eq. (8.25), it is inferred that any control point corresponds to a unique basis function. Therefore, if a basis function is defined only on part of the interval, it indicates the local nature of control points, i.e., only controlling part of the interval.

As shown in Fig. 8.7, $s_{1,3}$ represents the basis function corresponding to the control point $\mathbf{p}_1$. After decomposing through the dashed boxes layer by layer, it can be found that it is non-zero in the interval $[u_1, u_5)$. This indicates that the basis function $s_{i,p}$ corresponding to the control point can only affect the curve in its

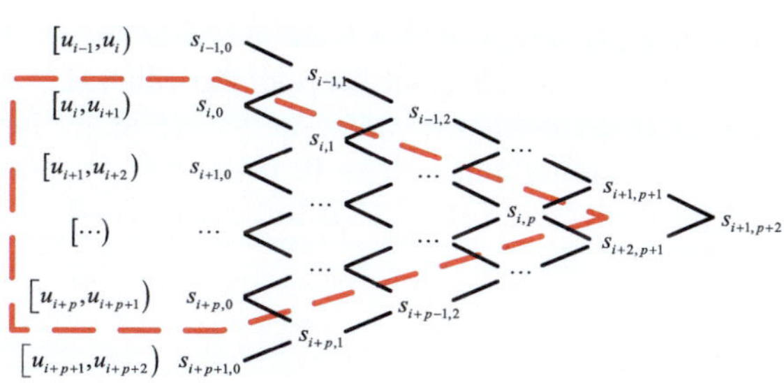

Fig. 8.6 Basis functions corresponding to the support intervals

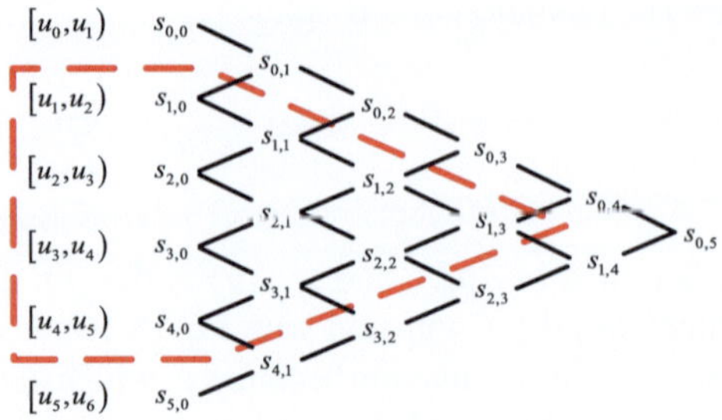

Fig. 8.7 Basis functions $s_{1,3}$ corresponding to the support intervals

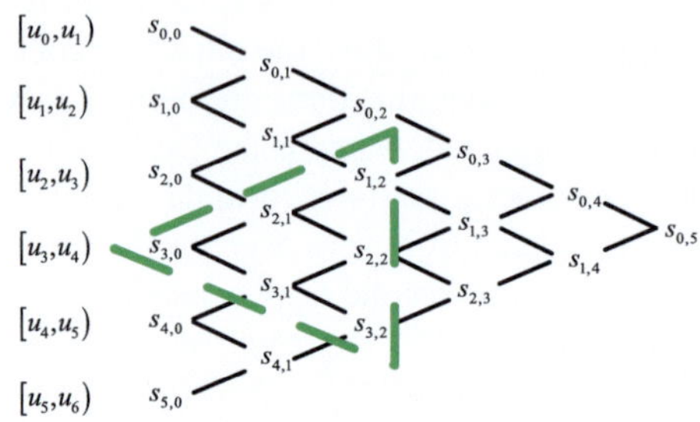

Fig. 8.8 Support intervals corresponding to basis functions and control points

corresponding interval and has no impact on other intervals. Conversely, the basis functions determining a particular interval are observed. Let $p = 2$, as shown in Fig. 8.8, the interval $[u_3, u_4)$ is influenced by $s_{1,2}$, $s_{2,2}$, and $s_{3,2}$, which means it is only affected by the control points corresponding to these basis functions. Therefore, in any given knot interval $\left[u_i, u_{i+1}\right)$, there exist at most $p + 1$ pth degree basis functions $s_{i-p,p}, s_{i-p+1,p}, \cdots, s_{i,p}$ that are non-zero, and the sum of these basis functions is 1.

According to Eq. (8.26), $\mathbf{u}$ is normalized during the calculation, so the basis functions are only related to the relative distances between a pair of knots and not to the absolute values of u_i. Based on this, the B-spline trajectory parameterized by u can be obtained, where $u \in \left[u_p, u_{m-p}\right]$. Therefore, Eq. (8.25) is written as

$$\mathbf{s}\left(u\right) = \sum_{i=j-p}^{j} \mathbf{p}_i s_{i,p}\left(u\right) \tag{8.27}$$

where $u \in [u_j, u_{j+1})$ $(p \leq j \leq n)$.

In numerical analysis, the De Boor algorithm [11] is a fast and numerically stable algorithm used to calculate spline curves in B-spline form. It is a generalization of the De Casteljau algorithm used for Bezier curves. Suppose the value of the spline curve with parameter $u \in [u_l, u_{l+1})$ is to be calculated, and $\mathbf{p}_i^{[0]} = \mathbf{p}_i$, $i = l - p, \ldots, l$, $k = 1, \ldots, p$, then

Fig. 8.9 An example to show B-spline properties

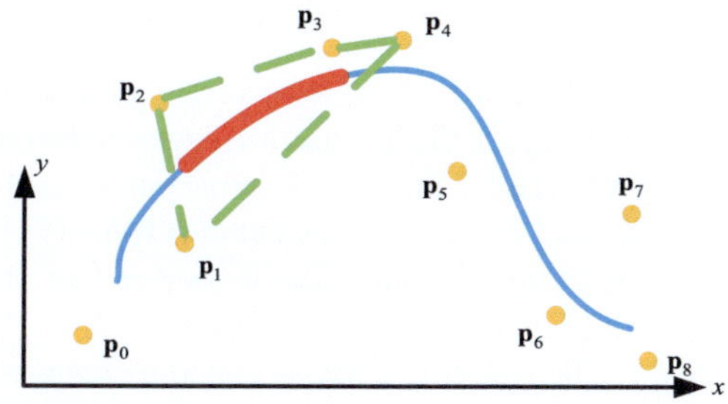

$$\mathbf{p}_i^{[k]} = \begin{cases} \left(1 - \alpha_{k,i}\right) \mathbf{p}_{i-1}^{[k-1]} + \alpha_{k,i} \mathbf{p}_i^{[k-1]} & , k = 1, 2, \ldots, p \\ \mathbf{p}_i & , k = 0 \end{cases}$$
$$\alpha_{k,i} = \frac{u - u_i}{u_{i+p+1-k} - u_i}. \tag{8.28}$$

Thus $\mathbf{s}\left(u\right) = \mathbf{p}_i^{[p]}, u \in [u_l, u_{l+1})$.

B-splines have different properties from Bezier curves:

(1) **Locality**: This is the most important difference compared to Bezier curves. The local support of B-spline basis functions determines the locality of B-splines. As described above and shown in Fig. 8.8, a point on a pth-degree curve is related to at most $p + 1$ control points and is independent of other control points. Therefore, moving the ith control point $\mathbf{p}_i$ may affect only the shape of the part of the curve defined in the interval corresponding to this point without affecting other parts of the curve.
(2) **Convex Hull**: Since an interval is determined by only $p + 1$ basis functions, whose sum equals 1, the basis functions act as weighting coefficients, and the points on the curve are a weighted sum of the corresponding control points. Consequently, B-spline curves lie within the convex hull formed by the control points $\mathbf{p}_i$, and the convex hull of B-spline curves is smaller than or equal to the convex hull defined by the same control points for Bezier curves. As shown in Fig. 8.9, the bold segments of the B-spline lie within a polygon consisting of four control points $\mathbf{p}_1, \mathbf{p}_2, \mathbf{p}_3, \mathbf{p}_4$.
(3) Unlike Bezier curves, B-spline curves do not necessarily pass through two endpoints.

8.1.4.3 B-Spline Trajectory Design

In B-spline trajectory planning, the shape of the curve can often be modified by adjusting the control point vector. The determining parameters include $n + 1$ control points, $m + 1$ knots, and a p-degree B-spline curve, where the parameters satisfy Eq. (8.24).

B-spline curves are classified into several types:

(1) Open B-spline: If the knot vector lacks a special structure, the resulting curve will not touch the first and last edges of the control polyline. This type of B-spline curve is called an *open* B-spline curve, as shown in Fig. 8.10.
(2) Clamped B-spline: If the curve is forced to be tangent to the first and last edges of the control points like Bezier curves, the first and last knots must have a multiplicity of $p + 1$, i.e., $u_0 = u_1 = \cdots = u_p$ and $u_{m-p} = u_{m-p+1} = \cdots = u_m$. The resulting curve is called a *clamped* B-spline curve. Clamped B-spline curves pass through the first and last control points, as shown in Fig. 8.11.
(3) Closed B-spline: By repeating specific knots and control points, the resulting curve will be a *closed* B-spline curve. In this case, the start and end points of the connection curve will form a closed loop, as shown in Fig. 8.12.
(4) Uniform B-spline curves: When the knots are uniformly spaced along the parameter axis, they are uniform B-spline curves, such as $\mathbf{u} = [0\ 1\ 2\ 3\ 4\ 5\ 6]^{\mathrm{T}}$. When p and m are fixed, the basis functions of uniform B-splines are periodic; all

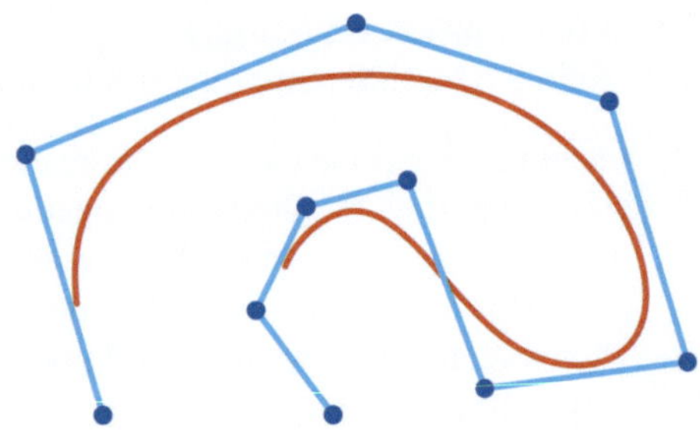

Fig. 8.10 An open B-spline curve, where the dots represent control points, and the curve represents the B-spline trajectory

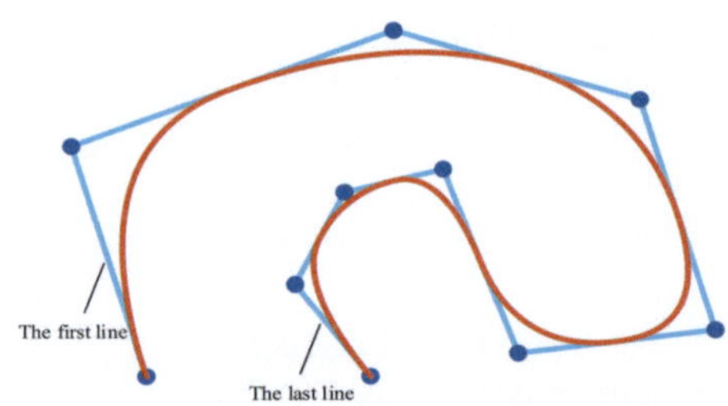

Fig. 8.11 A clamped B-spline curve, where the dots represent control points, and the curve represents the B-spline trajectory

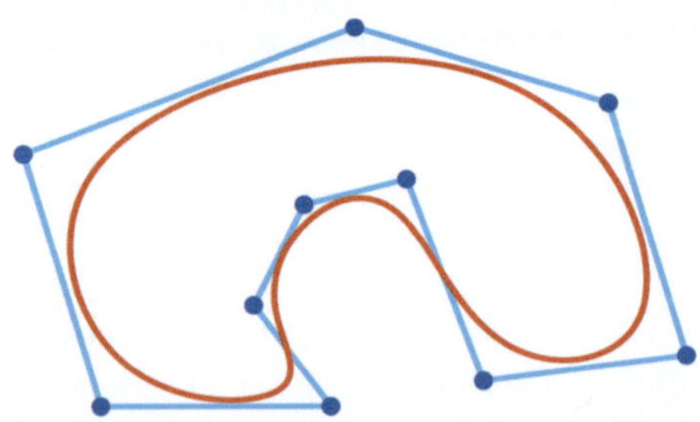

Fig. 8.12 A closed B-spline curve, where the dots represent control points, and the curve represents the B-spline trajectory

basis functions have the same shape, and each subsequent basis function is just a repetition of the previous one at a new position.

(5) Quasi-uniform B-spline curves: The knots at both ends have a multiplicity of $p + 1$, and the middle knots are uniformly spaced non-decreasing sequences, such as $\mathbf{u} = [0\ 0\ 0\ 1\ 2\ 3\ 4\ 5\ 5\ 5]^{\mathrm{T}}$. Quasi-uniform B-spline curves are the most widely used.

(6) Non-uniform B-spline curves: B-spline curves with non-uniform knot spacing.

8.1.4.4 B-Spline Trajectory Planning

During trajectory design, obstacle avoidance is often a consideration. Active obstacle avoidance is a method of achieving obstacle avoidance by designing trajectories as feedforward. The convex hull property of B-splines is utilized in the obstacle avoidance process. Considering grid-based mapping, as shown in Fig. 8.13, it is necessary to ensure that $d_h > 0$, where d_h represents the distance between the obstacle-occupied grid and any point $\mathbf{p}_h$ within the convex hull of the control points.

Let $\mathbf{p}_h$ be the distance to any control point (such as $\mathbf{p}_4$), r_h be the distance, and d_c be the distance from the control point to the obstacle-occupied grid. According to the triangle inequality, $d_h > d_c - r_h$. Let the sequential distances between control points (taking $p = 3$ as an example) be r_{12}, r_{23}, r_{34}. Therefore, it follows that $r_h < r_{12} + r_{23} + r_{34}$. Thus,

$$d_h > d_c - (r_{12} + r_{23} + r_{34}) \tag{8.29}$$

is derived. Therefore, ensuring

$$d_c > 0, \quad r_{i,i+1} < d_c/3, \quad i \in \{1, 2, 3\} \tag{8.30}$$

achieves collision-free trajectories [12].

For uniform or quasi-uniform B-spline trajectories, the knots are equidistant. Thus, let $\Delta t = u_{i+1} - u_i$. By

$$\mathbf{v}_i = \frac{1}{\Delta t}\left(\mathbf{p}_{i+1} - \mathbf{p}_i\right), \quad \mathbf{a}_i = \frac{1}{\Delta t}\left(\mathbf{v}_{i+1} - \mathbf{v}_i\right) \tag{8.31}$$

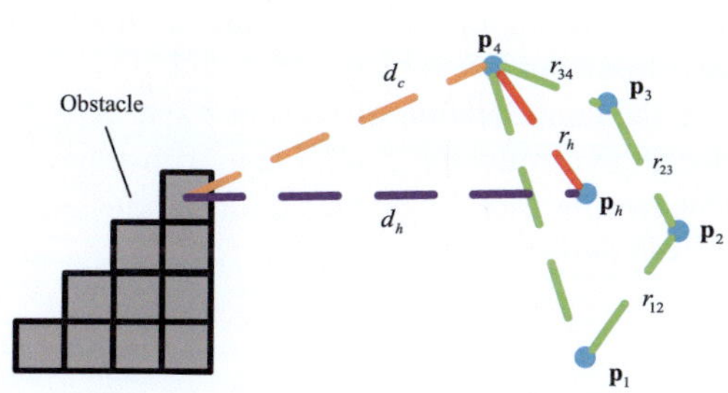

Fig. 8.13 B-Spline obstacle avoidance characteristics [12]

the velocity control point vectors $\mathbf{v}$ and acceleration control point vectors $\mathbf{a}$ are obtained. These control point vectors can also form B-splines and satisfy convex hull properties and other properties. Therefore, simply setting $\mathbf{v}_i \in [-v_{\max}, v_{\max}]^3$ and $\mathbf{a}_i \in [-a_{\max}, a_{\max}]^3$ achieves constraints on velocity and acceleration.

During trajectory planning, maps are often obtained through perception or provided in advance. Discrete path points $\mathbf{g}_i$ are obtained using graph search or grid search on the given map, as shown in Fig. 8.2. In the planning process, obstacle avoidance assurance is achieved by implementing (8.30), or by making control points as close as possible to the path points $\mathbf{g}_i$, or by directly using the path points as control points for trajectory planning.

Based on the above process, by constructing a series of convex objective functions, such as smoothness and traversal time, control points or knot vectors are optimized to obtain optimized trajectories, which vary depending on the different objectives. For example, an optimization objective is set as [12]

$$\begin{aligned} \min_{\mathbf{p}_i, i=1,\cdots,n} \quad & f_p = \lambda_s f_s + \lambda_g f_g \\ & = \lambda_s \sum_{i=p-1}^{n-p+1} \|\mathbf{p}_{i+1} - 2\mathbf{p}_i + \mathbf{p}_{i-1}\|^2 + \lambda_g \sum_{i=p}^{n-p} \|\mathbf{p}_i - \mathbf{g}_i\| \end{aligned} \tag{8.32}$$

where $\mathbf{g}_i$ is the path search point corresponding to control point $\mathbf{p}_i$, f_s represents the smoothness optimization objective, f_g represents optimizing the trajectory to follow the path obtained from the search closely, and λ_s and λ_g represent corresponding weights. Optimized or designed B-spline trajectories can serve as feedforward and be represented using differential flatness methods in input space to achieve trajectory tracking control.

8.1.5 Related Development

Currently, fixed-wing UAV trajectory generation algorithms often employ kinematic models, thereby avoiding the complexities of aerodynamics. A common approach is to extend Dubins paths [13] to find time-optimal trajectories under curvature constraints. This approach results in sudden changes in acceleration, making it difficult for the generated path to be accurately tracked. However, feedback control can be used to maintain tracking errors within acceptable limits during flights. To facilitate the consideration of both fast and agile flight, it is necessary to account for aircraft dynamics and control input constraints in the trajectory generation process. This ensures that the generated trajectories are dynamically feasible and accurately tracked during flight.

In UAV trajectory planning, B-spline curves are favored by researchers due to their parametric flexibility [14]. B-spline curves have smooth and continuous characteristics, which are crucial for UAVs. Furthermore, the judicious selection of parameters and constraints facilitates the integration of B-spline trajectory planning, thereby addressing the dual imperatives of kinematics and dynamics. This enhances the versatility of its application, particularly in the context of trajectory planning for quadrotor UAVs, a domain in which its efficacy has garnered significant recognition. [15, 16]. With the improvement of computing power, researchers are developing B-spline trajectory generation methods that offer better real-time performance and computational efficiency.

Differential flatness theory was proposed by Fliess et al. in the 1990s [17]. This theory states that in specific dynamic systems, all states and inputs of the system are represented by a set of variables called *flat outputs* and their derivatives. This enables the transformation of complex nonlinear systems into more easily manageable linear systems, providing convenience for control problems. This theory has been widely applied in computing efficient trajectory generation and tracking for quadrotor UAVs [7]. The employment of differential flatness in the trajectory generation of fixed-wing UAVs is predominantly constrained to kinematic or simplified dynamic models. However, certain studies have explored the use of differentially flat-coordinated flight models for path generation and tracking. Additionally, the incorporation of aircraft dynamic models has been proposed for executing aerobatic maneuvers. For example, the algorithm proposed in [3] is based on the differentially flat coordinated flight model presented in [1], which combines Dubins paths with lateral polynomial offsets to obtain smooth trajectories.

Differential flatness offers numerous advantages in the realm of UAV trajectory tracking, including the simplification of complex models and adaptability in highly dynamic environments. However, this approach is not without its challenges, such as computational complexity, model complexity, and the integration of advanced technologies [18]. However, these challenges also present new opportunities for UAV systems, such as enhanced adaptability and robustness, broader model generalizations, and integration with technologies like artificial intelligence. Differential flatness-based trajectory tracking methods provide a strong foundation for improving UAV system performance. In the future, the development of differential flatness control in the UAV field will continue to focus on enhancing the adaptability, real-time performance, and computational efficiency of algorithms, while expanding its application scope to address complex, dynamic real-world environments better [19]. Therefore, this field is expected to achieve more breakthroughs in control theory and engineering practice.

8.2 Basic Experiment

8.2.1 Experimental Objective

1. Things to prepare

(1) Software: MATLAB R2020b or above.
(2) Program: Experimental package "e6-1". The package mainly includes: The initialization program file for model and control parameters "InitDatactrldf.m", which can import UAV parameters into the workspace when executed; the file containing the complete differential flatness control program module for fixed-wing coordinated flight model "SmallFixedWingUAVdf.slx"; simulation result plotting program file "DF_plot.m"; rotation matrix function files "ROTX.m", "ROTY.m", and "ROTZ.m" called in Simulink.

2. Objective

(1) Comprehend the principles of differential flatness control, construct a differential flatness module based on the coordinated flight model, and analyze the relationship between desired trajectories and control inputs.
(2) Develop the code of differential flatness control, conduct Software-In-the-Loop (SIL) simulation in Simulink to complete the basic experiment on differential flatness control of simple trajectories in the ENU (East-North-Up) coordinate system,[1] and plot relevant experimental curves. The position vector in the ENU coordinate system is $[x_E\ x_N\ x_U]^{\mathrm{T}}$, the velocity vector is $[v_E\ v_N\ v_U]^{\mathrm{T}}$, and the acceleration vector is $[a_E\ a_N\ a_U]^{\mathrm{T}}$.

8.2.2 Experimental Procedure

The basic experiment mainly involves building the framework of the trajectory reference input module and the differential flatness control module in the "SmallFixedWingUAVdf.slx" file and completing the basic process. The basic experiment is divided into three steps: building the framework of the trajectory reference input module, designing and building the differential flatness control module, and connecting the controller and model modules in sequence. Step 1 is to build the trajectory reference input module framework for simple time-varying trajectories. Step 2 is to write the differential flatness control method. Finally, in Step 3, the controller and model modules are connected to perform a simulation.

The overall module building structure is shown in Fig. 8.14. The model mainly includes four modules: dashed box ① "track", trajectory input module, used to output desired trajectories over time; dashed box ② "DF", differential flatness module, used

[1] In this section, this coordinate system is used to maintain consistency with the coordinated flight module in Simulink.

to obtain differential flatness control based on the desired trajectories as feedforward; dashed box ③ control conversion module, which converts the control output by the differential flatness module into model inputs; dashed box ④ coordinated flight module, which uses the "6th Order Point Mass (Coordinated Flight)" module in "Aerospace Blockset/Equations of Motion/Point Mass".

1. Step 1: Framework construction of the trajectory reference input module

Build the framework of the expected trajectory module in the "SmallFixedWingUAVdf.slx" file in the program "e6-1", as shown in the dashed box ① in Fig. 8.14. In subsequent experiments, this module is used for verifying reference trajectory tracking. In the basic experiment of this section, to verify the effectiveness of differential flatness control, a simple time-varying tracking trajectory is designed first, as shown in Code 8.1. The input of the module is time t ("`t`"), and the output of the module is the time-varying desired position $\mathbf{x}_d$ ("`Xd`"), velocity $\mathbf{v}_d$ ("`Vd`"), acceleration $\mathbf{a}_d$ ("`Ad`"), and jerk $\mathbf{j}_d$ ("`Jd`"), corresponding to lines 2–5 in Code 8.1.

Code 8.1 Simple time-varying tracking trajectory design, from "e6/e6-1/SmallFixedWingUAVdf.slx"

```
1 function [Xd,Vd,Ad,Jd] = fcn(t)
2 Xd = [4*t - cos(t), -0.5*cos(2*t), 10 + sin(t)]';
3 Vd = [4 + sin(t), sin(2*t), cos(t)]';
4 Ad = [cos(t), 2*cos(2*t), -sin(t)]';
5 Jd = [-sin(t), -4*sin(2*t), -cos(t)]';
6 end
```

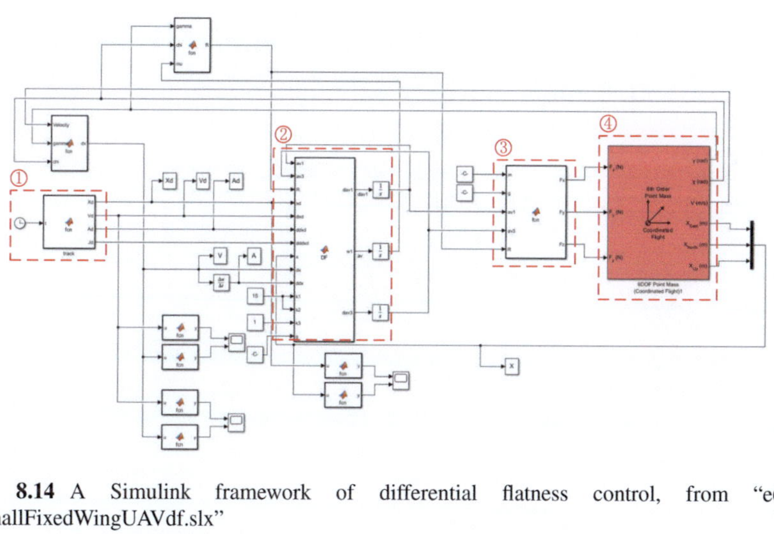

Fig. 8.14 A Simulink framework of differential flatness control, from "e6/e6-1/SmallFixedWingUAVdf.slx"

2. Step 2: Development and implementation of the differential flatness control module in MATLAB Function

To achieve trajectory tracking, according to the differential flatness control principle introduced earlier, build the framework of the differential flatness control module in the "SmallFixedWingUAVdf.slx" file (ENU coordinate system). The input of the module is the axial and normal accelerations a_{v1} ("av1") and a_{v3} ("av3") in the velocity coordinate system, the rotation matrix $\mathbf{R}_b^e$ ("R"), the output of the desired trajectory module from Step 1 $\mathbf{x}_d$ ("xd"), $\mathbf{v}_d$ ("dxd"), $\mathbf{a}_d$ ("ddxd"), $\mathbf{j}_d$ ("dddxd"), the current state of the fixed-wing $\mathbf{x}$ ("x"), $\mathbf{v}$ ("dx"), $\mathbf{a}$ ("ddx"), and the coefficients k_1 ("k1"), k_2 ("k2"), k_3 ("k3"), and gravity g ("g") in Eq. (8.10). The output of the module is $\dot{a}_{v1}$ ("dav1"), ω_x ("w1"), and $\dot{a}_{v3}$ ("dav3"). The example code is shown in Code 8.2. In the code, lines 7–9 represent Eq. (8.5) for calculating ω_y and ω_z, and since the coordinate system is ENU, $\mathbf{g} = [0\ 0\ \ -g]^{\mathrm{T}}$. Line 11 represents Equation (8.10) for calculating the tracking law. Lines 13–15 represent Eq. (8.9) for calculating the control commands $\dot{a}_{v1}$, ω_x, and $\dot{a}_{v3}$. Lines 18–26 handle potential errors due to division by zero, avoiding program crashes.

Code 8.2 Framework of the differential flatness control module, from "e6/e6-1/SmallFixedWingUAVdf.slx"

```
1  function [dav1, w1, dav3] = DF(av1, av3, R, xd, dxd, ddxd,
       dddxd, x, dx, ddx, k1, k2, k3, g)
2  dav1 = zeros(1, 1);
3  w1 = zeros(1, 1);
4  dav3 = zeros(1, 1);
5
6  V = norm(dx, 2);
7  gv = R' * [0; 0; -g];
8  w2 = -(av3 + gv(3)) / V;
9  w3 = gv(2) / V;
10
11 u = dddxd + k1 * (ddxd - ddx) + k2 * (dxd - dx) + k3 * (xd -
       x);
12
13 a = [-w2 * av3; w3 * av1 / av3; w2 * av1];
14 B = [1, 0, 0; 0, -1/av3, 0; 0, 0, 1];
15 out = a + B * R' * u;
16
17 dav1 = out(1); w1 = out(2); dav3 = out(3);
18 if isnan(dav1)
19 dav1 = 0;
20 end
21 if isnan(w1)
22 w1 = 0;
23 end
24 if isnan(dav3)
25 dav3 = 0;
26 end
27 end
```

3. Step 3: Framework construction of differential flatness control based on coordinated flight model and connecting modules

(1) First, run the module initialization file "InitDatactrldf.m" to initialize the model parameters. The model initialization conditions are defined as shown in Code 8.3, mainly explaining the position, mass, velocity, and attitude parameters in the ENU coordinate system. Configure the initial state of the fixed-wing UAV, create the parameter structure "uav.ic", where "uav.ic.Gam" and "uav.ic.Chi" are the initial pitch and yaw angles, "uav.ic.V" is the initial airspeed, "uav.ic.East", "uav.ic.North", and "uav.ic.Up" are the positions in the ENU coordinate system, "uav.ic.Mass" is the mass, and "uav.ic.g" is the gravitational acceleration.

Code 8.3 Initialization of the coordinated flight model, from "e6/e6-1/InitDatactrldf.m"

```
%% UAV
uav = struct;
%% Initial Condition
uav.ic = struct;
uav.ic.Gam = 0;
uav.ic.Chi = 0;
uav.ic.V = 4;
uav.ic.East = -1;
uav.ic.North = -0.5;
uav.ic.Up = 10;
uav.ic.Mass = 1;
uav.ic.g = 9.8;
```

(2) After completing the above steps, click into the Simulink model file "SmallFixedWingUAVdf.slx" to connect the trajectory input and differential flatness control modules and connect them to the input of the fixed-wing coordinated flight model. Then, the overall module construction of the basic experiment is completed.

(3) Run the Simulink simulation. After the simulation is completed, run the plot file "DF_plot.m" to plot the simulation results of position, velocity, and acceleration tracking, as shown in Figs. 8.15, 8.16, and 8.17. It is observed that although the initial position tracking error is significant, it gradually decreases over time. The tracking performance of velocity and acceleration also gradually improves over time, indicating that trajectory tracking is achieved. It is observed that due to slight differences in the initial values of acceleration and velocity, errors occur at the beginning of the simulation; however, they gradually converge over time. In segments with abrupt changes in the trajectory's curvature, errors may occur due to the limitations of fixed-wing dynamics, highlighting the importance of effective path planning and trajectory optimization for fixed-wing navigation.

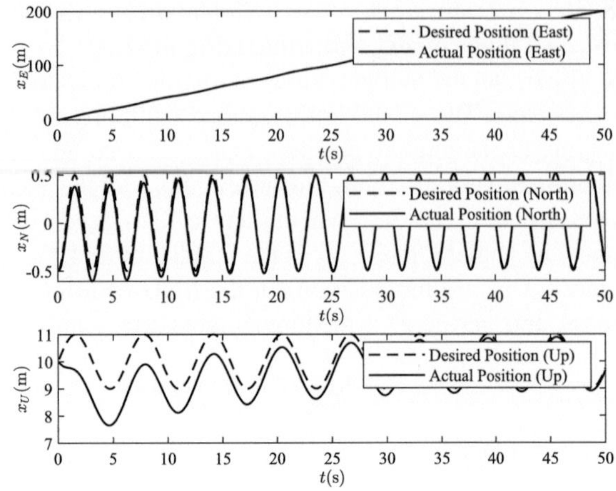

Fig. 8.15 Position tracking performance

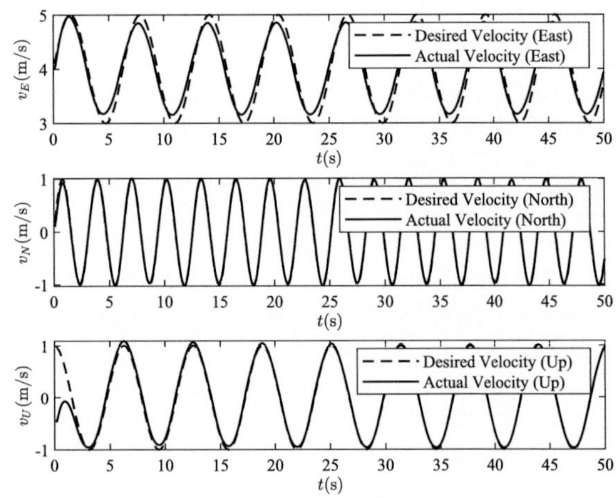

Fig. 8.16 Velocity tracking performance

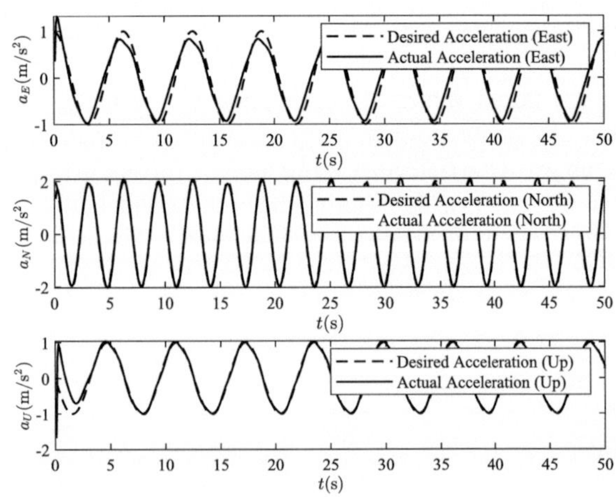

Fig. 8.17 Acceleration tracking performance

8.3 Analysis Experiment

8.3.1 Experimental Objective

1. Things to prepare

(1) Software: MATLAB R2020b or above.
(2) Program: Experimental package "e6-2". The package mainly includes: the Bezier curve generation program folder "bezir" and the B-spline curve generation program folder "B_spline".
The folder "bezir" contains the Bezier recursive equation program file "generate_poly.m" and the trajectory generation program submodule "bezir_n.m"; the folder "B_spline" includes function files for De Boor iteration "bspline_deboor.m" and derivative calculation "bspline_deriv.m", files for generating B-spline curves "Bspline_main.m", and files for validating the impact of control points and knot vectors on trajectories "spline.m".

2. Objective

(1) Understand the principles of Bezier curves and write MATLAB programs for generating Bezier curve trajectories.
(2) Understand the definition and design principles of B-splines and write MATLAB programs for generating B-spline trajectories.
(3) Analyze the impact of control point vectors and knot vectors on the design of B-spline trajectories.

8.3.2 Experimental Procedure

The analysis experiment focuses on the corresponding properties based on the design principles of Bezier curves and B-splines. The analysis experiment is divided into three steps: writing the Bezier curve trajectory generation function, writing the B-spline trajectory generation function, and analyzing the influence of different control point vectors and knot vectors on B-spline trajectories. First, understand the design principles of Bezier curves through Step 1, and further understand the principles of B-spline curves in Step 2. Finally, a comprehensive understanding of the properties of B-spline trajectories and the impact of control point vectors and knot vectors are gained through Step 3.

1. Step 1: Writing the Bezier curve trajectory generation function
(1) In alignment with the theoretical part, write the Bezier curve generation function in the "bezir_n.m" file located in the "bezir" folder within the package "e6-2". The function input is the control point vector "points", and the function output is the Bezier curve trajectory "x", "y", and the time parameter "t". The example code

is shown in Code 8.4. In this code, lines 6–13 are used to calculate the binomial coefficients in Eq. (8.23), and lines 15–29 represent Eq. (8.23)

Code 8.4 Bezier curve generation, from "e6/e6-2/bezir/bezir_n.m"

```
1  function [x, y, t] = bezir_n(points)
2  syms t;
3  % check points size
4  n = length(points) - 1;
5  
6  % calculate k
7  k = [1 1];
8  for i = 1 : (n - 1)
9  a = zeros(1, i+2);
10 a = conv(k, [1 1]);
11 clear k;
12 k = a;
13 end
14 
15 % calculate t
16 for i = 1 : (n + 1)
17 if i == 1
18 j = i - 1;
19 ft = (1 - t).^(n - j).*(t.^j);
20 x = points(1, i) * k(i) * ft;
21 y = points(2, i) * k(i) * ft;
22 else
23 j = i - 1;
24 ft = (1 - t).^(n - j).*(t.^j);
25 x = x + points(1, i) * k(i) * ft;
26 y = y + points(2, i) * k(i) * ft;
27 end
28 end
29 end
```

(2) After designing the control points of the Bezier curve, call "bezir_n.m" to generate the corresponding Bezier trajectory. In the "generate_poly.m" file located in the "bezir" folder of the program "e6-2", design the control point vector and use it as the input of the function "bezir_n.m", as shown in Code 8.5.

Code 8.5 Bezier curve generation, from "e6/e6-2/bezir/generate_poly.m"

```
points = [0 30 50 90 125 150 180 200 260 270;
0 3 3 0 -3 -3 0 3.5 2.8 0];
[x, y, t] = bezir_n(points);
```

To calculate the derivative of the Bezier function, the "diff" function is used. For ease of calculation and timestamp indexing, use the "matlabFunction" function to convert the generated "sym" form function into an anonymous function like $@(t)t^2$ (the content inside the parentheses is the independent variable, followed by the corresponding expression), as shown in Code 8.6.

Code 8.6 Conversion of bezier curve output, from "e6/e6-2/bezir/bezir_n.m"

```
dy = diff(y,t);
dx = diff(x,t);
fx = matlabFunction(x);
fy = matlabFunction(y);
```

(3) Run the file "generate_poly.m" to generate Bezier curves and plot them. The designed Bezier trajectory is shown in Fig. 8.18. It is observed that the designed Bezier trajectory passes through the first and last control points.

2. Step 2: Writing the B-spline trajectory generation function

(1) Based on the theoretical part, write the B-spline De Boor iteration algorithm function in the "bspline_deboor.m" file located in the "B_spline" folder within package "e6-2". The function input is the degree of the designed B-spline "n", input knot vector "t", control point vector "P", and sampling time sequence "U". The function output is the trajectory position sequence, denoted as "C", corresponding to the sampling time sequence, denoted as "U". The example code is shown in Code 8.7. Lines 2–12 initialize the parameters and function inputs, lines 14–22 calculate the control points associated with the corresponding segment, and lines 24–38 correspond to the specific iteration process of the De Boor algorithm represented by Eq. (8.28). Additionally, to obtain the derivative of the time-varying trajectory, such as velocity, or further obtain acceleration and jerk, the derivative function of the B-spline must be written. Here, readers can write on their own or refer to the function "bspline_deriv.m" in the software package to understand and learn.

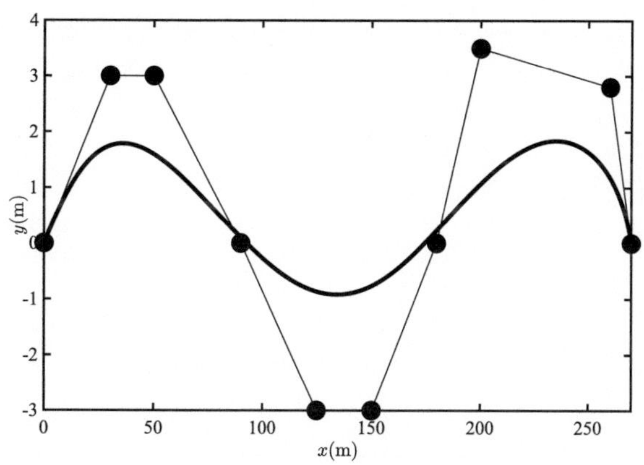

Fig. 8.18 A Bezier trajectory, where the circles represent control points and the curve represents the Bezier trajectory

Code 8.7 B-spline De Boor iteration algorithm, from "e6/e6-2/B_spline/bspline_deboor.m"

```
function [C,U] = bspline_deboor(n,t,P,U)
d = n;
m = size(P,1); % control points dimension
t = t(:).';    % knot vector
U = U(:);
S = sum(bsxfun(@eq, U, t), 2); % multiplicity of u in t (0 <=
    s <= d+1)
I = bspline_deboor_interval(U,t);

Pk = zeros(m,d+1,d+1);
a = zeros(d+1,d+1);

C = zeros(size(P,1), numel(U));
for j = 1 : numel(U)
u = U(j);
s = S(j);
ix = I(j);
Pk(:) = 0;
a(:) = 0;

% identify d+1 relevant control points
Pk(:, (ix-d):(ix-s), 1) = P(:, (ix-d):(ix-s));
h = d-s;

if h > 0
% de Boor recursion formula
for r = 1 : h
q = ix-1;
for i = (q-d+r) : (q-s)
a(i+1,r+1) = (u-t(i+1)) / (t(i+d-r+1+1)-t(i+1));
Pk(:,i+1,r+1) = (1-a(i+1,r+1)) * Pk(:,i,r) + a(i+1,r+1) *
    Pk(:,i+1,r);
end
end
C(:,j) = Pk(:,ix-s,d-s+1); % extract value from triangular
    computation scheme
elseif ix == numel(t) % last control point is a special case
C(:,j) = P(:,end);
else
C(:,j) = P(:,ix-d);
end
end
end
```

(2) After designing the control points and knot vectors of the B-spline, call "bspline_deboor.m" to generate the B-spline trajectory. As shown in Code 8.8, generate B-spline curves and obtain trajectory derivatives. In this code, lines 2–4 define the control point vector and knot vector, lines 6–8 calculate the polynomial degree, line 10 calculates the sampling time, and lines 11–12 call the function to calculate the B-spline trajectory and its derivatives.

Code 8.8 B-spline trajectory generation, from "e6/e6-2/B_spline/Bspline_main.m"

```
1  clear
2  P=[0,10,25,38,80,99,105,150,180,200;
3  0,22,26,12,46,31,15,10,37,50];
4  t=[0 0 0 0 0 1 2 3 4 5 6 6 6 6 6]*3;
5
6  m = size(t,2)-1;
7  n = size(P,2)-1;
8  p=m-n-1;
9
10 tq = linspace(t(p+1),t(m-p+1),1000); % just tq =
       linspace(0,1,1000) if normalized
11 Y = bspline_deboor(p,t,P,tq);
12 dY = bspline_deboor_deriv(p,t,P,tq);
```

Run the "Bspline_main.m" in "B_spline" folder to complete the generation and plotting of B-spline trajectories, as shown in Fig. 8.19. It can be found that the curve is a clamped B-spline, verifying that the parameters $m = 14$, $n = 9$, $p = 4$, and the knot multiplicity as shown in line 4 of Code 8.8 is $p + 1 = 5$, satisfying the conditions of a clamped B-spline.

3. Step 3: Analyzing the effects of different control point vectors and knot vectors on B-spline trajectories

(1) Use the "spmak" function in MATLAB/Curve Fitting Toolbox to generate quickly B-spline trajectories based on control point vectors and knot vectors. Here, the input of the "spmak" function is the control point vector and knot vector, and the output is the B-spline structure. Then, the derivative function "fnder" is used to realize derivative functionality, and the function "fnval" is used to obtain spline trajectory values based on the input time knots. The example code is shown in Code 8.9, where "knots" represents the knot vector, "ctrlpoints" represents the control point vector, "pp" is the B-spline order, "p", "dp", and "ddp" respectively represent the generated B-spline trajectory and its derivative trajectories, as shown in the "spline.m" file.

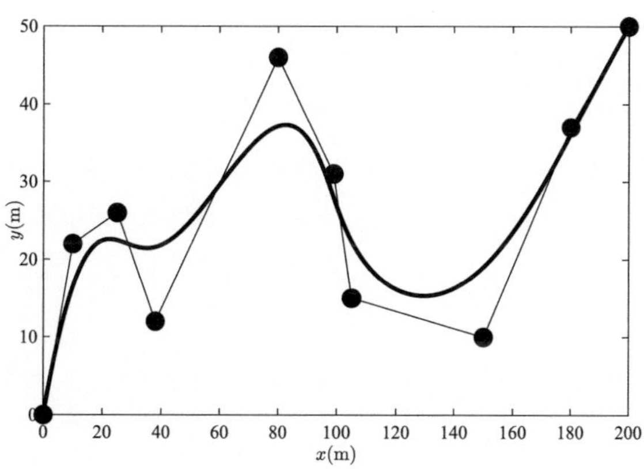

Fig. 8.19 A B-spline trajectory, where the circles represent control points and the curve represents the B-spline trajectory

Code 8.9 B-spline trajectory parameter calculation and output, from "e6/e6-2/B_spline/spline.m"

```
m = length(knots)-1; % length of knots vector - 1
n = size(ctrlpoints,2)-1; % length of ctrlpoints vector - 1
pp = m-n-1; % spline order
p = spmak(knots,ctrlpoints); % generate B-spline
t = linspace(knots(pp),knots(m+1-pp),10000);
dp = fnder(p); % derivate
ddp = fnder(dp);
Xd = fnval(p, t);

xt = Xd(1,:);
yt = Xd(2,:);
```

(2) As shown in Code 8.10, to explore the generation conditions of clamped B-splines, modify the control point vector and knot vector, and plot the clamped B-spline of the curve shown in Fig. 8.20.

Code 8.10 Control point and knot vectors for a clamped B-spline, from "e6/e6-2/B_spline/spline.m"

```
1 ctrlpoints=[0,6,4,-12,-6,10,30,26,36,32;
2 0,5,15,20,40,44,35,18,10,6];
3 knots=[0 0 0 1 2 3 4 5 6 7 8 8 8];% clamped B-spline
```

In the above example, the control point vector and knot vector have $m = 12$, $n = 9$, $p = m - n - 1 = 2$, and the knot multiplicity as shown in line 4 of Code 8.10 is $p + 1 = 3$, which satisfies the clamped condition. Therefore, the clamped B-spline curve passes through the first and last control points.

(3) As shown in Code 8.11, to explore the generation conditions of open B-splines, modify the control point vector and knot vector, and plot the open B-spline of the curve shown in Fig. 8.21.

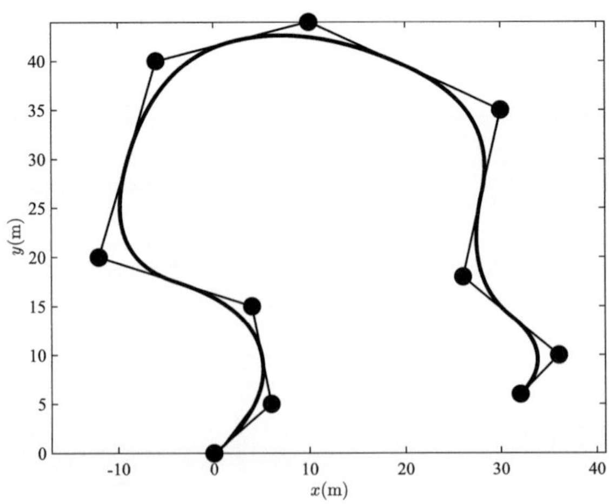

Fig. 8.20 A clamped B-spline trajectory, where the circles represent control points and the curve represents the B-spline trajectory

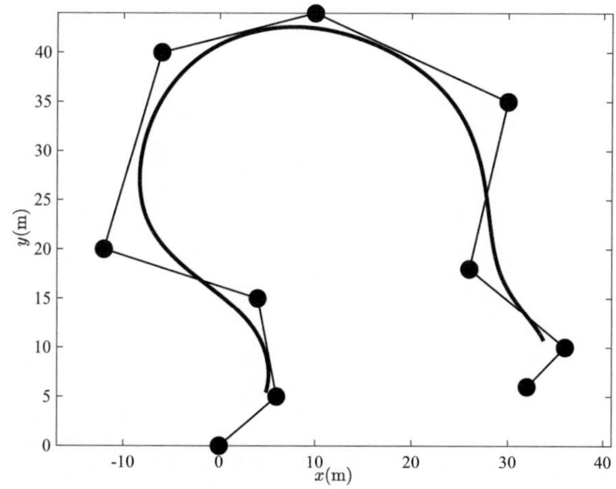

Fig. 8.21 An open B-spline trajectory position curve, where the circles represent control points and the curve represents the B-spline trajectory

Code 8.11 Control point and knot vectors for an open B-spline, from "e6/e6-2/B_spline/spline.m"

```
ctrlpoints=[0,6,4,-12,-6,10,30,26,36,32;
0,5,15,20,40,44,35,18,10,6];
knots=[1 5 9 12 15 22 28 31 35 44 49 55 59 66]; % open B-spline
```

In the above example, the control point vector and knot vector have $m = 14$, $n = 9$, $p = m - n - 1 = 4$, which is consistent with the generation conditions of open B-splines, so an open B-spline is obtained.

(4) As shown in Code 8.12, to explore the generation conditions of closed B-splines, modify the control point vector and knot vector, and plot the closed B-spline of the curve shown in Fig. 8.22. There are several methods for generating closed curves. A simple method is to "wrap" the control points. The specific operation of "wrapping" control points is as follows: first, design a uniform knot sequence containing $m + 1$ knots, and then "wrap" the first p and last p control points, as shown in the code below, i.e., $\mathbf{p}_0 = \mathbf{p}_{n-p+1}, \cdots, \mathbf{p}_{p-1} = \mathbf{p}_n$.

Code 8.12 Control point and knot vectors for a closed B-spline, from "e6/e6-2/B_spline/spline.m"

```
ctrlpoints=[0,6,4,-12,-6,10,30,26,36,32,0,6,4;
0,5,15,20,40,44,35,18,10,6,0,5,15];
knots = [0 1 2 3 4 5 6 7 8 9 10 11 12 13 14 15 16]; % closed
    B-spline
```

Through the above process, it is observed that given the control point vector and the length of the control point vector $n + 1$, when the degree of the B-spline trajectory p is determined, the length of the knot vector satisfies $m + 1 = n + p + 2$. For a knot vector of length $m + 1$, if the starting and ending knot multiplicity is $p + 1$, a clamped B-spline is designed. Simultaneously using the first p control points as the last p control points, a closed B-spline is obtained.

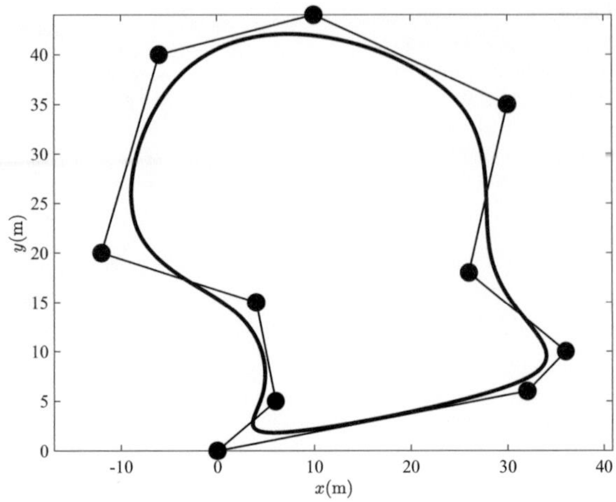

Fig. 8.22 A closed B-spline trajectory position curve, where the circles represent control points and the curve represents the B-spline trajectory

8.4 Design Experiment

8.4.1 Experimental Objective

1. Things to Prepare

(1) Software: MATLAB R2022b or above, RflySim toolchain. The RflySim toolchain provides a Hardware-In-the-Loop (HIL) simulation environment for micro-small fixed-wing UAVs.

(2) Program: Experimental packages "e6-3" and "e6-4". In package "e6-3", there are: a folder named "software" containing simulation programs based on basic experimental models, and a folder named "point mass" containing simulation programs based on dynamic system models of fixed-wing UAVs. Both folders contain the model and control parameter initialization program file "InitDatactrl.m", which imports UAV parameters into the workspace when executed, files containing trajectory planning and differential flatness control program modules "SmallFixedWingUAVdf.slx", simulation result plotting program file "DF_plot.m", rotation matrix function files "ROTX.m", "ROTY.m", and "ROTZ.m" called in Simulink, path setting and search file "path_search.m", B-spline trajectory generation function file "bspline.m", artificial potential field search function file "APF.m", and file "iter_first.m" is used to call the artificial potential field search function.

In package "e6-4", there are: a folder named "software" containing pre-software programs for testing the feasibility of hardware in advance, and two folders named "hitl" for HIL simulation programs. The folders contain model and control parameter initialization program files "Init.m" and "InitDatactrl.m", HIL

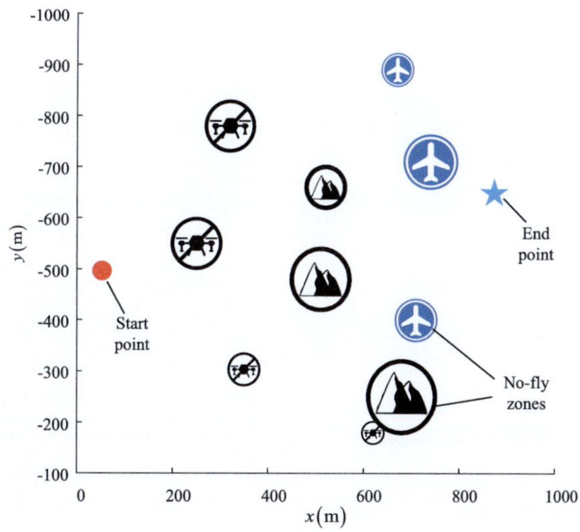

Fig. 8.23 Mission map

simulation control interface program file "SmallFixedWingUAVupperHIL.slx", and other files with the same meaning as those in package "e6-3".

(3) Hardware: a CubePilot/Pixhawk autopilot, a remote controller, and a receiver.

2. Objective

Set the current small fixed-wing UAV to complete a low-altitude cruise mission, as shown in Fig. 8.23. The map data in the North-East-Down (NED) coordinate system[2] are shown in Table 8.1, where the x-axis corresponds to the north direction, and the y-axis corresponds to the east direction. The position vector in the NED coordinate system is $\mathbf{p} = [x_N \ x_E \ x_D]^{\mathrm{T}}$, the velocity vector is $\mathbf{v} = [v_N \ v_E \ v_D]^{\mathrm{T}}$, and the acceleration vector is $\mathbf{a} = [a_N \ a_E \ a_D]^{\mathrm{T}}$. During the mission execution, the following task objectives need to be ensured:

- Start the task from the takeoff point and end it by landing at the recovery point;
- Avoid mountains, airports, and no-fly zones during the flight;
- The UAV should arrive within the specified time.

8.4.2 Experimental Design

The experimental design requires developing a differential flatness module tailored to the given interface, based on the contents of the basic experiment and analysis experiment, and completing the algorithm design for the task. The design phase

[2] This coordinate system is used in the experiments in this section to maintain consistency with the previous model in the book.

Table 8.1 Map data in the NED coordinate system

Type	Data
Start point	$x = 50$ m, $y = -500$ m
End point	$x = 900$ m, $y = -650$ m
Specified time	$t = 100$ s
Obstacles (No-fly Zones)	$x_1 = 350$ m, $y_1 = -310$ m, $r_1 = 30$ m
	$x_2 = 250$ m, $y_3 = -\ 550$ m, $r_2 = 50$ m
	$x_3 = 320$ m, $y_3 = -\ 780$ m, $r_3 = 50$ m
	$x_4 = 620$ m, $y_4 = -\ 180$ m, $r_4 = 20$ m
Obstacles (Mountains)	$x_5 = 520$ m, $y_5 = -\ 660$ m, $r_5 = 40$ m
	$x_6 = 510$ m, $y_6 = -\ 480$ m, $r_6 = 60$m
	$x_7 = 680$ m, $y_7 = -\ 250$ m, $r_7 = 70$m
Obstacles (Airports)	$x_8 = 670$ m, $y_8 = -\ 890$ m, $r_8 = 30$ m
	$x_9 = 710$ m, $y_9 = -400$ m, $r_9 = 40$ m
	$x_{10} = 740$ m, $y_{10} = -\ 710$ m, $r_{10} = 50$ m

encompasses the development of the low-level control interface, the path search and desired trajectory design utilizing the artificial potential field method, and the design of the feedforward-feedback composite controller.

8.4.2.1 Design of Low-Level Control Interface

The low-level flight control of the fixed-wing aircraft in this section is similar to Fig. 6.1 is mainly divided into two levels, namely, attitude control and altitude acceleration control. The closed-loop control framework is shown in Fig. 8.24. The fixed-wing system is an underactuated system with three virtual input variables (i.e., height H_{d}, longitudinal flight acceleration $\dot{V}_{\mathrm{ad}}$, and roll angle ϕ_{d}), nine output variables (position $\mathbf{p} \in \mathbb{R}^3$, velocity $\mathbf{v} \in \mathbb{R}^3$, and attitude $\boldsymbol{\Theta} \in \mathbb{R}^3$), and four control variables (elevator deflection δ_{e}, aileron deflection δ_{a}, rudder deflection δ_{r}, and throttle thrust T).

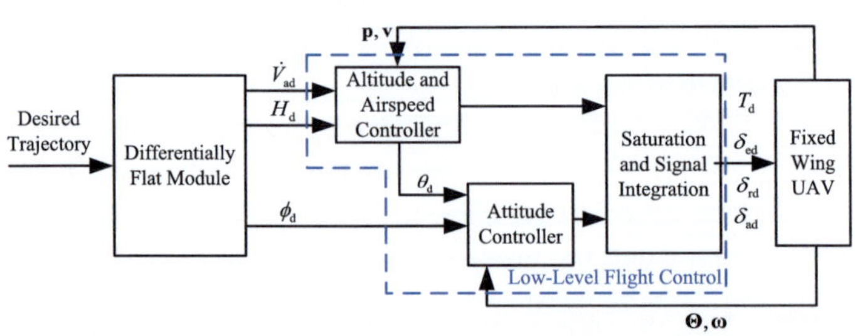

Fig. 8.24 Differential flatness control structure based on the low-level flight controller

Referring to Eq. (6.43) in Chap. 6,

$$m\dot{V}_{\mathrm{a}} = T\cos\alpha - f_{\mathrm{a}x} - mg\sin\gamma. \tag{8.33}$$

Assuming that the aerodynamic drag $f_{\mathrm{a}x}$ remains constant during flight and the angle of attack α and flight path angle γ are small, according to Eq. (8.33), the thrust feedforward is directly obtained as

$$T_{\mathrm{d}} = m\dot{V}_{\mathrm{ad}} \tag{8.34}$$

where $\dot{V}_{\mathrm{ad}}$ is represented by the result a_{v1} of differential flatness calculation in Eq. (8.11).

8.4.2.2 Desired Trajectory Design

As shown in Fig. 8.2, for the obstacle map obtained by the perception layer, path planning is completed using algorithms such as the artificial potential field method, A*, and RRT to achieve further trajectory planning. In this experiment, the artificial potential field method [20, 21] is used to construct the path search process. The artificial potential field $U(\mathbf{x})$ contains two types of force fields: the attractive field $U_{\mathrm{att}}(\mathbf{x})$ formed by the motion target position and the repulsive field $U_{\mathrm{rep}}(\mathbf{x})$ formed by obstacles. The attractive field is given by

$$U_{\mathrm{att}}(\mathbf{x}) = \begin{cases} k_{\mathrm{att}}\|\mathbf{x}_{\mathrm{g}} - \mathbf{x}\|^2 & \|\mathbf{x}_{\mathrm{g}} - \mathbf{x}\| \le d_{\mathrm{g}} \\ 2k_{\mathrm{att}}d_{\mathrm{g}}\|\mathbf{x}_{\mathrm{g}} - \mathbf{x}\| - k_{\mathrm{att}}\|\mathbf{x}_{\mathrm{g}} - \mathbf{x}\|^2 & \|\mathbf{x}_{\mathrm{g}} - \mathbf{x}\| > d_{\mathrm{g}} \end{cases} \tag{8.35}$$

where $\mathbf{x}$ represents the current position of the UAV, $\mathbf{x}_{\mathrm{g}}$ represents the position of the target point, k_{att} is the attraction coefficient, and d_{g} is the threshold to prevent excessive attraction when far from the target. The repulsive field is given by

$$U_{\mathrm{rep}}(\mathbf{x}) = \begin{cases} k_{\mathrm{rep}}\left(\frac{1}{D(\mathbf{x})} - \frac{1}{Q}\right)^2 & D(\mathbf{x}) \le Q \\ 0 & D(\mathbf{x}) > Q \end{cases} \tag{8.36}$$

where k_{rep} is the repulsion coefficient, $D(\mathbf{x})$ represents the distance between the UAV and the obstacles, and Q represents the range of repulsion action, within which obstacles generate repulsive force. The superposition of the two potential fields forms an artificial potential field, namely

$$U(\mathbf{x}) = U_{\mathrm{att}}(\mathbf{x}) + U_{\mathrm{rep}}(\mathbf{x}). \tag{8.37}$$

The negative gradient of the artificial potential field, i.e., $-\nabla U(\mathbf{x})$, yields the motion direction.

The control points of the B-spline are obtained through path search. After an artificial potential field search, a series of waypoints constitutes the route point vector. In this experiment, the route points are used as control points. After determining the B-spline order p, according to the design rules, the clamped B-spline trajectory is designed as the desired trajectory input for fixed-wing UAVs.

8.4.2.3 Controller Design

For the flight model provided in Chap. 4, before performing SIL simulation, it is necessary to apply the differential flatness control to provide input adapted to the interface. According to the properties of differential flatness control, the flat output can represent any state, so given the desired flat output, it is possible to represent any state variable or control input with feedforward. Therefore, given the desired output trajectory $\mathbf{y} = \mathbf{x}_\mathrm{d}$, it is necessary to represent the desired control inputs a_{v1}, a_{v3}, and ϕ_d.

The first column of the desired rotation matrix $\mathbf{R}_\mathrm{d}$ is obtained from $\mathbf{r}_1 = \dot{\mathbf{x}}_\mathrm{d}/\|\dot{\mathbf{x}}_\mathrm{d}\|$ as shown in Fig. 8.25. Thus, the axial acceleration is obtained through the projection

$$a_{v1} = \mathbf{r}_1^\mathrm{T} (\ddot{\mathbf{y}} - \mathbf{g}) . \tag{8.38}$$

Similar to (8.10), design the flat output as

$$\ddot{\mathbf{y}}_\mathrm{d} = \ddot{\mathbf{x}}_\mathrm{d} + k_4(\dot{\mathbf{x}}_\mathrm{d} - \dot{\mathbf{x}}) + k_5(\mathbf{x}_\mathrm{d} - \mathbf{x}) \tag{8.39}$$

where $k_4, k_5 > 0$. Based on this, Eq. (8.38) combines feedforward and feedback to enhance tracking performance.

As shown in Fig. 8.1, only the acceleration in the longitudinal plane is considered in the coordinated turn. The normal acceleration is obtained through

$$\begin{aligned} \mathbf{a}_\mathrm{n} &= \ddot{\mathbf{y}}_\mathrm{d} - \mathbf{g} - a_{v1}\mathbf{r}_1 \\ &= \left(\mathbf{I} - \mathbf{r}_1\mathbf{r}_1^\mathrm{T}\right)(\ddot{\mathbf{y}}_\mathrm{d} - \mathbf{g}) . \end{aligned} \tag{8.40}$$

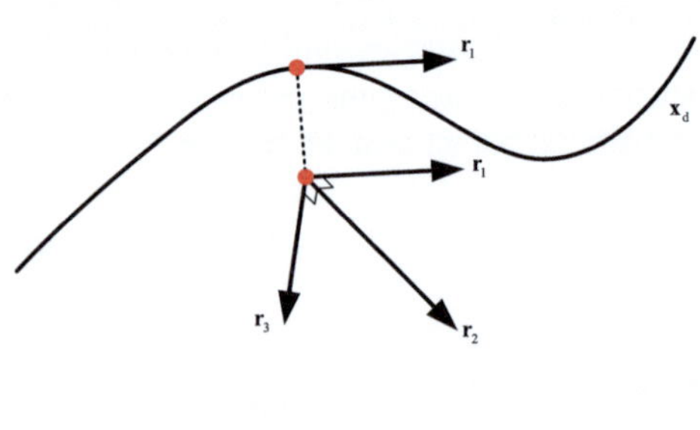

Fig. 8.25 Calculation of desired attitude

Therefore, $a_{v3} = \|\mathbf{a}_n\|$, then $\mathbf{r}_3 = \mathbf{a}_n / a_{v3}$. Thus, through $\mathbf{r}_2 = \mathbf{r}_3 \times \mathbf{r}_1$, $\mathbf{R}_d = [\mathbf{r}_1|\mathbf{r}_2|\mathbf{r}_3]$ is obtained, yielding the desired rotation matrix. Based on this, the desired roll angle feedforward is obtained

$$\phi_d = \text{atan2}\left(\mathbf{r}_2\left(3\right), \mathbf{r}_3\left(3\right)\right) \tag{8.41}$$

where $\mathbf{r}_2\left(3\right) = \sin\phi\cos\theta$, $\mathbf{r}_3\left(3\right) = \cos\phi\cos\theta$.

In this chapter, the differential flatness control of the fixed-wing UAV is based on the simplified model (8.2). Therefore, in the experimental design, using a more complex controlled model requires adding necessary feedback to compensate for the model. In this section, to achieve yaw stability tracking as the feedback of roll angle feedforward for fixed-wing heading, yaw feedback compensation is designed as roll angle compensation.

The lateral attitude motion of the fixed-wing UAV is mainly controlled by the aileron (roll control), and the rudder (yaw control) plays a role in eliminating sideslip. According to Eq. (4.63), when the desired roll angle ϕ_d is linearized, the change in roll angle and yaw rate is expressed as

$$\dot{\psi} \approx \dot{\psi}_d + \frac{g}{V_a}\left(\phi - \phi_d\right) \tag{8.42}$$

where ψ represents the current yaw angle, ψ_d represents the desired yaw angle, calculated as

$$\psi_d = \text{atan2}\left(\mathbf{r}_1\left(2\right), \mathbf{r}_1\left(1\right)\right) \tag{8.43}$$

where $\mathbf{r}_1\left(1\right) = \cos\theta\cos\psi$, $\mathbf{r}_1\left(2\right) = \cos\theta\sin\psi$. Therefore, when there is tracking error $\Delta\psi = \psi - \psi_d$, compensation is made by adding feedback to control the roll angle for yaw compensation, i.e.,

$$\phi_c = \phi_d + \Delta\phi_d \tag{8.44}$$

where ϕ_c represents the actual input roll angle control command, $\Delta\phi_d$ represents the roll angle compensation corresponding to yaw error, which is obtained according to

$$\Delta\dot{\psi} \approx \frac{g}{V_a}\Delta\phi_d \tag{8.45}$$

where $\Delta\dot{\psi} = \dot{\psi} - \dot{\psi}_d$ represents the yaw rate tracking error.

To ensure stable tracking of yaw, i.e., $\lim_{t\to\infty}\|\Delta\dot{\psi}(t)\| = 0$, it is necessary to design the compensation amount $\Delta\phi_d$ for the roll angle based on (8.45). As a result, the yaw rate satisfies the dynamic equation:

$$\Delta\dot{\psi} = -k_{\psi}\Delta\psi \tag{8.46}$$

where $k_\psi > 0$ represents yaw angle feedback. Based on this, combined with (8.45) and (8.46), the roll angle feedback compensation is obtained as

$$\Delta\phi_\mathrm{d} = -k_\psi \frac{V_\mathrm{a}}{g} \Delta\psi. \tag{8.47}$$

8.4.3 Experimental Procedure

The primary objective of the experimental design is to fulfill the specified task requirements. The experimental framework comprises four key steps: constructing the desired trajectory, refining the differential flatness module, executing SIL simulations, and conducting HIL simulations. First, in Step 1, the B-spline desired trajectory is constructed based on the path derived from the artificial potential field method, as depicted in Fig. 8.26. Subsequently, in Step 2, the differential flatness module is modified based on the theoretical analysis part of the experimental design. Moving forward to Step 3, a SIL simulation experiment is conducted to validate the algorithm's functionality. Finally, in Step 4, the RflySim simulation toolchain is utilized, together with hardware such as the CubePilot/Pixhawk autopilot, remote controller, and receiver, to conduct the HIL simulation experiment. This step aims to validate the feasibility and accuracy of the designed algorithm through the combination of actual hardware devices and simulation environments. The entire experimental process involves combining the SIL and HIL simulations to validate the effectiveness of the algorithm design comprehensively. Such a design scheme will help improve the performance of the control algorithm and provide reliable technical support, and guidance for practical applications.

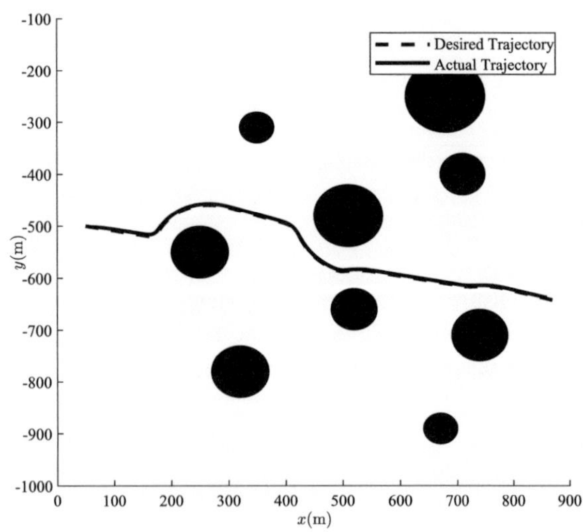

Fig. 8.26 Desired trajectory and actual trajectory

1. Step 1: Constructing the desired B-spline trajectory

In the experimental package "e6-3", use the artificial potential field method to complete the path search for the map situation mentioned above. The input of the artificial potential field method includes the start point "start", target point "target", obstacle position and size parameters "obs", step length "step", the maximum calculation count "countMax", attraction parameter "kAttr", repulsion parameter "kRep", obstacle scaling factor "kObs", etc., and the output is the searched path "result" and obstacle-related parameters "obs", as shown in Code 8.13. Refer to the "APF" file in the software package for the specific implementation details.

Code 8.13 Function call for APF method, from "e6/e6-3/APF.m"

```
[result, obs] = APF(start, target, obs, step, countMax, kAttr,
    kRep, kObs)
```

Based on the obtained search path, use the B-spline for trajectory planning, where "ctrlpoints" is the control point vector, "stoptime" is the trajectory stop time, "knots" is the knot vector, "p" is the generated B-spline trajectory, consistent with the previous code, and the desired trajectory is shown in Fig. 8.26.

Code 8.14 Generating B-spline trajectory based on APF method, from "e6/e6-3/bspline.m"

```
ctrlpoints=[xd';yd'];
n = size(ctrlpoints,2)-1;
pp = 3;
m = 1+pp+n;

stoptime = 200;
t1 = zeros([1,pp+1]);
t3 = stoptime*ones([1,pp+1]);
t2 = linspace(0,stoptime,m+1-2*pp-2);

knots=[t1, t2, t3]*3;

p = spmak(knots,ctrlpoints);% Generate B-spline function
```

Use the example code shown in Code 8.15 to obtain the time-varying desired trajectory "xts" and "yts" in Simulink, referring to the "bspline.m" file to get the derivative trajectory.

Code 8.15 Input of time-varying desired trajectory in Simulink, from "e6/e6-3/bspline.m"

```
t = linspace(knots(pp+1),knots(m-pp+1),10000);
Xd = fnval(p, t);

xt = Xd(1,:);
yt = Xd(2,:);
xts = timeseries(xt,t);
yts = timeseries(yt,t);
```

2. Step 2: Modifying the differential flatness module for the interface
After the derivation above, rewrite the code for the differential flatness control. An example code is shown in Code 8.16. The module takes the outputs of the desired trajectory module in Step 1 as input, denoted as "xd", "dxd", "ddxd", the current state of the fixed-wing denoted as "x", "dx", and the gravity acceleration denoted as "g", and outputs the acceleration a_{v1} (denoted as "av1"), altitude "Hz", yaw angle "yaw", and roll angle "roll". In the code, line 2 corresponds to the calculation of $\mathbf{r}_1$; line 3 corresponds to Eq. (8.38); lines 5 and 6 correspond to Eq. (8.40); lines 9–13 correspond to Eqs. (8.41) and (8.43) for the attitude angle calculation process. Similarly, lines 15–23 are designed to avoid invalid data ("Nan") caused by division by zero.

Code 8.16 Modified differential flatness control based on specific interface, from "e6/e6-3/SmallFixedWingUAVdf.slx"

```
1  function [av1,Hz,yaw,roll] = DF(xd,dxd,ddxd,x,dx,g)
2  r1 = (dxd)/norm(dxd);
3  av1 = r1'*(ddxd-[0;0;g]+3*(dxd-dx)+0.1*(xd-x));
4  
5  an = (eye(3)-r1*r1')*(ddxd-[0;0;g]);
6  av3 = -(norm(an));
7  
8  Hz = xd(3);
9  r3 = an/av3;
10 r2 = cross(r3,r1);
11 
12 yaw = atan2(r1(2),r1(1));
13 roll = atan2(r2(3),r3(3));
14 
15 if isnan(av1)
16 av1=0;
17 end
18 if isnan(Hz)
19 Hz=0;
20 end
21 if isnan(yaw)
22 yaw=0;
23 end
24 end
```

3. Step 3: SIL simulation in MATLAB/Simulink
(1) Open the "SmallFixedWingUAVdf.slx" file and connect the outputs of the modules from Step 2 to the model built in Chap. 4, as shown in Fig. 8.27. The model contains four modules: dashed box ① "track1", trajectory input module, used to output the desired trajectory over time, corresponding to Code 8.15; dashed box ② "DF", differential flatness module, based on the analysis of the experimental design, obtains the differential flatness control command according to the desired trajectory as feedforward at the given interface, corresponding to Code 8.16; dashed box ③ feedback

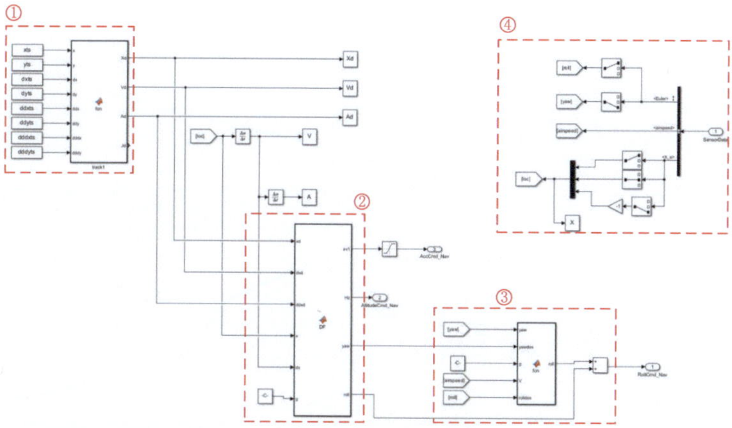

Fig. 8.27 Output module of differential flatness control, from "e6/e6-3/SmallFixedWingUAVdf.slx"

module, designed according to Eq. (8.42) to construct feedforward-feedback control; dashed box ④ model state acquisition, based on the fixed-wing model in Chap. 4, obtains the current state feedback of the UAV.

(2) Run the "InitDatactrl.m" file for initialization, including parameter initialization, path search, B-spline trajectory planning, etc.

(3) Open the "SmallFixedWingUAVdf.slx" file and run the SIL simulation.

(4) Run the "DF_plot.m" file for visualization of simulation results. The comparison between the actual trajectory and the tracked desired trajectory is shown in Fig. 8.26. The tracking performance of the position, velocity, and acceleration is shown in Figs. 8.28, 8.29, and 8.30, respectively. According to the comparison between the actual trajectory and the tracked desired trajectory, it is observed that the differential flatness control realizes the tracking of the desired trajectory.

Based on differential flatness control, small fixed-wing UAVs can perform trajectory tracking, which, compared to waypoint tracking control, achieves planning and control of waypoints in the time domain. Based on the above process, small fixed-wing UAVs can perform more maneuverable flight actions and achieve high-precision tracking of time-varying trajectories.

4. Step 4: HIL simulation experiment with RflySim toolchain

The HIL simulation layout is shown in Fig. 8.31.

(1) Open the "hitl" folder in the software experimental package "e6-4". First, run the initialization file "Init.m", open the "SmallFixedWingUAVupperHIL.slx" file, click "Model" in the "Model Browser" toolbar, as shown in Fig. 8.32.

(2) Click on "Model Workspace" as shown in Fig. 8.33, select "MAT File" in the model workspace, and click browse to select "e6-4/hitl/desire.mat", which is used to store the desired trajectory data for HIL implementation.

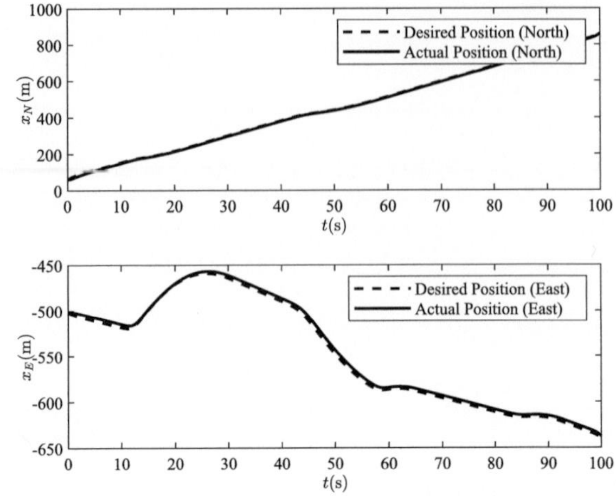

Fig. 8.28 Trajectory tracking performance

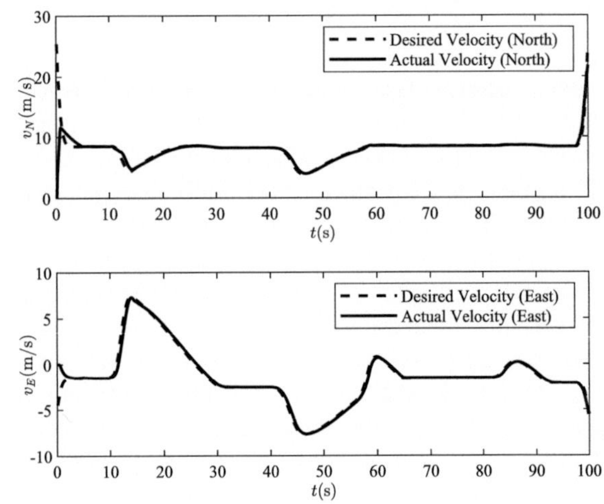

Fig. 8.29 Trajectory velocity tracking performance

(3) Click on "Simulink Parameter" in the toolbar as shown in Fig. 8.34, and rename the variable name in "Name" column to "x" and input "DD" in "value" column for assignment.
(4) Click "Save" on the Simulink page, and a confirmation box as shown in Fig. 8.35 will appear; click "Yes".
(5) Click on "Build" in the toolbar to compile the current program, as shown in Fig. 8.36.
(6) After successful compilation, connect the flight controller to the computer, click on "code" in the toolbar as shown in Fig. 8.36, select "PX4 PSP Upload code to PX4FMU", and burn the code to the CubePilot autopilot.

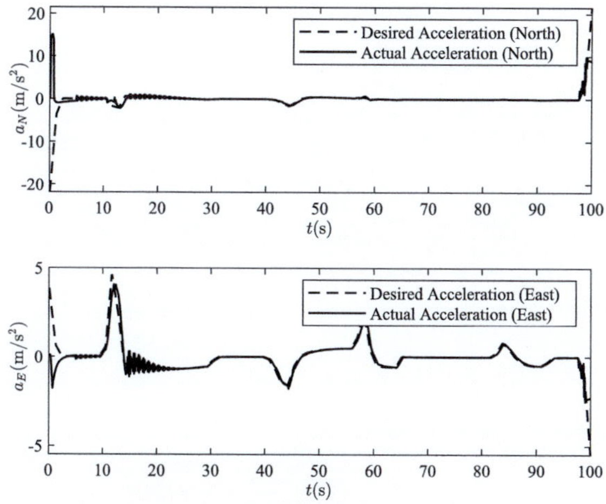

Fig. 8.30 Trajectory acceleration tracking performance

Fig. 8.31 Hardware layout of HIL simulation

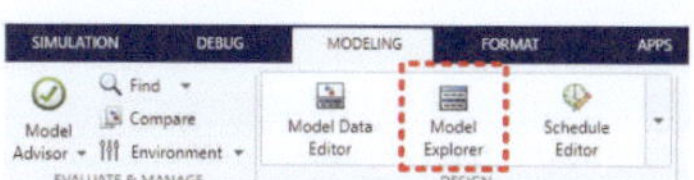

Fig. 8.32 Opening model browser

(7) Set the program to start automatic flight through "ch6" in the high position. Before starting the simulation, ensure that the remote controller's "ch6" channel is at a low position. After waiting for the HIL simulation environment to start, first unlock the UAV in QGC, and then turn "ch6" to the high position to start the automatic route flight. The flight process is shown in Fig. 8.37. The flight record in QGC is shown in Fig. 8.38.

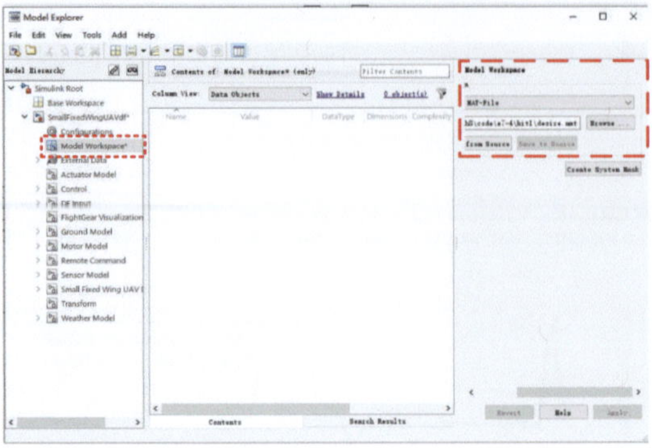

Fig. 8.33 Importing data from workspace

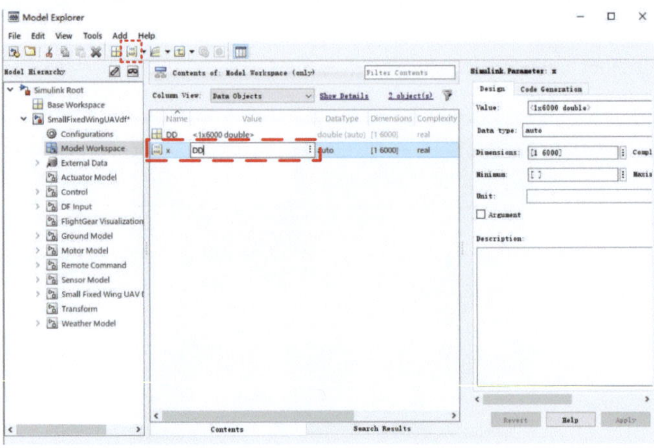

Fig. 8.34 Changing variable name

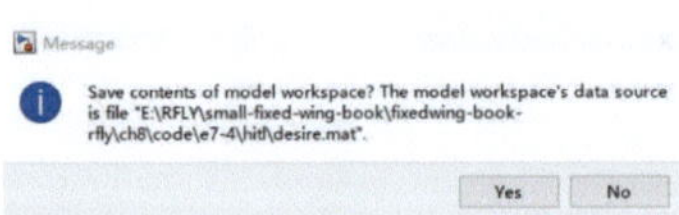

Fig. 8.35 Confirming to save data

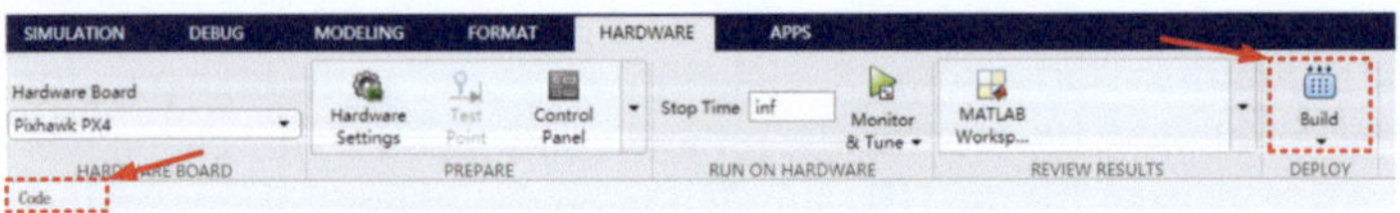

Fig. 8.36 Compiling the program

Fig. 8.37 Simulated view

Fig. 8.38 HIL flight route record

8.5 Exercise

1. Referring to [12], for experimental design, establish the optimization model as shown in Eq. (8.32), and then plan trajectories that satisfy the objective function. During the planning process, pay attention to the flying performance of the fixed-wing, such as ensuring that the curvature variation is not too large.
2. The existing implementation of the differential flatness algorithm exhibits some deficiencies in tracking performance. It is recommended that the control algorithm be refined to enhance tracking performance, for example, by improving Eq. (8.10).
3. In the preceding experiments and introductory material, a combination of feedback and feedforward was employed to control the fixed wing. Regarding the experimental design, it is recommended that Eqs. (8.34) or (8.47) be modified to consider performance with different feedback.
4. In addition to B-spline curves, classic polynomial trajectory planning algorithms such as Eq. (8.14) include methods such as minimum "Snap" trajectory optimization [7]. This type of trajectory often uses simple polynomials to represent curves and optimizes parameters to achieve optimization goals. Currently, this algorithm is mostly applied to multicopters, while the form of minimum "Snap" trajectory optimization is compatible with the flight characteristics of fixed-wing UAVs. Try designing optimized trajectories based on [7], and complete the experimental design again.

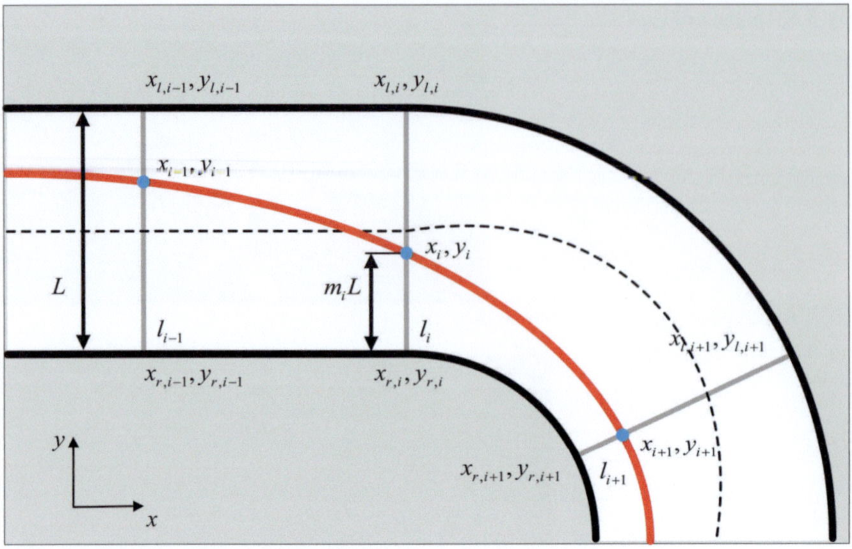

Fig. 8.39 Representation of the passable area

5. Based on the knowledge from Chap. 7, Dubins curves are more suitable for fixed-wing flight. Try to perform trajectory planning based on Dubins curves according to the reference [3], and complete the differential flatness control experiment.
6. This chapter has simplified the fixed-wing model to some extent, which may inevitably lead to significant errors in actual control. Based on the characteristics of fixed-wing UAVs, paths with slight curvature are planned to avoid substantial changes in the controller. As shown in Fig. 8.39, by sampling the track, $[l_1\ l_2\ l_3\ \cdots l_n]^{\mathrm{T}} \in \mathbb{R}^n$ are obtained, and the lengths of segments are equal. For the ith sampling point, the position on the track is expressed as:

$$\begin{aligned}\mathbf{p}_i &= [x_{ri} + m_i\,(x_{li} - x_{ri})\ \ y_{ri} + m_i\,(y_{li} - y_{ri})]^{\mathrm{T}} \\ &= [x_{ri} + m_i \Delta x_i\ \ y_{ri} + m_i \Delta y_i]^{\mathrm{T}}\end{aligned} \tag{8.48}$$

where $m_i \in [0, 1]$ represents the proportion of the passable area width L occupied by the given position. Since the path is connected by a series of discrete points, obtaining the desired path (minimum curvature path) is simplified to get the vector $\overline{\mathbf{m}} = [m_1\ m_2\ \cdots\ m_n]^{\mathrm{T}}$.
During the solving process, use quadratic spline curves for trajectory approximation to avoid discontinuities in curvature, represented as:

$$\begin{aligned}x_i\,(t) &= a_{i,x} + b_{i,x} t + c_{i,x} t^2 \\ y_i\,(t) &= a_{i,y} + b_{i,y} t + c_{i,y} t^2\end{aligned} \tag{8.49}$$

where $t = 2(l - l_i)/(l_{i+1} - l_{i-1}) \in [-1, 1)$ is the trajectory parameter. Since the square of curvature is proportional to the square of the second derivative of the trajectory, the problem of minimizing curvature is defined as:

$$\begin{aligned} &\min_{\overline{\mathbf{m}}} \sum_{i=2}^{n-1} c_{i,x}^2 + c_{i,y}^2 \\ &\text{s.t. } 0 \le m_i \le 1. \end{aligned} \tag{8.50}$$

Try to solve the minimum curvature path based on the above definition and complete the experimental design again.

References

1. J. Hauser, R. Hindman, Aggressive flight maneuvers, in *Proceedings of the 36th IEEE Conference on Decision and Control*, vol. 5 (IEEE, 1997), pp. 4186–4191
2. M.J. Van Nieuwstadt, R.M. Murray, Real-time trajectory generation for differentially flat systems. Int. J. Robust Nonlinear Control: IFAC-Affiliated J. **8**(11), 995–1020 (1998)
3. A. Bry, C. Richter, A. Bachrach, N. Roy, Aggressive flight of fixed-wing and quadrotor aircraft in dense indoor environments. Int. J. Robot. Res. **34**(7), 969–1002 (2015)
4. D. Hanover, A. Loquercio, L. Bauersfeld, A. Romero, R. Penicka, Y. Song, G. Cioffi, E. Kaufmann, D. Scaramuzza, Autonomous drone racing: a survey. IEEE Trans. Robot. **40**, 3044–3067 (2024)
5. H. Choset, K.M. Lynch, S. Hutchinson, G.A. Kantor, W. Burgard, *Principles of Robot Motion: Theory, Algorithms, and Implementations* (MIT Press, 2005)
6. A. Kouba, H. Bennaceur, I. Chaari, S. Trigui, A. Ammar, M.-F. Sriti, M. Alajlan, O. Cheikhrouhou, Y. Javed, *Robot Path Planning and Cooperation*, vol. 772 (Springer, 2018)
7. D. Mellinger, V. Kumar, Minimum snap trajectory generation and control for quadrotors, in *2011 IEEE International Conference on Robotics and Automation* (IEEE, 2011), pp. 2520–2525
8. S. Liu, M. Watterson, K. Mohta, K. Sun, S. Bhattacharya, C.J. Taylor, V. Kumar, Planning dynamically feasible trajectories for quadrotors using safe flight corridors in 3-D complex environments. IEEE Robot. Autom. Lett. **2**(3), 1688–1695 (2017)
9. G. Farin, D. Hansford, *The Essentials of CAGD* (CRC Press, 2000)
10. L. Piegl, W. Tiller, *The NURBS Book* (Springer Science & Business Media, 1996)
11. C. de Boor, *A Practical Guide to Splines* (Springer, 1978)
12. B. Zhou, F. Gao, L. Wang, C. Liu, S. Shen, Robust and efficient quadrotor trajectory generation for fast autonomous flight. IEEE Robot. Autom. Lett. **4**(4), 3529–3536 (2019)
13. L. Dubins, On curves of minimal length with a constraint on average curvature, and with prescribed initial and terminal positions and tangents. Am. J. Math. **79**, 497–516 (1957)
14. D. Jung, P. Tsiotras, On-line path generation for unmanned aerial vehicles using B-spline path templates. J. Guidance Control Dyn. **36**(6), 1642–1653 (2013)
15. J. Tordesillas, B.T. Lopez, M. Everett, J.P. How, Faster: fast and safe trajectory planner for navigation in unknown environments. IEEE Trans. Robot. **38**(2), 922–938 (2021)
16. W. Ding, W. Gao, K. Wang, S. Shen, An efficient B-spline-based kinodynamic replanning framework for quadrotors. IEEE Trans. Robot. **35**(6), 1287–1306 (2019)
17. M. Fliess, J.L. vine, P. Martin, P. Rouchon, Flatness and defect of non-linear systems: Introductory theory and examples. Int. J. Control **61**(6), 1327–1361 (1995)
18. T. Lee, Robust adaptive attitude tracking on SO(3) with an application to a quadrotor UAV. IEEE Trans. Control Syst. Technol. **21**(5), 1924–1930 (2012)

19. E. Tal, G. Ryou, S. Karaman, Aerobatic trajectory generation for a VTOL fixed-wing aircraft using differential flatness. IEEE Trans. Robot. **39**(6), 4805–4819 (2023)
20. O. Khatib, Real-time obstacle avoidance for manipulators and mobile robots. Int. J. Robot. Res. **5**(1), 90–98 (1986)
21. Y. Gao, C. Bai, R. Fu, Q. Quan, A non-potential orthogonal vector field method for more efficient robot navigation and control. Robot. Auton. Syst. **159**, 104291 (2023)

Chapter 9
Image-Based Visual Servo Docking Control Experiment

Runxiao Liu, Hanli Chen, Xunhua Dai, and Quan Quan

Image-Based Visual Servoing (IBVS) utilizes feedback information from two-dimensional images to control or adaptively adjust the robot's behavior. This method can be applied to achieve automatic aerial refueling docking and automatic recovery docking for micro-small fixed-wing Unmanned Aerial Vehicles (UAVs). This chapter aims to introduce IBVS control experiments, divided into four parts: a basic experiment, an analysis experiment, a design experiment, and a Hardware-In-the-Loop (HIL) experiment. In the basic experiment, readers will learn how to use IBVS to accomplish automatic aerial docking tasks. The analysis experiment will focus on the impact of key parameters in IBVS and compare the differences between IBVS and Position-Based Visual Servo (PBVS). Additionally, in the design experiment, readers will design an IBVS controller based on the upper-level control interface of fixed-wing UAVs to complete the task of automatic aerial refueling docking. Finally, a HIL experiment is conducted. For more information on this chapter, please refer to the link

https://rflysim.com/doc/en/C/5.FW_Book.html.

9.1 Preliminary

For cost reasons, some high-value UAVs need to be recovered by the mothership and reused at the end of their missions. The standard method of airborne recovery involves releasing a cable from the mothership with a drogue at its end. In Fig. 9.1, the UAV is then controlled to insert its probe into the drogue, completing the locking process. The cable is then retracted to a predetermined position, allowing the UAV to be mechanically recovered by the mothership's robotic arm, as shown in Fig. 9.2. The task of docking control for fixed-wing aircraft is similar to that of aerial refueling; however, both face significant challenges due to atmospheric disturbances and the

Q. Quan et al., *Micro-Small Fixed-Wing UAV Control*,
https://doi.org/10.1007/978-981-95-1954-5_9

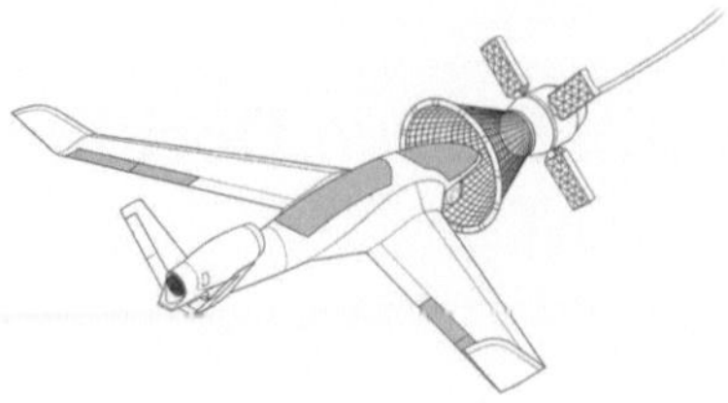

Fig. 9.1 Schematic diagram of aerial recovery device

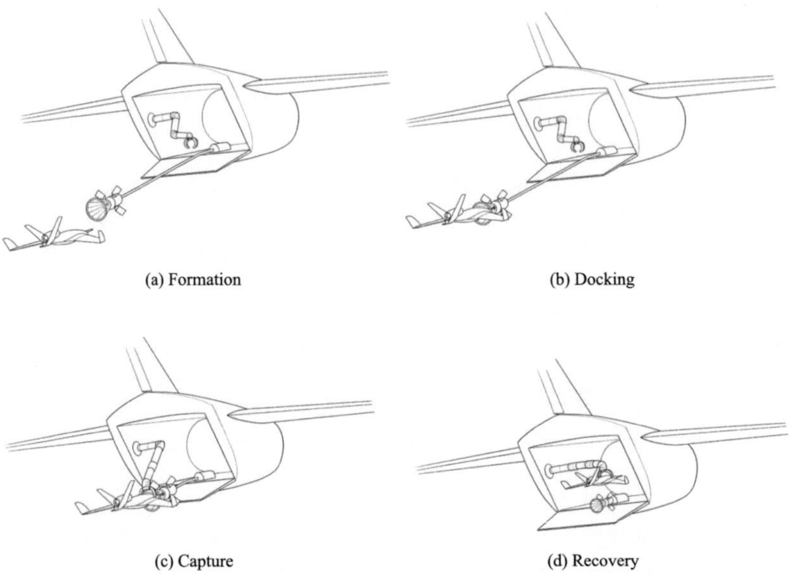

Fig. 9.2 Schematic diagram of recovery process

high precision requirements necessary for successful docking. This chapter introduces IBVS control technology for aerial refueling of fixed-wing UAVs, including target recognition, positioning, and docking control.

9.1.1 Fundamental Principles of Image-Based Visual Servo Control

The IBVS in this chapter utilizes camera data to control robot motion. The camera is mounted on the robot, causing it to move in tandem with the robot and produce changing images. The objective of IBVS control is to minimize the image tracking error, where the image tracking error $\mathbf{e}(t) \in \mathbb{R}^2$ is defined as:

$$\mathbf{e}(t) = \mathbf{s}(\mathbf{m}(t), \mathbf{a}) - \mathbf{s}^*. \tag{9.1}$$

Equation (9.1) represents a generalized form, where the vector $\mathbf{m}(t) \in \mathbb{R}^2$ represents the image measurement information (e.g., coordinates of target points in the image coordinate system or coordinates of the centroid of target objects). This image measurement information is used to compute visual features $\mathbf{s}(\mathbf{m}(t), \mathbf{a})$, where $\mathbf{a}$ represents parameters characterizing prior knowledge of the system (e.g., camera intrinsic matrix or 3D model of objects), and $\mathbf{s}^* \in \mathbb{R}^2$ represents the desired values of these visual features.

First, the relationship between visual feature $\mathbf{s}$ and camera motion velocity $\overline{\mathbf{v}}_c = [\mathbf{v}_c^T \ \omega_c^T]^T \in \mathbb{R}^6$ is given by

$$\dot{\mathbf{s}} = \mathbf{L_s}\overline{\mathbf{v}}_c \tag{9.2}$$

where $\mathbf{L_s} \in \mathbb{R}^{2\times6}$ is called the *feature Jacobian matrix* associated with $\mathbf{s}$, $\mathbf{v}_c$ is the instantaneous linear velocity of the camera coordinate system origin, and ω_c is the instantaneous angular velocity of the camera coordinate system.

According to Eqs. (9.1) and (9.2), the relationship between the error $\mathbf{e}$ and the camera motion velocity $\overline{\mathbf{v}}_c$ is given by:

$$\dot{\mathbf{e}} = \mathbf{L_e}\overline{\mathbf{v}}_c \tag{9.3}$$

where $\mathbf{L_e} = \mathbf{L_s}$. By using $\overline{\mathbf{v}}_c$ as the input to the robot's controller, if the objective is to make error $\mathbf{e}$ converge exponentially (i.e., $\dot{\mathbf{e}} = -\lambda\mathbf{e}$), one set:

$$\overline{\mathbf{v}}_c = -\lambda\mathbf{L_e^+}\mathbf{e} \tag{9.4}$$

where $\mathbf{L_e^+} = (\mathbf{L_e^T}\mathbf{L_e})^{-1}\mathbf{L_e^T}$.

In practical applications, it is necessary to derive or estimate $\mathbf{L_s}$ to complete the IBVS design. The derivation of $\mathbf{L_s}$ in classic IBVS problems is discussed next.

In Fig. 9.3, $o_c x_c y_c z_c$ represents the camera coordinate system, denoted as "c", and $o_i x_i y_i$ represents the image coordinate system, denoted as "i". The coordinates of the target point (referred to as "d" in this chapter, indicating the drogue) in the camera coordinate system are defined as $\mathbf{p}_d^c = [x_d^c \ y_d^c \ z_d^c]^T$, and it is projected onto the image as a two-dimensional point with coordinates $\mathbf{p}_d^i = [x_d^i \ y_d^i]^T$. The image tracking error $\mathbf{e}$ in the image is defined as

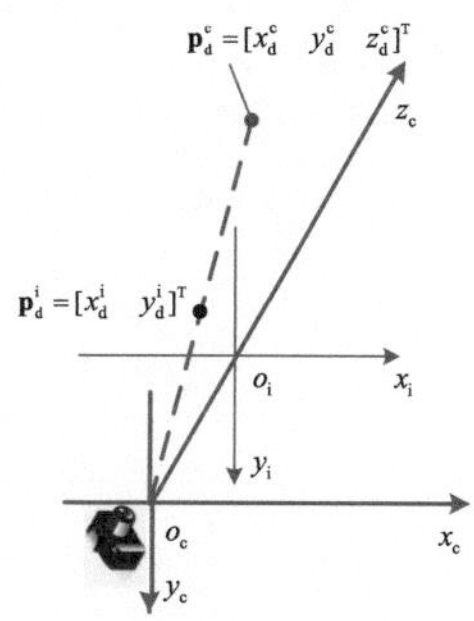

Fig. 9.3 Coordinate systems related to camera imaging

$$\mathbf{e} = \begin{bmatrix} e_x \\ e_y \end{bmatrix} = \begin{bmatrix} x_{\mathrm{d}}^{\mathrm{i}} - x_{\mathrm{o}}^{\mathrm{i}} \\ y_{\mathrm{d}}^{\mathrm{i}} - y_{\mathrm{o}}^{\mathrm{i}} \end{bmatrix} \tag{9.5}$$

where the point $\left(x_{\mathrm{o}}^{\mathrm{i}}, y_{\mathrm{o}}^{\mathrm{i}}\right)$ is a fixed point in the image, referred to as the *convergence point* of the image. Differentiating Eq. (9.5), one obtains:

$$\dot{\mathbf{e}} = \begin{bmatrix} \dot{e}_x \\ \dot{e}_y \end{bmatrix} = \begin{bmatrix} \dot{x}_{\mathrm{d}}^{\mathrm{i}} \\ \dot{y}_{\mathrm{d}}^{\mathrm{i}} \end{bmatrix}. \tag{9.6}$$

According to the pinhole imaging principle, one has:

$$\begin{cases} x_{\mathrm{d}}^{\mathrm{i}} = \dfrac{x_{\mathrm{d}}^{\mathrm{c}}}{z_{\mathrm{d}}^{\mathrm{c}}} \\ y_{\mathrm{d}}^{\mathrm{i}} = \dfrac{y_{\mathrm{d}}^{\mathrm{c}}}{z_{\mathrm{d}}^{\mathrm{c}}} \end{cases}. \tag{9.7}$$

Differentiating Eq. (9.7) yields:

$$\begin{cases} \dot{x}_{\mathrm{d}}^{\mathrm{i}} = \dfrac{\dot{x}_{\mathrm{d}}^{\mathrm{c}}}{z_{\mathrm{d}}^{\mathrm{c}}} - \dfrac{x_{\mathrm{d}}^{\mathrm{c}}\dot{z}_{\mathrm{d}}^{\mathrm{c}}}{z_{\mathrm{d}}^{\mathrm{c}\,2}} = \dfrac{\dot{x}_{\mathrm{d}}^{\mathrm{c}} - x_{\mathrm{d}}^{\mathrm{i}}\dot{z}_{\mathrm{d}}^{\mathrm{c}}}{z_{\mathrm{d}}^{\mathrm{c}}} \\ \dot{y}_{\mathrm{d}}^{\mathrm{i}} = \dfrac{\dot{y}_{\mathrm{d}}^{\mathrm{c}}}{z_{\mathrm{d}}^{\mathrm{c}}} - \dfrac{y_{\mathrm{d}}^{\mathrm{c}}\dot{z}_{\mathrm{d}}^{\mathrm{c}}}{z_{\mathrm{d}}^{\mathrm{c}\,2}} = \dfrac{\dot{y}_{\mathrm{d}}^{\mathrm{c}} - y_{\mathrm{d}}^{\mathrm{i}}\dot{z}_{\mathrm{d}}^{\mathrm{c}}}{z_{\mathrm{d}}^{\mathrm{c}}} \end{cases}. \tag{9.8}$$

The following relationship holds as

$$\dot{\mathbf{p}}_{\mathrm{d}}^{\mathrm{c}} = -\mathbf{v}_{\mathrm{c}}^{\mathrm{d}} - \omega_{\mathrm{c}}^{\mathrm{d}} \times \mathbf{p}_{\mathrm{d}}^{\mathrm{c}} \tag{9.9}$$

where

$$\begin{cases} \dot{x}_{\mathrm{d}}^{\mathrm{c}} = -v_{\mathrm{d},x}^{\mathrm{c}} - \omega_{\mathrm{d},y}^{\mathrm{c}} z_{\mathrm{d}}^{\mathrm{c}} + \omega_{\mathrm{d},z}^{\mathrm{c}} y_{\mathrm{d}}^{\mathrm{c}} \\ \dot{y}_{\mathrm{d}}^{\mathrm{c}} = -v_{\mathrm{d},y}^{\mathrm{c}} - \omega_{\mathrm{d},z}^{\mathrm{c}} x_{\mathrm{d}}^{\mathrm{c}} + \omega_{\mathrm{d},x}^{\mathrm{c}} z_{\mathrm{d}}^{\mathrm{c}} \\ \dot{z}_{\mathrm{d}}^{\mathrm{c}} = -v_{\mathrm{d},z}^{\mathrm{c}} - \omega_{\mathrm{d},x}^{\mathrm{c}} y_{\mathrm{d}}^{\mathrm{c}} + \omega_{\mathrm{d},y}^{\mathrm{c}} x_{\mathrm{d}}^{\mathrm{c}} \end{cases}. \tag{9.10}$$

Substituting Eqs. (9.10) into (9.8) and combining with Eq. (9.6) yield the relationship between $\dot{\mathbf{e}}$ and $\mathbf{v}_{\mathrm{d}}^{\mathrm{c}}$, $\omega_{\mathrm{d}}^{\mathrm{c}}$ as follows [1]:

$$\dot{\mathbf{e}} = \underbrace{\begin{bmatrix} -\frac{1}{z_{\mathrm{d}}^{\mathrm{c}}} & 0 & \frac{x_{\mathrm{d}}^{\mathrm{i}}}{z_{\mathrm{d}}^{\mathrm{c}}} & x_{\mathrm{d}}^{\mathrm{i}} y_{\mathrm{d}}^{\mathrm{i}} & -(1 + x_{\mathrm{d}}^{\mathrm{i}\,2}) & y_{\mathrm{d}}^{\mathrm{i}} \\ 0 & -\frac{1}{z_{\mathrm{d}}^{\mathrm{c}}} & \frac{y_{\mathrm{d}}^{\mathrm{i}}}{z_{\mathrm{d}}^{\mathrm{c}}} & 1 + y_{\mathrm{d}}^{\mathrm{i}\,2} & -x_{\mathrm{d}}^{\mathrm{i}} y_{\mathrm{d}}^{\mathrm{i}} & -x_{\mathrm{d}}^{\mathrm{i}} \end{bmatrix}}_{\mathbf{L_s}} \begin{bmatrix} v_{\mathrm{d},x}^{\mathrm{c}} \\ v_{\mathrm{d},y}^{\mathrm{c}} \\ v_{\mathrm{d},z}^{\mathrm{c}} \\ \omega_{\mathrm{d},x}^{\mathrm{c}} \\ \omega_{\mathrm{d},y}^{\mathrm{c}} \\ \omega_{\mathrm{d},z}^{\mathrm{c}} \end{bmatrix}. \tag{9.11}$$

9.1.2 Target Recognition and Localization

Target detection and localization refer to determining the location of the target. Currently, standard methods include recognition and localization based on landmark detection and neural network learning. Visual navigation based on landmark detection requires landmarks to be placed on the target in advance. In aerial refueling scenarios, LED lights or 850 nm narrow band infrared light sources can be used as landmarks. Combining optical filtering with algorithms can effectively extract landmarks.

It is assumed that the drogue LED is arranged in a circular shape, as shown in Fig. 9.4, and its projection in the image satisfies the expression

$$(x - a)^2 + (y - b)^2 = r^2 \tag{9.12}$$

where the coordinates (a, b) represent the center position of the drogue, and r is the radius of the drogue. It is assumed that N landmarks can be detected $p_i = (x_i, y_i) \in \mathscr{B}$, where $\mathscr{B}$ is the set of coordinates of all circular landmarks. Therefore, the corresponding residuals ε_i is expressed as

$$\varepsilon_i = (x_i - a)^2 + (y_i - b)^2 - r^2. \tag{9.13}$$

Taking the square of the residuals as the cost function to be fitted here yields

$$J = \sum_{i=1,\ldots,N} \varepsilon_i^2 = \sum_{i=1,\ldots,N} \left((x_i - a)^2 + (y_i - b)^2 - r^2\right)^2. \tag{9.14}$$

According to the principle of least squares, it can be inferred that when the fitting result is optimal, the partial derivatives of the cost function should be zero, namely

$$\frac{\partial J}{\partial a} = \frac{\partial J}{\partial b} = \frac{\partial J}{\partial r} = 0. \tag{9.15}$$

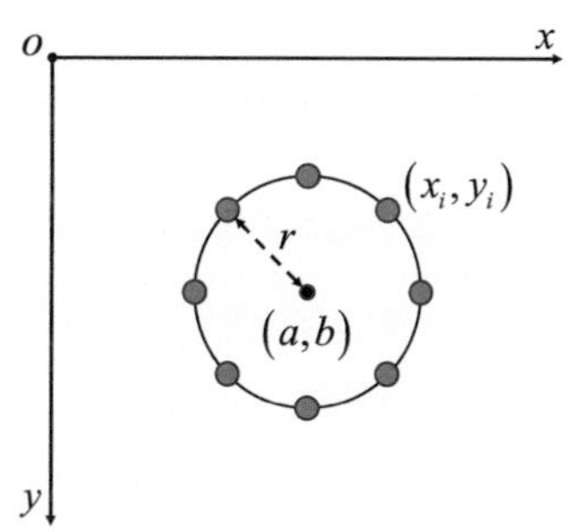

Fig. 9.4 Schematic diagram of drogue LED layout

After arranging the terms, one has

$$\begin{cases} (\overline{x}^2 - \overline{x^2})a + (\overline{x} \cdot \overline{y} - \overline{xy})b = \frac{1}{2}(\overline{x^2} \cdot \overline{x} + \overline{x} \cdot \overline{y^2} - \overline{x^3} - \overline{xy^2}) \\ (\overline{x} \cdot \overline{y} - \overline{xy})a + (\overline{y}^2 - \overline{y^2})b = \frac{1}{2}(\overline{x^2} \cdot \overline{y} + \overline{y} \cdot \overline{y^2} - \overline{x^2y} - \overline{y^3}) \end{cases} \tag{9.16}$$

where

$$\overline{x^m y^n} = \frac{\sum_{i=1}^{N} x_i^m y_i^n}{N} \qquad m, n \in 0, 1, 2. \tag{9.17}$$

Combining the above equations, the coordinates of the probe's center and its radius are obtained as follows

$$\begin{cases} a = \frac{(\overline{x^2}\cdot\overline{x}+\overline{x}\cdot\overline{y^2}-\overline{x^3}-\overline{xy^2})(\overline{y}^2-\overline{y^2})-(\overline{x^2}\cdot\overline{y}+\overline{y}\cdot\overline{y^2}-\overline{x^2y}-\overline{y^3})(\overline{x}\cdot\overline{y}-\overline{xy})}{2(\overline{x}^2-\overline{x^2})(\overline{y}^2-\overline{y^2})-2(\overline{x}\cdot\overline{y}-\overline{xy})^2} \\ b = \frac{(\overline{x^2}\cdot\overline{y}+\overline{y}\cdot\overline{y^2}-\overline{x^2y}-\overline{y^3})(\overline{x}^2-\overline{x^2})-(\overline{x^2}\cdot\overline{x}+\overline{x}\cdot\overline{y^2}-\overline{x^3}-\overline{xy^2})(\overline{x}\cdot\overline{y}-\overline{xy})}{2(\overline{x}^2-\overline{x^2})(\overline{y}^2-\overline{y^2})-2(\overline{x}\cdot\overline{y}-\overline{xy})^2} \\ r = \sqrt{a^2 - 2\overline{x}a + b^2 - 2\overline{y}b + \overline{x^2} + \overline{y^2}} \end{cases}. \tag{9.18}$$

Subsequently, based on the pinhole imaging principle and prior knowledge of the drogue, depth information can be calculated.

Recognition and positioning based on neural networks can utilize the YOLO neural network [2] for drogue recognition, eliminating the need to set markers. However, it requires a large amount of training data to be collected in advance, and it consumes a considerable amount of time and resources for training. It also requires high computational power for edge computing platforms when in use. Visual navigation based on the YOLO neural network can compute depth information based on the pinhole imaging principle and prior knowledge of the drogue, or it can directly integrate depth map information to obtain distance. The examples provided in this chapter all use the YOLO neural network for visual navigation.

9.1.3 Docking Control

Docking control addresses the problem of utilizing the information provided by visual navigation to achieve docking. The development of docking control algorithms requires model building and controller design. Figure 9.5 shows the control diagram for this chapter.

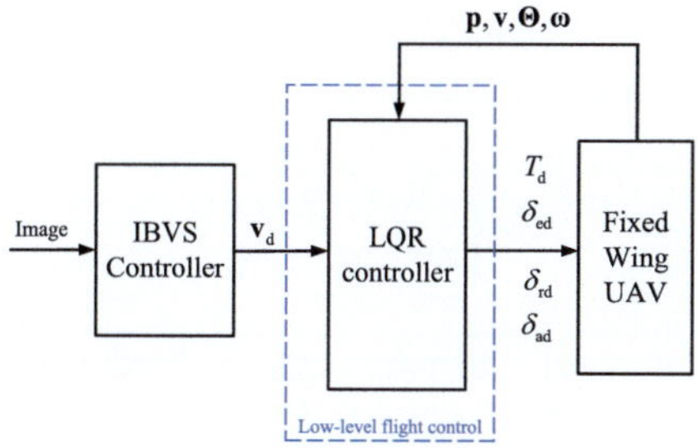

Fig. 9.5 Control diagram

9.1.3.1 Model

1. Coordinate System

When establishing the system model for the docking phase, it is necessary to define six coordinate systems: the ground coordinate system ($o_g x_g y_g z_g$), the receiver aircraft coordinate system ($o_r x_r y_r z_r$), the tanker aircraft coordinate system ($o_t x_t y_t z_t$), the drogue coordinate system ($o_d x_d y_d z_d$), the camera coordinate system ($o_c x_c y_c z_c$), and the relative coordinate system ($o_{re} x_{re} y_{re} z_{re}$), as shown in Fig. 9.6. Here, the origin of the relative coordinate system coincides with the origin of the tanker aircraft coordinate system, and its axes align with those of the camera coordinate system. During the docking phase, the relative position and attitude between the tanker aircraft and the receiver aircraft are mainly considered. The tanker aircraft coordinate system (a moving inertial coordinate system) is more convenient for modeling and analysis compared to the ground coordinate system (a fixed inertial coordinate system).

To further simplify the coordinate transformations among different coordinate systems, the direction of the axes in the ground coordinate system is also chosen to align with the horizontal movement direction of the tanker aircraft. Additionally, $\mathbf{R}_t^r$ represents the rotation matrix describing the angular relationship between the

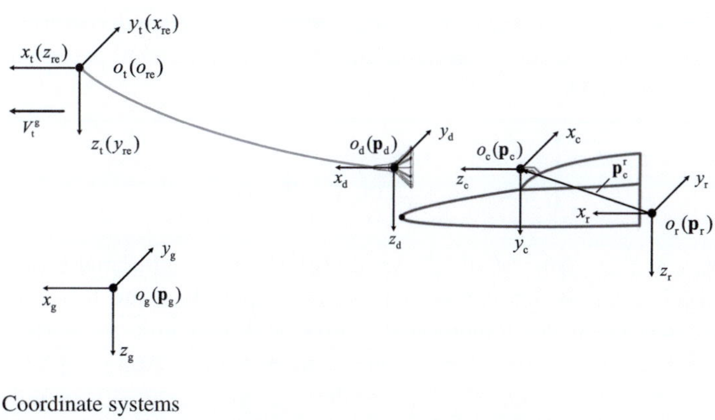

Fig. 9.6 Coordinate systems

receiver aircraft coordinate system 'r' and the tanker aircraft coordinate system 't'. The symbolic operation rules here are defined as:

$$\begin{cases} \mathbf{R}_j^i \mathbf{x}_k^j = \mathbf{x}_k^i \\ \mathbf{x}_k^j - \mathbf{x}_i^j = \mathbf{x}_k^i \end{cases}$$

where $i, j \in \{\mathrm{g, r, t, d, c, re}\}$ denotes the coordinate system set, and $k \in \{\mathrm{g, r, t, d, c, re}\}$ denotes the object set. For example, V_t^g represents the forward flight velocity of the tanker aircraft in the ground coordinate system, $\mathbf{v}_\mathrm{d}^\mathrm{c}$ represents the velocity of the drogue in the camera coordinate system. In contrast, $\mathbf{p}_\mathrm{c}^\mathrm{r}$ represents the vector difference from the receiver aircraft coordinate system to the camera coordinate system, namely $\mathbf{p}_\mathrm{c}^\mathrm{r} = \mathbf{p}_\mathrm{c}^\mathrm{t} - \mathbf{p}_\mathrm{r}^\mathrm{t}$ as illustrated in Fig. 9.6.

2. Receiver Model

During the docking phase, the pitch angle of the receiver changes very slightly. The reference state is defined as level flight at a constant speed. The attitude of the reference state satisfies the condition of coordinated turn, yielding roll angle and sideslip angle satisfying $\phi = \beta = 0$; angle of attack α, flight path angle γ, and pitch angle θ satisfy $\theta = \gamma + \alpha$. In this case, the motion equations of the receiver in the tanker's coordinate system can be decoupled into the lateral motion and the longitudinal motion.

According to Chap. 4, the motion state equation of the receiver is simplified as

$$\dot{\mathbf{x}}_\mathrm{r} = \mathbf{f}(\mathbf{x}_\mathrm{r}, \mathbf{u}_\mathrm{r}). \tag{9.19}$$

Here, $\mathbf{x}_\mathrm{r} = [\, x_\mathrm{r}\ y_\mathrm{r}\ h_\mathrm{r}\ \phi\ \theta\ \psi\ V_\mathrm{r}^\mathrm{g}\ \alpha\ \beta\ \omega_{x_\mathrm{b}}\ \omega_{y_\mathrm{b}}\ \omega_{z_\mathrm{b}}\,]^\mathrm{T}$ represents the state of the receiver, including position, Euler angles, aerodynamic angles, and angular rates. The vector $\mathbf{u}_\mathrm{r} = [T\ \delta_\mathrm{e}\ \delta_\mathrm{a}\ \delta_\mathrm{r}]^\mathrm{T}$ represents the control inputs, including throttle, elevator, aileron, and rudder. Similar to Chap. 4, one obtains

$$\Delta\dot{\mathbf{x}}_\mathrm{r} = \mathbf{A}\Delta\mathbf{x}_\mathrm{r} + \mathbf{B}\Delta\mathbf{u}_\mathrm{r}. \tag{9.20}$$

Here, $\Delta\mathbf{x}_\mathrm{r} = \mathbf{x}_\mathrm{r} - \mathbf{x}_\mathrm{r}^*$, $\mathbf{x}_\mathrm{r}^*$ is the trimmed state, and $\Delta\mathbf{x}_\mathrm{r}$ is the small perturbation state away from the trimmed state. Based on this, the decoupled small perturbation state equations are obtained:

$$\begin{cases} \Delta\dot{\mathbf{x}}_\mathrm{rlon} = \mathbf{A}_\mathrm{rlon}\Delta\mathbf{x}_\mathrm{rlon} + \mathbf{B}_\mathrm{rlon}\Delta\mathbf{u}_\mathrm{rlon} \\ \Delta\dot{\mathbf{x}}_\mathrm{rlat} = \mathbf{A}_\mathrm{rlat}\Delta\mathbf{x}_\mathrm{rlat} + \mathbf{B}_\mathrm{rlat}\Delta\mathbf{u}_\mathrm{rlat} \end{cases}. \tag{9.21}$$

Here, $\Delta\mathbf{x}_\mathrm{rlon} = [\Delta x_\mathrm{r}\ \Delta h_\mathrm{r}\ \Delta\theta\ \Delta V_\mathrm{r}^\mathrm{g}\ \Delta\alpha\ \Delta\omega_{y_\mathrm{b}}]^\mathrm{T}$ is the longitudinal channel state, $\Delta\mathbf{u}_\mathrm{rlon} = [\Delta T\ \Delta\delta_\mathrm{e}]^\mathrm{T}$ is the longitudinal channel input, $\mathbf{A}_\mathrm{rlon}$ and $\mathbf{B}_\mathrm{rlon}$ are the corresponding matrices of the longitudinal channel state equation. Similarly, $\Delta\mathbf{x}_\mathrm{rlat} = [\Delta y_\mathrm{r}\ \Delta\phi\ \Delta\psi\ \Delta\beta\ \Delta\omega_{x_\mathrm{b}}\ \Delta\omega_{z_\mathrm{b}}]^\mathrm{T}$ is the lateral channel state, $\Delta\mathbf{u}_\mathrm{rlat} = [\Delta\delta_\mathrm{a}\ \Delta\delta_\mathrm{r}]^\mathrm{T}$ is the lateral channel input, $\mathbf{A}_\mathrm{rlat}$ and $\mathbf{B}_\mathrm{rlat}$ are the corresponding matrices of the lateral channel state equation. Note that the equations and state variables here are not the

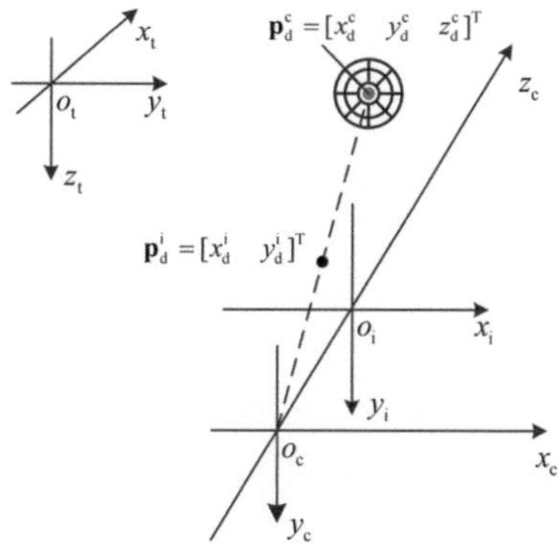

Fig. 9.7 Relationship among coordinate systems

same as those in Chap. 4. This is because the docking control based on visual servoing requires not only the state variables selected for trimming and linearization in Chap. 4, but also position-related state variables.

3. IBVS Model

In Fig. 9.7, $o_t x_t y_t z_t$ represents the tanker coordinate system, denoted as "t"; $o_c x_c y_c z_c$ represents the camera coordinate system, denoted as "c". The transformation matrix from "t" to "c" is denoted as $\mathbf{R}_t^c$; $o_i x_i y_i$ represents the image coordinate system, denoted as "i". Equation (9.11) yields

$$\dot{\mathbf{e}} = \underbrace{\begin{bmatrix} -\frac{1}{z_d^c} & 0 & \frac{x_d^i}{z_d^c} & x_d^i y_d^i & -(1+{x_d^i}^2) & y_d^i \\ 0 & -\frac{1}{z_d^c} & \frac{y_d^i}{z_d^c} & 1+{y_d^i}^2 & -x_d^i y_d^i & -x_d^i \end{bmatrix}}_{\mathbf{L}_s} \begin{bmatrix} v_{d,x}^c \\ v_{d,y}^c \\ v_{d,z}^c \\ \omega_{d,x}^c \\ \omega_{d,y}^c \\ \omega_{d,z}^c \end{bmatrix}, \tag{9.22}$$

where

$$\mathbf{v}_d^c = [v_{d,x}^c v_{d,y}^c v_{d,z}^c]^T = \mathbf{R}_t^{re}(\mathbf{v}_d^t - \mathbf{v}_c^t) = \mathbf{v}_d^{re} - \mathbf{v}_c^{re} \tag{9.23}$$

is the relative velocity, $\omega_d^c = [\ \omega_{d,x}^c \ \omega_{d,y}^c \ \omega_{d,z}^c]^T$ is the relative angular velocity, $\mathbf{R}_t^{re} = \mathbf{R}_t^c$ is the transformation matrix from the tanker coordinate system to the relative coordinate system, $\mathbf{v}_c^{re} = [\ v_{c,x}^{re} \ v_{c,y}^{re} \ v_{c,z}^{re}\ ]^T$ and $\mathbf{v}_d^{re}$ are the velocities of the camera and the drogue in the relative coordinate system. As shown in Fig. 9.8, z_d^c is the distance along $o_c z_c$ axis in the camera coordinate system from the camera to the center plane of the drogue (where the LED is mounted), also known as the *depth*. Equation (9.22) corresponds to the IBVS model for docking control, as stated in Eq. (9.3).

To simplify the calculation, by ignoring some unimportant variables and then decoupling Eq. (9.22) into lateral and longitudinal channels, the following equations are obtained.

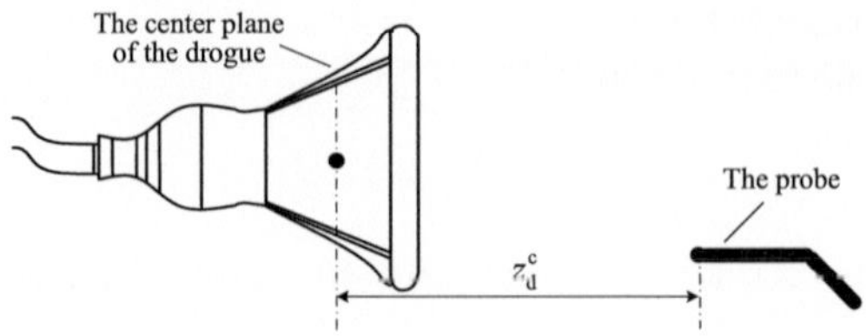

Fig. 9.8 Depth diagram

(1) In the $x_{\mathrm{c}} - z_{\mathrm{c}}$ plane, three degrees of freedom are $v_{\mathrm{d},x}^{\mathrm{c}}$, $v_{\mathrm{d},z}^{\mathrm{c}}$, $\omega_{\mathrm{d},y}^{\mathrm{c}}$, while $v_{\mathrm{d},y}^{\mathrm{c}} = 0$, $\omega_{\mathrm{d},x}^{\mathrm{c}} = 0$. Thus, one has

$$\dot{e}_x = -\frac{v_{\mathrm{d},x}^{\mathrm{c}}}{z_{\mathrm{d}}^{\mathrm{c}}} + \frac{e_x v_{\mathrm{d},z}^{\mathrm{c}}}{z_{\mathrm{d}}^{\mathrm{c}}} - (1 + e_x^2)\omega_{\mathrm{d},y}^{\mathrm{c}}. \tag{9.24}$$

(2) In the $y_{\mathrm{c}} - z_{\mathrm{c}}$ plane, three degrees of freedom are $v_{\mathrm{d},y}^{\mathrm{c}}$, $v_{\mathrm{d},z}^{\mathrm{c}}$, $\omega_{\mathrm{d},x}^{\mathrm{c}}$, while $v_{\mathrm{d},x}^{\mathrm{c}} = 0$, $\omega_{\mathrm{d},y}^{\mathrm{c}} = 0$. Thus, one has

$$\dot{e}_y = -\frac{v_{\mathrm{d},y}^{\mathrm{c}}}{z_{\mathrm{d}}^{\mathrm{c}}} + \frac{e_y v_{\mathrm{d},z}^{\mathrm{c}}}{z_{\mathrm{d}}^{\mathrm{c}}} - (1 + e_y^2)\omega_{\mathrm{d},x}^{\mathrm{c}}. \tag{9.25}$$

Suppose the tanker is treated as a rigid body. In that case, the linear velocity of the camera in the tanker coordinate system is equal to the vector sum of the linear velocity of the tanker center of mass and the velocity of rotation of the camera about the center of mass:

$$\mathbf{v}_{\mathrm{c}}^{\mathrm{t}} = \omega_{\mathrm{r}}^{\mathrm{t}} \times \mathbf{p}_{\mathrm{c}}^{\mathrm{r}} + \mathbf{v}_{\mathrm{r}}^{\mathrm{t}} \tag{9.26}$$

where $\mathbf{v}_{\mathrm{c}}^{\mathrm{t}} = [v_{\mathrm{c},x}^{\mathrm{t}}\ v_{\mathrm{c},y}^{\mathrm{t}}\ v_{\mathrm{c},z}^{\mathrm{t}}]^{\mathrm{T}}$ is the velocity of the camera, $\mathbf{v}_{\mathrm{r}}^{\mathrm{t}} = [v_{\mathrm{r},x}^{\mathrm{t}}\ v_{\mathrm{r},y}^{\mathrm{t}}\ v_{\mathrm{r},z}^{\mathrm{t}}]^{\mathrm{T}}$ is the linear velocity of the receiver, $\omega_{\mathrm{r}}^{\mathrm{t}} = [\omega_{\mathrm{r},x}^{\mathrm{t}}\ \omega_{\mathrm{r},y}^{\mathrm{t}}\ \omega_{\mathrm{r},z}^{\mathrm{t}}]^{\mathrm{T}}$ is the angular velocity of the receiver, and $\mathbf{p}_{\mathrm{c}}^{\mathrm{r}} = [x_{\mathrm{c}}^{\mathrm{r}}\ y_{\mathrm{c}}^{\mathrm{r}}\ z_{\mathrm{c}}^{\mathrm{r}}]^{\mathrm{T}}$ is the coordinates of the camera in the receiver coordinate system.

The transformation matrix from the receiver's coordinate system to the tanker's coordinate system is given by

$$\mathbf{R}_{\mathrm{r}}^{\mathrm{t}} = \mathbf{I}_3. \tag{9.27}$$

Therefore, Eq. (9.26) becomes

$$\begin{aligned} \mathbf{v}_{\mathrm{c}}^{\mathrm{t}} &= \omega_{\mathrm{r}}^{\mathrm{t}} \times \mathbf{p}_{\mathrm{c}}^{\mathrm{r}} + \mathbf{v}_{\mathrm{r}}^{\mathrm{t}} \\ &= \mathbf{R}_{\mathrm{r}}^{\mathrm{t}}\omega_{\mathrm{r}}^{\mathrm{r}} \times \mathbf{p}_{\mathrm{c}}^{\mathrm{r}} + \mathbf{R}_{\mathrm{r}}^{\mathrm{t}}\mathbf{v}_{\mathrm{r}}^{\mathrm{r}} \\ &= \omega_{\mathrm{r}}^{\mathrm{r}} \times \mathbf{p}_{\mathrm{c}}^{\mathrm{r}} + \mathbf{v}_{\mathrm{r}}^{\mathrm{r}} \end{aligned} \tag{9.28}$$

where $\omega_{\rm r}^{\rm r} = [\ \Delta\omega_{x_{\rm b}}\ \Delta\omega_{y_{\rm b}}\ \Delta\omega_{z_{\rm b}}\ ]^{\rm T}$ represent the angular velocity components of $\omega_{\rm r}^{\rm t}$ along the axes of the receiver's coordinate system, $\mathbf{v}_{\rm r}^{\rm r} = [\ \Delta u\ \Delta v\ \Delta w\ ]^{\rm T}$ represent the velocity components of $\mathbf{v}_{\rm r}^{\rm t}$ along the axes of the receiver's coordinate system.

Based on Eq. (9.28), the camera's linear velocity is expressed as

$$\begin{aligned}\mathbf{v}_{\rm c}^{\rm re} &= \mathbf{R}_{\rm t}^{\rm re}\mathbf{v}_{\rm c}^{\rm t} \\ &= \mathbf{R}_{\rm t}^{\rm re}(\boldsymbol{\omega}_{\rm r}^{\rm r} \times \mathbf{p}_{\rm c}^{\rm r} + \mathbf{v}_{\rm r}^{\rm r}).\end{aligned} \tag{9.29}$$

Since

$$\begin{cases}\Delta u = \Delta V_{\rm r}^{\rm g} \cos \Delta\alpha \cos \Delta\beta \\ \Delta v = \Delta V_{\rm r}^{\rm g} \sin \Delta\beta \\ \Delta w = \Delta V_{\rm r}^{\rm g} \sin \Delta\alpha \cos \Delta\beta\end{cases} \tag{9.30}$$

Equation (9.29) can be rewritten as:

$$\begin{cases}v_{{\rm c},x}^{\rm re} = \Delta V_{\rm r}^{\rm g} \sin \Delta\beta + x_{\rm c}^{\rm r}\Delta\omega_{z_{\rm b}} - z_{\rm c}^{\rm r}\Delta\omega_{x_{\rm b}} \\ v_{{\rm c},y}^{\rm re} = \Delta V_{\rm r}^{\rm g} \sin \Delta\alpha \cos \Delta\beta + y_{\rm c}^{\rm r}\Delta\omega_{x_{\rm b}} - x_{\rm c}^{\rm r}\Delta\omega_{y_{\rm b}} \\ v_{{\rm c},z}^{\rm re} = \Delta V_{\rm r}^{\rm g} \cos \Delta\alpha \cos \Delta\beta + z_{\rm c}^{\rm r}\Delta\omega_{y_{\rm b}} - y_{\rm c}^{\rm r}\Delta\omega_{z_{\rm b}}\end{cases}. \tag{9.31}$$

Ignoring higher-order terms, Eq. (9.31) simplifies to

$$\begin{cases}v_{{\rm c},x}^{\rm re} = x_{\rm c}^{\rm r}\Delta\omega_{z_{\rm b}} - z_{\rm c}^{\rm r}\Delta\omega_{x_{\rm b}} \\ v_{{\rm c},y}^{\rm re} = -x_{\rm c}^{\rm r}\Delta\omega_{y_{\rm b}} \\ v_{{\rm c},z}^{\rm re} = \Delta V_{\rm r}^{\rm g} + z_{\rm c}^{\rm r}\Delta\omega_{y_{\rm b}}\end{cases}. \tag{9.32}$$

Combining Eq. (9.21) yields the following two models.

(1) The longitudinal channel controlled model is

$$\begin{cases}\Delta\dot{\mathbf{x}}_{\rm rlon} = \mathbf{A}_{\rm rlon}\Delta\mathbf{x}_{\rm rlon} + \mathbf{B}_{\rm rlon}\Delta\mathbf{u}_{\rm rlon} \\ \mathbf{v}_{\rm rlon}^{\rm re} \triangleq \begin{bmatrix} v_{{\rm c},y}^{\rm re} \\ v_{{\rm c},z}^{\rm re}\end{bmatrix} = \underbrace{\begin{bmatrix} 0\ 0\ 0\ 0\ 0 & -x_{\rm c}^{\rm r} \\ 0\ 0\ 0\ 1\ 0 & z_{\rm c}^{\rm r}\end{bmatrix}}_{\mathbf{C}_{\rm rlon}}\Delta\mathbf{x}_{\rm rlon}\end{cases}. \tag{9.33}$$

(2) The lateral channel controlled model is

$$\begin{cases}\Delta\dot{\mathbf{x}}_{\rm rlat} = \mathbf{A}_{\rm rlat}\Delta\mathbf{x}_{\rm rlat} + \mathbf{B}_{\rm rlat}\Delta\mathbf{u}_{\rm rlat} \\ v_{\rm rlat}^{\rm re} \triangleq v_{{\rm c},x}^{\rm re} = \underbrace{\begin{bmatrix} 0\ 0\ 0\ 0 & -z_{\rm c}^{\rm r} & x_{\rm c}^{\rm r}\end{bmatrix}}_{\mathbf{C}_{\rm rlat}}\Delta\mathbf{x}_{\rm rlat}\end{cases}. \tag{9.34}$$

The control objective is: for the IBVS models (9.24) and (9.25) and the controlled models (9.33), (9.34), to design appropriate input signals $\Delta\mathbf{u}_{\rm rlon}$, $\Delta\mathbf{u}_{\rm rlat}$ such that as time t tends to infinity, the tracking error in the image $\mathbf{e}$ converges to 0 while the depth $z_{\mathrm{d}}^{\mathrm{c}}$ approaches to the desired value of 0.

9.1.3.2 Controller Design

The controller design consists of the outer loop controller design and the inner loop controller design. The outer loop controller design aims to develop the desired relative velocity $\mathbf{v}_{\mathrm{d}}^{\mathrm{c}}$ to converge the tracking error in the image $\mathbf{e}$ to 0 while the depth $z_{\mathrm{d}}^{\mathrm{c}}$ converges to the desired value of 0. On the other hand, the inner loop controller design aims to design the appropriate desired input signal $\Delta\mathbf{u}_{\mathrm{r}}$ to make $\mathbf{v}_{\mathrm{c}}^{\mathrm{re}}$ track $\mathbf{v}_{\mathrm{c,des}}^{\mathrm{re}}$ and reject complex disturbances.

1. Outer Loop Control

The outer loop controller design aims to design the desired relative velocity $\mathbf{v}_{\mathrm{d}}^{\mathrm{c}}$ to converge the tracking error in the image $\mathbf{e}$ to 0 while the depth $z_{\mathrm{d}}^{\mathrm{c}}$ converges to the desired value of 0. This includes the lateral control channel design and the longitudinal control channel design.

(1) Lateral Control Channel Design

Ignoring $\omega_{\mathrm{d},y}^{\mathrm{c}}$, the desired velocity of $v_{\mathrm{d},x}^{\mathrm{c}}$ is designed as

$$v_{\mathrm{d},x\mathrm{des}}^{\mathrm{c}} = k_1 e_x. \tag{9.35}$$

If $v_{\mathrm{d},x}^{\mathrm{c}} = v_{\mathrm{d},x\mathrm{des}}^{\mathrm{c}}$, then Eq. (9.24) becomes

$$\dot{e}_x = -\lambda_1 e_x \tag{9.36}$$

where $\lambda_1 = \frac{k_1 - v_{\mathrm{d},z}^{\mathrm{c}}}{z_{\mathrm{d}}^{\mathrm{c}}}$, k_1 is an adjustable parameter satisfying $k_1 > \max(v_{\mathrm{d},z}^{\mathrm{c}})$, and $\lim_{t\to\infty} |e_x(t)| = 0$ [3].

(2) Longitudinal Control Channel Design

Ignoring $\omega_{\mathrm{d},x}^{\mathrm{c}}$, the desired velocity of $v_{\mathrm{d},y}^{\mathrm{c}}$ is designed as

$$v_{\mathrm{d},y\mathrm{des}}^{\mathrm{c}} = k_2 e_y. \tag{9.37}$$

If $v_{\mathrm{d},y}^{\mathrm{c}} = v_{\mathrm{d},y\mathrm{des}}^{\mathrm{c}}$, then Eq. (9.25) becomes

$$\dot{e}_y = -\lambda_2 e_y \tag{9.38}$$

where $\lambda_2 = \frac{k_2 - v_{\mathrm{d},z}^{\mathrm{c}}}{z_{\mathrm{d}}^{\mathrm{c}}}$, k_2 is an adjustable parameter satisfying $k_2 > \max(v_{\mathrm{d},z}^{\mathrm{c}})$, and $\lim_{t\to\infty} \left|e_y(t)\right| = 0$ [3]. The desired velocity design for the relative speed along the depth direction $v_{\mathrm{d},z}^{\mathrm{c}}$ is

$$v_{\mathrm{d},z\mathrm{des}}^{\mathrm{c}} = -k_3 z_{\mathrm{d}}^{\mathrm{c}}. \tag{9.39}$$

In this case, if $v^{\mathrm{c}}_{\mathrm{d},z} = v^{\mathrm{c}}_{\mathrm{d},z\mathrm{des}}$ and $k_3 > 0$, then $\lim_{t\to\infty} \left|z^{\mathrm{c}}_{\mathrm{d}}(t)\right| = 0$. Additionally, to prevent overshooting due to the rapid speed of the tanker, an adjustment term "$-k_4 \left|e_x\right| - k_5 \left|e_y\right|$" is added in Eq. (9.39) to regulate the speed of the tanker [3]:

$$v^{\mathrm{c}}_{\mathrm{d},z\mathrm{des}} = -k_3 z^{\mathrm{c}}_{\mathrm{d}} - k_4 \left|e_x\right| - k_5 \left|e_y\right|. \tag{9.40}$$

If $\left|e_x\right|$, $\left|e_y\right|$ are large, this term can slow down the speed to avoid overshooting.

In summary, in the outer loop controller, the desired velocity $\mathbf{v}^{\mathrm{c}}_{\mathrm{d}}$ is

$$\mathbf{v}^{\mathrm{c}}_{\mathrm{d,des}} = \left[v^{\mathrm{c}}_{\mathrm{d},x\mathrm{des}}\ v^{\mathrm{c}}_{\mathrm{d},y\mathrm{des}}\ v^{\mathrm{c}}_{\mathrm{d},z\mathrm{des}}\right]^{\mathrm{T}}. \tag{9.41}$$

Actually, one can only control $\mathbf{v}^{\mathrm{re}}_{\mathrm{c}}$. The relationship between $\mathbf{v}^{\mathrm{re}}_{\mathrm{c}}$ and $\mathbf{v}^{\mathrm{re}}_{\mathrm{d}}$ is $\mathbf{v}^{\mathrm{c}}_{\mathrm{d}} = \mathbf{v}^{\mathrm{re}}_{\mathrm{d}} - \mathbf{v}^{\mathrm{re}}_{\mathrm{c}}$, where $\mathbf{v}^{\mathrm{re}}_{\mathrm{d}}$ is considered as disturbance. Thus $\mathbf{v}^{\mathrm{re}}_{\mathrm{c,des}} = -\mathbf{v}^{\mathrm{c}}_{\mathrm{d,des}} = \mathbf{v}^{\mathrm{d}}_{\mathrm{c,des}}$ is obtained. Then, decoupling it results in

$$\mathbf{v}^{\mathrm{re}}_{\mathrm{c,des}} = \begin{bmatrix} \mathbf{v}^{\mathrm{re}}_{\mathrm{rlon,des}} \\ v^{\mathrm{re}}_{\mathrm{rlat,des}} \end{bmatrix} \tag{9.42}$$

where $\mathbf{v}^{\mathrm{re}}_{\mathrm{rlon,des}} = [v^{\mathrm{d}}_{\mathrm{c},y\mathrm{des}}\ v^{\mathrm{c}}_{\mathrm{d},z\mathrm{des}}\ ]^{\mathrm{T}}$, $v^{\mathrm{re}}_{\mathrm{rlat,des}} = v^{\mathrm{d}}_{\mathrm{c},x\mathrm{des}}$.

2. Inner Loop Control

The design of the inner loop controller aims to design appropriate desired input signals $\Delta\mathbf{u}_{\mathrm{r}}$ to make $\mathbf{v}^{\mathrm{re}}_{\mathrm{c}}$ track $\mathbf{v}^{\mathrm{re}}_{\mathrm{c,des}}$ and reject complex disturbances. This includes the design of the lateral control channel and the longitudinal control channel. The inner loop controllers employ optimal control methods, such as the Linear Quadratic Regulator (LQR).

LQR Fundamentals.

Consider a linear time-invariant system:

$$\dot{\mathbf{x}} = \mathbf{A}\mathbf{x} + \mathbf{B}\mathbf{u} \tag{9.43}$$

where $\mathbf{x} \in \mathbb{R}^n$ is the n-dimensional state and $\mathbf{u} \in \mathbb{R}^m$ is the m-dimensional input. This system is controllable or at least stabilizable. The performance cost is defined as:

$$J = \frac{1}{2}\int_0^{\infty} \left(\mathbf{x}(t)^{\mathrm{T}}\mathbf{Q}\mathbf{x}(t) + \mathbf{u}(t)^{\mathrm{T}}\mathbf{R}\mathbf{u}(t)\right)\mathrm{d}t \tag{9.44}$$

where $\mathbf{u}$ is unconstrained, and $\mathbf{Q}$, $\mathbf{R}$ are generally constant symmetric positive definite matrices (sometimes relaxed to symmetric semi-positive definite matrices). The optimal control that minimizes J is given by:

$$\mathbf{u} = -\mathbf{R}^{-1}\mathbf{B}^{\mathrm{T}}\mathbf{P}\mathbf{x} \tag{9.45}$$

where $\mathbf{P} \in \mathbb{R}^{n\times n}$ is a positive definite symmetric matrix satisfying the Riccati equation:

$$-\mathbf{PA} - \mathbf{A}^{\mathrm{T}}\mathbf{P} + \mathbf{PBR}^{-1}\mathbf{B}^{\mathrm{T}}\mathbf{P} - \mathbf{Q} = 0. \tag{9.46}$$

This equation can be conveniently solved using the "lqr" function in MATLAB.

(1) Lateral Control Channel Design

The tracking error is defined as $\mathbf{e}_{\mathrm{rlon}} = \mathbf{v}^{\mathrm{re}}_{\mathrm{rlon}} - \mathbf{v}^{\mathrm{re}}_{\mathrm{rlon,des}}$. Here, an integrator is used to reject complex disturbances, defined as:

$$\mathbf{q}_{\mathrm{rlon}} = \int_0^t \mathbf{e}_{\mathrm{rlon}}(\tau)\mathrm{d}\tau = \int_0^t (\mathbf{v}^{\mathrm{re}}_{\mathrm{rlon}}(\tau) - \mathbf{v}^{\mathrm{re}}_{\mathrm{rlon,des}}(\tau))\mathrm{d}\tau. \tag{9.47}$$

Substituting Eqs. (9.47) into (9.24) yields:

$$\begin{bmatrix} \Delta\dot{\mathbf{x}}_{\mathrm{rlon}} \\ \dot{\mathbf{q}}_{\mathrm{rlon}} \end{bmatrix} = \begin{bmatrix} \mathbf{A}_{\mathrm{rlon}} & \mathbf{0}_{6\times 2} \\ \mathbf{C}_{\mathrm{rlon}} & \mathbf{0}_{2\times 2} \end{bmatrix} \begin{bmatrix} \Delta\mathbf{x}_{\mathrm{rlon}} \\ \mathbf{q}_{\mathrm{rlon}} \end{bmatrix} + \begin{bmatrix} \mathbf{B}_{\mathrm{rlon}} \\ \mathbf{0}_{2\times 2} \end{bmatrix} \Delta\mathbf{u}_{\mathrm{rlon}} - \begin{bmatrix} \mathbf{0}_{6\times 2} \\ \mathbf{I}_{2\times 2} \end{bmatrix} \mathbf{v}^{\mathrm{re}}_{\mathrm{rlon,des}}. \tag{9.48}$$

The controller is designed as:

$$\Delta\mathbf{u}_{\mathrm{rlon}} = -\mathbf{K}_{x1}\Delta\mathbf{x}_{\mathrm{rlon}} - \mathbf{K}_{e1}\mathbf{q}_{\mathrm{rlon}} \tag{9.49}$$

where $\mathbf{K}_{x1} \in \mathbb{R}^{2\times 6}$, $\mathbf{K}_{e1} \in \mathbb{R}^{2\times 2}$. To compute $\mathbf{K}_{x1}$ and $\mathbf{K}_{e1}$, the following cost function is designed:

$$J(\Delta\mathbf{u}_{\mathrm{rlon}}) = \underset{\mathbf{K}_{x1},\mathbf{K}_{e1}}{\operatorname{argmin}} \int_0^{\infty} \left(\left[\Delta\mathbf{x}_{\mathrm{rlon}}(t)^{\mathrm{T}} \ \mathbf{q}_{\mathrm{rlon}}(t)^{\mathrm{T}}\right] \mathbf{Q}_{\mathrm{rlon}} \begin{bmatrix} \Delta\mathbf{x}_{\mathrm{rlon}}(t) \\ \mathbf{q}_{\mathrm{rlon}}(t) \end{bmatrix} + \Delta\mathbf{u}_{\mathrm{rlon}}(t)^{\mathrm{T}}\mathbf{R}_{\mathrm{rlon}}\Delta\mathbf{u}_{\mathrm{rlon}}(t) \right) \mathrm{d}t \tag{9.50}$$

where $\mathbf{Q}_{\mathrm{rlon}} \geq 0$ and $\mathbf{R}_{\mathrm{rlon}} > 0$ are weighting matrices.

(2) Longitudinal Channel Controller Design

Similarly, define the tracking error as $e_{\mathrm{rlat}} = v^{\mathrm{re}}_{\mathrm{rlat}} - v^{\mathrm{re}}_{\mathrm{rlat,des}}$, and the error integration is

$$q_{\mathrm{rlat}} = \int_0^t e_{\mathrm{rlat}}(\tau)\mathrm{d}\tau = \int_0^t (v^{\mathrm{re}}_{\mathrm{rlat}}(\tau) - v^{\mathrm{re}}_{\mathrm{rlat,des}}(\tau))\mathrm{d}\tau. \tag{9.51}$$

Substituting Eqs. (9.51) into (9.25) yields:

$$\begin{bmatrix} \Delta\dot{\mathbf{x}}_{\mathrm{rlat}} \\ \dot{q}_{\mathrm{rlat}} \end{bmatrix} = \begin{bmatrix} \mathbf{A}_{\mathrm{rlat}} & \mathbf{0}_{6\times 2} \\ \mathbf{C}_{\mathrm{rlat}} & \mathbf{0}_{2\times 2} \end{bmatrix} \begin{bmatrix} \Delta\mathbf{x}_{\mathrm{rlat}} \\ q_{\mathrm{rlat}} \end{bmatrix} + \begin{bmatrix} \mathbf{B}_{\mathrm{rlat}} \\ \mathbf{0}_{2\times 2} \end{bmatrix} \Delta\mathbf{u}_{\mathrm{rlat}} - \begin{bmatrix} \mathbf{0}_{6\times 2} \\ \mathbf{I}_{2\times 2} \end{bmatrix} v^{\mathrm{re}}_{\mathrm{rlat,des}}. \tag{9.52}$$

The controller is designed as

$$\Delta \mathbf{u}_{\text{rlat}} = -\mathbf{K}_{x2} \Delta \mathbf{x}_{\text{rlat}} - \mathbf{K}_{e2} q_{\text{rlat}} \tag{9.53}$$

where $\mathbf{K}_{x2} \in \mathbb{R}^{2\times 6}$, $\mathbf{K}_{e2} \in \mathbb{R}^{2}$. The cost function is defined as

$$J(\Delta \mathbf{u}_{\text{rlat}}) = \underset{\mathbf{K}_{x2},\mathbf{K}_{e2}}{\operatorname{argmin}} \int_0^{\infty} \left(\left[\Delta \mathbf{x}_{\text{rlat}}(t)^{\text{T}} \; q_{\text{rlat}}(t)^{\text{T}} \right] \mathbf{Q}_{\text{rlat}} \begin{bmatrix} \Delta \mathbf{x}_{\text{rlat}}(t) \\ q_{\text{rlat}}(t) \end{bmatrix} + \Delta \mathbf{u}_{\text{rlat}}(t)^{\text{T}} \mathbf{R}_{\text{rlat}} \Delta \mathbf{u}_{\text{rlat}}(t) \right) \text{d}t \tag{9.54}$$

where $\mathbf{Q}_{\text{rlat}} \geq 0$ and $\mathbf{R}_{\text{rlat}} > 0$ are weighting matrices.

3. Controller Synthesis
In summary, the IBVS controller is synthesized as

$$\begin{cases} v_{\text{d},x\text{des}}^{\text{c}} = k_1 e_x \\ v_{\text{d},y\text{des}}^{\text{c}} = k_2 e_y \\ v_{\text{d},z\text{des}}^{\text{c}} = -k_3 z_{\text{d}}^{\text{c}} - k_4 \left| e_x \right| - k_5 \left| e_y \right| \\ \Delta \mathbf{u}_{\text{rlon}} = -\mathbf{K}_{x1} \Delta \mathbf{x}_{\text{rlon}} - \mathbf{K}_{e1} \mathbf{q}_{\text{rlon}} \\ \Delta \mathbf{u}_{\text{rlat}} = -\mathbf{K}_{x2} \Delta \mathbf{x}_{\text{rlat}} - \mathbf{K}_{e2} q_{\text{rlat}} \end{cases} \tag{9.55}$$

where $k_i > 0$, $i = 1, 2, ..., 5$ are control gains.

The structure diagram of the IBVS control is shown in Fig. 9.9.

9.1.4 Related Development

In recent years, IBVS has gained significant attention in the field of UAVs. UAVs use onboard cameras for environmental perception and achieve precise motion control based on image information. This enables them to obtain richer environmental information and higher control accuracy when performing complex tasks. IBVS relies primarily on cameras for visual feedback, reducing reliance on Global Positioning Systems (GPS) and other external navigation aids. This increases environmental adaptability and system reliability.

Visual servoing technology has a wide range of applications for multirotor UAVs, including precision landing, object tracking, airborne target interception, and autonomous obstacle avoidance. UAVs can use visual servoing to achieve safe and precise autonomous landing of the moving platform by identifying specific markers or features in landing areas and automatically adjusting attitude and speed [4, 5]. In recent years, more sophisticated applications have emerged that utilize multirotor UAVs to capture and track moving targets using onboard cameras [6, 7]. In addition, UAVs can autonomously intercept targets in the air [8–10]. In addition, UAVs can detect obstacles using their cameras and adjust their flight paths accordingly to avoid collisions [11].

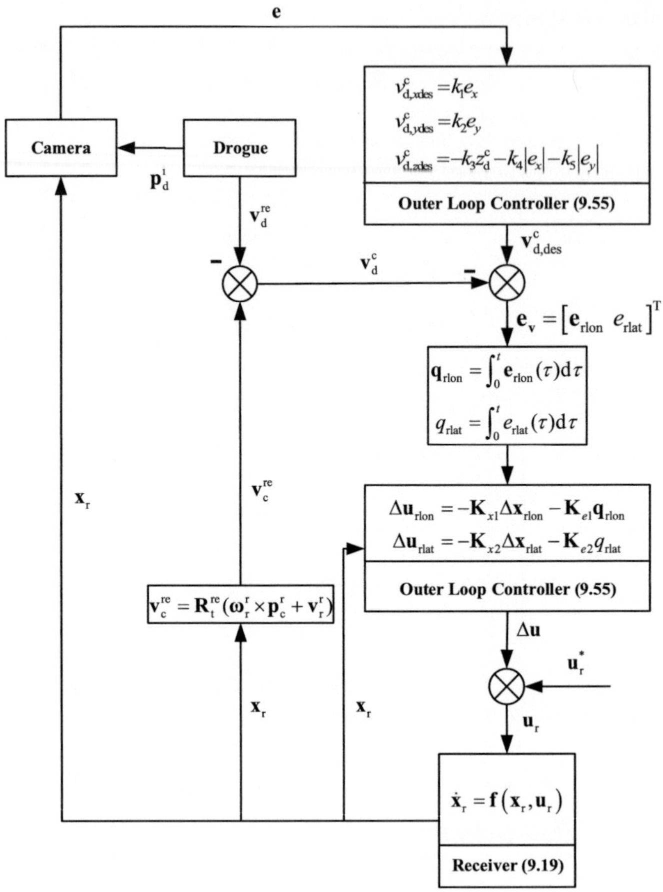

Fig. 9.9 IBVS control diagram

For fixed-wing UAVs, IBVS also plays an important role in areas such as autonomous interception and aerial docking. In the military domain, the autonomous interception capability of fixed-wing UAVs has become a hot research topic in recent years. This capability enables UAVs to identify and intercept enemy targets independently, reducing direct involvement and risks to personnel on the battlefield. It represents an important direction for future UAV technology research. In aerial docking tasks, fixed-wing UAVs capture images of docking targets through their onboard cameras, and the control system calculates guidance directly based in the image information, enabling precise flight adjustments [3, 12].

The difficulty of docking control, whether for UAV recovery or aerial refueling, is compounded by the presence of various atmospheric perturbations. Consequently, automatic docking control using image servoing has emerged as a prominent area of research in recent years. This is mainly attributable to the fact that the use of image-based servoing for aerial refueling docking offers several notable advantages, particularly in automated or semi-automated operational contexts. Image servoing enables the attainment of high-precision visual tracking, which is crucial for executing intricate docking operations. The use of image servo control reduces the reliance on external sensors, particularly in environments where GPS signals are limited or unavailable. Furthermore, it enhances the system's resilience in the context of adversarial electronic warfare. IBVS streamlines the design and implementation of the system, facilitating a reduction in the quantity and complexity of hardware necessary on the vehicle, thereby decreasing weight and cost [13]. In conclusion, IBVS provides a precise, adaptable, and resilient solution for aerial refueling docking, particularly in intricate, dynamic, and high-fidelity flight environments that require a high degree of autonomy.

9.2 Basic Experiment

9.2.1 Experimental Objective

1. Things to prepare

(1) Software: MATLAB2022b or above, Python, RflySimAAR.[1]
(2) Program: Experimental package "e7/e7-1", including "Simulink" and "Python" folders, where "Simulink" folder mainly contains the aerial refueling docking simulation "AAR_IBVS.slx" based on IBVS, the "F16_Init.m" initialization file, and other functions called in the platform. "Python" folder mainly contains the drogue recognition script "detect-realtime.py" and its called functions, etc.

2. Objective
Understand and master the principle of the aerial refueling docking simulation platform based on IBVS, familiarize with the functions of each module, write IBVS code, run the simulation to complete the aerial refueling docking experiment, and analyze the experimental results.

9.2.2 Experimental Procedure

The experimental procedure is divided into three steps: first, understand the composition of the IBVS aerial refueling docking simulation platform in Step 1; then, implement the IBVS control algorithm in Step 2; finally, run the simulation and analyze the results of the aerial refueling docking experiment in Step 3.

---

[1] https://github.com/RflySim/RflySimAAR/tree/master.

Step 1. Understand the composition of the IBVS aerial refueling docking simulation platform

As shown in Fig. 9.10, the aerial refueling docking simulation platform is mainly composed of five parts: "Tanker System", "Receiver System","Drogue&Probe System", "Environment", and "Scene".

① In the "Tanker System" module, the tanker's acceleration is set to zero to maintain a constant speed level flight, which is achieved through a PI controller, as shown in Fig. 9.11. The output of this module is "Tanker States", representing the state of the tanker.

② The input to the "Receiver System" module includes: "ImageInput" indicating image information input, including the estimated position of the drogue, tracking error in the image, and depth; "WindTurb_earth", representing wind disturbance in the ground coordinate system; "Tanker States", representing the state of the tanker. The outputs are: "Pos_pr2tx", representing the position of the drogue relative to the tanker; "recAttitude", representing the attitude of the receiver. This module comprises three main components: the receiver model, the controller, and the coordinate transformation module. The receiver model is similar to the Simulink model of a small UAV discussed in Chap. 4, a six-degree-of-freedom full-state model, and based on it, trimming and some data conversions are performed to facilitate docking control design. The controller module uses the results of visual navigation to implement docking control based on IBVS, with the IBVS code implemented in this part. The coordinate transformation module completes the calculation of coordinate conversions among different coordinate systems.

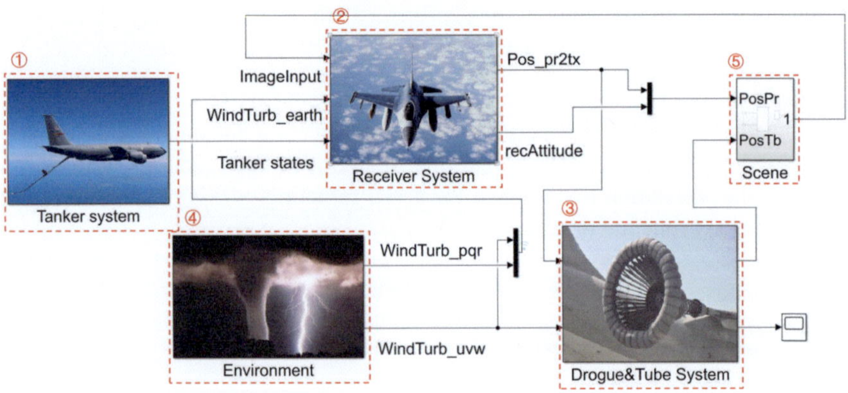

Fig. 9.10 Air refueling docking simulation platform

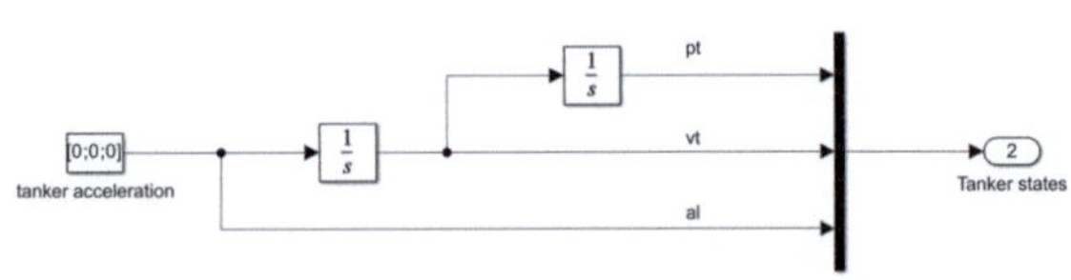

Fig. 9.11 Simulink implementation of the tanker state

③ "Drogue&Probe System" module calculates the pose of the docking drogue, with inputs: the position of the probe relative to the tanker "Pos_pr2tx" and the wind linear velocity "WindTurb_uvw". The output is the pose of each hose and drogue "PosTb".

④ "Environment" module calculates atmospheric data such as wind speed based on airspeed and altitude, with outputs being the wind disturbance linear velocity "WindTurb_uvw" and the wind disturbance angular velocity "WindTurb_pqr".

⑤ "Scene" module completes the scene display of the docking, with inputs being the pose of each hose and drogue "PosTb" and the pose of the receiver "PosPr". The implementation principle of this module is to communicate with RflySim and Python terminal through UDP broadcast, sending the pose information of each hose, drogue, receiver, and tanker to RflySim3D for display, and receiving the visual processing results returned from Python yields the estimated position of the drogue, tracking error in the image, and depth, as shown in Fig. 9.12. The scene display performance is shown in Fig. 9.13.

Step 2. Implementing IBVS Control Algorithm

Building upon the aerial refueling docking simulation platform, with reference to the experimental principles set forth in Eq. (9.55), the IBVS control algorithm is implemented. Under the path "AAR_IBVS\Receiver System\Controller\IBVS Controller", a MATLAB Function module for the outer loop controller has been reserved. This module takes the tracking error in the image, depth estimation results, and the state of the tanker as input, and produces the desired velocity of the camera in the relative coordinate system. Corresponding to Eq. (9.55), the key code is as follows in Code 9.1. The first line reverses the tracking error in the image along the y-axis; the second line sets the compensation value for depth error to control the depth of insertion into the drogue; lines 3–6 set control parameters; lines 7–9 compute the desired camera velocity relative to the drogue.

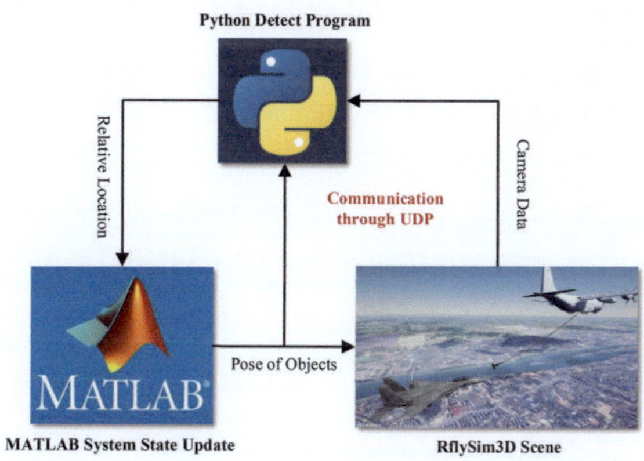

Fig. 9.12 Information interaction structure diagram

Fig. 9.13 Docking scene

Code 9.1 IBVS controller, from "e7/e7-1/Simulink/AAR_IBVS.slx"

```
ey = -ey; % reverse ey
Z_d = 1.0; % compensation Z
kz = 0.5; % set control parameters
ky = 3;
kx = 1;
k1 = 3; k2 = 1;
u_vcz = kz*(Z-Z_d)-k1*(abs(ex))-k2*(abs(ey)); % calculate
    desired velocity of camera
u_vcy = ky*(ey-0);
u_vcx = kx*(ex-0);
```

The inner loop controller is implemented using LQR control. Taking the longitudinal channel as an example, the key code is as shown in Code 9.2. Lines 1–4 set $\mathbf{A}_{\text{rlon}}, \mathbf{B}_{\text{rlon}}, \mathbf{C}_{\text{rlon}}$ as in Eq. (9.48). Lines 6–7 augment the system by introducing the integral of the tracking error, corresponding to Eq. (9.48). Lines 9–10 set $\mathbf{Q}$ matrix and $\mathbf{R}$ matrix for LQR. Line 12 solves the LQR problem using the "lqr" function in MATLAB. Lines 13–14 split the feedback gain matrix of the augmented system into $\mathbf{K}_{x2}, \mathbf{K}_{e2}$ as in Eq. (9.53).

Code 9.2 LQR controller, from "e7/e7-1/Simulink/F16_Init.m"

```
1 A_lon=A([1,3,5,7,8,11,13,14],[1,3,5,7,8,11,13,14]); % set
      State equation matrix
2 B_lon=B([1,3,5,7,8,11,13,14],1:2);
3 C_lon_xd_h=[1 0 zeros(1,6);
4             0 1 zeros(1,6)];
5 % augment the system
6 AA_lon=[A_lon,zeros(8,2);-C_lon_xd_h,zeros(2,2)];
7 BB_lon=[B_lon;zeros(2,2)];
8 % set lqr parameters
9 Q_lon=diag([4 10 10 1 10 10 0 0 1 3]) %
```

```
    R_lon=diag([50 50]);
    % calculate K of lqr
    [k_lon,p_lon,e_lon]=lqr(AA_lon,BB_lon,Q_lon,R_lon);
    Kp_lon=k_lon(:,1:8);
    Ki_lon=-k_lon(:,9:10);
```

After computing "Ki" and "Kp", the IBVS controller is established in Simulink, as shown in Fig. 9.14.

Step 3. Running Simulation and Analyzing Air Refueling Docking Experiment Results

Once the IBVS algorithm has been fully realized, as previously described, it is then possible to proceed with the execution of the Simulink simulation. It is essential to note that the RflySim3D application must be opened before launching the Simulink file. Once this is done, the file should be paused within a five-second timeframe. Subsequently, execute the "detect-realtime.py" script and await the appearance of a confirmation window before continuing with the execution of the Simulink file. If Python returns a file path error, it may be necessary to copy the platform to the C drive and then proceed with its execution. Upon completion of the simulation, the "AnalysisVisualError.mlx" script can be executed to compute the docking error during the docking process and plot the results, resulting in the following docking experiment results.

Figure 9.15 illustrates the visualization of the docking process, indicating that the IBVS algorithm has successfully accomplished the docking task. In Fig. 9.16, "**e**" represents the tracking error in the image plane during the docking process, indicating the difference between the center point of the drogue and the origin of the image coordinate system. It can be observed from the curve that the tracking error in the image converges to zero within 100 s.

"$z_{\mathrm{d}}^{\mathrm{c}}$" in Fig. 9.16 shows the depth error response during the docking process, representing the distance between the probe tip and the drogue. From the curve, it can be observed that the smaller the depth error, the slower the convergence rate of the depth error, which is consistent with the design of the corresponding outer loop controller (Eq. (9.41)).

"$\mathbf{p}_{\mathrm{d}}^{\mathrm{p}}$" in Fig. 9.16 shows the 3D docking error in the relative coordinate system during the docking process, namely the position difference between the probe tip

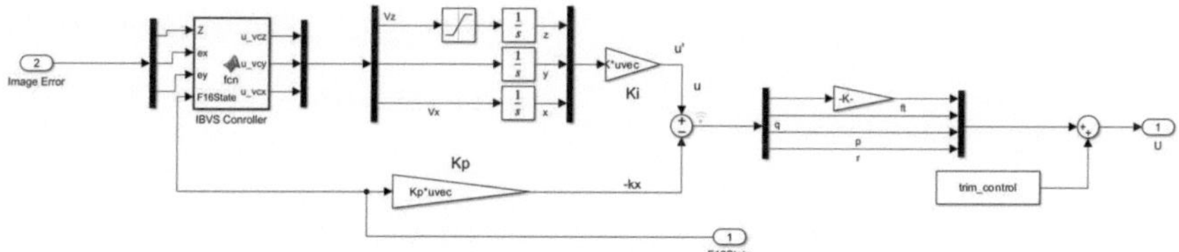

Fig. 9.14 IBVS controller, from "e7/e7-1/Simulink/AAR_IBVS.slx"

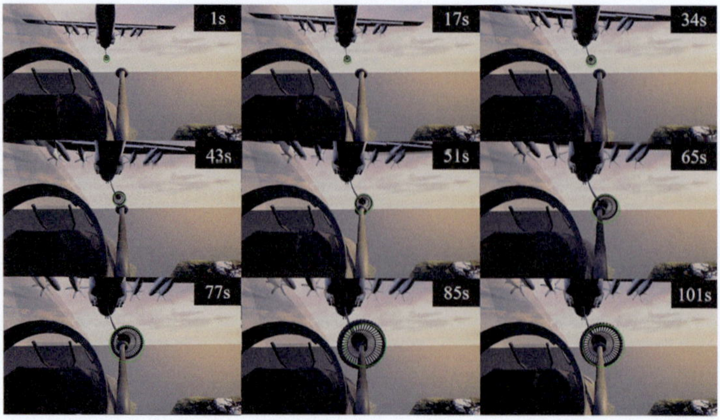

Fig. 9.15 Visualization of the docking process

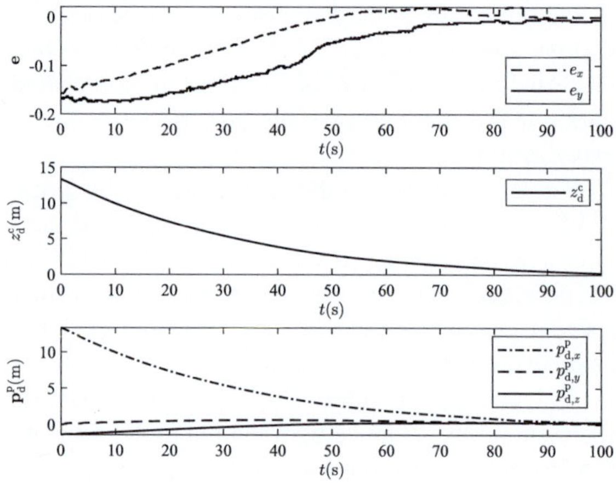

Fig. 9.16 Docking error response

and the center of the drogue. From the curve, it is observed that in the lateral and longitudinal directions, the probe tip approaches the center of the drogue at a faster speed, while in the depth direction, due to controller design, the speed at which the probe tip approaches the center of the drogue gradually decreases as the depth error is decreasing. Eventually, 3D docking errors converge to zero, implying successful docking.

9.3 Analysis Experiment

9.3.1 Experimental Objective

1. Things to prepare

(1) Software: MATLAB 2022b or above, Python, RflySimAAR.
(2) Program: Experimental package "e7/e7-2", mainly including the IBVS docking control program folder "IBVS" and the PBVS docking control program folder "PBVS". Taking the "IBVS" folder as an example, it includes two subfolders: "Simulink" and "Python". The "Simulink" folder mainly contains the simulation for aerial refueling based on IBVS, "AAR_IBVS.slx", the "F16_Init.m" initialization file, and other functions called in the platform. The "Python" folder mainly includes the drogue recognition script "detect-realtime.py" and other functions it calls. Note that "detect-realtime.py" in the "IBVS" folder and "PBVS" folder are not the same. The difference lies in the fact that the former returns the pixel position and depth of the drogue center in the image coordinate system, while the latter returns the estimated three-dimensional position of the drogue.

2. Objective
Compare and analyze the impact of different turbulence intensities on the aerial refueling docking algorithm based on IBVS, as well as different control parameters under different turbulence conditions. Compare it with the PBVS docking algorithm and analyze the difference in their disturbance rejection capabilities due to camera installation errors.

9.3.2 Experimental Procedure

The experimental procedure is divided into three steps: first, Step 1 compares the docking performance under different turbulence intensities, then, Step 2 compares the docking performance of IBVS and PBVS.

Step 1. Comparing Docking Performances Under Different Turbulence Intensities
Turbulence intensity is a function of altitude and the probability of exceedance intensity. Under the premise of constant altitude, turbulence intensity is negatively correlated with the probability of exceedance intensity. In the Aerospace Toolbox of Simulink, there is a module called "Dryden Wind Turbulence Model". In the "Probability of exceedance of high-altitude intensity" option, the probability of exceedance intensity can be set. In general, exceedance intensity probability of 10^{-1} corresponds to "Class I turbulence", 10^{-2} corresponds to "Class II turbulence", and 10^{-3} corresponds to "Class III turbulence". In the aerial refueling docking simulation platform,

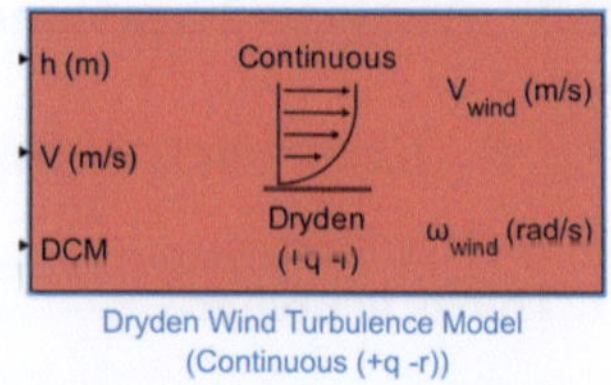

Fig. 9.17 "Dryden Wind Turbulence Model" module

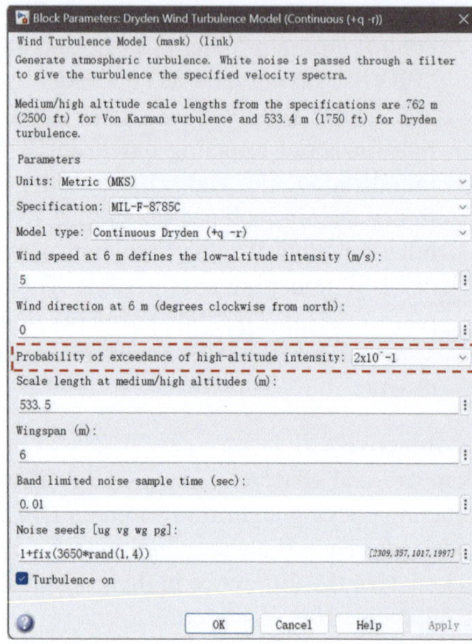

Fig. 9.18 Turbulence intensity setting

the "Dryden Wind Turbulence Model" module can be found in the environment module, as shown in Fig. 9.17. Modify the exceedance probability intensity in Fig. 9.18 to adjust the turbulence intensity.

Set the turbulence intensity to 0, I, and II, and run the simulation without changing the control parameters. Observe the simulation results, as shown below.

(1) Comparing Figs. 9.19 and 9.20, it can be observed that as the wind disturbance intensity increases, although there is some fluctuation in the trajectory of the tanker during the docking process, the image error $\mathbf{e}$, depth error z_d^c, and docking error $\mathbf{p}_d^p$ still converge to nearly zero within 100 s. This implies that the docking is still successful under Class I turbulence, and the docking error curve remains relatively smooth, showing that the controller has a certain suppression function for wind disturbances and exhibits good robustness against minor wind disturbances.
(2) From Fig. 9.21, it can be observed that under Class II turbulence, the tracking error in the image curve shows more pronounced fluctuations, especially after 40 s. This is because when the camera is closer to the probe, a small move-

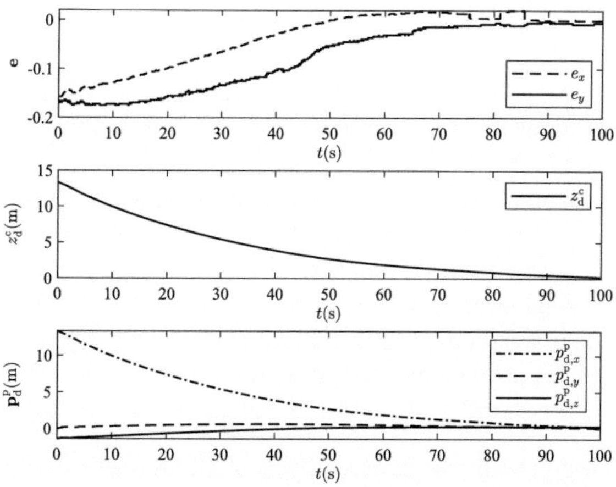

Fig. 9.19 Docking error response without turbulence

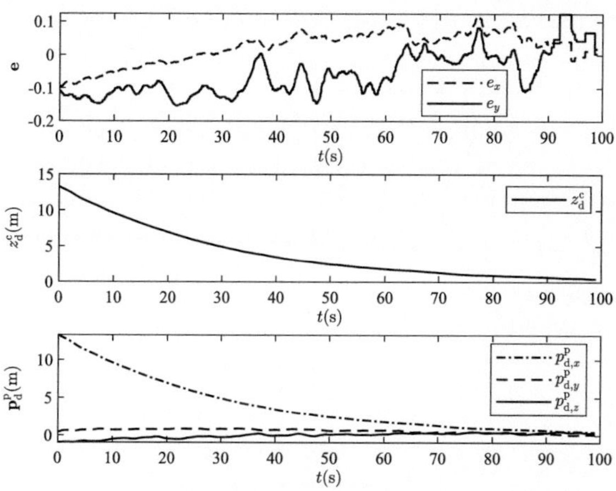

Fig. 9.20 Docking error response under Class I turbulence

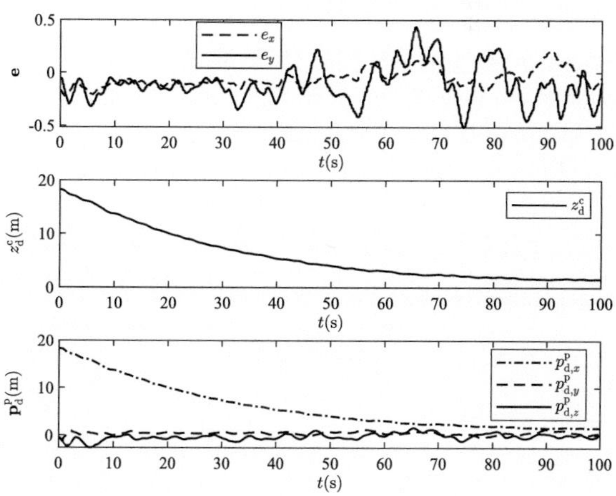

Fig. 9.21 Docking error response under Class II turbulence

ment of the probe in the coordinate system of the refueling aircraft will result in a significant change in the image. The depth error curve also shows more noticeable fluctuations compared to Fig. 9.20. The docking error curve has not converged to near zero even after 110 s. The final error along $p^{\mathrm{p}}_{\mathrm{d},x}$ direction is about 1 m, and $p^{\mathrm{p}}_{\mathrm{d},y}$ and $p^{\mathrm{p}}_{\mathrm{d},z}$ are about 0.5 m, indicating that the probe has not been inserted into the drogue. Therefore, the docking is unsuccessful, which is a normal phenomenon. Under the influence of strong wind disturbances, the stability of the tanker itself will also be greatly disturbed, not to mention the higher precision requirements for aerial refueling docking control. In practice, unmanned aerial refueling and docking operations are subject to regulations regarding atmospheric conditions. When the wind disturbance exceeds a certain limit, the docking task is not allowed to be executed due to safety concerns.

Step 2. Comparative Analysis of IBVS and PBVS for Docking

The difference between the airborne refueling docking platforms based on IBVS("AAR_IBVS.slx") and PBVS("AAR_PBVS.slx") lies in the controller part of the refueling system, as shown in Fig. 9.22.

It is assumed that calibration issues have caused some error in the camera installation position, such as $\begin{bmatrix} 1 & 0 & -0.5 \end{bmatrix}^{\mathrm{T}}$, modifications are made in the initialization file "F16_Init.m" for the PBVS platform, as shown in Code 9.3. Lines 1–2 are commented out to represent the scenario without camera installation position measurement error, while lines 4–5 account for the inclusion of camera installation position error.

Code 9.3 Injecting camera installation position measurement error in PBVS platform, from "e7-2/PBVS/F16_Init.m"

```
1 %Without camera location error
2 % d_cam_rc = [4.7;0.54;-1.45];
3
4 %With camera location error
5 d_cam_rc = [4.7+1;0.54;-1.45-0.5];
```

Modifications are also made in the IBVS platform's initialization file "F16_Init.m", as shown in Code 9.4. Lines 1–4 are commented out to represent the scenario without camera installation position measurement error, while lines 6–8 account for the inclusion of camera installation position error.

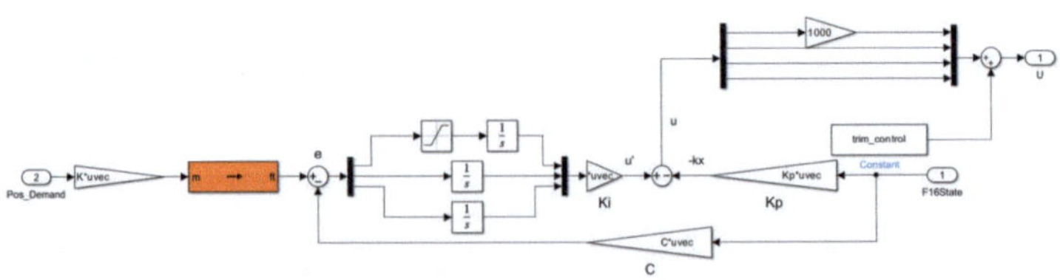

Fig. 9.22 A PBVS Controller, from "e7/e7-2/PBVS/Simulink/AAR_PBVS.slx"

Code 9.4 Injecting camera installation position measurement error in IBVS platform, from "e7-2/IBVS/F16_Init.m"

```
%Without camera location error
% Lx = 4.7;
% Ly = 0.54;
% Lz = -1.45;

%With camera location error
Lx = 4.7+1;
Ly = 0.54;
Lz = -1.45-0.5;
```

Simulations are run separately for the airborne refueling docking platforms based on IBVS, "AAR_IBVS.slx", with its corresponding "detect-realtime.py", and for those based on PBVS, "AAR_PBVS.slx", with its corresponding "detect-realtime.py." If Python reports a file path error, the platform needs to be copied to the C drive before running. The simulation results in Figs. 9.23 and 9.24 are obtained. The plotting script "AnalysisError.mlx" can be run after the simulation to obtain the docking error.

In Fig. 9.23, the period from 0 to 15 s represents the approaching phase, with the distance between the drogue and the receptacle approximately 2 m. The interval between 15 and 30 s represents the phase of holding position and estimating the drogue position. The interval between 30 and 50 s represents the initial attempt at docking. It is evident that, due to measurement error in the camera position, the docking error does not converge to zero and remains significant, particularly along the $p_{\mathrm{d},x}^{\mathrm{p}}$ direction. Following a failed docking attempt, the drogue position is re-estimated after a period of 50–70 s, after which a second attempt is made. This also fails. Figure 9.24 illustrates that the docking error stabilizes to zero within 160 s, although the docking process itself is longer. It can be observed that in the presence of camera position measurement error, the IBVS controller is capable of achieving

Fig. 9.23 Docking error response with camera position measurement error in PBVS

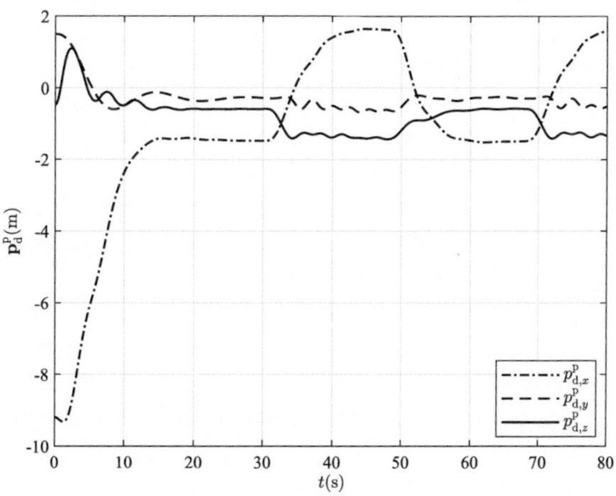

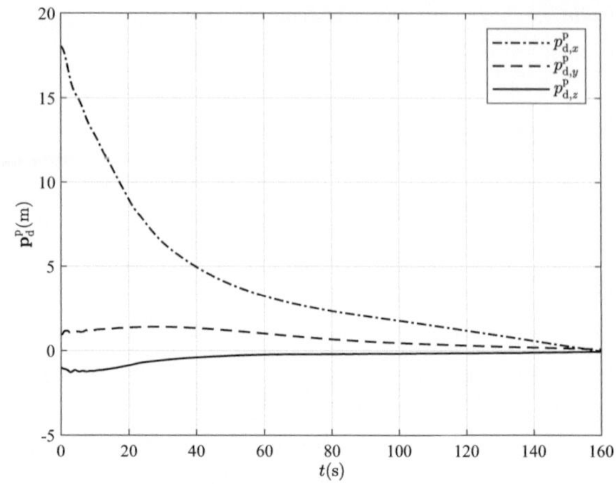

Fig. 9.24 Docking error response with camera position measurement error in IBVS

successful docking, whereas the PBVS controller is not. The following analysis will examine this phenomenon in greater detail.

1. In the case of the IBVS controller, the camera position measurement is only employed for the decoupled longitudinal and lateral IBVS models. This, in turn, serves to augment the longitudinal and lateral channel state equations, which are utilized in the design of the inner-loop stabilization controller. Consequently, the measurement of camera position only affects the inner-loop controller of IBVS. From another perspective, for the IBVS controller, errors such as 3D model reconstruction of the camera, position measurement, camera calibration, and incorrect installation position of the camera directly contribute to tracking errors in the image and depth errors. Image error is insensitive to these errors and can be eliminated by the outer-loop controller. Depth error is sensitive to these errors, but depth only determines the forward velocity during docking and does not significantly affect docking success compared to lateral error.
2. In order for the PBVS controller to function correctly, the inner-loop controller must track the drogue position in the tanker coordinate system. This is achieved by the outer-loop controller using position servoing. Therefore, for position servoing, the camera's measurement directly affects the outer-loop controller in accurately locating the drogue. From another perspective, errors associated with the camera's 3D reconstruction, position measurement, calibration, and installation position directly contribute to 3D position estimation errors, significantly affecting the estimation of position and, in turn, the success of docking.
3. In the event of uncertainties, such as measurement errors, occurring in the context of IBVS control, these uncertainties exert an influence solely on the inner-loop stabilization, specifically on the augmented state equations employed for inner-loop LQR control. This has the consequence that the outer-loop controller remains unaffected and is able to maintain precise tracking. In contrast, in PBVS control, such uncertainties directly affect the precise retrieval of drogue position in the

outer-loop control, thereby directly influencing the effectiveness of docking. Consequently, IBVS controllers exhibit superior robustness compared to traditional PBVS controllers, thereby validating the advantage of IBVS control.

9.4 Design Experiment

9.4.1 Experimental Objective

1. Things to prepare

(1) Software: MATLAB R2022b or above, Python, RflySimAAR.
(2) Program: Experimental package "e7/e7-3", which includes IBVS aerial refueling docking simulation under velocity mode, named "AAR_IBVS_Vmode.slx", initialization file "F16_In_new.m", and other functions called in the platform. The Python script for drogue recognition can be reused from the basic experiment.

2. Objectives

1. Design an IBVS controller with "ground speed" as the upper-level control interface and validate the docking control in RflySimAAR.
2. Furthermore, utilize the CubePilot/Pixhawk autopilot with "receiver's waypionts" as the control interface to validate docking control in the HIL simulation.

9.4.2 Experimental Design

To design an IBVS controller with "ground speed" as the control interface, the relationship among the tanker's motion model, guidance model, and IBVS control model needs to be considered. This is because the linear velocity, attitude, and angular velocity of the tanker affect the pose of the camera, thereby influencing the tracking error in the image in the IBVS model.

Step 1. Establishing Guidance Model Based on Velocity Control
The low-level flight control under velocity control mode can be implemented using LQR, model predictive control and feedback linearization ("MPC+FL"), or differential flatness methods. The guidance model for the receiver with the velocity control mode is described as follows:

$$\begin{cases} \dot{\mathbf{p}}_{\mathrm{r}}^{\mathrm{g}} = \mathbf{v}_{\mathrm{r}}^{\mathrm{g}} \\ \underbrace{\begin{bmatrix} v_{\mathrm{r},x}^{\mathrm{g}} \\ v_{\mathrm{r},y}^{\mathrm{g}} \\ v_{\mathrm{r},z}^{\mathrm{g}} \end{bmatrix}}_{\mathbf{v}_{\mathrm{r}}^{\mathrm{g}}} = \begin{bmatrix} G_x & & \\ & G_y & \\ & & G_z \end{bmatrix} \underbrace{\begin{bmatrix} v_{\mathrm{r},x\mathrm{ref}}^{\mathrm{g}} \\ v_{\mathrm{r},y\mathrm{ref}}^{\mathrm{g}} \\ v_{\mathrm{r},z\mathrm{ref}}^{\mathrm{g}} \end{bmatrix}}_{\mathbf{v}_{\mathrm{r,ref}}^{\mathrm{g}}} \end{cases} \tag{9.56}$$

where $\mathbf{p}_{\mathrm{r}}^{\mathrm{g}} = \left[p_{\mathrm{r},x}^{\mathrm{g}}\ p_{\mathrm{r},y}^{\mathrm{g}}\ p_{\mathrm{r},z}^{\mathrm{g}}\right]^{\mathrm{T}}$ represents the position of the tanker in the ground coordinate system, $\mathbf{v}_{\mathrm{r}}^{\mathrm{g}} = \left[v_{\mathrm{r},x}^{\mathrm{g}}\ v_{\mathrm{r},y}^{\mathrm{g}}\ v_{\mathrm{r},z}^{\mathrm{g}}\right]^{\mathrm{T}}$ represents the velocity of the tanker in the ground coordinate system, $\mathbf{v}_{\mathrm{r,ref}}^{\mathrm{g}} = \left[v_{\mathrm{r},x\mathrm{ref}}^{\mathrm{g}}\ v_{\mathrm{r},y\mathrm{ref}}^{\mathrm{g}}\ v_{\mathrm{r},z\mathrm{ref}}^{\mathrm{g}}\right]^{\mathrm{T}}$ represents the reference velocity of the tanker in the ground coordinate system, and G_x, G_y, G_z are transfer functions from $\mathbf{v}_{\mathrm{r,ref}}^{\mathrm{g}}$ to $\mathbf{v}_{\mathrm{r}}^{\mathrm{g}}$. The transfer functions G_x, G_y, G_z can be obtained through system identification.

Step 2. Establishing Camera Motion Model under Velocity Control Mode
To fully consider the influence of attitude and angular velocity on tracking error in the image in velocity control mode, unlike the basic and analysis experiments, a more comprehensive derivation of Eq. (9.11) is performed here, resulting in:

$$\dot{e}_x = -\frac{v_{\mathrm{d},x}^{\mathrm{c}}}{z_{\mathrm{d}}^{\mathrm{c}}} + \frac{e_x v_{\mathrm{d},z}^{\mathrm{c}}}{z_{\mathrm{d}}^{\mathrm{c}}} + e_x e_y \omega_{\mathrm{d},x}^{\mathrm{c}} - (1 + e_x^2)\omega_{\mathrm{d},y}^{\mathrm{c}} + e_y \omega_{\mathrm{d},z}^{\mathrm{c}} \tag{9.57}$$

$$\dot{e}_y = -\frac{v_{\mathrm{d},y}^{\mathrm{c}}}{z_{\mathrm{d}}^{\mathrm{c}}} + \frac{e_y v_{\mathrm{d},z}^{\mathrm{c}}}{z_{\mathrm{d}}^{\mathrm{c}}} + (1 + e_y^2)\omega_{\mathrm{d},x}^{\mathrm{c}} - e_x e_y \omega_{\mathrm{d},y}^{\mathrm{c}} - e_x \omega_{\mathrm{d},z}^{\mathrm{c}}. \tag{9.58}$$

Next, the relationship between the velocity of the refueling aircraft in the ground coordinate system and the velocity of the camera relative to the tanker (considering the velocity, attitude, and angular velocity of the refueling aircraft) is derived. From Eqs. (9.28) and (9.29), it is known that:

$$\begin{cases} v_{\mathrm{c},x}^{\mathrm{re}} = v_{\mathrm{r},y}^{\mathrm{t}} + x_{\mathrm{c}}^{\mathrm{r}} \Delta\omega_{z_{\mathrm{b}}} - z_{\mathrm{c}}^{\mathrm{r}} \Delta\omega_{x_{\mathrm{b}}} \\ v_{\mathrm{c},y}^{\mathrm{re}} = v_{\mathrm{r},z}^{\mathrm{t}} + y_{\mathrm{c}}^{\mathrm{r}} \Delta\omega_{x_{\mathrm{b}}} - x_{\mathrm{c}}^{\mathrm{r}} \Delta\omega_{y_{\mathrm{b}}} \\ v_{\mathrm{c},z}^{\mathrm{re}} = v_{\mathrm{r},x}^{\mathrm{t}} + z_{\mathrm{c}}^{\mathrm{r}} \Delta\omega_{y_{\mathrm{b}}} - y_{\mathrm{c}}^{\mathrm{r}} \Delta\omega_{z_{\mathrm{b}}} \end{cases}. \tag{9.59}$$

Considering the motion relationship between the refueling aircraft and the tanker, one has:

$$\mathbf{v}_{\mathrm{r}}^{\mathrm{t}} = \mathbf{v}_{\mathrm{r}}^{\mathrm{g}} - \mathbf{v}_{\mathrm{t}}^{\mathrm{g}}. \tag{9.60}$$

Therefore, Eq. (9.59) is expressed as:

$$\begin{cases} v_{\mathrm{c},x}^{\mathrm{re}} = v_{\mathrm{r},y}^{\mathrm{g}} + x_{\mathrm{c}}^{\mathrm{r}}\Delta\omega_{z_{\mathrm{b}}} - z_{\mathrm{c}}^{\mathrm{r}}\Delta\omega_{x_{\mathrm{b}}} \\ v_{\mathrm{c},y}^{\mathrm{re}} = v_{\mathrm{r},z}^{\mathrm{g}} + y_{\mathrm{c}}^{\mathrm{r}}\Delta\omega_{x_{\mathrm{b}}} - x_{\mathrm{c}}^{\mathrm{r}}\Delta\omega_{y_{\mathrm{b}}} \\ v_{\mathrm{c},z}^{\mathrm{re}} = v_{\mathrm{r},x}^{\mathrm{g}} - V_{\mathrm{t}}^{\mathrm{g}} + z_{\mathrm{c}}^{\mathrm{r}}\Delta\omega_{y_{\mathrm{b}}} - y_{\mathrm{c}}^{\mathrm{r}}\Delta\omega_{z_{\mathrm{b}}} \end{cases}. \tag{9.61}$$

Step 3. IBVS Controller Design

For Eq. (9.57), the desired velocity for this channel is designed as

$$v_{\mathrm{d},x\mathrm{d}}^{\mathrm{c}} = k_1 e_x + z_{\mathrm{d}}^{\mathrm{c}} e_x e_y \omega_{\mathrm{d},x}^{\mathrm{c}} - z_{\mathrm{d}}^{\mathrm{c}}(1+e_x^2)\omega_{\mathrm{d},y}^{\mathrm{c}} + z_{\mathrm{d}}^{\mathrm{c}} e_y \omega_{\mathrm{d},z}^{\mathrm{c}}. \tag{9.62}$$

If $v_{\mathrm{d},x}^{\mathrm{c}} = v_{\mathrm{d},x\mathrm{d}}^{\mathrm{c}}$, then Eq. (9.57) becomes

$$\dot{e}_x = -\frac{k_1 - v_{\mathrm{d},z}^{\mathrm{c}}}{z_{\mathrm{d}}^{\mathrm{c}}} e_x. \tag{9.63}$$

Choosing $k_1 > \max(v_{\mathrm{d},z}^{\mathrm{c}})$ ensures $\lim\limits_{t\to\infty} |e_x(t)| = 0$.

Considering Eq. (9.58), the desired velocity for this channel is designed as

$$v_{\mathrm{d},y\mathrm{d}}^{\mathrm{c}} = k_2 e_y - z_{\mathrm{d}}^{\mathrm{c}} e_x e_y \omega_{\mathrm{d},y}^{\mathrm{c}} + z_{\mathrm{d}}^{\mathrm{c}}(1+e_y^2)\omega_{\mathrm{d},x}^{\mathrm{c}} - z_{\mathrm{d}}^{\mathrm{c}} e_x \omega_{\mathrm{d},z}^{\mathrm{c}}. \tag{9.64}$$

If $v_{\mathrm{d},y}^{\mathrm{c}} = v_{\mathrm{d},y\mathrm{d}}^{\mathrm{c}}$, then Eq. (9.58) becomes

$$\dot{e}_y = -\frac{k_2 - v_{\mathrm{d},z}^{\mathrm{c}}}{z_{\mathrm{d}}^{\mathrm{c}}} e_y. \tag{9.65}$$

Choosing $k_2 > \max(v_{\mathrm{d},z}^{\mathrm{c}})$ ensures $\lim\limits_{t\to\infty} \left|e_y(t)\right| = 0$.

The desired velocity in the depth direction refers to Eq. (9.40). In this experiment, the angular velocity of the drogue can be ignored. It is assumed $\omega_{\mathrm{c}}^{\mathrm{d}} = \omega_{\mathrm{c}}^{\mathrm{re}} = \mathbf{R}_{\mathrm{re/t}}\omega_{\mathrm{c}}^{\mathrm{t}} \approx [\omega_{y_{\mathrm{b}}}\ \omega_{z_{\mathrm{b}}}\ \omega_{x_{\mathrm{b}}}]^{\mathrm{T}}$. Therefore, combining the designs for all three channels, the IBVS controller is:

$$\begin{cases} v_{\mathrm{d},x\mathrm{d}}^{\mathrm{c}} = k_1 e_x + z_{\mathrm{d}}^{\mathrm{c}} e_x e_y \Delta\omega_{y_{\mathrm{b}}} - z_{\mathrm{d}}^{\mathrm{c}}(1+e_x^2)\Delta\omega_{z_{\mathrm{b}}} + z_{\mathrm{d}}^{\mathrm{c}} e_y \Delta\omega_{x_{\mathrm{b}}} \\ v_{\mathrm{d},y\mathrm{d}}^{\mathrm{c}} = k_2 e_y - z_{\mathrm{d}}^{\mathrm{c}} e_x e_y \Delta\omega_{z_{\mathrm{b}}} + z_{\mathrm{d}}^{\mathrm{c}}(1+e_y^2)\Delta\omega_{y_{\mathrm{b}}} - z_{\mathrm{d}}^{\mathrm{c}} e_y \Delta\omega_{x_{\mathrm{b}}} \\ v_{\mathrm{d},z\mathrm{d}}^{\mathrm{c}} = k_3 z_{\mathrm{d}}^{\mathrm{c}} + k_4 |e_x| + k_5 \left|e_y\right| \end{cases}. \tag{9.66}$$

If $\mathbf{v}_{\mathrm{d}}^{\mathrm{c}} = \mathbf{v}_{\mathrm{d,d}}^{\mathrm{c}}$, then $\lim\limits_{t\to\infty} |e_x(t)| = 0$, $\lim\limits_{t\to\infty} \left|e_y(t)\right| = 0$, $\lim\limits_{t\to\infty} \left|z_{\mathrm{d}}^{\mathrm{c}}(t)\right| = 0$. Here, $\mathbf{v}_{\mathrm{d,d}}^{\mathrm{c}}$ represents the desired velocity of the drogue in the camera coordinate system.

In practice, since $\mathbf{v}_{\mathrm{d}}^{\mathrm{c}} = \mathbf{v}_{\mathrm{d}}^{\mathrm{re}} - \mathbf{v}_{\mathrm{c}}^{\mathrm{re}}$, one can only control $\mathbf{v}_{\mathrm{c}}^{\mathrm{re}}$ and not $\mathbf{v}_{\mathrm{d}}^{\mathrm{re}}$, because $\mathbf{v}_{\mathrm{d}}^{\mathrm{re}}$ represents the motion velocity of the drogue. Therefore, $\mathbf{v}_{\mathrm{c,d}}^{\mathrm{re}} = \mathbf{v}_{\mathrm{d,d}}^{\mathrm{c}}$ is set, treating $\mathbf{v}_{\mathrm{d}}^{\mathrm{re}}$ as disturbance. According to Eq. (9.61), the desired velocity of $\mathbf{v}_{\mathrm{r}}^{\mathrm{g}}$ relative to the ground is obtained as:

$$\begin{cases} v_{\mathrm{r},x\mathrm{d}}^{\mathrm{g}} = -k_3 z_{\mathrm{d}}^{\mathrm{c}} - k_4 \left| e_x \right| - k_5 \left| e_y \right| + V_{\mathrm{t}}^{\mathrm{g}} - z_{\mathrm{c}}^{\mathrm{r}} \Delta\omega_{y_{\mathrm{b}}} + y_{\mathrm{c}}^{\mathrm{r}} \Delta\omega_{z_{\mathrm{b}}} \\ v_{\mathrm{r},y\mathrm{d}}^{\mathrm{g}} = k_1 e_x + z_{\mathrm{d}}^{\mathrm{c}} e_x e_y \Delta\omega_{y_{\mathrm{b}}} - z_{\mathrm{d}}^{\mathrm{c}} (1 + e_x^2) \Delta\omega_{z_{\mathrm{b}}} + z_{\mathrm{d}}^{\mathrm{c}} e_y \Delta\omega_{x_{\mathrm{b}}} - x_{\mathrm{c}}^{\mathrm{r}} \Delta\omega_{z_{\mathrm{b}}} + z_{\mathrm{c}}^{\mathrm{r}} \Delta\omega_{x_{\mathrm{b}}} \\ v_{\mathrm{r},z\mathrm{d}}^{\mathrm{g}} = k_2 e_y - z_{\mathrm{d}}^{\mathrm{c}} e_x e_y \Delta\omega_{z_{\mathrm{b}}} + z_{\mathrm{d}}^{\mathrm{c}} (1 + e_y^2) \Delta\omega_{y_{\mathrm{b}}} - z_{\mathrm{d}}^{\mathrm{c}} e_y \Delta\omega_{x_{\mathrm{b}}} - y_{\mathrm{c}}^{\mathrm{r}} \Delta\omega_{x_{\mathrm{b}}} + x_{\mathrm{c}}^{\mathrm{r}} \Delta\omega_{y_{\mathrm{b}}} \end{cases}. \tag{9.67}$$

In general, the small UAV has a fast dynamic response in velocity control mode, which can meet the requirements of docking control. Therefore, $\mathbf{v}_{\mathrm{r,d}}^{\mathrm{g}}$ can generally be directly used as the reference velocity, namely,

$$\mathbf{v}_{\mathrm{r,ref}}^{\mathrm{g}} = \mathbf{v}_{\mathrm{r,d}}^{\mathrm{g}}. \tag{9.68}$$

Furthermore, for HIL experiments with "receiver's waypoints" as the control interface, one can obtain:

$$\begin{cases} V_{\mathrm{r}}^{\mathrm{g}} = \|\mathbf{v}_{\mathrm{r,ref}}^{\mathrm{g}}\| \\ \mathbf{p}_{\mathrm{r}}^{\mathrm{g}} = \int_0^t \mathbf{v}_{\mathrm{r,ref}}^{\mathrm{g}}(\tau)\mathrm{d}\tau \end{cases}. \tag{9.69}$$

9.4.3 Experimental Procedure

The experimental procedure is divided into two steps: first, implementing the IBVS controller under the velocity control mode in Step 1; then, in Step 2, running simulations and analyzing the performance of the aerial refueling docking experiment.

Step 1. Implementing IBVS Controller under Velocity Control Mode
Under the path "AAR_IBVS_Vmode\Receiver System\Controller", use "Matlab Function" to implement the IBVS controller under velocity control mode, where the inputs are tracking error in the images, depth estimation results, and the state of the receiver. The output is the velocity reference of the receiver in the ground coordinate system. Corresponding to Eq. (9.67), the key code is shown in Code 9.5, where line 2 reverses $o_{\mathrm{i}}y_{\mathrm{i}}$ axis direction of the tracking error in the image; line 3 sets the compensation value for depth error to control the probe; lines 4–8 set the control parameters; lines 9–11 compute the expected distance of the camera relative to the drogue. It should be noted that since the platform is designed to calculate the speed and position of the receiver relative to the tanker first, it then superimposes these values on the receiver's speed and position for display. The forward speed by the image servo controller here is the relative speed.

Code 9.5 IBVS controller under velocity control mode, from "e7/e7-3/AAR_IBVS_Vmode.slx"

```
d_pr_rc = [6.06;0.54;-0.86]; % the distance between the camera and the receiver
ey = -ey; % reverse ey
Z_d = 1.0; % compensation Z
kz = 0.5; % set control parameters
ky = 3;
kx = 1;
k1 = 3;
k2 = 1;
vgz = kz*(Z-Z_d)-k1*(abs(ex))-k2*(abs(ey))-d_pr_rc(3)*q+d_pr_rc(2)*r; % calculate desired velocity of receiver
vgy = ky*(ey-0)+Z*ex*ey*q-Z(1+ex^2)r+Z*ey*p-d_pr_rc(1)*r+d_pr_rc(3)*p;
vgx = kx*(ex-0)-Z*ex*ey*r+Z(1+ey^2)q-Z*ex*p-d_pr_rc(2)*p+d_pr_rc(1)*q;
```

The low-level controller adopts the LQR method to implement the velocity control mode. The primary difference from the basic experiment's LQR controller is that there is no need to augment the system to compensate for visual tracking errors. Instead, the velocity of the receiver in the ground coordinate system is directly set as the control target.

Step 2. Running Simulation and Analyzing the performance of Aerial Refueling Docking

After implementing the IBVS algorithm above, the Simulink simulation can be executed. Figure 9.25 presents the visual display of the aerial refueling docking.

Figure 9.26 illustrates the error curves during the docking process without disturbance, including "Image Error $\mathbf{e}$", "Depth Error $z_{\mathrm{d}}^{\mathrm{c}}$", and "Docking Error $\mathbf{p}_{\mathrm{d}}^{\mathrm{p}}$". The tracking error in the image, representing the difference between the center point of the drogue and the origin of the image coordinate system, converges to zero within 67 s. From the depth curve, it can be observed that the smaller the depth error, the slower the convergence rate of the depth error. The curves of the relative coordinate system in three axes show that the lateral and longitudinal relative distances converge rapidly, while the depth direction, due to controller design, exhibits a convergence rate inversely related to the depth error. Eventually, all three-axis docking errors converge to zero, indicating successful docking. Figure 9.27 shows the docking process.

Figures 9.28, 9.29, and 9.30 show the error under disturbances including: level I turbulence, bow wave effect, and refueling aircraft wake turbulence; level I turbulence, bow wave effect, refueling aircraft wake turbulence, and discrete gust dis-

Fig. 9.25 Docking scene

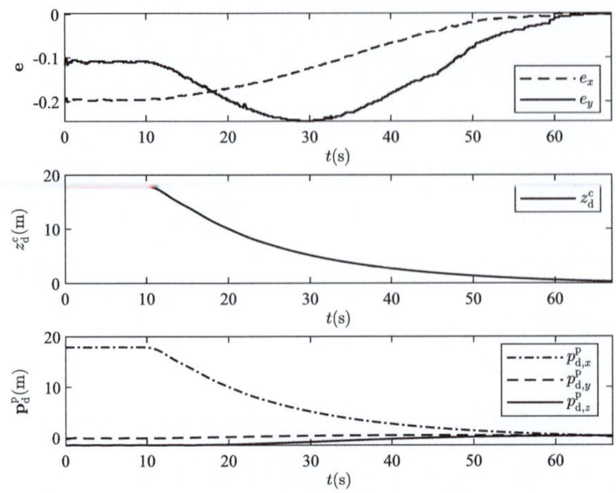

Fig. 9.26 Docking error response without disturbance

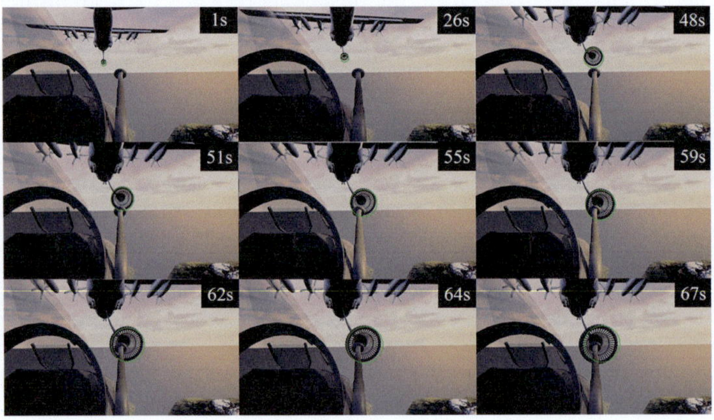

Fig. 9.27 Docking process

turbances; level II turbulence, bow wave effect, refueling aircraft wake turbulence, and discrete gust disturbances, respectively. Compared Figs. 9.28 with 9.29, it can be observed that a discrete gust has no significant impact on the docking result, indicating that the designed IBVS controller can resist disturbances from a discrete gust. A comparison of Figs. 9.29 and 9.30 reveals that, while the disturbances gradually increase, the fluctuation of the tracking error in the image and docking error curves during the docking process also increases. However, these fluctuations eventually lead to successful docking, indicating that the designed controller possesses robust disturbance rejection capabilities.

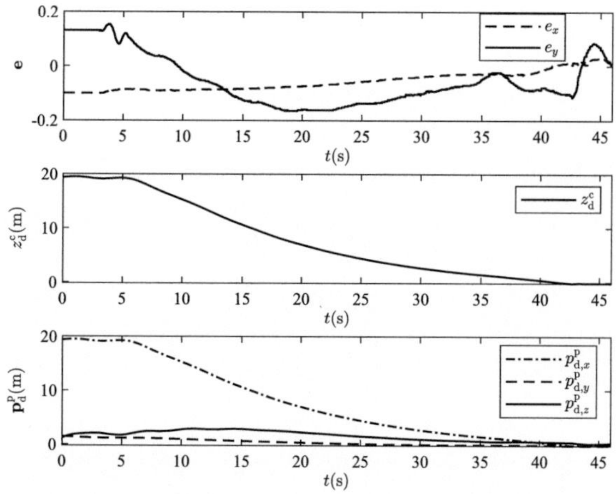

Fig. 9.28 Docking error response under Class I turbulence, bow wave effect, and wake of the tanker

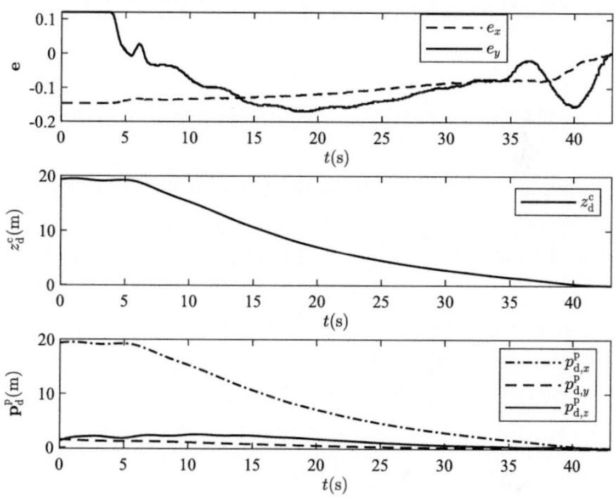

Fig. 9.29 Docking error response under Class I turbulence, bow wave effect, wake of the tanker, and discrete gust

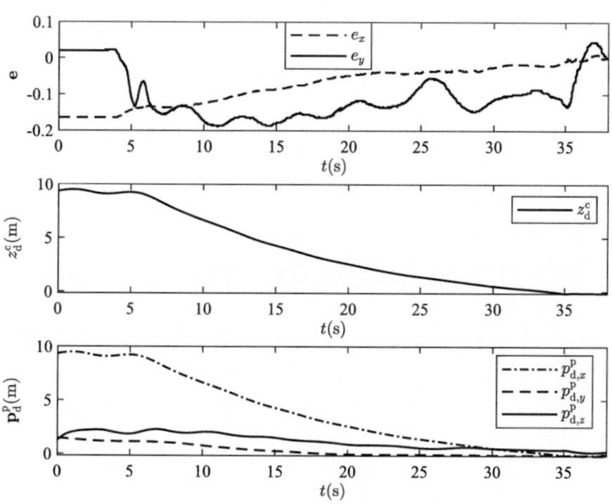

Fig. 9.30 Docking error response under Class II turbulence, bow wave effect, wake of the tanker, and discrete gust

9.5 HIL Experiment

9.5.1 Experimental Objective

1. Things to prepare

(1) Software: MATLAB R2022b or above, Python, RflySimAAR, RflySim toolchain. The RflySim toolchain provides a HIL simulation environment for micro-small fixed-wing UAVs. It can send commands to the fixed-wing UAV in "Offboard" mode via MAVLINK using a Python terminal. The commands support "airspeed + altitude + course angle" control mode.[2]
(2) Program: Experimental package "e7/e7-4". The experimental package mainly includes: the "CopterSimulink_HITL" folder containing the HIL simulation program for autonomous aerial docking, the "Simulink" folder for simulating mothership and drogue movements, and the "module" folder for storing RflySim3D models of mothership, UAV, and drogue. The "CopterSimulink_HITL" folder mainly includes the UAV control script "Mytrajectory.py", the drogue detection script "detect-realtime.py", the one-click start RflySim batch file "PX4SimHITL.bat", the Python environment configuration list file "requirements.txt", a folder containing Python environment installation packages "packages", a folder containing flight control firmware "Firmware", a file saving image errors "error_data.mat", the error analysis plotting file "AnalysisVisualError.m", and other functions called. The "Simulink" folder mainly includes the file for running mothership and drogue movement simulation "AirRefueling_Platform.slx" and functions called in other platforms. The "module" folder mainly includes the mothership model "WestTransportC130J", UAV model "MQ-9Reaper", and drogue model "ProbeLightScene".
(3) Hardware: a CubePilot/Pixhawk autopilot.

2. Objective

Based on the experiment design, using CubePilot/Pixhawk autopilot with the "airspeed + altitude + course angle" control interface, a complete HIL simulation for docking control needs to complete.

9.5.2 Experimental Design

In the experimental design, the Simulink simulation model of the flight controller can execute commands under the "ground speed" control mode, whereas the autopilot only supports commands under the "airspeed + altitude + course angle" control mode. Therefore, to complete the HIL simulation, it is necessary to unify the two control modes and design an IBVS controller under the "airspeed + altitude + course

[2] The control mode is designed in Simulink and compiled into ".px4" PX4 autopilot software firmware file using the PSP toolbox.

angle" control mode. To design an IBVS controller with "airspeed + altitude + course angle" as the control interface, it is necessary to establish the relationship between the guidance models based on "airspeed + altitude + course angle" control and ground speed control. The "ground speed" interface is used to represent the "airspeed + altitude + course angle" interface, so that the IBVS controller based on the ground speed obtained in the design experiment can be derived into an IBVS controller based on the "airspeed + altitude + course angle" guidance model.

Step 1. Establishing the Relationship between "Ground Speed" Guidance Model and "Airspeed + Altitude + Course Angle" Guidance Model
The position of the UAV in the ground coordinate system corresponds to its position in the North-East-Down coordinate system and can be represented as:

$$\begin{bmatrix} p_{\mathrm{r},x}^{\mathrm{g}} \\ p_{\mathrm{r},y}^{\mathrm{g}} \\ p_{\mathrm{r},z}^{\mathrm{g}} \end{bmatrix} = \begin{bmatrix} p_{x_{\mathrm{e}}} \\ p_{y_{\mathrm{e}}} \\ H \end{bmatrix} \tag{9.70}$$

where $\mathbf{p}_{\mathrm{r}}^{\mathrm{g}} = \left[p_{\mathrm{r},x}^{\mathrm{g}}\ p_{\mathrm{r},y}^{\mathrm{g}}\ p_{\mathrm{r},z}^{\mathrm{g}}\right]^{\mathrm{T}}$ represents the position of the UAV in the ground coordinate system; $p_{x_{\mathrm{e}}}$ represents the northward position of the UAV in the North-East-Down coordinate system; $p_{y_{\mathrm{e}}}$ represents the eastward position of the UAV in the North-East-Down coordinate system; H represents the actual height of the UAV relative to the ground.

In Chap. 5, the guidance model based on "airspeed + altitude + course angle" control has been derived, which can be represented by

$$\begin{cases} \dot{p}_{x_{\mathrm{e}}} = V_{\mathrm{a}} \cos \psi + v_{\mathrm{w}x} \\ \dot{p}_{y_{\mathrm{e}}} = V_{\mathrm{a}} \sin \psi + v_{\mathrm{w}y} \\ \psi = \chi + f\left(v_{\mathrm{w}x}, v_{\mathrm{w}y}\right) \\ V_{\mathrm{a}} = G_{V_{\mathrm{a}}} V_{\mathrm{ad}} \\ \chi = G_{\chi} \chi_{\mathrm{d}} \\ H = G_H H_{\mathrm{d}} \end{cases} \tag{9.71}$$

where V_{a} represents the airspeed of the UAV (scalar); ψ is the yaw angle of the UAV; χ is the course angle of the UAV; $v_{\mathrm{w}x}, v_{\mathrm{w}y}$ represent the northward and eastward wind speeds, respectively; V_{ad} is the desired airspeed of the UAV (scalar); χ_{d} is the desired course angle of the UAV; H_{d} is the desired altitude of the UAV relative to the ground; $G_{V_{\mathrm{a}}}, G_{\chi}, G_H$ represent the transfer functions from the desired airspeed, desired course angle, and desired altitude of the UAV to the airspeed, course angle, and altitude of the UAV, respectively.

The guidance model based on the speed control has been derived in this chapter's design experiment. From Eqs. (9.56), (9.70), and (9.71), one has:

$$\begin{cases} v_{\mathrm{r},x\mathrm{ref}}^{\mathrm{g}} = \dfrac{1}{G_x}(V_\mathrm{a}\cos\psi + v_{\mathrm{w}x}) \\ v_{\mathrm{r},y\mathrm{ref}}^{\mathrm{g}} = \dfrac{1}{G_y}(V_\mathrm{a}\sin\psi + v_{\mathrm{w}y}) \\ v_{\mathrm{r},z\mathrm{ref}}^{\mathrm{g}} = \dfrac{1}{G_z}\dot{H} \\ \psi = \chi + f\left(v_{\mathrm{w}x}, v_{\mathrm{w}y}\right) \\ V_\mathrm{a} = G_{V_\mathrm{a}} V_\mathrm{ad} \\ \chi = G_\chi \chi_\mathrm{d} \\ H = G_H H_\mathrm{d} \end{cases} \tag{9.72}$$

where $\mathbf{v}_{\mathrm{r,ref}}^{\mathrm{g}} = \left[v_{\mathrm{r},x\mathrm{ref}}^{\mathrm{g}}\ v_{\mathrm{r},y\mathrm{ref}}^{\mathrm{g}}\ v_{\mathrm{r},z\mathrm{ref}}^{\mathrm{g}}\right]^\mathrm{T}$ represents the reference speed of the UAV in the ground coordinate system; G_x, G_y, G_z represent the transfer functions from the reference ground speed $\mathbf{v}_{\mathrm{r,ref}}^{\mathrm{g}}$ to the ground speed $\mathbf{v}_\mathrm{r}^\mathrm{g}$. In the HIL experiment, the wind speeds $v_{\mathrm{w}x}, v_{\mathrm{w}y}$ and $f\left(v_{\mathrm{w}x}, v_{\mathrm{w}y}\right)$ can be ignored as they are small. The docking speed of the UAV in the experiment is slow, and its dynamic process can be ignored. $G_x, G_y, G_z, G_{V_\mathrm{a}}, G_\chi, G_H$ are all considered as 1. Therefore, Eq. (9.72) is simplified as:

$$\begin{cases} v_{\mathrm{r},x\mathrm{ref}}^{\mathrm{g}} = V_\mathrm{ad}\cos\chi_\mathrm{d} \\ v_{\mathrm{r},y\mathrm{ref}}^{\mathrm{g}} = V_\mathrm{ad}\sin\chi_\mathrm{d} \\ v_{\mathrm{r},z\mathrm{ref}}^{\mathrm{g}} = \dot{H}_\mathrm{d} \end{cases}. \tag{9.73}$$

The following section presents the relationship for the desired airspeed, desired course angle, desired altitude, and the UAV's desired ground speed. According to Eq. (9.68), the UAV's desired ground speed $\mathbf{v}_{\mathrm{r,d}}^{\mathrm{g}}$ is considered as the UAV's ground reference speed $\mathbf{v}_{\mathrm{r,ref}}^{\mathrm{g}}$. Considering the relationship between the two guidance models and simple trigonometric relations, the "ground speed" interface can be expressed as "airspeed + altitude + course angle" interface, leading to:

$$\begin{cases} V_\mathrm{ad} = \sqrt{{v_{\mathrm{r},x\mathrm{d}}^{\mathrm{g}}}^2 + {v_{\mathrm{r},y\mathrm{d}}^{\mathrm{g}}}^2} \\ \chi_\mathrm{d} = \arcsin\dfrac{v_{\mathrm{r},y\mathrm{d}}^{\mathrm{g}}}{V_\mathrm{ad}} = \arcsin\dfrac{1}{\sqrt{\dfrac{v_{\mathrm{r},x\mathrm{d}}^{\mathrm{g}}}{v_{\mathrm{r},y\mathrm{d}}^{\mathrm{g}}}^2 + 1}} \\ H_\mathrm{d} = \displaystyle\int_0^t v_{\mathrm{r},z\mathrm{d}}^{\mathrm{g}}(\tau)\mathrm{d}\tau + H_0 \end{cases} \tag{9.74}$$

where H_0 represents the initial altitude of the UAV before entering IBVS control.

Step 2. Design of IBVS Controller under "Airspeed + Altitude + Course Angle" Control Mode

The ground speed IBVS controller, as obtained in the design experiment, is then improved upon and combined with Eq. (9.74), to generate an IBVS controller that is capable of controlling airspeed, altitude, and course angle.

The "airspeed + altitude + course angle" control interface used in the HIL simulation experiment is an upper-level interface encapsulated within the PX4 library in Python. In contrast with the findings of the design experiment, it is impossible to realize stabilizing control on the low-level control interface through the design of an inner-loop LQR controller. It is challenging to eliminate disturbances through the design of an outer-loop proportional controller, leading to substantial oscillations during docking. Accordingly, an improvement is made in the IBVS controller within the design experiment through the incorporation of differential damping into the outer-loop control, with the objective of mitigating oscillations during docking.

For Eq. (9.62), after introducing the damping, the desired velocity for this channel becomes:

$$v^{\mathrm{c}}_{\mathrm{d},x\mathrm{d}} = k_{xp}e_x + k_{xd}\dot{e_x} + z^{\mathrm{c}}_{\mathrm{d}}e_x e_y \omega^{\mathrm{c}}_{\mathrm{d},x} - z^{\mathrm{c}}_{\mathrm{d}}(1+e_x^2)\omega^{\mathrm{c}}_{\mathrm{d},y} + z^{\mathrm{c}}_{\mathrm{d}}e_y\omega^{\mathrm{c}}_{\mathrm{d},z} \tag{9.75}$$

where k_{xp} is the proportional coefficient, and k_{xd} is the differential coefficient. If $v^{\mathrm{c}}_{\mathrm{d},x} = v^{\mathrm{c}}_{\mathrm{d},x\mathrm{d}}$, Eq. (9.57) becomes

$$\dot{e}_x = -\frac{k_{xp} - v^{\mathrm{c}}_{\mathrm{d},z}}{z^{\mathrm{c}}_{\mathrm{d}} + k_{xd}} e_x. \tag{9.76}$$

By choosing $k_{xp} > \max(v^{\mathrm{c}}_{\mathrm{d},z})$, it ensures $\lim_{t\to\infty} |e_x(t)| = 0$, indicating that the image error for this channel converges to 0.

Considering Eq. (9.64), after introducing the damping, the desired velocity for this channel becomes:

$$v^{\mathrm{c}}_{\mathrm{d},y\mathrm{d}} = k_{yp}e_y + k_{yd}\dot{e_y} - z^{\mathrm{c}}_{\mathrm{d}}e_x e_y \omega^{\mathrm{c}}_{\mathrm{d},y} + z^{\mathrm{c}}_{\mathrm{d}}(1+e_y^2)\omega^{\mathrm{c}}_{\mathrm{d},x} - z^{\mathrm{c}}_{\mathrm{d}}e_x\omega^{\mathrm{c}}_{\mathrm{d},z} \tag{9.77}$$

where k_{yp} is the proportional coefficient, and k_{yd} is the derivative coefficient. If $v^{\mathrm{c}}_{\mathrm{d},y} = v^{\mathrm{c}}_{\mathrm{d},y\mathrm{d}}$, Equation (9.58) becomes

$$\dot{e}_y = -\frac{k_{yp} - v^{\mathrm{c}}_{\mathrm{d},z}}{z^{\mathrm{c}}_{\mathrm{d}} + k_{yd}} e_y. \tag{9.78}$$

By choosing $k_{yp} > \max(v^{\mathrm{c}}_{\mathrm{d},z})$, it ensures $\lim_{t\to\infty} \left|e_y(t)\right| = 0$, indicating that the image error for this channel converges to 0.

The design of the desired velocity in the depth direction is consistent with the design experiment, and the angular velocity of the drogue is also ignored. It is assumed that $\omega^{\mathrm{c}}_{\mathrm{d}} = \omega^{\mathrm{re}}_{\mathrm{c}} = \mathbf{R}_{\mathrm{re/t}}\omega^{\mathrm{t}}_{\mathrm{c}} \approx [\omega_{y_\mathrm{b}}\ \omega_{z_\mathrm{b}}\ \omega_{x_\mathrm{b}}]^{\mathrm{T}}$. Treating $\mathbf{v}^{\mathrm{re}}_{\mathrm{d}}$ as disturbance, let $\mathbf{v}^{\mathrm{re}}_{\mathrm{c,d}} = \mathbf{v}^{\mathrm{c}}_{\mathrm{d,d}}$ and $\mathbf{v}^{\mathrm{c}}_{\mathrm{d}} = \mathbf{v}^{\mathrm{c}}_{\mathrm{d,d}}$.

According to Eq. (9.61), the improved desired velocity of the UAV relative to the ground, denoted as $\mathbf{v}^{\mathrm{g}}_{\mathrm{r,d}}$, is obtained as follows:

$$\begin{cases} v^{\mathrm{g}}_{\mathrm{r},xd} = -k_z z^{\mathrm{c}}_{\mathrm{d}} - k_1 \left|e_x\right| - k_2 \left|e_y\right| + V^{\mathrm{g}}_{\mathrm{t}} - z^{\mathrm{r}}_{\mathrm{c}} \Delta\omega_{y_{\mathrm{b}}} + y^{\mathrm{r}}_{\mathrm{c}} \Delta\omega_{z_{\mathrm{b}}} \\ v^{\mathrm{g}}_{\mathrm{r},yd} = k_{xp} e_x + k_{xd} \dot{e_x} + z^{\mathrm{c}}_{\mathrm{d}} e_x e_y \Delta\omega_{y_{\mathrm{b}}} - z^{\mathrm{c}}_{\mathrm{d}} (1 + e_x^2) \Delta\omega_{z_{\mathrm{b}}} + z^{\mathrm{c}}_{\mathrm{d}} e_y \Delta\omega_{x_{\mathrm{b}}} - x^{\mathrm{r}}_{\mathrm{c}} \Delta\omega_{z_{\mathrm{b}}} + z^{\mathrm{r}}_{\mathrm{c}} \Delta\omega_{x_{\mathrm{b}}} \\ v^{\mathrm{g}}_{\mathrm{r},zd} = k_{yp} e_y + k_{yd} \dot{e_y} - z^{\mathrm{c}}_{\mathrm{d}} e_x e_y \Delta\omega_{z_{\mathrm{b}}} + z^{\mathrm{c}}_{\mathrm{d}} (1 + e_y^2) \Delta\omega_{y_{\mathrm{b}}} - z^{\mathrm{c}}_{\mathrm{d}} e_y \Delta\omega_{x_{\mathrm{b}}} - y^{\mathrm{r}}_{\mathrm{c}} \Delta\omega_{x_{\mathrm{b}}} + x^{\mathrm{r}}_{\mathrm{c}} \Delta\omega_{y_{\mathrm{b}}} \end{cases} \tag{9.79}$$

where k_z is the proportional coefficient in the depth direction; k_1 and k_2 are feedback coefficients for adjusting the UAV velocity, slowing down the speed when the image error is too large to avoid overshooting.

Combining Eqs. (9.74) and (9.79), the IBVS controller is obtained with the ability to control airspeed, altitude, and course angle as

$$\begin{cases} v^{\mathrm{g}}_{\mathrm{r},xd} = -k_z z^{\mathrm{c}}_{\mathrm{d}} - k_1 \left|e_x\right| - k_2 \left|e_y\right| + V^{\mathrm{g}}_{\mathrm{t}} - z^{\mathrm{r}}_{\mathrm{c}} \Delta\omega_{y_{\mathrm{b}}} + y^{\mathrm{r}}_{\mathrm{c}} \Delta\omega_{z_{\mathrm{b}}} \\ v^{\mathrm{g}}_{\mathrm{r},yd} = k_{xp} e_x + k_{xd} \dot{e_x} + z^{\mathrm{c}}_{\mathrm{d}} e_x e_y \Delta\omega_{y_{\mathrm{b}}} - z^{\mathrm{c}}_{\mathrm{d}} (1 + e_x^2) \Delta\omega_{z_{\mathrm{b}}} + z^{\mathrm{c}}_{\mathrm{d}} e_y \Delta\omega_{x_{\mathrm{b}}} - x^{\mathrm{r}}_{\mathrm{c}} \Delta\omega_{z_{\mathrm{b}}} + z^{\mathrm{r}}_{\mathrm{c}} \Delta\omega_{x_{\mathrm{b}}} \\ v^{\mathrm{g}}_{\mathrm{r},zd} = k_{yp} e_y + k_{yd} \dot{e_y} - z^{\mathrm{c}}_{\mathrm{d}} e_x e_y \Delta\omega_{z_{\mathrm{b}}} + z^{\mathrm{c}}_{\mathrm{d}} (1 + e_y^2) \Delta\omega_{y_{\mathrm{b}}} - z^{\mathrm{c}}_{\mathrm{d}} e_y \Delta\omega_{x_{\mathrm{b}}} - y^{\mathrm{r}}_{\mathrm{c}} \Delta\omega_{x_{\mathrm{b}}} + x^{\mathrm{r}}_{\mathrm{c}} \Delta\omega_{y_{\mathrm{b}}} \\ V_{\mathrm{ad}} = \sqrt{(v^{\mathrm{g}}_{\mathrm{r},xd})^2 + (v^{\mathrm{g}}_{\mathrm{r},yd})^2} \\ \chi_{\mathrm{d}} = \arcsin \dfrac{v^{\mathrm{g}}_{\mathrm{r},yd}}{V_{\mathrm{ad}}} = \arcsin \dfrac{1}{\sqrt{(\frac{v^{\mathrm{g}}_{\mathrm{r},xd}}{v^{\mathrm{g}}_{\mathrm{r},yd}})^2 + 1}} \\ H_{\mathrm{d}} = \displaystyle\int_0^t v^{\mathrm{g}}_{\mathrm{r},zd}(\tau) \mathrm{d}\tau + H_0 \end{cases} \tag{9.80}$$

9.5.3 *Experimental Procedure*

The experimental procedure includes: first, completing the configuration of the PyTorch environment in Step 1; then, implementing the IBVS controller in "airspeed + altitude + course angle" control mode in Step 2; finally, running the HIL simulation in Step 3 to analyze the performance of autonomous aerial docking.

Step 1. Completing PyTorch Environment Configuration
Running YOLO image recognition in Python requires configuring PyTorch and a series of related packages. First, the software "Anaconda" needs to be installed on the system. After installation, open the command line and input "conda-V" to check if it is installed and view the current version of conda, as shown in Fig. 9.31. Then, use the command "conda create -n AAR_plat python = 3.9.6" to create a virtual environment named "AAR_plat" with Python version 3.9.6. "AAR_plat" folder can be found in the "envs" folder under the "Anaconda" installation directory. Next,

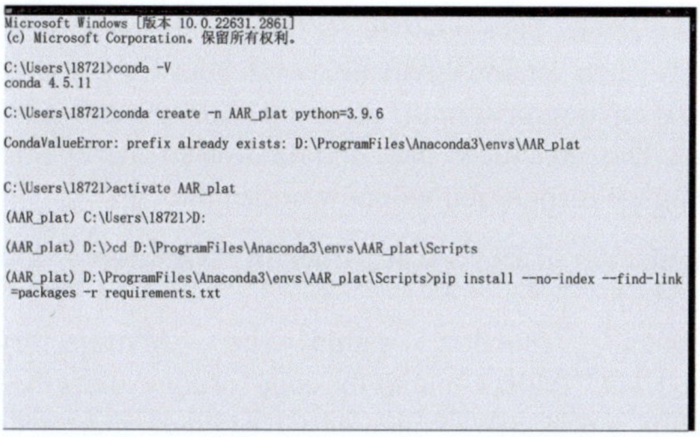

Fig. 9.31 Python environment configuration

use the "activate AAR_plat" command to activate the virtual environment named "AAR_sim", and use the "cd Anaconda/envs/AAR_plat/Scripts" command to open the folder for configuring the virtual environment in the command line. Place the "requirements.txt" file and the "packages" folder from the experimental package "e7-4" in the "envs/AAR_plat/Scripts" folder of the virtual environment, and enter the command "pip install SPSVERBc1no-index SPSVERBc1find-links=packages -r requirements.txt" in the command line. The virtual environment will automatically install all the packages in the "requirements.txt" file offline. If the user's computer does not have an Nvidia graphics card, the PyTorch environment configuration is completed, and the drogue recognition program based on YOLO will be calculated by the CPU. Users with Nvidia graphics cards need to complete the CUDA configuration further so that the drogue recognition program can use GPU computing to achieve better recognition results.

Below is the configuration for CUDA. As shown in Fig. 9.32, first, check the CUDA version supported by your GPU. Enter the command "nvidia-smi" in the command line to obtain the highest CUDA version supported by the GPU, as shown in the dashed box. Download and install the corresponding version of CUDA from

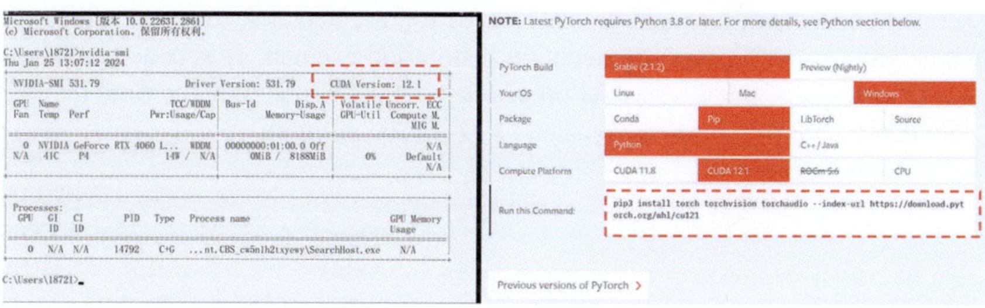

Fig. 9.32 Checking CUDA version

the Nvidia official website. Then, visit the PyTorch official website to find the download link for the PyTorch version compatible with your CUDA. After activating the "AAR_sim" virtual environment in the command line, paste the link and install the PyTorch package. This concludes the configuration of the PyTorch environment. Running YOLO image recognition will now utilize the GPU.

Step 2. Implementation of IBVS Controller in "Airspeed + Altitude + Course Angle" Control Mode

The "SimpleRoute()" function within the "Mytrajectory.py" file in "CopterSimulink_HITL" folder contains the code for controlling the UAV. The "SimpleRoute()" function is primarily divided into five phases, indicated by the flag variable. In the first phase, when "flag=0", the UAV is unlocked. In the second phase, when "flag=1", the UAV takes off and enters "Offboard" mode. In the third phase, when "flag=2", it enters the attitude control mode and stabilizes the climbing attitude. In the fourth phase, when "flag=3", it enters the "airspeed + altitude + course angle" control mode, maintaining a constant altitude and course angle while following the mothership. In the fifth phase, when "flag=4", the YOLO image recognition is activated, implementing the IBVS controller in the "airspeed + altitude + course angle" control mode. In this fifth phase, the IBVS controller takes the image error, depth estimation result, and the UAV's angular velocity as inputs and outputs the desired airspeed, desired course angle, and desired altitude. This is executed by the function "SendVelYawAlt", which represents the upper-level control interface for "airspeed + altitude + course angle". Corresponding to Eq. (9.80), the key code is shown in Code 9.6. The first two lines normalize the image error; lines 4–6 set the initial differential process deviation to 0; line 7 represents the position of the camera relative to the UAV center; line 8 represents the flight speed of the mothership; lines 9–15 set the control parameters; lines 16–20 determine whether the UAV has successfully docked with the mothership, and if so, the UAV accelerates to dock the recovery device; lines 21–27 correspond to Eq. (9.80), first obtaining the desired ground speed, then calculating the desired airspeed, desired course angle, and desired altitude; line 28 executes the desired command for "airspeed + altitude + course angle" upper-level controller; lines 29–30 record the previous image error, thereby calculating the differential process. It should be noted that the program uses PD control, and the differential process can be represented as $k_d \mathrm{d}e(t)/\mathrm{d}t$. Since the interval between each program execution instruction is equal, $\mathrm{d}t$ is constant. Then, the coefficient k_d of the differential process can be combined with the time interval $\mathrm{d}t$ into one coefficient, namely constant "k_d" in the program.

Code 9.6 IBVS controller under "airspeed + altitude + course angle" mode,from "e7/e7-4/CopterSimulink_HITL/Mytrajectory_HITL.py"

```
ex = (obj_xx - px) / 640 # unitisation of image error
ey = (obj_yy - py) / 360
ez = obj_zz
if flagIII == 1:
    ex_last = ex # first difference is zero
    ey_last = ey
d_pr_rc = [6.5, 0, -2.1] # the distance between the camera and the receiver
Vgt = 14.7 # mothership speed
k1 = 0.002 # set control parameters
k2 = 0.002
kx_p = 3.6
kx_d = 0.12
ky_p = 0.1
ky_d = 0.1
kz = 0.03
if abs(obj_xx - px) <= 30 and abs(obj_yy - py) <= 30 and obj_zz <= 0.5:
    Va = 15 # docking success when less image error and depth
    Chi = -mav.uavPosNED[1] / 50
    H = -92.2
    flag = 5
else:
    Vgx = kz*(ez)-k1*abs(ex)-k2*abs(ey)+Vgt-d_pr_rc[2]*q+d_pr_rc[1]*r # calculate desired velocity of receiver
    Vgy = kx_p*(ex)+kx_d*(ex-ex_last)+ez*ex*ey*q-ez*(1+ex**2)*r+ez*ey*
        p-d_pr_rc[0]*r+d_pr_rc[2]*p
    Vgz = ky_p*(ey)+ky_d*(ey-ey_last)-ez*ex*ey*r+ez*(1+ey**2)*q-ez*ey*
        p-d_pr_rc[1]*p+d_pr_rc[0]*q
    Va = math.sqrt(Vgx**2 + Vgy**2) # calculate desired velocity, angle and height of receiver
    Chi = math.asin(Vgy/Va)
    H = Vgz + H
mav.SendVelYawAlt(Va, Chi, H) # controller send commands
ex_last = ex # calculate the difference
ey_last = ey
```

Step 3. Running the HIL Simulation and Analyzing the Results of the Aerial Docking Experiment

Before starting the experiment, it is necessary to import the models into the RflySim3D model library. Copy the mothership model "WestTransportC130J", UAV model "MQ-9Reaper", and drogue model "ProbeLightScene" from the "module" folder to the "PX4PSP_RflySim3D_RflySim3D_Content" folder. In the HIL simulation experiment, the upper-level interface of the IBVS controller is changed to "velocity + altitude + course angle," and the flight control firmware must also be modified accordingly and then re-burned into the autopilot through the QGroundControl ground station. The modified flight control firmware corresponding to the model is stored in the "CopterSimulink_HITL/Firmware" folder, which includes two

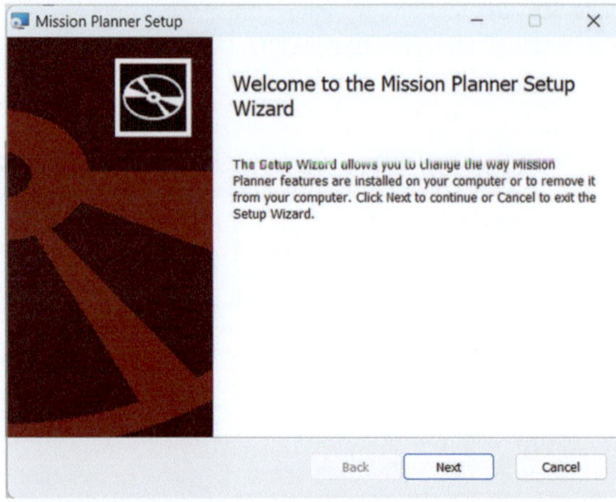

Fig. 9.33 Installation of the driver software

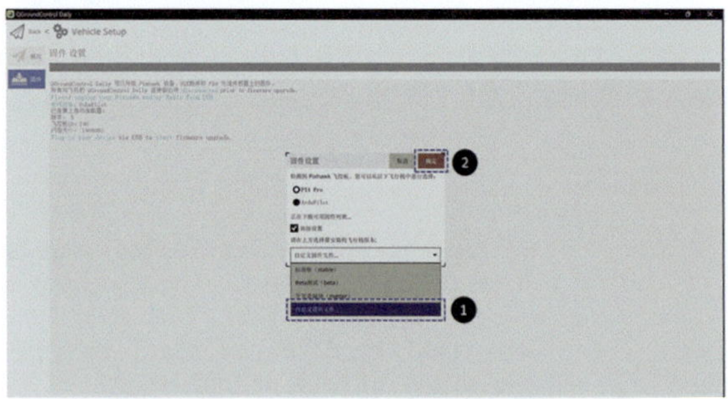

Fig. 9.34 Flashing the flight controller firmware

types: Cube series autopilot and autopilot using the "PX4_fmu-v5" firmware. Taking CubeOrange autopilot as an example, first, install the driver software for the Cube autopilot. Run the "MissionPlanner-1.3.76.msi" installation program in the "CopterSimulink_HITL/Firmware" folder, as shown in Fig. 9.33, to complete the installation of the driver software. After successful installation, burn the flight control firmware to the autopilot through the QGroundControl ground station. As shown in Fig. 9.34, open the QGroundControl ground station, select "Firmware Setup - Advanced Settings - Custom Firmware File - OK" in the aircraft settings, then in the pop-up file manager, find the "CopterSimulink_HITL/Firmware/Cubepilot" folder, and select the "cubepilot_cubeorange_default.px4" file. Then the QGroundControl ground station will perform the flight control firmware burning. When prompted with "Upgrade Complete", the burning is completed, as shown in Fig. 9.35.

After completing the firmware burning, proceed with the HIL simulation experiment. As shown in Fig. 9.36, connect the CubePilot/Pixhawk autopilot to the computer. Then, open the "CopterSimulink_HITL" folder within the experimental package "e7-4". Run the "PX4SimHITL.bat" batch file. A window similar to Fig. 9.37 will pop up. Enter the recommended serial port number in the terminal window and press "Enter" to start the RflySim toolchain in one click.

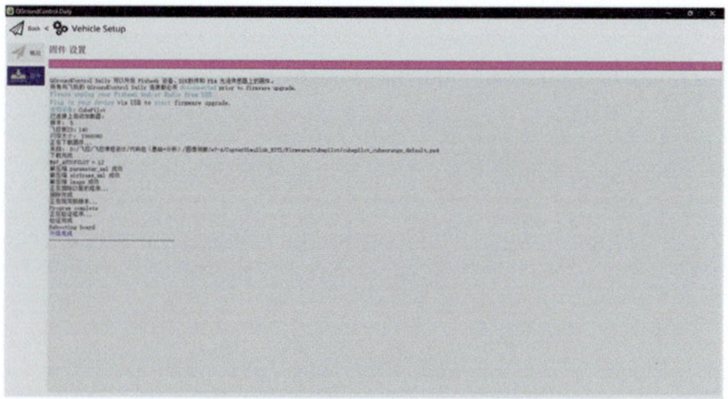

Fig. 9.35 Firmware flashing success

Fig. 9.36 Hardware layout of HIL simulation

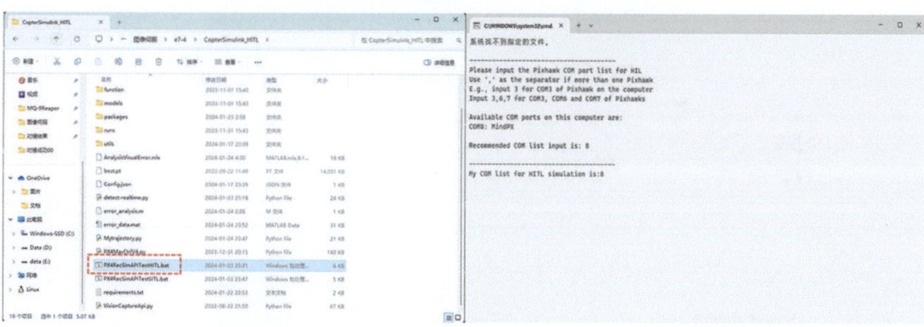

Fig. 9.37 Setting up RflySim HIL simulation platform

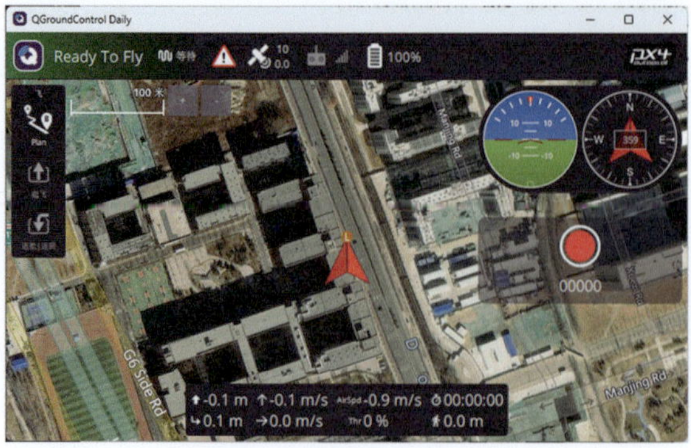

Fig. 9.38 Ready to fly

Switch to the QGC ground station interface, as shown in Fig. 9.38. After the top left corner of the ground station displays the aircraft in the "Ready to fly" state, run the mothership motion simulation platform "AirRefueling_Platform.slx" in the "Simulink" folder. As shown in Fig. 9.39, the mothership motion simulation platform mainly consists of four parts: ① refueling module "Tanker system", ② drogue module "Hose-drogue system", ③ environment setting module "Natural wind model", and ④ mothership trajectory planning module "3D rendering engine". The refueling module, drogue module, and environment setting module are basically the same as those introduced in the basic experiment. The motion speed and trajectory of the mothership can be set in the mothership trajectory planning module, yielding a path for "AirRefueling_Platform/3D rendering engine", using the "MATLAB Function" module. The input of the "MATLAB Function" module is time, and the output is

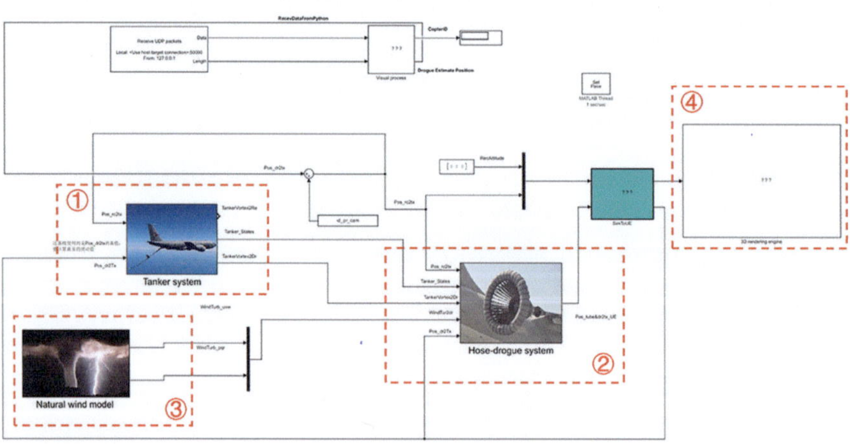

Fig. 9.39 Mothership motion Simulink simulation, from "e7/e7-4/Simulink/AirRefueling_Platform.slx"

the position of the mothership. The specific code is as follows in Code 9.7. In this experiment, the mothership is set to fly in a straight line at a constant altitude, with starting coordinates of $(-100, -119, -100)$ and a speed of 15 m/s. Note that the coordinate system here is the North-East-Down coordinate system, so the mothership flies in the direction along the north at a constant altitude of 100 m. Then, run the UAV control program "Mytrajectory.py" in the "CopterSimulink_HITL" folder first. Wait for the compiler terminal to output "Waiting for client to connect to named pipe..." and then run the drogue recognition program "detect-realtime.py". Note that, to save computer memory, image processing will only start when the UAV is close enough to the mothership. The onboard camera will not be activated in the first four phases.

Code 9.7 Setting the mothership speed and motion trajectory

```
function y = fcn(t)
    y = [-100 -119 -100]; % initial position
    v1 = 15; % velocity
    y(1) = y(1) + v1 * t; % trajectory
end
```

During the UAV flight process, the aircraft's coordinates, three-dimensional velocity, and attitude angle can be observed in real-time through CopterSim, as shown in the figure. At the same time, aircraft motion can be observed in real time through RflySim3D. Figure 9.40 shows the visual display of UAV docking with the mothership.

The onboard camera's visual display of the IBVS docking process is shown in Fig. 9.41, indicating the successful completion of the docking task between the UAV and the mothership. Throughout the UAV-mothership docking process, the "Mytrajectory.py" program continuously records image errors, depth errors, and position errors, saving them in the "error_data.mat" file. Run error analysis plotting program "AnalysisVisualError.mlx", and then generate error curves for the docking process, as shown in Fig. 9.42.

The image error represents the difference between the center of the drogue and the position of the UAV probe in the image coordinate system. The position error represents the difference between the absolute position of the drogue and the UAV probe in the world coordinate system. From the curves, it can be observed that both the image error and the position error converge to zero within 40 s. From the depth curve,

Fig. 9.40 Docking visual display

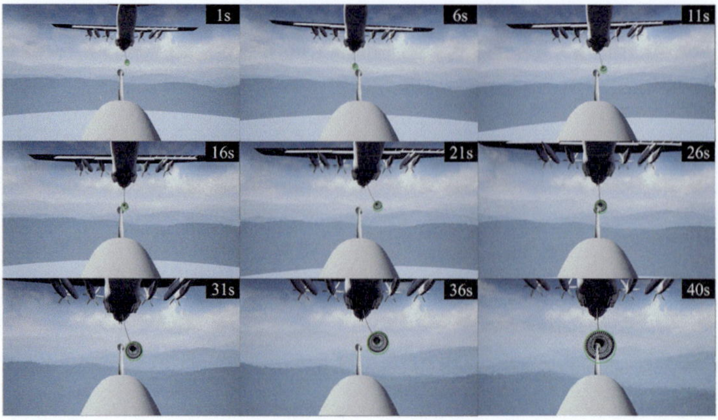

Fig. 9.41 Docking process

Fig. 9.42 Error response in HIL simulation

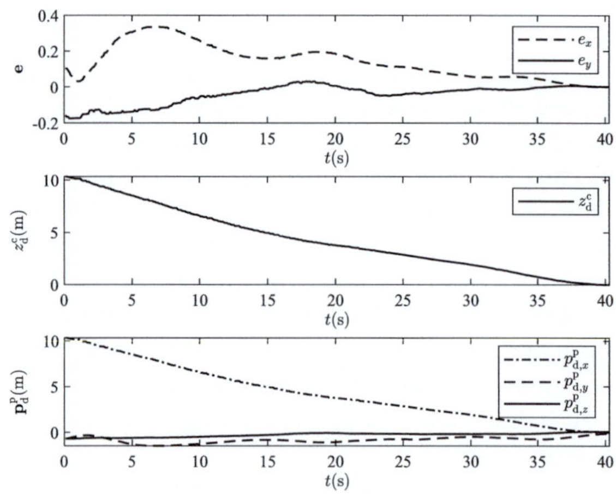

it can be observed that the smaller the depth error, the slower the rate of convergence of the depth error, consistent with the design of the IBVS controller in the depth direction. Although oscillations appear in the curves of image error and position error, these oscillations continuously converge, and eventually the errors converge to zero. This implies that the improved PD controllers in the lateral and longitudinal directions effectively attenuate UAV's oscillations and swings, demonstrating good disturbance rejection capability. Even without the stabilizing performance of the inner-loop LQR method, the docking error can still converge to zero, successfully completing the docking task.

9.6 Exercise

1. The drogue identification and localization in this chapter are based on the YOLO neural network implementation. Please attempt to use the target-based method provided in Sect. 9.1.1 to achieve the goal of identification and localization.
2. Based on RflySimAAR presented in this chapter, try to use other control methods, such as iterative learning control based on position servo, as mentioned in [14].
3. The LQR control is utilized for the inner loop control in this chapter. Please explore advanced control algorithms to replace the LQR control for the inner loop and analyze its performance.
4. How would the uncertainty caused by incorrect intrinsic camera calibration be reflected in the model? What are the different impacts on PBVS and IBVS? Please analyze in detail.

References

1. F. Chaumette, S. Hutchinson, Visual servo control. I. basic approaches. IEEE Robot. Autom. Mag. **13**(4), 82–90 (2006)
2. G. Jocher, YOLOv5 by ultralytics, version 7.0, May 2020. https://doi.org/10.5281/zenodo.3908559. [Online]. Available: https://github.com/ultralytics/yolov5
3. Q. Quan, R. Liu, H. Liu, Z. Ma, J. Ren, An image based visual servo method for probe-and-drogue autonomous aerial refueling. J. Guidance Control Dyn. **48**(7), 1689–1698 (2025). https://doi.org/10.2514/1.G008889
4. D. Lee, T. Ryan, H.J. Kim, Autonomous landing of a VTOL UAV on a moving platform using imagebased visual servoing, in *2012 IEEE International Conference on Robotics and Automation* (IEEE, 2012), pp. 971–976
5. J.S. Wynn, *Visual Servoing for Precision Shipboard Landing of an Autonomous Multirotor Aircraft System* (Brigham Young University, 2018)
6. Y. Chen, Y. Wu, Z. Zhang, Z. Miao, H. Zhong, H. Zhang, Y. Wang, Image-based visual servoing of unmanned aerial manipulators for tracking and grasping a moving target. IEEE Trans. Ind. Inform. **19**(8), 8889–8899 (2022)
7. C. Qin, Q. Yu, H.H. Go, H.H.-T. Liu, Perception-aware image-based visual servoing of aggressive quadrotor UAVs. IEEE/ASME Trans. Mechatron. **28**(4), 2020–2028 (2023)
8. K. Yang, Q. Quan, An autonomous intercept drone with image-based visual servo, in *2020 IEEE International Conference on Robotics and Automation (ICRA)* (IEEE, 2020), pp. 2230–2236
9. H. Tao, D. Lin, S. He, T. Song, R. Jin, Optimal terminal-velocity-control guidance for intercepting non-cooperative maneuvering quadcopter. J. Field Robot. **39**(4), 457–472 (2022)
10. G. Wang, J. Qin, Q. Liu, Q. Ma, C. Zhang, Image-based visual servoing of quadrotors to arbitrary flight targets. IEEE Robot. Autom. Lett. **8**(4), 2022–2029 (2023)
11. J. Ning, H. Zhang, Q. Quan, Dynamic obstacle avoidance of quadcopters with monocular camera based on image-based visual servo, in *2022 International Conference on Unmanned Aircraft Systems (ICUAS)* (IEEE, 2022), pp. 150–156
12. H. Duan, Y. Sun, Y. Shi, Bionic visual control for probe-and-drogue autonomous aerial refueling. IEEE Trans. Aerosp. Electron. Syst. **57**(2), 848–865 (2020)
13. J. Valasek, K. Gunnam, J. Kimmett, M. Tandale, J. Junkins, D. Hughes, Vision-based sensor and navigation system for autonomous air refueling. J. Guidance Control Dyn. **28**(5), 979–989 (2005)

14. X. Dai, Q. Quan, J. Ren, Z. Xi, K.-Y. Cai, Terminal iterative learning control for autonomous aerial refueling under aerodynamic disturbances. J. Guidance Control Dyn. **41**(7), 1577–1584 (2018)

Chapter 10
Swarm Flight Control Experiment

Yue Li, QingPeng Chen, Xunhua Dai, and Quan Quan

In recent years, Unmanned Aerial Vehicles (UAVs) have undergone rapid development and are now widely used in various fields. The research focus on UAV-related technologies has gradually shifted from single-aircraft flight to distributed coordination, driven by advancements in technology and hardware. A UAV swarm is a group system composed of multiple agents that execute collective tasks through various instructions. UAV swarms offer convenient and efficient solutions to various problems, addressing issues such as the low fault tolerance and limited efficiency of single UAVs. This chapter introduces the principles of fixed-wing UAV swarm flight and commonly used control methods. For more information on this chapter, please refer to the link

https://rflysim.com/doc/en/C/5.FW_Book.html.

10.1 Preliminary

10.1.1 Basic Concepts

In nature, many organisms exhibit characteristics of group behavior. These organisms have simple organizational structures but can accomplish complex tasks through communication and cooperation, such as foraging in bee colonies, transporting in ant colonies, migrating birds, and schooling fish [1–3]. The UAV swarm comprises numerous uncomplicated agents that can achieve desired tasks through local interaction or perception [4–6]. Flight control of UAV swarms can mainly be categorized into centralized and distributed approaches.

The centralized approach involves processing all information in the UAV swarm through a central hub, which can result in comprehensive information and potentially optimal results. However, this approach has several drawbacks, including poor robustness and a heavy computational load. If the central node fails, the entire multi-

Q. Quan et al., *Micro-Small Fixed-Wing UAV Control*,
https://doi.org/10.1007/978-981-95-1954-5_10

UAV system will fail. Additionally, communication resources are burdened due to the required links between the central node and other UAVs. High reliability is essential for a large amount of information exchange to avoid potential conflicts. As the swarm scales up, the sharply increased computational load becomes challenging to handle. The multi-UAV system in the distributed scheme does not require a central node. In some cases, distributed schemes can operate autonomously without wireless communication or identification of other UAVs. Distributed control only requires that UAVs acquire essential environmental information, such as the positions, velocities, and attitudes of different UAVs, as well as the presence of obstacles or threats in the environment. The use of distributed control enables high flexibility and robustness for the UAV swarm, allowing them to complete assigned tasks even with partial node failures or unreliability. However, the distributed structure of UAVs can lead to locally optimal decisions rather than globally optimal ones, potentially causing the entire swarm to fail to exhibit the desired behavior.

As depicted in Fig. 10.1, a scene from the movie "Angel Has Fallen" showcases a combat scenario involving a swarm of micro-small fixed-wing UAVs. This innovative attack strategy offers a new perspective and is likely to become a common force in future warfare, potentially reshaping traditional combat methods. The capability to rapidly and safely navigate to the target location is crucial in assessing the feasibility of a UAV swarm in accomplishing its combat objective. This is also a pressing issue that requires immediate attention in the practical implementation of UAV swarm combat. The United States' Defense Advanced Research Projects Agency (DARPA) has introduced a novel concept known as "Mosaic Warfare". Mosaic Warfare is an approach that considers combat as an emerging complex system. It utilizes low-cost UAV swarms and other electronic and network effects to overwhelm opponents. The approach employs distributed control, which relies only on the state information of nearby neighboring UAVs. This results in high task completion reliability, making it suitable for large-scale, dense UAV swarms.

Fig. 10.1 Scene from the movie "Angel Has Fallen"

10.1.2 Controlled Model

For fixed-wing UAV swarms, it is quite challenging to analyze and control using the six-degree-of-freedom full-state equation. Simplified models of fixed-wing UAVs, commonly referred to as *guidance models*, are frequently employed in the development of swarm control algorithms. In Chap. 5, Eq. (5.1) represents a guidance model of fixed-wing UAVs. When there is no wind or wind speed is low, v_{wx}, v_{wy}, and $f\left(v_{\mathrm{wx}}, v_{\mathrm{wy}}\right)$ can all be ignored, yielding

$$\begin{cases} \dot{p}_{x_{\mathrm{e}}} = V_{\mathrm{e}} \cos \psi \\ \dot{p}_{y_{\mathrm{e}}} = V_{\mathrm{e}} \sin \psi \\ \psi = \chi \\ V_{\mathrm{e}}(s) = G_{V_{\mathrm{e}}}(s) V_{\mathrm{ed}}(s) \\ \chi(s) = G_{\chi}(s) \chi_{\mathrm{d}}(s) \\ H(s) = G_H(s) H_{\mathrm{d}}(s) \end{cases} \tag{10.1}$$

where $p_{x_{\mathrm{e}}}$ is the northward position of the fixed-wing UAV in the North-East-Down coordinate system; $p_{y_{\mathrm{e}}}$ is the eastward position of the fixed-wing UAV in the North-East-Down coordinate system; V_{e} is the forward flight speed of the fixed-wing UAV in the North-East-Down coordinate system; ψ is the yaw angle of the fixed-wing UAV, χ is the course angle; H is the actual altitude of the fixed-wing UAV relative to the ground; H_{d} is the desired altitude of the fixed-wing UAV relative to the ground; V_{ed} is the desired forward flight speed of the fixed-wing UAV relative to the ground; χ_{d} is the desired course angle of the fixed-wing UAV.

As shown in Eq. (10.1), the guidance model of the fixed-wing UAV has three control channels, namely speed, course angle (or yaw), and altitude. Define the position of the fixed-wing UAV in the two-dimensional plane as $\mathbf{p} = [p_{x_{\mathrm{e}}} \quad p_{y_{\mathrm{e}}}]^{\mathrm{T}}$, and the velocity as $\mathbf{u} = [v_x \quad v_y]^{\mathrm{T}}$. In the controller design process, a velocity controller for the point mass model in the two-dimensional plane is defined as

$$\dot{\mathbf{p}} = \mathbf{u} \tag{10.2}$$

and then map it to the guidance model.

If the control input $\mathbf{u}$ tends to $\mathbf{u}_{\mathrm{d}} = [v_{x\mathrm{d}} \quad v_{y\mathrm{d}}]^{\mathrm{T}}$, then the position $\mathbf{p}$ tends to $\mathbf{p}_{\mathrm{d}}$. The desired forward flight speed V_{ed} is solved as

$$V_{\mathrm{ed}} = \sqrt{V_{\mathrm{ed}}^2 \left(\cos^2\psi_{\mathrm{d}} + \sin^2\psi_{\mathrm{d}}\right)} = \sqrt{v_{x\mathrm{d}}^2 + v_{y\mathrm{d}}^2} \tag{10.3}$$

and the desired yaw angle ψ_{d} as

$$\psi_{\mathrm{d}} = \arctan\left(\frac{v_{x\mathrm{d}}}{v_{y\mathrm{d}}}\right). \tag{10.4}$$

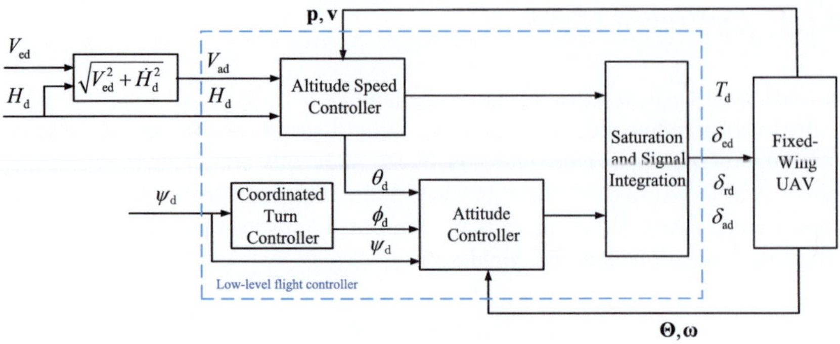

Fig. 10.2 Guidance control structure utilizing a low-level flight controller

Therefore, a controller for the point mass model can be designed to obtain the desired velocity $\mathbf{u}_d$. Then it can be mapped to the fixed-wing guidance model to obtain the desired forward flight speed V_{ed} and desired yaw angle ψ_d, along with the input of the desired altitude H_d, resulting in the control for the fixed-wing guidance model (10.1). Figure 10.2 is the guidance and control structure based on the low-level flight controller shown in Chap. 6; the input to the low-level flight controller is the desired airspeed V_{ad}, the desired yaw angle ψ_d, and the desired altitude H_d, and the output of the low-level flight controller is four control commands (elevator deflection δ_{ed}, aileron deflection δ_{ad}, rudder deflection δ_{rd} and throttle thrust T_d).

10.1.3 Swarm Control Method

1. Formation control

In the formation control strategy of UAVs, a simple approach is to use the "leader-follower" method [7]. The leader follows a planned path, and each follower controls its relative position to the leader to form a formation. Typically, the leader is not necessarily a real UAV. It can also be a virtual point. As the number of UAVs increases, multi-level structures may appear in the formation. For example, the leader is considered the priority and transmits information to second-level followers. Similarly, each level of followers also sends information to its lower-level followers. Based on this principle, UAVs can eventually form a formation to execute tasks. However, this approach has a clear drawback: the formation structure is highly dependent on the leader and the communication between UAVs, which can amplify errors at each level.

(1) Leader controller design

The leader controller design problem mainly focuses on guiding UAVs through a series of waypoints to complete the entire path. This part of the design can be considered from two aspects: the lateral position error of UAVs from the trajectory and

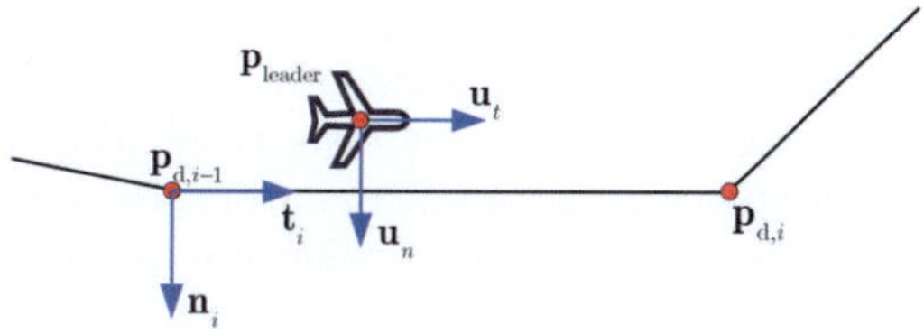

Fig. 10.3 A UAV flies along a series of desired paths

the speed error from the desired speed along the trajectory. As shown in Fig. 10.3, $\mathbf{p}_{\text{leader}}$ represents the position of the leader UAV, $\mathbf{p}_{\text{d},i-1}$ represents the $(i-1)$th waypoint, and $\mathbf{p}_{\text{d},i}$ represents the ith waypoint. The segment from the $(i-1)$th waypoint to the ith waypoint is referred to as the ith segment, and the unit vector in the tangent direction on this segment is defined as $\mathbf{t}_i$, and the unit vector in the normal direction is $\mathbf{n}_i$.

For the leader adopting a point mass model as

$$\dot{\mathbf{p}}_{\text{leader}} = \mathbf{v}_{\text{leader}} \tag{10.5}$$

the lateral position error of the ith segment is

$$\mathbf{e}_{\text{n}} = \left(\left(\mathbf{p}_{\text{leader}} - \mathbf{p}_{\text{d},i}\right)^{\text{T}} \mathbf{t}_i\right)\mathbf{t}_i - \left(\mathbf{p}_{\text{leader}} - \mathbf{p}_{\text{d},i}\right). \tag{10.6}$$

Since the position of the ith waypoint $\mathbf{p}_{\text{d},i}$ is a constant, its derivative is 0. By taking the derivative of the position error $\mathbf{e}_{\text{n}}$, one gets

$$\dot{\mathbf{e}}_{\text{n}} = \left(\mathbf{v}_{\text{leader}}^{\text{T}}\mathbf{t}_i\right)\mathbf{t}_i - \mathbf{v}_{\text{leader}}. \tag{10.7}$$

The deviation of the leader's tangential velocity from the desired velocity on the trajectory is described as

$$\dot{\mathbf{e}}_{\text{t}} = \mathbf{v}_{\text{d},i} - \left(\mathbf{v}_{\text{leader}}^{\text{T}}\mathbf{t}_i\right)\mathbf{t}_i. \tag{10.8}$$

The leader's tangential and normal control commands are designed as

$$\mathbf{u}_{\text{t}} = K_{\text{tp}}\dot{\mathbf{e}}_{\text{t}} + K_{\text{ti}}\int \dot{\mathbf{e}}_{\text{t}} \tag{10.9}$$

$$\mathbf{u}_{\text{n}} = K_{\text{np}}\mathbf{e}_{\text{n}} + K_{\text{nd}}\dot{\mathbf{e}}_{\text{n}} + K_{\text{ni}}\int \mathbf{e}_{\text{n}} \tag{10.10}$$

where, $K_{\text{tp}}, K_{\text{ti}}, K_{\text{np}}, K_{\text{nd}}, K_{\text{ni}} \in \mathbb{R}$. Therefore, the controller for the leader is finally designed as

$$\mathbf{v}_{\text{leader}} = \mathbf{u}_{\text{t}} + \mathbf{u}_{\text{n}}. \tag{10.11}$$

(2) Follower controller design

Compared to the leader, the follower controller also needs to maintain the formation. The control system can control the position difference between the follower and the leader, ensuring that the follower maintains a certain distance and angle with respect to the leader. The follower's point mass model is

$$\dot{\mathbf{p}}_{\text{follower},j} = \mathbf{v}_{\text{follower},j}. \tag{10.12}$$

The position difference between the follower and the leader is described as

$$\mathbf{p}_{\text{leader},j} = \mathbf{p}_{\text{leader}} - \mathbf{p}_j. \tag{10.13}$$

The control term for the follower to follow the leader is designed as

$$\mathbf{u}_{\text{f}} = k_{\text{p}} \left(\mathbf{d}_j - \mathbf{p}_{\text{leader},j} \right) + k_{\text{d}} \mathbf{v}_{\text{leader},j}. \tag{10.14}$$

Here, $\mathbf{d}_j$ is the desired relative position between the follower and the leader, k_{p} and k_{d} are the proportional and derivative coefficients in the PD control, respectively; and $k_{\text{d}}\mathbf{v}_{\text{leader},j}$ corresponds to the derivative term of this tracking process. Adding $\mathbf{u}_{\text{f}}$ to the leader's navigation control, the follower controller can be shown as

$$\mathbf{v}_{\text{follower}} = \mathbf{u}_{\text{t}} + \mathbf{u}_{\text{n}} + \mathbf{u}_{\text{f}}. \tag{10.15}$$

The advantage of UAV formation control using the leader-follower architecture lies in the well-defined geometric relationships for each UAV, making the selection of a leader straightforward and efficient.

2. Control within a virtual tube

The formation control introduced in the previous section is a commonly used method for controlling a UAV swarm. However, as the environment becomes more complex and the scale of the UAV swarm increases, the formation structure will become somewhat fragile, and its adaptability and stability evidently may not meet the requirements. For example, a single UAV's abnormal change can easily lead to confusion in the formation. In addition to this, another approach to UAV swarm control is path planning for multiple UAVs, which considers the simultaneous path planning of all UAVs. Although this method can effectively enhance the adaptability and flexibility of the swarm to its environment, it faces challenges such as low computational efficiency and difficulties in conflict resolution as the number of UAVs increases. Furthermore, this approach requires extensive communication among UAVs. In this context, inspired by fluid motion in nature and traffic flow in vehicles, a new framework based on "virtual tubes" has been proposed [8, 9]. A virtual tube serves only as a constraint and guidance for the swarm. Moreover, the controller treats neighboring UAVs only as obstacles without requiring their ID information, solving the communication denial problem.

Fig. 10.4 A virtual tube with a generating curve [9]

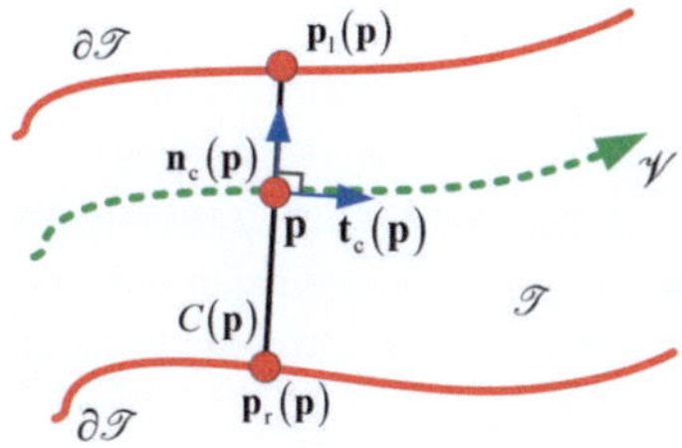

As shown in Fig. 10.4, a virtual tube is considered as generated on a curve $\mathscr{V} \in \mathbb{R}^2$. For a point $\mathbf{p}$ on the tube, its tangent direction unit vector is $\mathbf{t}_c(\mathbf{p})$, and its normal direction unit vector is $\mathbf{n}_c(\mathbf{p})$. The core of generating virtual tubes lies in the generating curve.

The mathematical definitions of the relevant concepts in virtual tubes are as follows:

- Curve: For any point $\mathbf{p} \in \mathscr{V}$ on the curve, define the vector $\mathbf{t}_c(\mathbf{p}) \in \mathbb{R}^2$ as the unit tangent vector pointing in the direction of progress along the curve at that point, and this vector is continuously differentiable. The direction of progress of the curve is the direction of flight for the UAV swarm inside the virtual tube. Similarly, define $\mathbf{n}_c(\mathbf{p}) \in \mathbb{R}^2$ as the unit tangent vector pointing in the direction orthogonal to the direction of progress of the curve at that point. It can be concluded that

$$\mathbf{t}_c^{\mathrm{T}}(\mathbf{p})\mathbf{n}_c(\mathbf{p}) = 0. \tag{10.16}$$

- Cross-section: For any point $\mathbf{p} \in \mathscr{V}$, the cross-section of the tube passing through that point is defined as

$$\mathscr{C}(\mathbf{p}) = \left\{\mathbf{p} + \lambda(\mathbf{p})\mathbf{n}_c(\mathbf{p}) \in \mathbb{R}^2 | \lambda_l(\mathbf{p}) \leq \lambda(\mathbf{p}) \leq \lambda_r(\mathbf{p}), \lambda_l(\mathbf{p}), \lambda_r(\mathbf{p}) \in \mathbb{R}\right\}. \tag{10.17}$$

For any point $\mathbf{p}' \in \mathscr{C}(\mathbf{p})$, define $\mathscr{C}(\mathbf{p}')$ as the cross-section passing through point $\mathbf{p}'$. Clearly, $\mathscr{C}(\mathbf{p}') = \mathscr{C}(\mathbf{p})$. For $\mathbf{x}_1 \in \mathscr{C}(\mathbf{p}_1)$, $\mathbf{x}_2 \in \mathscr{C}(\mathbf{p}_2)$ and $\mathbf{p}_1, \mathbf{p}_2 \in \mathscr{V}$, if any point can move from $\mathbf{p}_1$ to $\mathbf{p}_2$ along the unit tangent vector $\mathbf{t}_c(\cdot)$, it indicates that $\mathbf{x}_1$ or $\mathscr{C}(\mathbf{p}_1)$ is behind $\mathbf{x}_2$ or $\mathscr{C}(\mathbf{p}_2)$. Furthermore, the two endpoints of the cross-section are defined as

$$\begin{aligned} \mathbf{p}_l(\mathbf{p}) &= \mathbf{p} + \lambda_l(\mathbf{p})\mathbf{n}_c(\mathbf{p}) \\ \mathbf{p}_r(\mathbf{p}) &= \mathbf{p} + \lambda_r(\mathbf{p})\mathbf{n}_c(\mathbf{p}). \end{aligned} \tag{10.18}$$

These two endpoints are the boundary points of the curved tube corresponding to the point $\mathbf{p} \in \mathscr{V}$. Next, the midpoint $\mathbf{p}_m \in \mathscr{C}(\mathbf{p})$ of the cross-section, defined as a function of the point $\mathbf{p} \in \mathscr{V}$, is represented as

$$\mathbf{p}_m = m(\mathbf{p}) = \frac{1}{2}\left(\mathbf{p}_l(\mathbf{p}) + \mathbf{p}_r(\mathbf{p})\right). \tag{10.19}$$

The width of the cross-section $C(\mathbf{p})$ is denoted as $2r_{\mathrm{t}}(\mathbf{p})$, where $r_{\mathrm{t}}(\mathbf{p}) = \frac{1}{2}\left|\lambda_{\mathrm{l}}(\mathbf{p}) - \lambda_{\mathrm{r}}(\mathbf{p})\right|$.
Based on the definitions of curve and cross-section, a definition for the two-dimensional curved tube is given here.

- Two-dimensional Curved Tube: For a two-dimensional curved tube, the tube is composed of cross-sections at various points on the curve, and it is represented as

$$\mathscr{T} = \cup_{\mathbf{p}\in\mathscr{V}}\mathscr{C}(\mathbf{p}). \tag{10.20}$$

The boundary of the two-dimensional curved tube $\mathscr{T}$ is represented as

$$\partial\mathscr{T} = \{\mathbf{x}|\mathbf{x} = \mathbf{p}_{\mathrm{l}}(\mathbf{p}) \cup \mathbf{p}_{\mathrm{r}}(\mathbf{p}), \mathbf{p} \in \mathscr{V}\}. \tag{10.21}$$

For the ith UAV, the velocity control based on the virtual tube $\mathbf{v}_{\mathrm{c},i}$ consists of three components: the forward flight control term $\mathbf{v}_{\mathrm{f},i}$, the collision avoidance control term $\mathbf{v}_{\mathrm{m},i}$, and the term to maintain within the virtual tube $\mathbf{v}_{\mathrm{t},i}$:

$$\mathbf{v}_{\mathrm{c},i} = \mathrm{sat}\left(\mathbf{v}_{\mathrm{f},i} + \mathbf{v}_{\mathrm{m},i} + \mathbf{v}_{\mathrm{t},i}, v_{\max}\right). \tag{10.22}$$

Here, $v_{\max}$ represents the maximum flight velocity of the UAV. The forward flight control term is used to guide each UAV to move forward along the virtual tube, the collision avoidance control term is used for inter-UAV collision avoidance, and the term to maintain within the virtual tube is used to constrain all UAVs within the tube. The specific design methods for these three components will be introduced in the experimental design.

10.2 Basic Experiment

10.2.1 Experimental Objective

1. Things to prepare

(1) Software: MATLAB R2022b or above.
(2) Program: Experimental package "e8/e8-1" includes a fixed-wing UAV guidance model, an example of the basic experiment of fixed-wing UAV swarm flight. For example, "GuidanceModel.m" is the fixed-wing UAV guidance model, "initFormation.m" is the initialization file for simulation, and "formation_8_1.slx" is the simulation file for fixed-wing UAV formation flight.

2. Objective

(1) Understand the fixed-wing UAV guidance model and master the method of fixed-wing UAV formation flight.
(2) UAV flight paths and complete fixed-wing UAV formation flight using Simulink.

10.2.2 Experimental Procedure

The objective of the basic experiment is to enhance understanding of the core principles of formation flight, with an emphasis on designing the formation controller and developing guidance models using S-functions for simulation. The basic experiment is divided into four steps: defining the initial condition, designing a fixed-wing formation controller, developing a fixed-wing guidance model, and integrating the controller with the model. Step 1 defines initial conditions by assigning values to parameters in the simulation. Subsequently, Steps 2 and 3 write the controller and model, which are ultimately used in fixed-wing formation tasks in Step 4.

Step 1: Initial condition definition

In the simulation, the initial state of the fixed-wing UAV is defined by the following variables: position along the x-coordinate, position along the y-coordinate, altitude, altitude change rate, speed, yaw angle, and yaw rate. Although altitude and its derivative are included in the model, only two-dimensional control is considered in the experiment. In this basic experiment, the UAV swarm is assumed to have a three-level structure. The first UAV in the sequence is the first priority, the second and third UAVs are the second priority, and the remaining UAVs are the third priority.

First, initialize the state of the fixed-wing UAV swarm. In the "e8-1" folder, there is an initialization file, "initFormation.m", which specifies the number of UAVs and simulation step size, and defines the initial positions, speeds, yaw angles, and corresponding initial states of the UAVs. The example code is shown in Code 10.1.

Code 10.1 Initialization of a fixed-wing swarm, from "e8/e8-1/initFormation.m"

```
global N UAV_Initial ts
N=7;%Number of UAVs
ts=0.01;%Simulation step size
UAV_PInit=[0 0 0;-5 0 0;-5 -5 0;-5 -10 0;-10 0 0;-10 -5 0;-10 -10 0];%Initial location of the UAV
UAV_VInit=[1;1;1;1;1;1;1];%Initial speed of the UAV
UAV_YawInit=[0;0;0;0;0;0;0];%Initial yaw angle of the UAV
UAV_Initial=[UAV_PInit(:,1);UAV_PInit(:,2);UAV_PInit(:,3);zeros(N,1);UAV_VInit;UAV_YawInit;zeros(N,1)];
%Initial state of the UAV[x,y,h,dot_h,V,yaw,dot_yaw]
```

An example code is provided in Code 10.2 to generate the trajectory for the leader UAV. The flight trajectory of the UAV is designed in the form of a sine function. A series of points on a two-dimensional plane are defined as waypoints $\mathbf{p}_{d,i}$ in Eq. (10.6), and written into "point", as shown in lines 1–3 of the code. On each path segment, the desired speed "vdes" for the fixed-wing UAV is defined as line 5. Constraints are also applied to the speed of the fixed-wing, with its minimum speed defined as "v_{min}" and maximum speed as "v_{max}".

Code 10.2 Generating desired waypoints and setting speeds, from "e8/e8-1/initFormation.m"

```
1 pointx=0.1:4:100.1;%The x-axis coordinate of the path point
2 pointy=10*sin(pointx/4/pi);%The y-axis coordinate of the path point
3 point=[pointx' pointy']%The coordinate of the path point
4 len=length(point(:,1));%The number of waypoints
5 vdes=5;%The desired speed
6 v_min=1;%The minimum speed
7 v_max=8;%The maximum speed
```

Step 2: Writing a fixed-wing formation controller

A fixed-wing formation controller is designed to guide the formation along a desired path. It employs various control strategies for both the leader and the followers. The leader's controller focuses on waypoint tracking, managing both lateral and longitudinal control in each segment. This is illustrated in Eqs. (10.9) and (10.10), where tangential input $\mathbf{u}_t$ and normal input $\mathbf{u}_n$ are presented. The follower controller is primarily concerned with tracking the leader and maintaining the formation structure. Code 10.3 shows the PID control for the leader UAV. The calculation of PID parameters is shown in lines 1–3. For simplicity in tuning, the integral coefficient in the PID parameters given here is set to 0.

It should be noted that the criterion for switching to the following path segment is to check whether the angle between the line connecting the current position and the current waypoint is obtuse or not. When the angle is acute, it indicates that the current waypoint has been reached, and the UAV should switch to the following path. Figure 10.6 intuitively illustrates this process. Correspondingly, line 4 calculates the dot product of the vector formed by the current position and the current waypoint with the direction vector of the next path segment. Its sign can be used to determine whether the angle is acute or obtuse, thereby determining whether to enter the next path segment. When an integral term is present, continued integration may lead to saturation. It needs to be cleared at different paths, as shown in lines 8–11.

Code 10.3 PID control of Leader UAV, from "e8/e8-1/formation_8_1.slx"

```
1  en=sum((p-point_now).*ti)*ti-(p-point_now);
2  den=sum(v.*ti)*ti-v;
3  det=vdes*ti-sum(v.*ti)*ti;
4  f=sum((p-point_next).*ti);
5  if flg==flglast
6  eni=eni+en*ts;
7  eti=eti+det*ts;
8  else
9  eni=zeros(1,2);
10 eti=zeros(1,2);
11 end
```

Code 10.4 calculates the control command for both leader and follower UAVs, where the two-dimensional matrix "d" represents the vector of position differences between each pair of UAVs, and "k1" and "k2" are the PID parameters for follower control. As shown in Fig. 10.5, the UAV formation is divided into three levels, with the 1st UAV at the first level, the 2nd and 3rd UAVs at the second level, and the 4th to 7th UAVs at the third level. UAV 2 and UAV 3 follow UAV 1, making UAV 1 the leader and UAV 2 and UAV 3 the followers. Similarly, the UAV 4 and UAV 5 follow UAV 2, and the UAV 6 and UAV 7 follow UAV 3. Therefore, based on Eqs. (10.11) and (10.15), UAV control commands are calculated separately based on the index "i", computed from lines 4–15 in the following code.

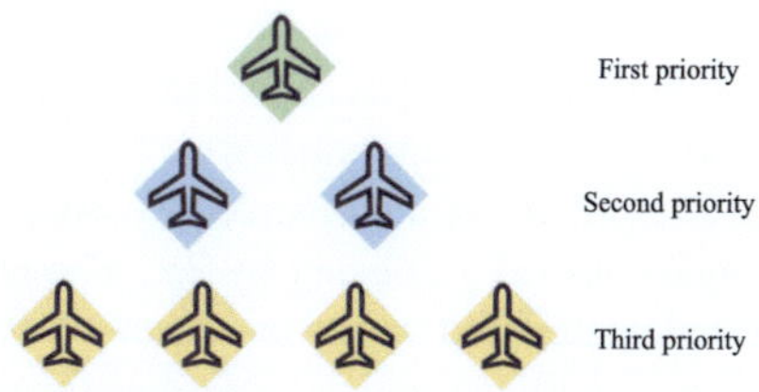

Fig. 10.5 Fixed-wing UAV formation structure

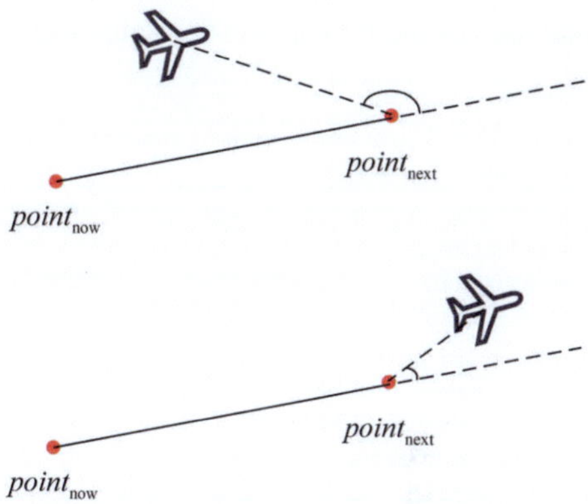

Fig. 10.6 Waypoint switching judgment

Code 10.4 UAV controllers for leader and follower, from"e8/e8-1/formation_8_1.slx"

```
d=[0 0;-3 5;-3 -5;-3 3;-3 -3;-3 3;-3 -3];%Vector of position difference between each UAV
k1=0.5;%PID parameters of followers
k2=0.1;
for i=1:N
u(i,:)=mysat2(knp*en+kni*eni+knd*den+ktd*det+kti*eti,v_min,v_max);%PID control
if i>1 && i<=3
u(i,:)=u(i,:)+k1*(Pcur(1,1:2)+d(i,:)-Pcur(i,1:2))+k2*(Vcur(1,1:2)-Vcur(i,1:2));%Leader-follower control
end
if i>3 && i<=5
u(i,:)=u(i,:)+k1*(Pcur(2,1:2)+d(i,:)-Pcur(i,1:2))+k2*(Vcur(2,1:2)-Vcur(i,1:2));%Leader-follower control
end
if i>5
u(i,:)=u(i,:)+k1*(Pcur(3,1:2)+d(i,:)-Pcur(i,1:2))+k2*(Vcur(3,1:2)-Vcur(i,1:2));%Leader-follower control
end
end
```

After obtaining the control commands for velocity, they need to be projected onto the three channels of the fixed-wing guidance model: velocity, altitude, and yaw, as shown in Code 10.5. Since the fixed-wing guidance model is a second-order model, its altitude and yaw derivatives are also state variables in the model, which are calculated in the program for iterative computation of the fixed-wing guidance model. Since this experiment only considers a two-dimensional plane, the altitude and its derivatives are set to zero.

There is a discontinuity issue when the change in the desired yaw angle switches at the critical transition points of π and $-\pi$, causing the fixed-wing UAV to be unable to turn smoothly in the nearest direction. For example, if the current yaw angle of the fixed-wing UAV is $5/6\pi$, it will continue to rotate counterclockwise by $1/3\pi$ in the next moment. If the value is constrained within the range of $-\pi$ to π, the desired yaw angle at the next moment will be $-5/6\pi$, with a change of $-5/3\pi$, corresponding to a clockwise rotation of $5/3\pi$. Therefore, an angle conversion is applied to the desired yaw angle, as shown in lines 9–14 of the code. Assuming the UAV can turn clockwise or counterclockwise for up to 10 laps, i.e., setting "k" to traverse from 1 to 21, "kPsid" is incremented by $2(k-11)\pi$ based on "Psid", obtaining a series of angles "kPsid" differing by 2π. The absolute error of each "kPsid" from "Psicur" is calculated and stored in "error". "kPsid" corresponding to the minimum value of "error" is found, which represents the desired yaw angle for the next moment of the UAV. Simultaneously, the derivative of the desired yaw angle, "dPsid", is calculated and used as input to update the state of the fixed-wing UAV in the guidance model.

Code 10.5 Projecting control commands to fixed-wing guidance model, from"e8/e8-1/formation_8_1.slx"

```
1  %% Convert to three-channel control
2  ex=[1;0];
3  ey=[0;1];
4  for i=1:N
5  Vd(i)=norm(u(i,:));
6  Psid(i)=atan2(ey'*u(i,:)',ex'*u(i,:)');
7  kPsid =zeros(21,1);
8  error = zeros(21,1);
9  for k=1:21
10 kPsid(k) = Psid(i) + (k-11)*2*pi;
11 error(k) = abs(kPsid(k) - Psicur(i));
12 end
13 [~,index]=min(error);
14 Psid(i)=kPsid(index);
15 dPsid(i)=(Psid(i)-Psidlast(i))/ts;
16 end
```

Step 3: Writing a fixed-wing guidance model

The fixed-wing guidance model is written using the S-function (System function) module in Simulink. S-functions can describe dynamic systems, including discrete, continuous, and composite systems. In the simulation, the fixed-wing guidance model is named "GuidanceModel", corresponding to Eq. (10.1). The main code for the model is shown in Code 10.6.

Code 10.6 mdlDerivatives function in the fixed-wing guidance model S-function, from "e8/e8-1/GuidanceModel.m"

```
function sys=mdlDerivatives(t,x,u)
global N
sys(1:N)      = x(4*N+1:5*N).*cos(x(5*N+1:6*N)); %dot_x=Vcos(psi)
sys(N+1:2*N)  = x(4*N+1:5*N).*sin(x(5*N+1:6*N)); %dot_y=Vsin(psi)
sys(2*N+1:3*N) = x(3*N+1:4*N);                   %dot_h=hdot;
sys(3*N+1:4*N) = -1.83*x(3*N+1:4*N)-1.166*x(2*N+1:3*N)-0.306*u(3*N+1:4*N)+1.166*u(N+1:2*N);
      %dot_dot_h=-1.83*hdot-1.166*h-0.306*hd_dot+1.166*hd
sys(4*N+1:5*N) = -0.8958*x(4*N+1:5*N)+0.8517*u(1:N); %dot_V=-0.8958*V+0.8517*Vd
sys(5*N+1:6*N) = x(6*N+1:7*N);                   %dot_psi=psi_dot
sys(6*N+1:7*N) = -1.709*x(6*N+1:7*N)-1.109*x(5*N+1:6*N)-0.289*u(4*N+1:5*N)+1.105*u(2*N+1:3*N);
      %dot_dpsi=-1.709*psi_dot-1.109*psi-0.289*psid_dot+1.105*psid
```

Based on the fixed-wing guidance model in Eq. (10.1), the state vector in the S-function is defined as a 7N-dimensional vector, representing the x-coordinate position, y-coordinate position, altitude, altitude derivative, speed, yaw angle, and yaw

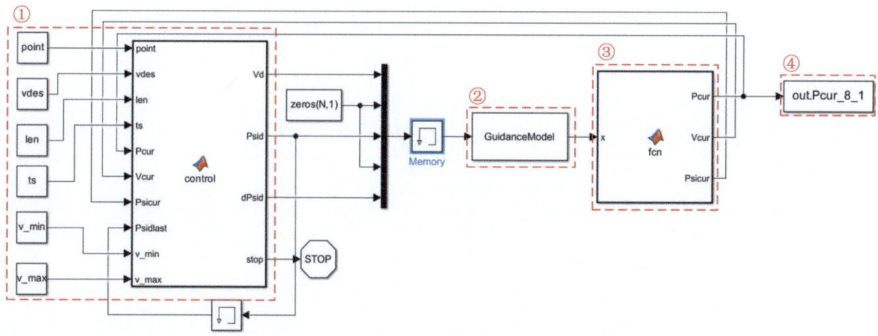

Fig. 10.7 Simulink simulation structure, from "e8/e8-1/formation_8_1.slx"

rate. The control input is a $5N$-dimensional vector, representing the desired speed, desired altitude, desired yaw angle, the derivative of desired altitude, and the derivative of the desired yaw angle.

Step 4: Connecting the controller with the model
In Simulink, the instructions outputted by the controller are connected to the model, while feedback information is fed to the controller, forming a closed loop. This is illustrated in Fig. 10.7.

This model consists of four modules:

1. Dashed box ①—"control", the controller module, including the contents of Code 10.3 to Code 10.5, to achieve UAV control.
2. Dashed box ②—"GuidanceModel", the fixed-wing simulation module, including the contents of Code 10.6, used to receive control output from the controller and output the current state of the fixed-wing UAV.
3. Dashed box ③—The UAV state data conversion module. This module takes the UAV states transmitted by "GuidanceModel" as input and outputs the current position "Pcur", velocity "Vcur", and yaw "Psicur".
4. Dashed box ④—The position "Pcur" is outputted using a "To workspace" module as "out.Pcur_8_1".

It should be noted that the "Memory" module used between the dashed box ② and dashed box ③ for the delay in Simulink is added to prevent an "algebraic loop" error in simulation.

After running the simulation program, corresponding animations can be drawn for observation. Typing "Draw(out.Pcur_8_1,point)" in the MATLAB command line allows observation of the movement of UAVs in the fixed-wing UAV formation at each moment during the simulation. Additionally, by running the "Draw_traj.m" file in the "e8-1" folder, trajectory plots of the UAVs in the fixed-wing UAV formation can be generated.

Upon examination of the simulation results, it becomes evident that these fixed-wing UAVs have formed a formation flight. In Fig. 10.8, the solid line represents the flight trajectory of each UAV. The solid circles on the trajectory depict the position

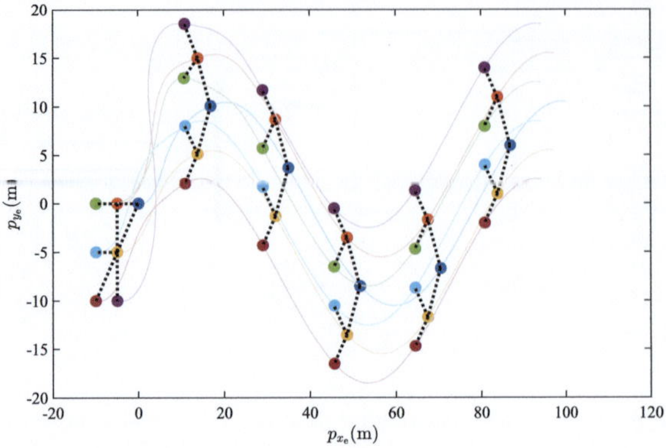

Fig. 10.8 Trajectories of a fixed-wing UAV swarm

of the UAVs at intervals of 12 s, and the dotted lines show the formation of the UAV swarm. The controllers designed above have enabled the fixed-wing UAV swarm to fly in the desired formation in the basic experiment.

10.3 Analysis Experiment

10.3.1 Experimental Objective

1. Things to prepare

(1) Software: MATLAB R2022b or above.
(2) Program: Experimental package "e8/e8-2" is available. The package contains the fixed-wing UAV guidance model "GuidanceModel.m", the initialization file for simulation "initFormation.m", simulation files for a fixed-wing formation flight such as "formation_8_1.slx", "formation_8_2_1.slx" for a fixed-wing formation flight with added noise, "formation_8_2_2.slx" for a fixed-wing formation flight with added delay, and "formation_8_2_3.slx" for a fixed-wing formation flight with added noise and delay.

2. Objective

Considering factors such as noise and delay in the formation flight of fixed-wing UAVs, observe the formation flight behavior and analyze the limitations of the applied methods.

10.3.2 Experimental Procedure

The purpose of the analysis experiment is to understand the impact of external disturbances on the formation flight of UAVs. By changing relevant parameters in the UAV swarm, analyze how noise and delay affect formation flight. This experiment focuses on comparative analysis under different scenarios and learns how to simulate and verify the effects of formation control. The analysis experiment is divided into four steps: initial condition definition, simulation scene design, performing a Simulink simulation, and drawing and analysis of results. Step 1 defines initial variables by assigning values to parameters in the simulation. In Step 2, different simulation scenarios are designed to observe and analyze the effects of noise, packet loss, and delay on formation control in Steps 3 and 4.

Step 1: Initial condition definition
The initialization file "initFormation.m" exists in the "e8-2" folder. The definition of initial variables is the same as in Sect. 10.2. In the simulation, the formation is structured into three levels: the first UAV in the sequence has the first priority, the second and third UAVs are designated as the second priority, and the remaining UAVs are classified as the third priority.

Step 2: Design of simulation scenarios
The controllers designed here are similar to those mentioned in Sect. 10.2. Only parameters such as noise, packet loss, and delay in UAV formation are changed for comparison with the ideal simulation.

(1) Situation 1: Noise with no delay
In the simulation, two variables are added to represent the estimated position and velocity of fixed-wing UAVs, denoted as "Pcur_est" and "Vcur_est", respectively. These estimated positions and velocities will be sent by the UAVs to other companions for formation. The variables "Pcur_est" and "Vcur_est" are obtained by adding random noise to the true position and velocity of the fixed-wing UAVs, as shown in Code 10.7.

Code 10.7 Noise in UAV formation, from "e8/e8-2/initFormation.m"

```
%Add noise to position and speed
bp=2;%The magnitude of the added noise at the location
bv=2;%The magnitude of the noise added to the speed
Pcur_est=Pcur+[bp*(rand(N,2)-0.5)*2 zeros(N,1)];
Vcur_est=Vcur+[bv*(rand(N,2)-0.5)*2 zeros(N,1)];
```

In MATLAB, "rand" represents uniformly distributed random numbers. This MATLAB function returns a random scalar obtained from a uniform distribution over the interval (0, 1). Since the noise added to position and velocity should have a mean of 0, appropriate processing is done to obtain random numbers in the interval $\left(-b_p, b_p\right)$ and $(-b_v, b_v)$.

(2) Situation 2: No noise but with delay
In Simulink, a module called "Transport Delay" is available for modeling the delay between UAVs, as shown in Fig. 10.9.

By double-clicking on the "Transport Delay" module, the time delay parameter can be adjusted. In this simulation, the time delay is set to 0.8 (Fig. 10.10).

Fig. 10.9 "Transport Delay" module in Simulink

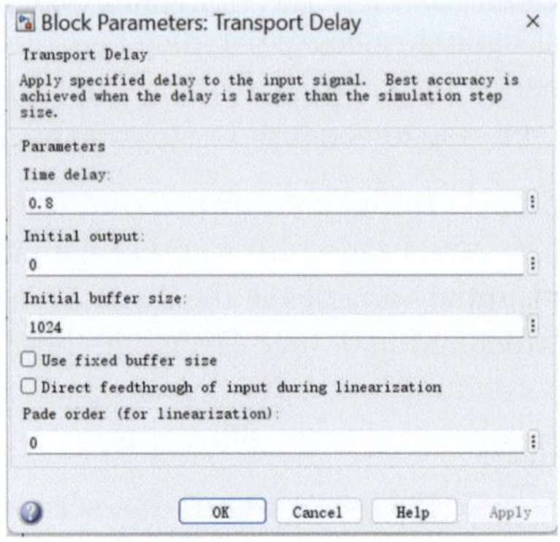

Fig. 10.10 Parameter configuration for "Transport Delay" module

(3) Situation 3: With noise and delay
In the third scenario, both noise and delay factors are considered simultaneously in the simulation to observe the formation flight of fixed-wing UAVs.

Step 3: Constructing Simulink simulation
The fixed-wing guidance model used here is the same S-function written in Sect. 10.2, and the structure of the Simulink simulation is the same as in the basic experiment. In Fig. 10.11, noise is added to the actual position and velocity to obtain the estimated position and velocity.

This model contains four major modules:

1. Dashed box ①—Controller module "control" is used to design the controller to control the UAVs.
2. Dashed box ②—Fixed-wing simulation module "GuidanceModel" is used to receive the control commands outputted by the controller and output the current state of the fixed-wing UAVs.
3. Dashed box ③—UAV state data conversion module, which takes the state variable "x" from "GuidanceModel" and then outputs the current position "Pcur", velocity "Vcur", yaw "Psicur", estimated position "Pcur_est", and estimated velocity "Vcur_est".
4. Dashed box ④—A "To workspace" module is used to save the position data "Pcur" as "out.Pcur_8_2_1".

In Fig. 10.12, a delay component is added to the feedback loop of real position and velocity to obtain the estimated position and velocity.

This model consists of five modules:

1. Dashed box ①—"control", the controller module used to design the controller to control the UAVs.

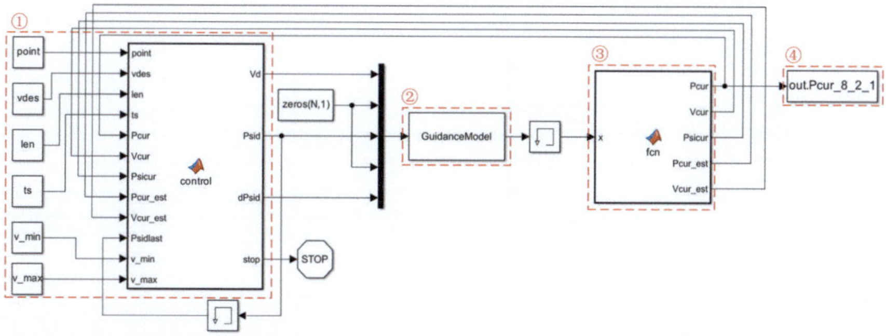

Fig. 10.11 Situation 1: Simulation of fixed-wing UAV formation flight with added noise, from "e8/e8-2/formation_8_2_1.slx"

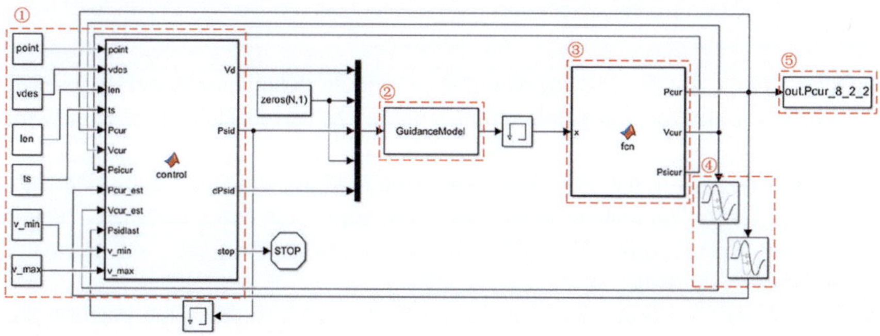

Fig. 10.12 Situation 2: Simulation of fixed-wing UAV formation flight with added delay, from "e8/e8-2/formation_8_2_2.slx"

2. Dashed box ②—"GuidanceModel", the fixed-wing simulation module used to receive the control commands outputted by the controller and output the current state of the fixed-wing UAVs.
3. Dashed box ③—The UAV state data conversion module. This module takes the state variable "x" from "GuidanceModel" and then outputs the current position "Pcur", velocity "Vcur", and yaw "Psicur".
4. Dashed box ④—The state delay module, which is used to delay the position and velocity of the UAVs.
5. Dashed box ⑤—A "To workspace" module is used to save the position data "Pcur" as "out.Pcur_8_2_2".

In Fig. 10.13, noise and delay components are added to the real position and velocity to obtain the estimated position and velocity.

This model contains five major modules:

1. Dashed box ①—Controller module "control" is used to design the controller to control UAVs.

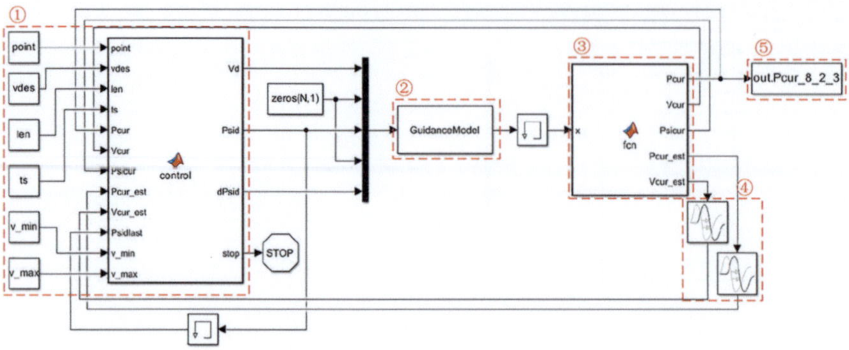

Fig. 10.13 Situation 3: Simulation of fixed-wing UAV formation flight with added noise and delay, from "e8/e8-2/formation_8_2_3.slx"

2. Dashed box ②—Fixed-wing simulation module "GuidanceModel" is used to receive the control commands and then output the current state of the fixed-wing UAVs.
3. Dashed box ③—UAV state data conversion module, which takes the state variable "x" from "GuidanceModel" and then outputs the current position "Pcur", velocity "Vcur", yaw "Psicur", estimated position "Pcur_est", and estimated velocity "Vcur_est".
4. Dashed box ④—State delay module, which is used to delay the position and velocity of the UAVs.
5. Dashed box ⑤—A "To workspace" module is used to save the position data "Pcur" as "out.Pcur_8_2_3".

Step 4: Result drawing and analysis
Open the "e8-2" folder and run the "Draw_traj.m" file. This file initializes and simulates various scenarios, plotting the simulation results in four subplots for comparison, as shown in Fig. 10.14. The dotted lines represent the topological structure of UAVs.

In addition to the trajectory comparison above, each scenario's Simulink file can be run separately. For example, first run "initFormation.m" for initialization, then run the "formation_8_2_1.slx" simulation file. After the simulation is complete, run the command "Draw(out.Pcur_8_2_1, point)" in MATLAB to dynamically plot the real-time states of the fixed-wing UAV formation.

The simulation results indicate that the inclusion of noise and delay influences the stability of the fixed-wing UAV formation. This is due to the inherent uncertainty associated with the position and velocity information received by the UAVs, which is compounded by the presence of noise and delay. The introduction of noise represents an inaccuracy in the acquired UAV positions. In contrast, the effect of delay implies that the UAVs may be tracking the expected positions from the previous time step. These interference factors inevitably affect the structure maintained by the UAV formation, resulting in the oscillations observed in the simulation results.

Fig. 10.14 Trajectories of fixed-wing UAV swarm flight in four scenarios, from "e8/e8-2/Draw_traj.m"

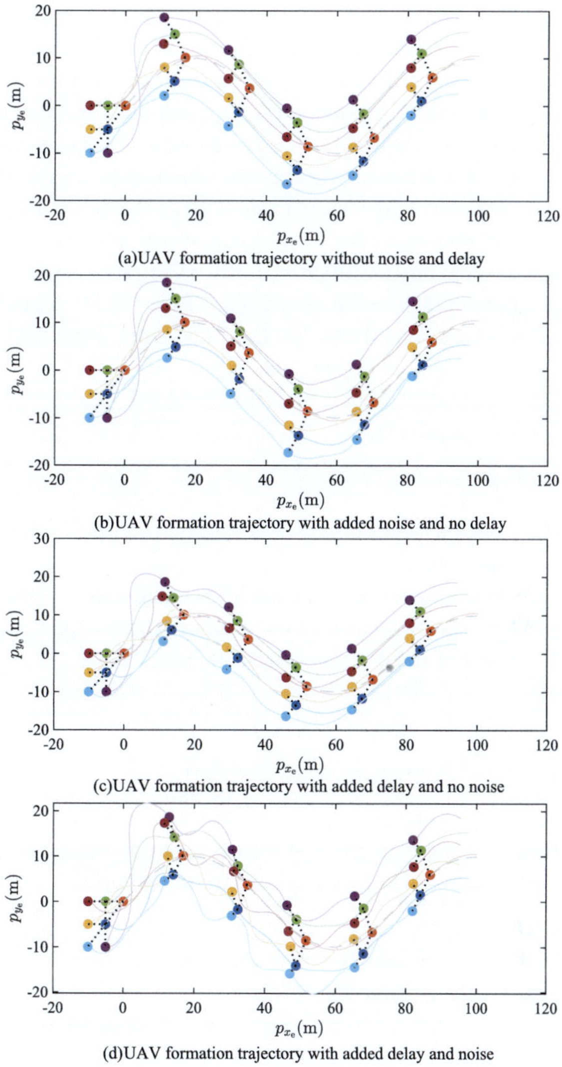

10.4 Design Experiment

10.4.1 Experimental Objective

1. Things to prepare

(1) Software: MATLAB R2022b or above, RflySim toolchain. The RflySim toolchain provides a Hardware-In-the-Loop (HIL) simulation environment for micro-small fixed-wing UAVs.

(2) Program: Experimental package "e8/e8-3" includes a guidance model for fixed-wing UAVs and design examples of fixed-wing UAV swarm flight. For example, "GuidanceModel.m" is the guidance model for fixed-wing UAVs, "initTube.m" is the initialization file for simulation, "Tube.slx" is the simulation file for fixed-wing swarm flight using virtual tubes, "Tube_noise.slx" is the simulation file for fixed-wing swarm flight with added noise, "Tube_delay.slx" is the simulation file for fixed-wing swarm flight with added delay, and "Tube_noise_delay.slx" is the simulation file for fixed-wing swarm flight with added noise and delay. "Small-FixedWingUAVnoctrlHIL.dll" is a model file for a fixed-wing UAV. The system parameter identification file is "Tube_HITL_identification.slx", the initialization files are "init_Tube_SITL.m" and "init_Tube_HITL.m", and the simulation files are "Tube_SITL.slx" and "Tube_HITL.slx".

(3) Hardware: Four CubePilot/PixHawk autopilots, four remote controllers, and four receivers.

2. Objective

(1) Design flight controllers based on virtual tubes to accomplish swarm flight missions.

(2) By taking noise and delay into consideration separately, observe and analyze the flight performance of fixed-wing swarms and compare them with the formation flight of fixed-wing swarms in Sect. 10.3.

10.4.2 Experimental Design

Based on the principles of virtual tubes discussed in Sect. 10.1.3, the detailed controller design will be developed in this section.

(1) Guiding the swarm to fly forward along a virtual tube
In the UAV swarm, the control component for the ith UAV to fly forward along the virtual tube is given by

$$\mathbf{v}_{\mathrm{f},i} = v_{\mathrm{f},i}\mathbf{t}_{\mathrm{c}}\left(\mathbf{p}_i\right) \tag{10.23}$$

where $v_{\mathrm{f},i}$ is the magnitude of the control component for flying forward along the virtual tube, i.e., the desired forward speed, and $\mathbf{t}_{\mathrm{c}}\left(\mathbf{p}_i\right)$ is the unit tangent vector in the forward direction at the current point of the ith UAV on the curve.

(2) Collision avoidance among UAVs
Define

$$\tilde{\mathbf{p}}_{ij} = \mathbf{p}_i - \mathbf{p}_j \tag{10.24}$$

where $\mathbf{p}_i$ and $\mathbf{p}_j$ are the positions of the ith and jth UAVs, respectively.

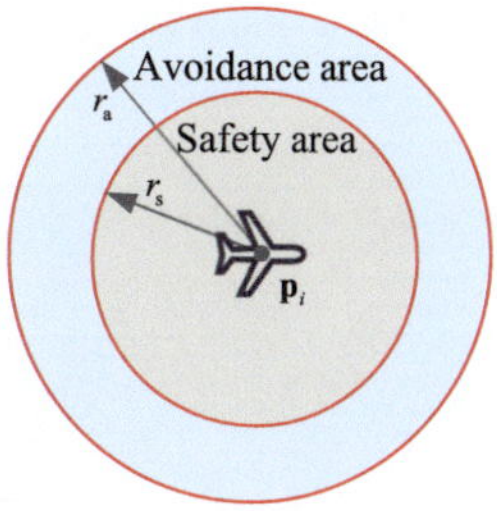

Fig. 10.15 Safety radius and avoidance radius for UAVs

Here, the concepts of safety radius and avoidance radius for UAVs, as shown in Fig. 10.15, are introduced below. To ensure that UAVs do not collide with obstacles or other UAVs in the environment, the safety area of the ith UAV is defined as

$$\mathscr{S}_i = \left\{ \mathbf{x} \in \mathbb{R}^2 \,|\, \|\mathbf{x} - \mathbf{p}_i\| \leq r_\text{s} \right\} \tag{10.25}$$

where r_s is the safety radius of the ith UAV. It should be noted that in the design of the safety area, not only is the current position of the UAV considered, but also its velocity [10]. For any pair of UAVs in the swarm, they do not collide if

$$\mathscr{S}_i \cap \mathscr{S}_j = \emptyset. \tag{10.26}$$

Based on the definition of the safety area for UAVs, the definition of the avoidance area for UAVs is given here. This area is used for collision avoidance control of UAVs. The avoidance area of the ith UAV is represented as

$$\mathscr{A}_i = \left\{ \mathbf{x} \in \mathbb{R}^2 \,|\, \|\mathbf{x} - \mathbf{p}_i\| \leq r_\text{a} \right\} \tag{10.27}$$

where r_a is the avoidance radius of the ith UAV. If $\mathscr{A}_i \cap \mathscr{S}_j \neq \emptyset$, i.e.,

$$\left\| \mathbf{p}_j - \mathbf{p}_i \right\| \leq r_\text{s} + r_\text{a}. \tag{10.28}$$

Then, the jth UAV should avoid the ith UAV. According to the definition of the avoidance area, the ith UAV should also avoid the jth UAV simultaneously. Additionally, to ensure that the avoidance area fully encompasses the safety area, the condition $r_\text{a} > r_\text{s}$ must be satisfied.

For collision avoidance between the ith and jth UAVs, a control barrier function is designed as

$$V_{\text{m},ij} = \frac{k_2 \sigma_\text{m} \left(\left\| \tilde{\mathbf{p}}_{ij} \right\| \right)}{(1 + \varepsilon_\text{m}) \left\| \tilde{\mathbf{p}}_{ij} \right\| - 2 r_\text{s} s \left(\frac{\|\tilde{\mathbf{p}}_{ij}\|}{2r_\text{s}}, \varepsilon_\text{s} \right)} \tag{10.29}$$

where $\sigma_\text{m}(x) = \sigma(x, 2r_\text{s}, r_\text{a} + r_\text{s})$, the bump function $\sigma(\cdot)$ and the approximate smooth saturating function $s(x, \varepsilon_\text{s})$ are defined in Appendix. When $r_\text{s} = 10, r_\text{a} = 20$, $\varepsilon_\text{m} = 10^{-6}$, and $k_2 = 1$, the function $V_{\text{m},ij}$ is shown in Fig. 10.16.

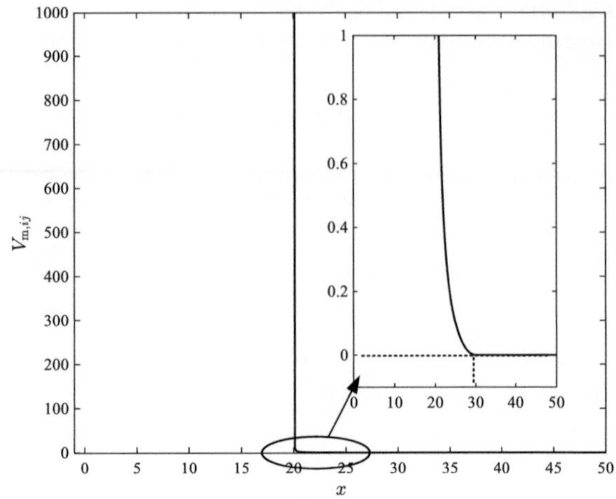

Fig. 10.16 Function of $V_{\mathrm{m},ij}$

Such a control barrier function has the following properties.

(1) $\partial V_{\mathrm{m},ij}/\partial \left\|\tilde{\mathbf{p}}_{ij}\right\| \leq 0$, $V_{\mathrm{m},ij}$ is a non-increasing function of $\left\|\tilde{\mathbf{p}}_{ij}\right\|$.
(2) If $\left\|\tilde{\mathbf{p}}_{ij}\right\| > r_{\mathrm{a}} + r_{\mathrm{s}}$, implying $\mathscr{A}_i \cap \mathscr{S}_j = \emptyset$ and $\mathscr{A}_j \cap \mathscr{S}_i = \emptyset$, then $V_{\mathrm{m},ij} = 0$ and $\partial V_{\mathrm{m},ij}/\partial \left\|\tilde{\mathbf{p}}_{ij}\right\| \leq 0$. If $V_{\mathrm{m},ij} = 0$, then $\left\|\tilde{\mathbf{p}}_{ij}\right\| > r_{\mathrm{a}} + r_{\mathrm{s}} > 2r_{\mathrm{s}}$.
(3) If $0 < \left\|\tilde{\mathbf{p}}_{ij}\right\| < 2r_{\mathrm{s}}$, indicating $\mathscr{S}_j \cap \mathscr{S}_i \neq \emptyset$ (collision may not occur in practice), then there exists a sufficiently small $\epsilon_{\mathrm{s}} > 0$ such that

$$V_{\mathrm{m},ij} = \frac{k_2}{\epsilon_{\mathrm{m}} \left\|\tilde{\mathbf{p}}_{ij}\right\|} \geq \frac{k_2}{2\epsilon_{\mathrm{m}} r_{\mathrm{s}}}. \tag{10.30}$$

The goal of designing velocity commands is to make $V_{\mathrm{m},ij}$ zero or as small as possible. According to property (2), this implies $\left\|\tilde{\mathbf{p}}_{ij}\right\| > 2r_{\mathrm{s}}$, i.e., the ith UAV will not collide with the jth UAV.

For the ith UAV, its collision avoidance control component is given by

$$\mathbf{v}_{\mathrm{m},i} = \sum_{j \in N_{\mathrm{m},i}} \frac{\partial V_{\mathrm{m},ij}}{\partial \left\|\tilde{\mathbf{p}}_{ij}\right\|} \frac{\tilde{\mathbf{p}}_{ij}}{\left\|\tilde{\mathbf{p}}_{ij}\right\|}. \tag{10.31}$$

(3) Confining swarm flight within a curved virtual tube
For the ith UAV, to confine its flight within curved virtual tubes, a barrier function is constructed here:

$$V_{\mathrm{t},i} = \frac{k_3 \sigma_{\mathrm{t}} \left(r_{\mathrm{t}}\left(\mathbf{p}_i\right) - \left\|\tilde{\mathbf{p}}_{\mathrm{t},i}\right\|\right)}{\left(r_{\mathrm{t}}\left(\mathbf{p}_i\right) - r_{\mathrm{s}}\right) - \left\|\tilde{\mathbf{p}}_{\mathrm{t},i}\right\| s\left(\frac{r_{\mathrm{t}}(\mathbf{p}_i) - r_{\mathrm{s}}}{\left\|\tilde{\mathbf{p}}_{\mathrm{t},i}\right\| + \varepsilon_{\mathrm{t}}}, \varepsilon_{\mathrm{s}}\right)} \tag{10.32}$$

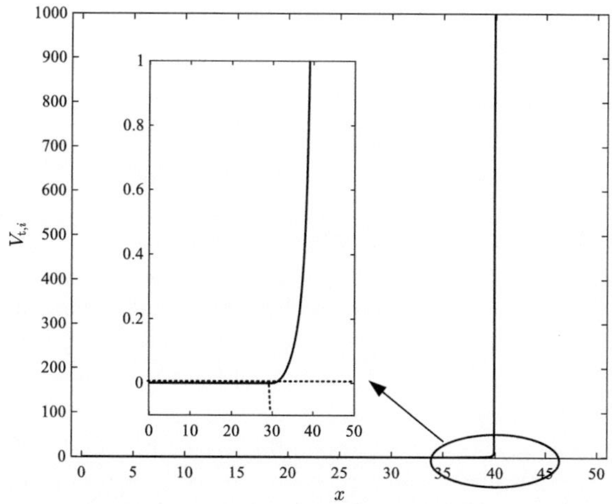

Fig. 10.17 Function of $V_{t,i}$

where ε_t is any small constant, $\left\|\tilde{\mathbf{p}}_{t,i}\right\|$ is the distance of the UAV's position from the centerline of the tube, r_t is half the width of the corresponding cross-section at the current location in the tube, and $\sigma_t(x) = \sigma(x, r_s, r_a)$. When $r_t = 50$, $r_s = 10$, $r_a = 20$, $\varepsilon_t = 10^{-6}$, and $k_3 = 1$, the function curve of $V_{t,i}$ is shown in Fig. 10.17.

The function $V_{t,i}$ possesses the following properties.

(1) $\partial V_{t,i}/\partial\left\|\tilde{\mathbf{p}}_{t,i}\right\| \geq 0$, where $V_{t,i}$ is a non-decreasing function of $\left\|\tilde{\mathbf{p}}_{t,ij}\right\|$.
(2) If $r_t - \left\|\tilde{\mathbf{p}}_{t,i}\right\| \geq r_a$, meaning the edge of the virtual tube is not within the avoidance region of the ith UAV, then $\sigma_t\left(r_t(\mathbf{p}_i) - \left\|\tilde{\mathbf{p}}_{t,i}\right\|\right) = 0$. Therefore, $V_{t,i} = 0$ and $\partial V_{t,i}/\partial\left\|\tilde{\mathbf{p}}_{t,i}\right\| = 0$.
(3) If $r_t - \left\|\tilde{\mathbf{p}}_{t,i}\right\| < r_s$, meaning the edge of the virtual tube enters the safety region of the ith UAV, then there exists a sufficiently small $\epsilon_s > 0$ such that

$$s\left(\frac{r_t(\mathbf{p}_i) - r_s}{\left\|\tilde{\mathbf{p}}_{t,i}\right\| + \epsilon_t}, \epsilon_s\right) = \frac{r_t(\mathbf{p}_i) - r_s}{\left\|\tilde{\mathbf{p}}_{t,i}\right\| + \epsilon_t} < 1. \tag{10.33}$$

Therefore, when ϵ_s is very small,

$$V_{t,i} = \frac{k_3\left(\left\|\tilde{\mathbf{p}}_{t,i}\right\| + \epsilon_t\right)}{\epsilon_t\left(r_t(\mathbf{p}_i) - r_s\right)} \tag{10.34}$$

becomes very large.

The goal of velocity controller design is to make $V_{t,i}$ equal to 0, which implies $r_t - \left\|\tilde{\mathbf{p}}_{t,i}\right\| \geq r_a$. According to property (2), the ith UAV will remain within the virtual tube.

The control component limiting the ith UAV's flight within the curved tube is given by

$$\mathbf{v}_{\mathrm{t},i} = \left(\mathbf{I}_2 - \frac{\partial m(\mathbf{p}_i)}{\partial \mathbf{p}_i}\right)^{\mathrm{T}} \left(\frac{\partial V_{\mathrm{t},i}}{\partial \left\|\tilde{\mathbf{p}}_{\mathrm{t},i}\right\|}\frac{\tilde{\mathbf{p}}_{\mathrm{t},i}}{\left\|\tilde{\mathbf{p}}_{\mathrm{t},i}\right\|} + \frac{\partial V_{\mathrm{t},i}}{\partial r_{\mathrm{t}}(\mathbf{p}_i)}\left(\frac{\partial r_{\mathrm{t}}(\mathbf{p}_i)}{\partial \mathbf{p}_i}\right)^{\mathrm{T}}\right). \tag{10.35}$$

(4) Distributed swarm controller design
After obtaining three control components, the final distributed swarm controller can be synthesized. For the ith UAV, its velocity command is given by

$$\mathbf{v}_{\mathrm{c},i} = \mathrm{sat}\left(\mathbf{v}_{\mathrm{f},i} + \mathbf{v}_{\mathrm{m},i} + \mathbf{v}_{\mathrm{t},i}, v_{\max}\right) \tag{10.36}$$

where $v_{\max}$ is the maximum flight speed of the UAV. For a more detailed proof of the controller design, please refer to [8, 9]. After designing, the velocity control command for the swarm can be further mapped to the guidance model of fixed-wing UAVs, following a similar process to that in Code 10.5.

10.4.3 Experimental Procedure

This experiment aims to help readers understand the swarm control method based on virtual tubes. Readers will learn how to design a swarm controller through the principles of the virtual tube method. The experiment focuses on the design ideas of the swarm control algorithm, learning how to enable UAVs to complete a swarm flight without the need to maintain a fixed structure. The experimental design is divided into four steps: defining initial conditions, generating virtual tubes, designing controllers, and setting up Simulink simulations. In Step 1, initial variables are described, and parameters in the simulation are set. In Step 2, virtual tubes are generated, and the flight constraints for fixed-wing UAV swarms are set. After designing the controller in Step 3, a Simulink simulation is performed in Step 4.

Step 1: Defining initial variables and generating a virtual tube
First, define the initial variables of the UAV swarm, including the initial states of UAVs and the number of UAVs, and set parameters such as the minimum speed, the maximum speed, the safety radius, and the avoidance radius. As shown in Code 10.8, to generate a virtual tube for the UAV swarm, the essential parameters of the tube are defined, including the centerline and corresponding left and right boundaries. The left and right boundaries of the tube are extended using the centerline "p_leader" and tube width "dis".

Code 10.8 Defining the basic parameters of the tube, from "e8/e8-3/initTube.m"

```
curve_x=(0.1:4:100.1)';
curve_y=10*sin(curve_x/4/pi);
curve_theta=atan(2.5/pi*cos(curve_x/4/pi));
dis=10*ones(length(curve_x),1);%Define the tube width
p_leader=[curve_x,curve_y];%Define the generating curve
p_left=[];
p_right=[];
for i=1:length(curve_x)%Define the left and right boundaries of the tube
    p_left(i,:)=(p_leader(i,:))+dis(i)*[cos(curve_theta(i)+pi/2),sin(curve_theta(i)+pi/2)];
    %p_left is the position of point i going forward vertically minus 90 degrees, and the distance between the two is dis(i)
    p_right(i,:)=(p_leader(i,:))+dis(i)*[cos(curve_theta(i)-pi/2),sin(curve_theta(i)-pi/2)];
    %p_right is the position where point i is forward 90 degrees perpendicular, and the distance between the two is dis(i)
end
```

Step 2: Writing controller

Based on the controller design for flying along the virtual tube, the code corresponding to this design needs to be written. According to Eq. (10.23), the method for calculating the control component $\mathbf{v}_{\mathrm{f},i}$ to guide the swarm forward along the tube needs to be obtained. First, it is necessary to find the unit tangent vector $\mathbf{t}_{\mathrm{c}}\left(\mathbf{p}_i\right)$ along the forward direction of the current UAV i corresponding to the point $\mathbf{p}_i$ on the curve. Code 10.9 provides instructions on identifying "leader_locate", which represents the position closest to the UAV on the generating curve of the virtual tube at the current moment. This position corresponds to $\mathbf{p}_i$ in the set of points from the generating curve of the virtual tube.

Code 10.9 Finding the closest position on the generating curve of the virtual tube, from"e8/e8-3/Tube.slx"

```
%Find the location closest to the UAV on the generating curve of the tube at the current moment
leader_locate=zeros(N,1);
for i=1:N
    dis_leader_min=10^8;
    for j=1:length(p_leader(:,1))
        dis_leader=norm(Pcur(i,1:2)-p_leader(j,:));%Calculates the distance between the current location and the point on the center of the tube
        if dis_leader < dis_leader_min%if the distance is smaller than previously recorded
            leader_locate(i)=j;%Update the record value
            dis_leader_min=dis_leader;%Update the record value
        end
    end
end
```

For the collision avoidance control component $\mathbf{v}_{\mathrm{m},i}$ and the boundary restriction component $\mathbf{v}_{\mathrm{t},i}$ of the tube, a barrier function similar to $V_{\mathrm{m},ij}$ in Eq. (10.29) and $V_{\mathrm{t},i}$ in Eq. (10.31) is used for approximate calculation in practical applications, as shown in Code 10.10. This function has similar functionality but is simpler and easier to use.

Code 10.10 Calculating the controller components, from"e8/e8-3/dmysigma2.m"

```
A =-(d2*em - d1*em + 2)/(d1 - d2)^3;
B =(- em*d1^2 - em*d1*d2 + 3*d1 + 2*em*d2^2 + 3*d2)/(d1 - d2)^3;
C =-(d2*(- 2*em*d1^2 + em*d1*d2 + 6*d1 + em*d2^2))/(d1 - d2)^3;
D =(d2*(- em*d1^2*d2 + em*d1*d2^2 + 3*d1*d2 - d2^2))/((d1 - d2)*(d1^2 - 2*d1*d2 + d2^2));
if x<=d1
    u = 1;
elseif d1<=x && x<=d2
    u = A*x^3 + B*x^2 + C*x + D ;
else
    u = 0;
end
```

The components of the controller are calculated according to the final velocity command in Eq. (10.36). The key code is provided in Code 10.11. The first three lines calculate the forward flight component "V1" of the fixed-wing UAV, denoted as $\mathbf{v}_{f,i}$. Lines 4–11 iterate over the positions of other jth UAV for the ith UAV and calculate the avoidance component "V2" denoted as $\mathbf{v}_{m,i}$. Here, "pmij" represents the position difference between the ith and jth UAVs, and "bil" is the coefficient of the avoidance component. Lines 12–18 calculate the restriction component "V3" denoted as $\mathbf{v}_{t,i}$ based on the distance between the UAV and the boundary of the tube. Here, "mpi" represents the position of the middle point of the tube cross-section, "pti" is the difference between the current position of the ith UAV and the middle point of the tube cross-section, "rtpi" is half the width of the tube cross-section, and "rtpi-norm(pti)" is the distance between the UAV and the boundary of the tube at the current time. Thus, the velocity component coefficient provided by the tube boundary "ci" and the direction "un" is calculated, and then the tube boundary restriction component "V3" is obtained.

Code 10.11 Calculation of controller components, from"e8/e8-3/Tube.slx"

```
1  v_line=4;
2  e1(i,:) = nc(leader_locate(i),:)*v_line;%Forward component
3  V1(i,:) = -mysat(k1*e1(i,:),vmax);%Saturate the speed
4  for j=1:N
5      if i~=j
6          em = 0;
7          pmij = Pcur(i,1:2) - Pcur(j,1:2);
8          bil = k2*dmysigma2(norm(pmij),2*rs,rs+ra,em);%Use a simpler form of barrier function to achieve similar results
9          V2(i,:) = V2(i,:)+ bil*(pmij/norm(pmij));%The avoidance component between UAV i and UAV j
10     end
11 end
12 et=0;
13 mpi=0.5*(tunnel_left(leader_locate(i),:)+tunnel_right(leader_locate(i),:));%The middle point of the tube cross-section
14 pti(i,:) = Pcur(i,1:2) - mpi;%The difference between the current position and the midpoint of the tube cross-section
15 rtpi(i)=0.5*norm(tunnel_left(leader_locate(i),:)-tunnel_right(leader_locate(i),:));%Half the width of the tube cross-section
16 ci=k3*dmysigma2(rtpi(i)-norm(pti(i,:)),rs,ra,et);%The velocity component factor provided by the tube boundary
17 un=pti(i,:)-(dot(pti(i,:),nc(leader_locate(i),:))*nc(leader_locate(i),:));%The direction of the velocity component provided by the tube boundary
18 V3(i,:)=-(un/norm(un))*ci;%Tube boundary limit components
```

Step 3: Building Simulink simulation

Connect the controller with the model in Simulink, as shown in Figs. 10.18, 10.19, and 10.20. The three simulations illustrate the control within virtual tubes of a fixed-wing swarm under different conditions: without additional influencing factors, with added noise, and with delay. After running the simulation, plots are shown to observe the flight behavior of the UAVs.

Figure 10.18 describes the control modules with virtual tubes of a fixed-wing swarm without additional influencing factors. Running this simulation provides a better understanding of control based on the virtual tube. The simulation of the fixed-wing swarm contains four major modules:

1. Dashed box ①—"mytunnel" contains Codes 10.9 to 10.11, which represent the controller for fixed-wing UAVs based on the virtual tube. This module calculates the three velocity components, namely the forward component $\mathbf{v}_{f,i}$, the collision

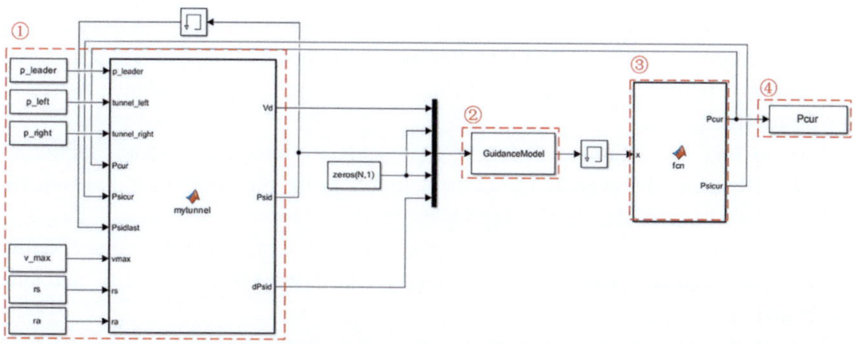

Fig. 10.18 Fixed-wing swarm closed-loop control simulation module, from "e8/e8-3/Tube.slx"

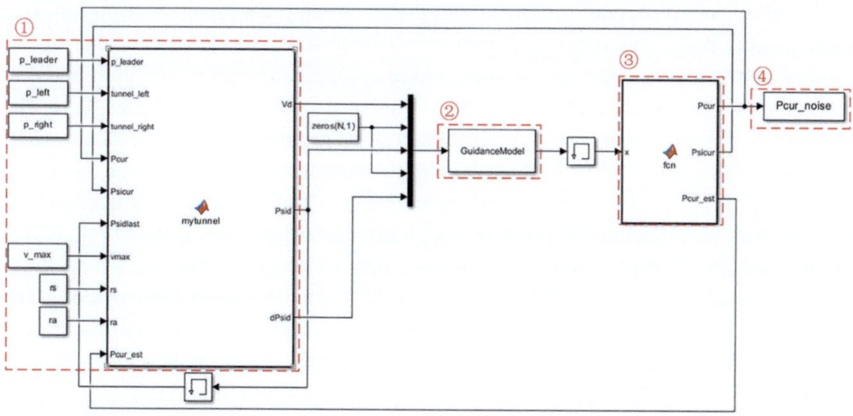

Fig. 10.19 Fixed-wing swarm closed-loop control simulation module subject to noise, from "e8/e8-3/Tube_noise.slx"

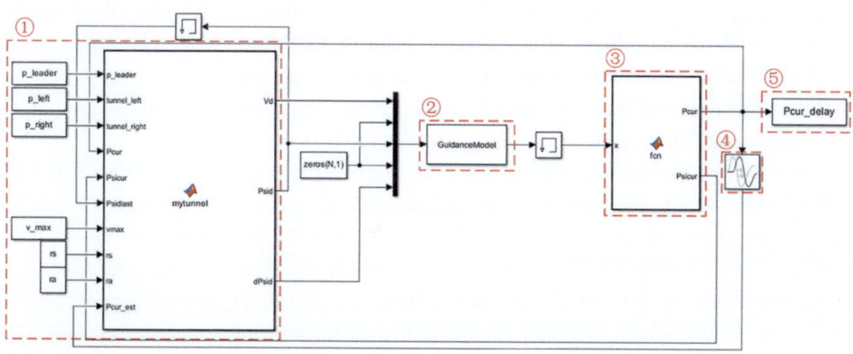

Fig. 10.20 Fixed-wing swarm closed-loop control simulation module subject to delay, from "e8/e8-3/Tube_delay.slx"

avoidance component $\mathbf{v}_{\mathrm{m},i}$, and the tube boundary restriction component $\mathbf{v}_{\mathrm{t},i}$. According to Eq. (10.36), the three terms are synthesized into the velocity command $\mathbf{v}_{\mathrm{c},i}$, which is then mapped to the control input of the guidance model to control the UAVs.
2. Dashed box ②—"GuidanceModel" represents the guidance model, including Code 10.6, which receives the control inputs from the controller and outputs the current state of the fixed-wing UAVs.
3. Dashed box ③—UAV state data conversion module, which takes the state variable "x" from "GuidanceModel" and then outputs the current position "Pcur" and yaw "Psicur".
4. Dashed box ④—A "To workspace" module is used to save the position data "Pcur".

Nevertheless, the actual circumstances under which the swarm functions are frequently suboptimal and subject to numerous disruptions. To investigate the impact of noise on the swarm, a series of closed-loop control simulations are conducted in which the fixed-wing swarm is subjected to noise. In comparison to Fig. 10.18, the simulation depicted in Fig. 10.19 incorporates noise factors, whereby the positions of the UAVs with respect to each other cannot be accurately estimated. This simulates the flight of the UAV swarm in the presence of noise. The simulation comprises four principal modules, the first two of which are identical to those described in Fig. 10.18. The dashed box ③ represents the UAV state data conversion module, which takes the state variable "x" from "GuidanceModel". Then it outputs the current position "Pcur", yaw "Psicur", and estimated position "Pcur_est". The position "Pcur" is output to a "To workspace" module, named "Pcur_noise", as shown in dashed box ④.

In practical applications, the delay is also a factor to consider in UAV swarm flight. Compared to Fig. 10.18, the simulation in Fig. 10.20 considers delay factors, i.e., the UAVs cannot obtain information in real-time, simulating the fixed-wing swarm closed-loop control subject to delay. The simulation comprises five major modules, with the first three being identical to those explained in Fig. 10.18. Dashed box ④—State delay module, which applies a delay to the UAV's position and velocity to obtain the delayed position; the position "Pcur" is output to a "To workspace" module, named "Pcur_delay", as shown in dashed box ⑤.

Considering the impact of both noise and delay, the simulation is performed as shown in Fig. 10.21. The explanations of the modules used in the simulation are the same as those in Figs. 10.19 and 10.20. To compare the four scenarios, trajectory plots are shown in Fig. 10.22.

As illustrated in Fig. 10.22, the seven lines represent the trajectories of seven fixed-wing UAVs in the swarm. At the same time, the solid circles indicate the UAVs' positions at a specific point in time. At the times indicated by the figure, namely 0, 12, and 24 s, the positions of each UAV in the swarm are shown. The solid lines on the sides represent the boundaries of the virtual tube, while the dotted line in the center represents the generating curve of the virtual tube. The simulation results demonstrate that fixed-wing UAVs can complete swarm flights with the assistance of

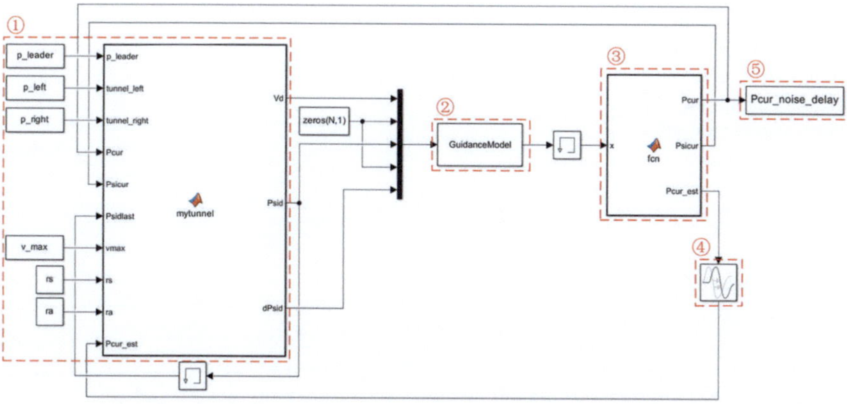

Fig. 10.21 Fixed-wing swarm closed-loop control simulation module subject to noise and delay, from "e8/e8-3/Tube_noise_delay.slx"

a virtual tube. There is little difference in the UAV flight trajectories when comparing the conditions of no noise and no delay to the three scenarios. This indicates that the control method based on the virtual tube is capable of effectively coping with noise and delay. This is because the control within the virtual tube is not based on a fixed formation structure for the swarm positions; instead, it is capable of maintaining the swarm flight along the tube through distributed control. The control within the virtual tube remains effective for a larger number of UAVs without assigning specific tracking targets to each UAV, thereby demonstrating robust performance. Conversely, the simulation results indicate that the trajectory of the UAV based on the virtual tube is smoother than that of formation control. This also implies that the energy consumption of the control within the virtual tube will be less than that of formation control, which is one of the advantages of this method.

In addition to the trajectory comparison above, the Simulink files for each scenario can also be run separately. For example, first run "initTube.m" for initialization, then run the "Tube_delay.slx" simulation file. After the simulation, running the command "Draw(Pcur_delay,p_leader)" in the MATLAB command window can dynamically plot the real-time state of the fixed-wing UAV swarm.

10.4.4 HIL Experiment

HIL simulation aims to validate the effectiveness of a previously designed control algorithm. By utilizing HIL flight control systems, readers can gain a deeper understanding of the process involved in performing real-time simulation and verification of control algorithms. This experiment particularly focuses on the utilization of multi-autopilot HIL simulation. HIL simulation is divided into four distinct phases: hardware configuration, parameter identification, SIL simulation, and HIL simula-

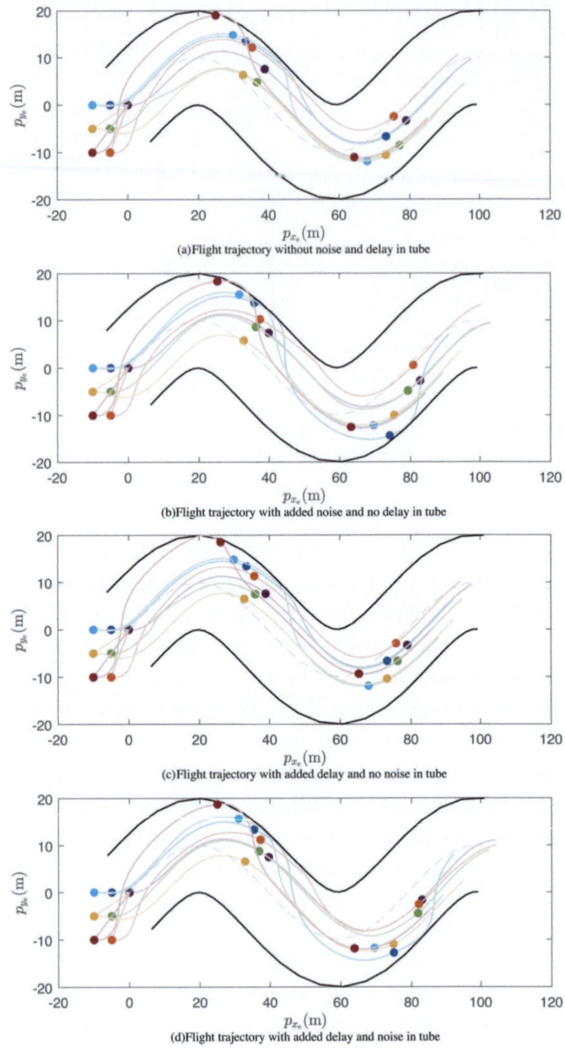

Fig. 10.22 Comparison of fixed-wing UAV control within virtual tubes under different conditions, from "e8/e8-3/Draw_traj.m"

tion. Step 1 involves configuring the hardware necessary for HIL simulation. Step 2 identifies the fixed-wing model to establish the guidance model. In Step 3, the model with identified parameters is used for SIL simulation. Finally, in Step 4, the hardware is connected to complete the HIL simulation.

Step 1: Hardware configuration

The HIL simulation experiment conducted in this book uses Cubepilots/PixHawks. Place the model file of the fixed-wing UAV named "SmallFixedWingUAVnoctrl-HIL.dll" under the path "\PX4PSP\Coptersim\external\model". The first step is to configure the hardware, as outlined in the third and fourth steps of Chap. 2. Open

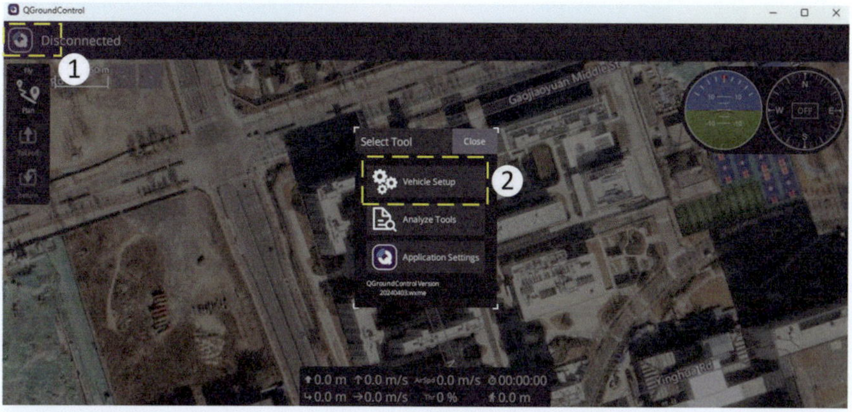

Fig. 10.23 Opening QGC

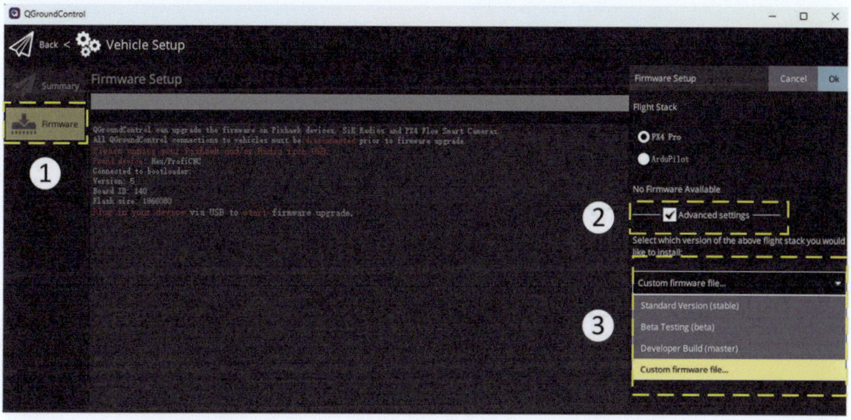

Fig. 10.24 Flashing the firmware

QGroundControl (QGC), as shown in Fig. 10.23, click the software icon in the upper left corner, and select "Vehicle Setup". As shown in Fig. 10.24, click "Firmware" in the left list. Follow the prompts to unplug and replug the flight control. Select "Advanced Settings" in the popup interface, as shown by ②. Then, choose "Custom firmware file" from the dropdown menu, as shown by ③. Import the file "cubepilot_cubeorange_default.px4".

As shown in Fig. 10.25, select "Parameters" in the left list as shown by ①, then click "Tool" in the upper right corner, and select "Load from file" as shown by ②. Import the pre-configured parameter file "pixhawk.params".

Subsequently, it is necessary to ascertain whether there is any issue with the frame settings. As illustrated in Fig. 10.26, the fixed-wing frame should be selected.

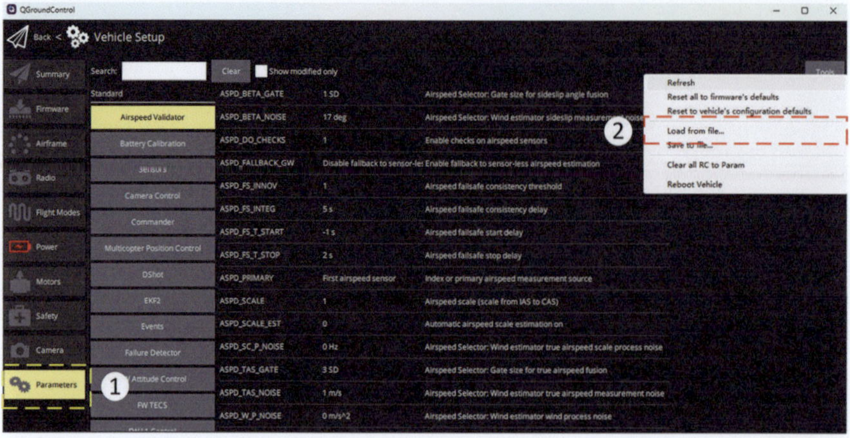

Fig. 10.25 Importing parameters

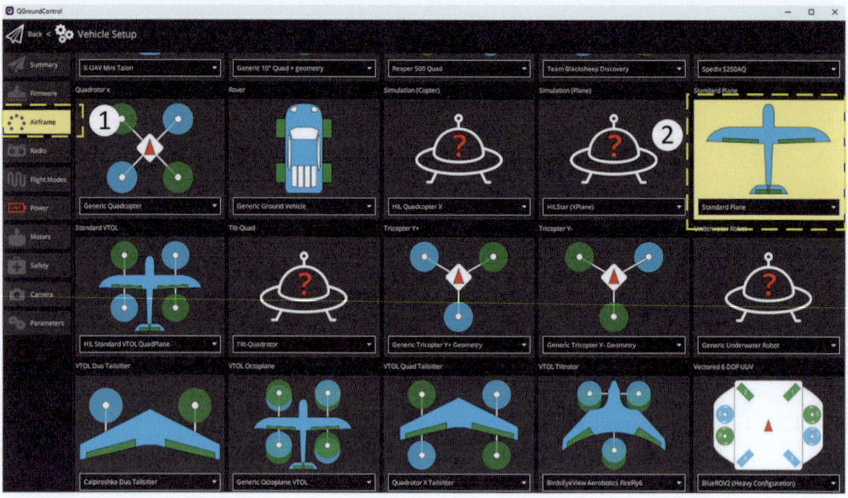

Fig. 10.26 Selecting frame

For each flight controller, change "MAV_SYS_ID" as shown in Fig. 10.27. Select "Parameters", search for "MAV_SYS_ID" in "Search", and set it to 1, 2, 3, and so on according to the number of flight controllers.

Then, all settings for the flight control hardware are completed.

Step 2: Parameter identification

Connect the flight control hardware and refer to the system identification experiment in Chap. 5 to perform parameter identification, thereby obtaining the guidance model. Here, due to the difficulty in obtaining desired yaw data from the flight control log file, Simulink input-output data is recorded to complete parameter identification.

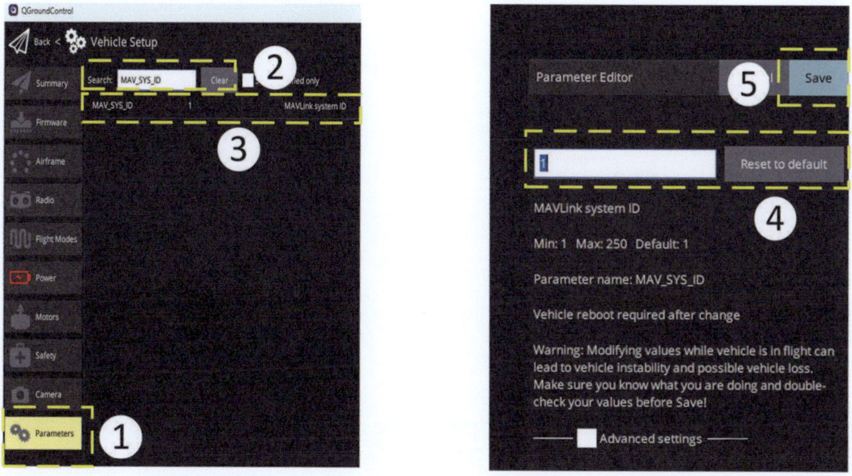

Fig. 10.27 Changing SYSID

Three flight modes are designed to collect data for altitude, yaw, and speed channels. When identifying a certain channel, the other two channels are kept unchanged to avoid coupling.

Open the simulation "Tube_HITL_identification.slx", as shown in Fig. 10.28, the desired speed "Vd", the desired altitude "hd", and the desired yaw "Psid" are designed as inputs to enter "Commander". The HIL simulation utilizes the RflyUdpFast interface of Simulink, employing the UltraSimple Mode. The usage of this interface can be found in Lecture 7 of the RflySim toolchain PPT (located at the RflySim installation path \PX4PSP\PPTs). The control structure sent by Simulink is shown in Code 10.12.

Code 10.12 Control structure

```
struct inOffboardShortData{
    int checksum; // Check digit 1234567890
    int ctrlMode; // Mode selection
    float controls[4]; //Four-digit control command
}
```

In Code 10.11, "checksum" needs to be set to "1234567890" as the data verification key. "ctrlMode" can have multiple protocols (need to change RflyUdpFast.cpp and CopterSim). Here, two modes are used. When "ctrlMode" is set to 11, it indicates the local takeoff command of MAVLink, where "controls[0]" to "controls[2]" represent the desired local takeoff coordinates. When "ctrlMode" is set to 13, it indicates velocity, waypoint, and altitude commands, where "controls[0]" to "controls[2]" represent the desired velocity, altitude, and heading.

Therefore, during takeoff, set the fixed-wing mode to "ctrlMode=11" in "Commander" and send the command "outCMDs=[11;posX;posY;posZ;0]" to RflyUdp-

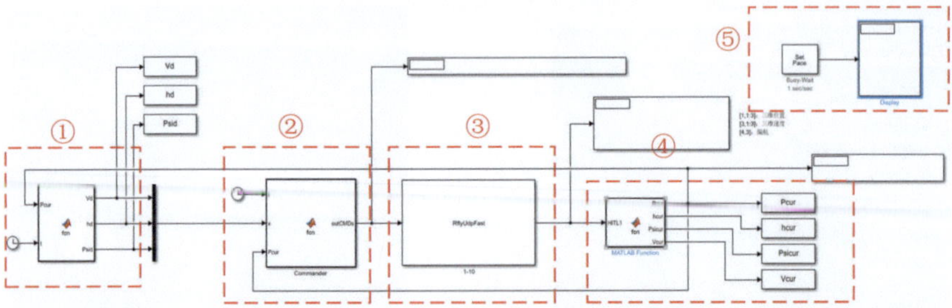

Fig. 10.28 System identification of the flight controller, from "e8/HITL/Tube_HITL_identification.slx"

Fast, where "posX", "posY", and "posZ" are the coordinates where the UAV is expected to fly to. After takeoff, "ctrlMode" can be set to 13, and the command sent is "outCMDs=[13;Vd;hd;Psid;0]", where "Vd" is the desired velocity, "hd" is the desired altitude, and "Psid" is the desired yaw.

Figure 10.28 describes the Simulink simulation for system identification of the fixed-wing flight control. The system identification simulation consists of five main modules: dashed box ①—generation of desired speed, desired altitude, and desired yaw; dashed box ②—generation of fixed-wing UAV control commands; dashed box ③—"RflyUdpFast" is the UDP communication module, which sends control commands externally and receives flight status data from the fixed-wing UAV; dashed box ④—records UAV status data. Use a "To workspace" module in Simulink to record the desired input and the actual output returned by the flight control; dashed box ⑤—"Set Pace" module, which allows the simulation to run at a step size of 1 s/s to ensure real-time simulation.

Connect the autopilot to the computer, run the one-click startup script "SmallFixedWingUAVnoctrlHITLRun.bat", and enter the flight control port number as prompted. After a short wait, when the window in the lower left corner of CopterSim displays "PX4: GPS 3D fixed & EKF initialization finished." and "PX4: Enter Auto Loiter Mode!", it indicates that the flight control has started successfully.

Run the Simulink simulation, and after it ends, follow the instructions for using the "System Identification" Toolbox in Chap. 5 to perform parameter identification on the flight control. It should be noted that when "timeseries" is the data format selected in the "To workspace" module, the data in the workspace needs to be converted to the "double" format first, and then it can be imported into the "System Identification" Toolbox. For the velocity channel, for instance, import the desired velocity as the input and the actual velocity as the output. Perform parameter identification using different configurations: 2 poles and 1 zero, 2 poles and 0 zeros, and 1 pole and 0 zeros. The resulting data is illustrated in Fig. 10.29.

Finally, select the transfer function of the velocity channel as shown in Fig. 10.30, which is $V_{\mathrm{e}}(s) = \frac{0.2882}{s+0.2731} V_{\mathrm{ed}}(s)$.

Fig. 10.29 Velocity channel parameter identification 1

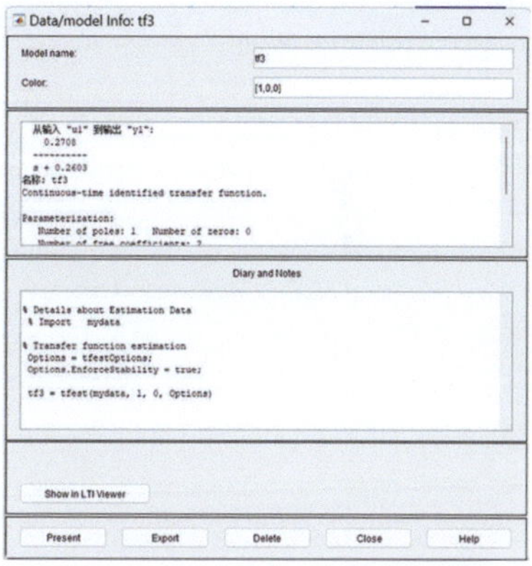

Fig. 10.30 Velocity channel parameter identification 2

Similarly, identification results for the altitude and yaw channels can be obtained as shown in Figs. 10.31 and 10.32, respectively. The transfer function of the altitude channel is

$$H(s) = \frac{0.0851 + 0.007894}{s^2 + 0.148s + 0.007868} H_{\mathrm{d}}(s) \tag{10.37}$$

and the transfer function of the yaw channel is

$$\chi(s) = \frac{0.3404\,s + 0.1109}{s^2 + 0.4581s + 0.111} \chi_{\mathrm{d}}(s) . \tag{10.38}$$

Thus, based on the identified results and the original Eq. (10.1), the guidance model of the fixed-wing UAV is organized as follows:

Fig. 10.31 Altitude channel parameter identification

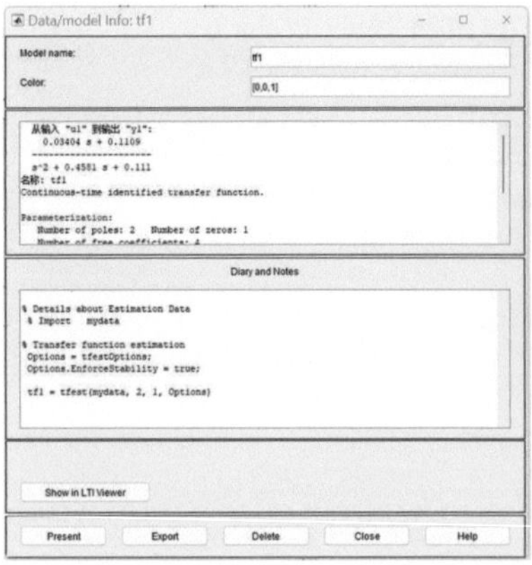

Fig. 10.32 Yaw channel parameter identification

$$\left\{\begin{aligned} \dot{p}_{x_e} &= V_e \cos \psi \\ \dot{p}_{y_e} &= V_e \sin \psi \\ \psi &= \chi \\ V_e(s) &= \frac{0.2708}{s + 0.2603} V_{ed}(s) \\ \chi(s) &= \frac{0.3404\, s + 0.1109}{s^2 + 0.4581 s + 0.111} \chi_d(s) \\ H(s) &= \frac{0.0851 + 0.007894}{s^2 + 0.148 s + 0.007868} H_d(s). \end{aligned}\right. \tag{10.39}$$

After obtaining the parameter identification results for the channels mentioned above, the guidance model of the fixed-wing UAV can be obtained. The guidance model is written as an S-function for subsequent software simulation, and its key code is shown in Code 10.13.

Code 10.13 Fixed-wing guidance model, from "e8/HITL/GuidanceModel.m"

```
function sys=mdlDerivatives(t,x,u)
global N
sys(1:N)      = x(4*N+1:5*N).*cos(x(5*N+1:6*N));
sys(N+1:2*N)  = x(4*N+1:5*N).*sin(x(5*N+1:6*N));
sys(2*N+1:3*N) = x(3*N+1:4*N);
sys(3*N+1:4*N) = -0.148*x(3*N+1:4*N)-0.007868*x(2*N+1:3*N)+0.0851*u(3*N+1:4*N)+0.007894*u(N+1:2*N);
sys(4*N+1:5*N) = -0.2603*x(4*N+1:5*N)+0.2708*u(1:N);
sys(5*N+1:6*N) = x(6*N+1:7*N);
sys(6*N+1:7*N) = -0.4581*x(6*N+1:7*N)-0.111*x(5*N+1:6*N)+0.03404*u(4*N+1:5*N)+0.1109*u(2*N+1:3*N);
```

Step 3: Software simulation

Following the design steps outlined in the experiment design, the controller for the fixed-wing swarm is designed, and its Simulink simulation structure is depicted in Fig. 10.18. Simply rewrite the S-function for fixed-wing UAVs. The initial conditions are designed for four fixed-wing UAVs, with initial positions set to "UAV_PInit=[0 0 0;0 10 0;10 0 0;10 10 0]+[-250,-119,0]". To match the coordinates of the 3D visual map, the parameters for the virtual tube are designed as shown in Code 10.14.

Code 10.14 Virtual tube parameters, from "e8/HITL/init_Tube_HITL.m"

```
curve_x=(0:4:400)';
curve_y=10*sin(curve_x/10/pi+pi/2);
curve_theta=atan(10/10/pi*cos(curve_x/10/pi+pi/2));

curve_x1=(-10:4:150)';
curve_y1=zeros(length(curve_x1),1);
curve_theta1=zeros(length(curve_x1),1);

curve_x=[curve_x1;curve_x+curve_x1(end)-curve_x(1)];
curve_y=[curve_y1;curve_y+curve_y1(end)-curve_y(1)];
curve_theta=[curve_theta1;curve_theta];
dis=20*ones(length(curve_x),1);
p_leader=[curve_x,curve_y]+[-250,-119+5];
```

After debugging the controller and achieving the desired simulation results, the final HIL simulation experiment can be conducted.

Step 4: HIL simulation

First, set up the Simulink simulation as shown in Fig. 10.33. The Simulink file for HIL simulation consists of five main modules:

1. Dashed box ①—Generation of desired speed, altitude and yaw.
2. Dashed box ②—Generation of control commands for fixed-wing UAVs. Each flight controller uses a separate control module to send commands to RflyUdpFast. Set "vehicle Number" to 4.
3. Dashed box ③—RflyUdpFast for UDP communication, responsible for sending control commands and receiving flight status data from fixed-wing UAVs.
4. Dashed box ④—Recording UAV status data using "To Workspace" module.
5. Dashed box ⑤—"Set Pace" module to ensure the simulation runs at a pace of 1 s per second for real-time simulation (Fig. 10.34).

Connect four autopilots to the computer, as shown in Fig. 10.35. Run the one-click startup script "SmallFixedWingUAVnoctrlHITLRun.bat" and enter the flight

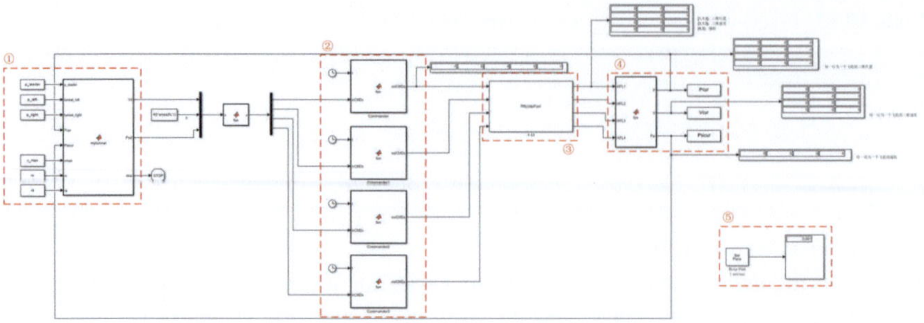

Fig. 10.33 HIL simulation of Simulink programs, from "e8/HITL/Tube_HITL.slx"

Fig. 10.34 Setting up the RflyUdpFast module

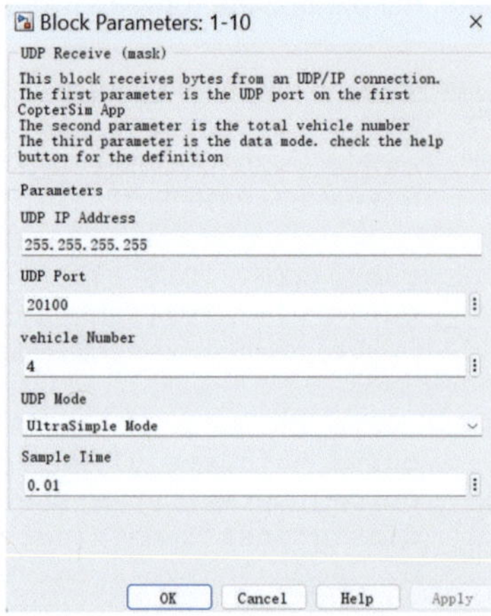

Fig. 10.35 Hardware layout of HIL simulation

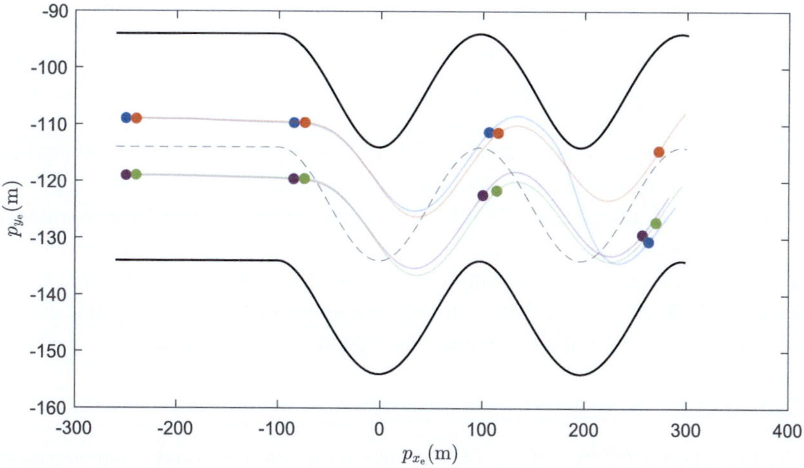

Fig. 10.36 Flight trajectories of a fixed-wing swarm within a virtual tube performed by HIL simulation, from "e8/HITL/Draw_traj_HITL.m"

controller port numbers as prompted. After a short wait, when the flight controller startup is successful, run the initialization file "init_Tube.m" and the Simulink simulation file. The fixed-wing swarm taking off smoothly and completing the tube flight can be observed.

Once the HIL simulation is complete, the corresponding trajectory can be plotted. Figure 10.36 illustrates the trajectories of four fixed-wing UAVs within the swarm. The solid circles represent the positions of the UAVs at specific moments. At times 0, 12, 24, and 36 s, the positions of the UAVs within the swarm are shown in the plot. The solid lines on the sides represent the two boundaries of the virtual tube, and the dotted line in the center represents the centerline of the virtual tube. The simulation results indicate that the fixed-wing UAVs can complete the swarm flight with the assistance of the virtual tube. Nevertheless, due to the inherent lag in the state of fixed-wing UAVs and the response of control inputs, it is possible to exceed the boundaries of the virtual tube briefly.

10.4.5 Related Development

The control objective of the virtual tube is to maintain the UAV swarm within the virtual tube and guide it to fly along it. In practical applications, the virtual tube can represent paths connecting two locations, or features such as gates, corridors, or windows. The control strategy used for the virtual tube is similar to the artificial potential field method, which involves designing a potential barrier function [11] as an artificial potential function. Each UAV employs distributed control laws to generate a vector field by superimposing the negative gradients of attractive and repulsive potential functions, ensuring that UAVs pass through the virtual tube while avoiding collisions with other UAVs. Unlike the formation control, with virtual tubes, the UAV swarm can freely pass through virtual tubes without maintaining a fixed geometry.

In the virtual-tube-based control method, UAV swarms employ distributed control. This structure utilizes only the state information of nearby neighbor UAVs, rather than global information, thereby enhancing mission reliability and suitability for large-scale, dense UAV swarms. Additionally, this method can accommodate heterogeneous UAV swarms; for example, high-speed UAVs and low-speed UAVs can collectively perform swarm flights in complex environments. During this process, some high-speed UAVs may overtake low-speed ones to execute their respective tasks.

Current research on virtual tubes has covered several aspects.

(1) Virtual tube modeling: In complex environments, accurately and reasonably modeling virtual tubes is crucial. Currently, virtual tubes are divided into three types: open virtual tubes, segmented open virtual tubes, and closed virtual tubes. They are further divided into straight virtual tubes [8], trapezoidal virtual tubes [12], quadrilateral spliced virtual tubes [13], circular virtual tubes [14], and curved virtual tubes [9, 15]. In [17], the authors proposed a model for a class of regular virtual tubes and designed a method to obtain regular virtual tubes based on trajectory planning and regular conditions. In [18], the authors further proposed a class of optimal virtual tubes and a practical virtual tube planning method.
(2) Distributed controller design for UAV swarms based on virtual tubes: Given the definition and the types of virtual tubes, distributed controllers must be designed for UAV swarms to ensure that they complete predetermined tasks smoothly without collisions while operating within the virtual tubes. In [10], the authors proposed a separation principle for safety radius and controller design in uncertain communication environments. Using this principle, the authors investigated safety radius design methods for both the design phase (with no uncertainty) and the flight phase (with uncertainty). In [15], the authors extended the distributed controller within curved virtual tubes in an ideal condition to all UAVs under self-positioning drift and accurate relative positioning conditions by introducing a flocking algorithm to reduce the adverse effects of self-positioning drift. Concretely, similar to the "many wrongs principle", cohesion behavior and velocity alignment behavior can respectively reduce the adverse impact of position measurement drift and velocity measurement errors.
(3) Optimization of virtual tube trafficability [19]: To guarantee the safe and smooth passage of the UAV swarm through the virtual tube, the authors proceeded to enhance the proposed distributed controller to optimize the trafficability of the virtual tube and facilitate the rapid passage of the UAV swarm. To achieve this, the authors improved upon the traditional artificial potential field method by proposing a method for orthogonalizing a potential-free vector field. This method maintains system stability while increasing the UAV swarm's trafficability. The

objective of the improvement strategy is to make the repulsive vector field orthogonal to the attractive vector field under certain conditions, thereby enabling UAVs to move toward the target with greater velocity.

For future research on virtual tubes, it is of paramount importance to address the issue of dealing with various uncertain conditions. In complex environments, UAV swarms often operate under unfavorable conditions of inaccurate positioning, which inevitably leads to drift in self-positioning information. This has a significant impact on the effectiveness of swarm flight control. Additionally, although virtual tubes theoretically do not contain any internal obstacles, in practice, there may be malfunctioning UAVs and small obstacles that have not been detected in advance inside virtual tubes. Therefore, it is crucial to explore methods for addressing the effects of various uncertainties to achieve effective control of UAV swarms within virtual tubes. For more information, readers can refer to the survey [20].

10.5 Exercise

1. A UAV swarm can be viewed as a collective system, comprising multiple individuals with limited capabilities that interact and collaborate to complete complex tasks. This mirrors the collective behaviors seen in nature, such as ant colonies, bee swarms, bird flocks, and fish schools. Many UAV swarm control strategies are inspired by swarm intelligence, where simple individuals exhibit complex behaviors through local interactions modeled by various algorithms. Explain the principles of bird flocking and their application to UAV control.
2. In collective systems, each individual possesses a simple structure yet collaborates to perform complex tasks such as foraging, transport, and migration. Ants, for example, can forage, coordinate transport, and build structures without central command. Explain the principles underlying these behaviors and their relevance to UAV control.
3. The virtual structure method treats a UAV formation as a rigid structure, with each UAV assigned to a specific position within it. Each UAV tracks its corresponding point on the virtual structure [21]. Based on this method, a UAV controller can be designed to achieve formation flight.
4. In the actual process of swarm flight, due to performance limitations and other constraints, UAVs are expected not to be too far apart or too close to each other. How to design a controller within virtual tubes under accurate relative positioning to ensure these performance criteria? Hint: Introduce the concept of fluid viscosity, considering the swarm as a group of liquid molecules to keep them together as a whole. The presence of fluid viscosity ensures that the fluid moves forward as a whole while maintaining an appropriate distance between fluid molecules.
5. As illustrated in Fig. 10.37, each aircraft can only detect its surroundings and the gate's position without data link communication. How can they cooperate to avoid collisions and pass through the gate efficiently?

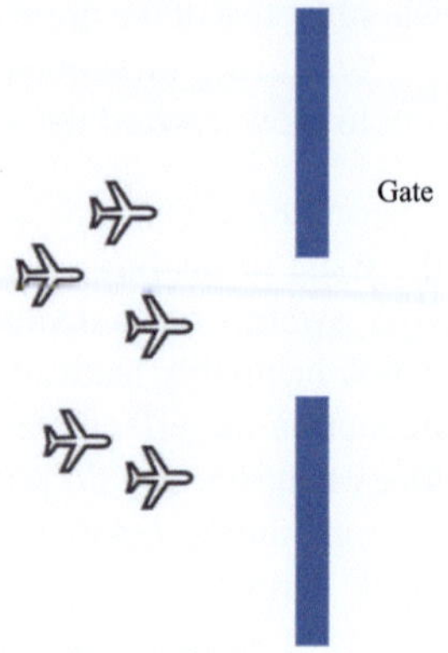

Fig. 10.37 Illustration for Exercise 5

10.6 Appendix

The design of the swarm controller involves two important smooth functions. The first smooth function is a continuously differentiable bump function as

$$\sigma(x, d_1, d_2) = \begin{cases} 1 & x \leq d_1 \\ Ax^3 + Bx^2 + Cx + D & d_1 \leq x \leq d_2 \\ 0 & d_2 \leq x \end{cases} \tag{10.40}$$

where $A = -2/(d_1 - d_2)^3$,$B = 3(d_1 + d_2)/(d_1 - d_2)^3$,$C = -6d_1d_2/(d_1 - d_2)^3$, $D = d_2^2(3d_1 - d_2)/(d_1 - d_2)^3$. Taking the derivative with respect to x yields:

$$\frac{\partial \sigma(x, d_1, d_2)}{\partial x} = \begin{cases} 0 & x \leq d_1 \\ 3Ax^2 + 2Bx + C & d_1 \leq x \leq d_2 \\ 0 & d_2 \leq x. \end{cases} \tag{10.41}$$

Another smooth function is an approximate smooth saturation function. The original saturation function is a piecewise function

$$\bar{s}(x) = \min(x, 1), x \geq 0. \tag{10.42}$$

Here, a smooth function to approximate the saturation function is defined as

$$s(x, \epsilon_s) = \begin{cases} x & 0 \leq x \leq x_1 \\ (1 - \epsilon_s) + \sqrt{\epsilon_s^2 - (x - x_2)^2} & x_1 \leq x \leq x_2 \\ 1 & x_2 \leq x \end{cases} \tag{10.43}$$

where $x_2 = 1 + \epsilon_s/\tan 67.5°$,$x_1 = x_2 - \sin 45°\epsilon_s$. Since $x_1 \geq 0$ it follows that $\epsilon_s \leq \frac{\tan 67.5°}{\tan 67.5° \sin 45° - 1}$. For any $\epsilon_s \in \left[0, \frac{\tan 67.5°}{\tan 67.5° \sin 45° - 1}\right]$, it can be shown that

$$s(x, \epsilon_s) \leq \bar{s}(x). \tag{10.44}$$

And

$$\lim_{\epsilon_s \to 0} \sup_{x \geq 0} |\bar{s}(x) - s(x, \epsilon_s)| = 0. \tag{10.45}$$

The derivative of the function $s(x, \epsilon_s)$ with respect to x is

$$\frac{\partial s(x, \epsilon_s)}{\partial x} = \begin{cases} 1 & 0 \leq x \leq x_1 \\ \frac{x_2 - x}{\sqrt{\epsilon_s^2 - (x - x_2)^2}} & x_1 \leq x \leq x_2 \\ 0 & x_2 \leq x. \end{cases} \tag{10.46}$$

For any $\epsilon_s > 0$, it can be shown that $\sup_{x \geq 0} \left| \frac{\partial s(x, \epsilon_s)}{\partial x} \right| \leq 1$ (Fig. 10.38).

Additionally, there is a directional vector saturation function sat $(\mathbf{x}, n)$, described as

$$\text{sat}(\mathbf{x}, n) = \begin{cases} \mathbf{x} & \|\mathbf{x}\| \leq n \\ n\frac{\mathbf{x}}{\|\mathbf{x}\|} & \|\mathbf{x}\| > n \end{cases}. \tag{10.47}$$

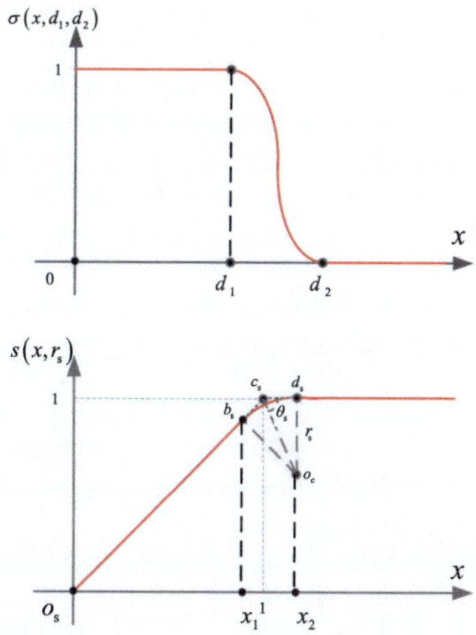

Fig. 10.38 Schematic diagram of two smooth functions

References

1. E. Bonabeau, M. Dorigo, G. Theraulaz, *Swarm Intelligence: From Natural to Artificial Systems*, vol. 2 (Oxford University Press, 1999), pp. 25–34
2. P. Vincent, I. Rubin, A framework and analysis for cooperative search using UAV swarms, in *Proceedings of the 2004 ACM Symposium on Applied Computing, ser. SAC 04* (Association for Computing Machinery, Nicosia, Cyprus, 2004), pp. 79–86
3. Y. Li, Y. Gao, B.-B. Yin, Q. Quan, Bee-dance-inspired UAV trajectory pattern design for target information transfer without direct communication, in *2021 40th Chinese Control Conference (CCC)* (IEEE, 2021), pp. 4279–4284
4. E. Sahin, Swarm robotics: From sources of inspiration to domains of application, in *International Workshop on Swarm Robotics* (Springer, 2004), pp. 10–20
5. Y. Li, Y. Gao, S. Yang, Q. Quan, Swarm robotics search and rescue: A bee-inspired swarm cooperation approach without information exchange, in *2023 IEEE International Conference on Robotics and Automation (ICRA)* (2023), pp. 1127–1133
6. Y. Jiang, Y. Gao, W. Song, Y. Li, Q. Quan, Bibliometric analysis of UAV swarms. J. Syst. Eng. Electron. **33**(2), 406–425 (2022)
7. H.G. Tanner, G.J. Pappas, V. Kumar, Leader-to-formation stability. IEEE Trans. Robot. Autom. **20**(3), 443–455 (2004)
8. Q. Quan, R. Fu, M. Li, D. Wei, Y. Gao, K.-Y. Cai, Practical distributed control for VTOL UAVs to pass a virtual tube. IEEE Trans. Intell. Veh. **7**(2), 342–353 (2022)
9. Q. Quan, Y. Gao, C. Bai, Distributed control for a robotic swarm to pass through a curve virtual tube. Robot. Auton. Syst. **162** (2023)
10. Q. Quan, R. Fu, K.-Y. Cai, How far two UAVs should be subject to communication uncertainties. IEEE Trans. Intell. Transp. Syst. **24**(1), 429–445 (2022)
11. L. Wang, A.D. Ames, M. Egerstedt, Safety barrier certificates for collisions-free multirobot systems. IEEE Trans. Robot. **33**(3), 661–674 (2017)
12. Y. Gao, C. Bai, Q. Quan, Distributed control within a trapezoid virtual tube containing obstacles for robotic swarms subject to speed constraints. IEEE Trans. Control Netw. Syst. **12**(1), 287–299 (2025)
13. Y. Gao, C. Bai, Q. Quan, Distributed control for a multiagent system to pass through a connected quadrangle virtual tube. IEEE Trans. Control Netw. Syst. **10**(2), 693–705 (2023)
14. Y. Gao, C. Bai, L. Zhang, Q. Quan, Multi-UAV cooperative target encirclement within an annular virtual tube. Aerosp. Sci. Technol. **128** (2022)
15. Y. Gao, C. Bai, Q. Quan, Robust distributed control within a curve virtual tube for a robotic swarm under self-localization drift and precise relative navigation. Int. J. Robust Nonlinear Control **33**(16), 9489–9513 (2023)
16. P. Mao, Q. Quan, Making robotics swarm flow more smoothly: a regular virtual tube model, in *2022 IEEE/RSJ International Conference on Intelligent Robots and Systems (IROS)* (IEEE, 2022), pp. 4498–4504
17. P. Mao, Q. Quan, Making robotics swarm flow more smoothly: a regular virtual tube model, in *2022 IEEE/RSJ International Conference on Intelligent Robots and Systems (IROS), IEEE* (2022), pp. 4498–4504
18. P. Mao, R. Fu, Q. Quan, Optimal virtual tube planning and control for swarm robotics. Int. J. Robot. Res. **43**(5), 602–627 (2024)
19. Y. Gao, C. Bai, L. Zhang, Q. Quan, Multi-UAV cooperative target encirclement within an annular virtual tube. Aerosp. Sci. Technol. **128**, 107800 (2022)
20. Y. Gao, C. Bai, Q. Quan, A survey on passing-through control of multi-robot systems in cluttered environments. J. Syst. Eng. Electron. **36**(4), 1037–1056 (2025)
21. A. Sadowska, H. Huijberts, D. Kostic, N. van de Wouw, H. Nijmeijer, Formation control of unicycle robots using the virtual structure approach, in *2011 15th International Conference on Advanced Robotics (ICAR)* (2011), pp. 365–370

MIX
Papier aus verantwortungsvollen Quellen
Paper from responsible sources
FSC® C105338

If you have any concerns about our products,
you can contact us on
ProductSafety@springernature.com

In case Publisher is established outside the EU,
the EU authorized representative is:
Springer Nature Customer Service Center GmbH
Europaplatz 3, 69115 Heidelberg, Germany

Printed by Libri Plureos GmbH
in Hamburg, Germany